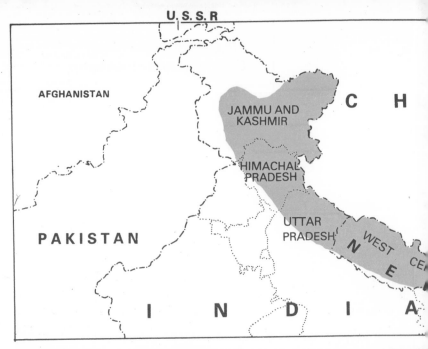

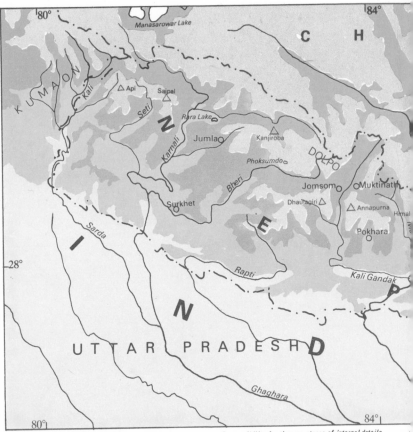

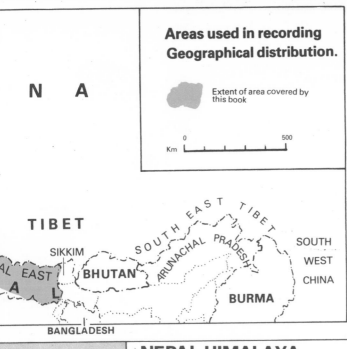

Areas used in recording Geographical distribution.

Extent of area covered by this book

Km 0 ——————— 500

N A

TIBET

SIKKIM

SOUTH EAST TIBET

BHUTAN

ARUNACHAL PRADESH

AL EAST
A L

SOUTH WEST CHINA

BURMA

BANGLADESH

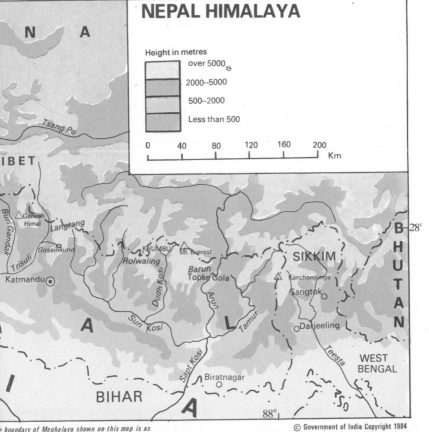

NEPAL HIMALAYA

Height in metres

over 5000

2000–5000

500–2000

Less than 500

Km 0 — 40 — 80 — 120 — 160 — 200

N A

Tsang Po

IBET

Buri Gandak
△ Ganesh Himal
Langtang
Trisuli
Gosainkund
KHUMBU △ Mt. Everest
Rolwaling
Barun
Topke Gola
Dudh Kosi
Katmandu ⊙
Arun
Sun Kosi
A
L
Tamur
SIKKIM
△ Kanchenjunga
Gangtok ○
Darjeeling ○
B H U T A N
28°

Sapt Kosi
Biratnagar ○

BIHAR
A
88°

Teesta

WEST BENGAL

I

FLOWERS OF THE HIMALAYA

FLOWERS OF THE HIMALAYA

OLEG POLUNIN
ADAM STAINTON

Drawings by Ann Farrer

DELHI
OXFORD UNIVERSITY PRESS
CALCUTTA CHENNAI MUMBAI
1997

Oxford University Press, Walton Street, Oxford OX2 6DP

Oxford   New York
Athens   Auckland   Bangkok   Calcutta
Cape Town   Chennai   Dar es Salaam   Delhi
Florence   Hong Kong   Istanbul   Karachi
Kuala Lumpur   Madrid   Melbourne   Mexico City
Mumbai   Nairobi   Paris   Singapore
Taipei   Tokyo   Toronto
and associates in
Berlin   Ibadan

Oxford is a trade mark of Oxford University Press

First published 1984
Seventh impression 1992

First published in Oxford India Paperbacks 1997

ISBN 0 19 564187 6

Printed in India
Text typeset at Sri Aurobindo Ashram Press, Pondicherry
and printed at Rekha Printers Pvt. Ltd., New Delhi 110020
Plates printed at Sahara India Mass Communication, Noida
Published by Manzar Khan, Oxford University Press
YMCA Library Building, Jai Singh Road, New Delhi 110001

This book is dedicated to Sir George Taylor F.R.S., an enthusiastic supporter of Himalayan botanical exploration for over fifty years, who gave the authors, in 1949 and 1954 respectively, their first opportunity to see the magnificence of the Himalaya and the richness of its flora.

Acknowledgements

A book such as this must inevitably be heavily indebted to the work of earlier botanists and collectors. For the last fifty years the Department of Botany in the British Museum (Natural History) has been a centre for work on the Himalayan flora, and recently it has published in conjunction with the University of Tokyo *An Enumeration of the Flowering Plants of Nepal*. This invaluable publication has greatly helped the authors in the task of preparing the first field guide to the flowers of the Himalaya.

Our thanks go first to Mr J. F. M. Cannon, Keeper of Botany at the Natural History Museum, and to his predecessors in office, who have allowed us to make use of the great collection of Himalayan specimens in the herbarium and of Himalayan literature in the library; also to all members of the staff there who over the years have always assisted us with unfailing courtesy.

We owe a special debt of gratitude to Mr L. H. J. Williams and to Mr A. O. Chater for much help which they have given to us during the periods they were editing the *Nepal Enumeration*; to Professor H. Hara during his welcome visits to London; and to Miss S. Y. Sutton during her long association with the same work.

The *Nepal Enumeration* and the herbarium at the Museum are arranged in the family order first adopted by Bentham and Hooker. All the older standard Indian botanical publications also follow this order. We have therefore chosen to follow the Bentham and Hooker family order in this book in preference to any of the other orders more recently proposed.

Many people have kindly allowed us to see their photographs with a view to selecting illustrations. They are too numerous for us to mention individually, but we thank them all, including the following whose photographs are reproduced in the plates: Sir C. Barclay Bart. 1223; Mrs S. Beer (widow of the late Mr L. Beer) 21, 22, 94, 196, 447, 476, 1105, 1220, 1248; Mr A. B. C. Harrison 1020; Mr A. Huxley 163, 852, 1333, 1401, 1425; Mr R. Lancaster 395, 723, 1374; Mr D. J. McCosh 76, 882; Mr D. Sayers 379; Mr G. Smith 990; Mr L. H. J. Williams 817, 842.

We thank particularly Dr. C. Grey-Wilson who has assisted us with *Clematis* and *Impatiens*, as well as lending us many photographs: 14, 19, 27, 70, 111, 184, 373, 475, 502, 542, 628, 641, 686, 711, 756, 772, 774, 781, 843, 847, 882, 936, 946, 1072, 1075, 1327, 1358, 1363, 1373, 1376, 1402, 1463, 1464.

We thank Ann Farrer for undertaking the exacting task of botanical illustration, and we hope that readers will find her drawings most useful in the identification of species. Ann Farrer was awarded a Churchill Travelling Fellowship to draw Himalayan flowers in the field, and she accompanied the authors during the summer of 1977 for that purpose.

Our thanks are also due to Mrs Kathleen Cooke for her expert typing of a difficult manuscript, and to Mrs Lorna Polunin for preparing the index.

Finally, we wish to thank all those collectors, camp assistants, porters and pony-men who have accompanied us on our Himalayan journeys. They have been many in number and varied in race, their homes ranging from Ladakh to Bhutan. Those who know the Himalaya will be able to appreciate how much these men have contributed to the successful conclusion of our fieldwork.

Contents

Maps

Contents

Introduction

Limits of the country covered by this book

From Nanga Parbat on the Indus the Himalaya stretch eastwards for 2250 km to Namcha Barwa on the bend of the Tsang-Po in South-East Tibet. This book covers only that part of the range which lies within Nepal and the Indian Western Himalaya, a distance of about 1450 km.

The Nepal-Sikkim border has been chosen as the eastern boundary of our area. The Eastern Himalaya beyond this point have a particularly rich and interesting flora, but at the present time few people can go to Bhutan, and virtually no non-Indian to Arunachal Pradesh. Sikkim too was closed while this book was in course of preparation, and although a limited part is now open to tourists from abroad most of it remains inaccessible.

The India-Pakistan border has been chosen as the western boundary of our area. Remnants of the Himalayan flora continue westwards into the mountains of Northern Pakistan, but since the mountain flora of Pakistan also includes many species with a West or Central Asian distribution it has seemed advisable to limit coverage here to Indian territory.

To the north our boundary is the Tibetan border. As will become apparent below, within the country covered by this book there are some quite extensive areas lying to the north of the main Himalayan range which have a climate and flora similar to that of Tibet, and yet are politically a part of Nepal or India. A number of species have therefore been included which are much more typical of the dry Tibetan plateau than of the monsoon Himalaya.

To the south our boundary is based not upon a political border but upon a line of altitude of approximately 1200 m. In order to keep this book within reasonable size it has been necessary to exclude most of the subtropical flora, though a few species have been included which are prominent in the lower valleys and the outer foothills which lie towards the Indian plains.

Even after excluding most of the subtropical species it has been necessary to deal with the flora selectively. In making our selection we have tried to include all those species which are both common and attractive. Opinions may differ as to the latter attribute, and we hope that in the judgement of the public we have not committed too many sins of omission. A certain number of much rarer plants have also been included, though here we have reluctantly been compelled to omit some interesting species, also all the sedges and

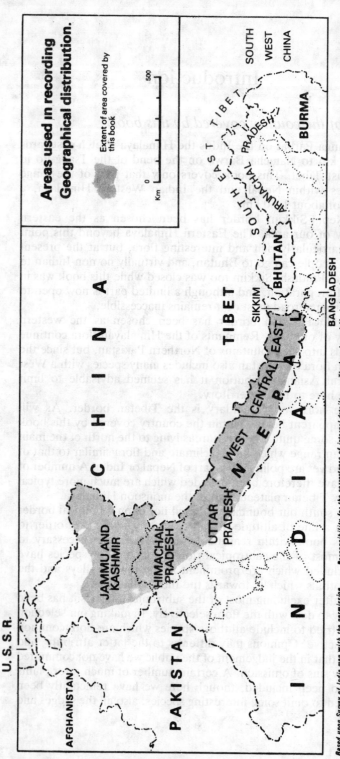

Areas used in recording Geographical distribution.

Extent of area covered by this book

Km   0   500

U.S.S.R.

AFGHANISTAN

PAKISTAN

CHINA

JAMMU AND
KASHMIR

HIMACHAL
PRADESH

UTTAR
PRADESH

I N D I A

N E P A L

WEST

CENTRAL

EAST

SIKKIM

BHUTAN

TIBET

ARUNACHAL
PRADESH

SOUTH
WEST
CHINA

BURMA

BANGLADESH

SOUTH EAST TIBET

grasses. There are many species in the Himalaya which, although not native, are naturalized and quite prominent in the flora. A number of these exotics have been included, because it is felt that many people who see them will be unaware that they are not truly Himalayan. Those who know the European flora will notice that some extremely familiar species have been included. They should bear in mind that these plants will probably be quite unknown to a visitor from the Indian plains or from other parts of the world. 1500 species are described in the text; 689 are illustrated photographically; and 316 are illustrated by line drawings.

Countries, states and geographical divisions which indicate distribution of species

Afghanistan

Pakistan Includes the mountain areas of Chitral and Swat; also those parts now subject to the de facto control of Pakistan which were formally part of the state of Kashmir, i.e. Gilgit, Hunza and Baltistan.

Kashmir The state of Jammu and Kashmir, including Ladakh, Zanskar and Kishtwar.

Himachal Pradesh This comparatively new state includes Chamba, Lahul, Spiti, Kulu, Kangra, Kinnaur and Simla. In older publications much of it often was referred to as the Punjab Himalaya.

Uttar Pradesh Much of this state lies in the Indian plains. In recording distribution, reference is intended only to that mountainous part of the state which lies between Himachal Pradesh and the Nepal border, and which formerly comprised Kumaon and the states of Garhwal.

West Nepal From the border with Uttar Pradesh to Long. 83°E., i.e. just to the west of Dhaulagiri.

Central Nepal From Long. 83°E. to Long. 86°30′E., i.e. just to the west of Everest.

East Nepal From Long. 86°30′E. to the Sikkim border.

Sikkim Records from the adjacent Darjeeling district of North Bengal have been referred to Sikkim.

Bhutan

Arunachal Pradesh Until 1972 this state was known as the North-East Frontier Agency. In older publications it was often called the Assam Himalaya.

Tibet Now known as the Xizang Autonomous Region of the People's Republic of China. It contains much high, dry plateau country.

S.E. Tibet That part of Tibet which lies adjacent to the borders of Arunachal Pradesh, Burma, Yunnan and Szechwan. Since its climate is much wetter and its flora much richer than that of the plateau it requires separate mention.

S.W. China The provinces of Yunnan and Szechwan, which include much mountain country extending eastwards from the Himalaya.

Other geographical divisions used to describe wider distribution are self-explanatory. The above divisions have been used on the grounds that they are the most likely to be understood. No political significance should be attached to the choice made, nor to the delineation of any frontier in any map appearing in this book.

The flora of the area covered by this book is now fairly well known to botanists. One may still be fortunate enough to find here unrecorded species, or even a species unknown to science, but the odds against doing so now hardly justify the labour and expense of making a large general collection. On the other hand a specialist can still find much work to do, and some taxonomic problems cannot satisfactorily be solved without further fieldwork.

The Himalayan flora, and some factors which influence its composition

The Himalaya have a rich flora. In Nepal, within boundaries which are about 150 km in breadth by 700 km in length, some 6500 species of seed-plant are known to occur. The number occurring in the Indian Western Himalaya is not known, but the total number in our whole area, including all of the subtropical flora, may be about 9000 species. Those parts of the world which possess an unusually rich flora generally owe this richness to the fact that the country has remained undisturbed by any cataclysm, such as glaciation or inundation, for a long period of time, thus enabling many species to evolve and old species to survive. In geological terms the Himalaya are a recent creation, and the richness of the flora is due primarily to the great variation in climate and habitat to be found here. Much of the charm of the Himalaya lies in the possibility of seeing very different types of flora within quite a short distance.

Altitude

Altitude is the most important of the various factors which combine to create contrasts in habitat, climate and flora.

In the deep valleys and the lower foothills of Nepal the flora is subtropical, the subtropical zone extending upwards to about 1800 m. Although much of the zone is under cultivation, a number of colourful and interesting plants can be found here. Many of the species have an Indian distribution, and in wetter eastern parts of Nepal the Burma-Malaya element is strong.

In the Western Himalaya the subtropical zone is not so extensive, nor is its flora so rich. Here the upper limit of this zone is at about 1400 m. Winters in the west are considerably colder than in Nepal, and perhaps it is not strictly correct to refer to the low-altitude flora here as subtropical. Elements of the flora of Pakistan and Afghanistan are also present, with species which can withstand both hot summers and cold winters.

Above the subtropical zone lies the temperate zone, which extends upwards to the tree-line. The altitude of the tree-line is much influenced by aspect, rainfall and the hand of man. In Nepal it lies roughly at 4000 m and in the west some 300 m lower. The lower part of the temperate zone is largely under cultivation, but higher up there are extensive forests of evergreen oaks, conifers, and birch. The shrub flora is rich and interesting. It is most strongly represented in the wetter eastern parts of our area, and its richness is particularly noticeable in the genus *Rhododendron*. Many species of this temperate flora range eastwards into China, though there are also a number of species with a much more limited distribution.

From the tree-line the alpine zone extends upwards until the zone of permanent snow and ice is reached. The lower parts of the alpine zone contain the summer grazing-grounds, with meadows bright with alpine flowers. The upper parts have a high-altitude flora, with species which are adapted to withstand the extremes of cold and desiccation. In Nepal glaciers descend to about 4400 m. In the west they descend somewhat lower; eg., to the north of Nun Kun in Ladakh a glacier reaches the Suru river at an altitude of about 3600 m. Obviously the general permanent snow-line is much higher than the above-mentioned altitudes, and in high summer slopes may be free from snow even at 6000 m. In Nepal isolated plants are quite commonly found up to 5500 m. The *Guiness Book of Records* gives the highest species as *Stellaria decumbens* at 6135 m, but A. F. R. Wollaston writing in *Mount Everest, The Reconnaissance, 1921*, awards the honour to *Arenaria bryophylla* at 6180 m.

The alpine flora as a whole is Sino-Himalayan in distribution, though it includes a much smaller number of species which are widely distributed throughout the North Temperate Zone. There are also many endemic species which occur only in quite limited areas.

Rainfall

Rainfall is the next most important factor after altitude to influence the composition of the Himalayan flora. Much of the precipitation comes in the form of summer monsoon rain, which reaches the Himalaya both from the Bay of Bengal and from the Arabian Sea. Monsoon conditions begin to affect the eastern end of Nepal early in June and gradually spread westwards, until by the beginning of July all those parts of the Western Himalaya which are not protected by mountain barriers will also be receiving heavy rainfall.

In Nepal there are normally four full monsoon months, from the beginning of June to the end of September, and this monsoon rain comprises much the larger part of the total precipitation received there. In nearby Darjeeling, out of an average annual rainfall of 3100 mm, only 518 mm fall outside the monsoon months. Nepal does receive some winter precipitation, which in the mountains falls as snow, but it varies quite widely from year to year both in quantity and in the time of its arrival.

In the Western Himalaya the total monsoon rainfall in general is less than in Nepal, for the monsoon does not reach the west until early July, and it begins to recede early in September. On the other hand the Western Himalaya receive a much heavier winter rainfall and snowfall, and unsettled weather often continues well into May. This winter precipitation in the west is therefore a much higher proportion of the total annual precipitation received there. At Srinagar in Kashmir, out of an annual average rainfall of 661 mm only 195 mm fall during the period June-September, whereas 331 mm fall during the period January-April. It should be pointed out however that monsoon rainfall in the Kashmir valley is considerably reduced by the protection of the Pir Panjal range which lies to the south. Monsoon rainfall in other more exposed parts of the Western Himalaya is much higher.

It can therefore be said that Nepal, with the exception of the rainshadow areas discussed below, has a true monsoon climate consisting of a wet summer and a dry winter. The Western Himalaya in general have a lighter summer monsoon rainfall and a wetter winter than Nepal, and those parts of it which are sheltered from the monsoon by mountain barriers tend towards the wet winter and dry summer type of climate which prevails in West Asia and the Mediterranean. This accounts for the presence in the Western Himalaya of elements of a flora more typical of West Asia, with species such as *Eremurus himalaicus, Fritillaria imperialis* and *Tulipa stellata*. These species do not extend as far east as Nepal. Conversely, much of the rich Eastern Himalayan flora which can be seen in wetter parts of Nepal does not extend into the Western Himalaya.

Latitude

It is a common misconception that the Himalaya run in an east-west direction. In fact that part of the range covered by this book runs south-east to north-west. In consequence there is a difference of about eight degrees of latitude between the two ends of our area, equal to the difference in latitude between London and Genoa. The climate of the Western Himalaya is more temperate than that of Nepal, whose south-eastern border lies close to the Tropic of Cancer. The tree-line in the west is lower than in Nepal, and so also is the snow-line, though in the latter respect it must be remembered that glaciers may descend lower not only because of a more temperate climate but also because of a higher winter snowfall. Snowfalls are common in Srinagar, but are unknown in Kathmandu which lies at a comparable altitude.

In view of its more temperate climate, it is hardly surprising that in the Western Himalaya plants with a widespread North Temperate or European distribution form a far more important element in the flora than they do in Nepal. Another consequence of the higher latitude of the west is that differences in length of daylight between summer and winter are more marked here than in Nepal, and also may help to influence the distribution of certain North Temperate species.

Rainshadow

Within the general pattern of rainfall described above there are wide local variations. The main Himalayan range forms a massive barrier which prevents most of the monsoon rain from reaching trans-Himalayan country. The effectiveness of the mountains as a barrier can be illustrated by some rainfall figures. Dalhousie in the Western Himalaya lies on the outer Dhaula Dhar range, and is fully exposed to the monsoon. It has an average annual rainfall of 2542 mm. Leh in Ladakh is only 240 km distant to the north, but it is separated from Dalhousie by much high mountain country which holds off almost all the rain. The annual average at Leh is less than 100 mm. The Dhaulagiri and Annapurna districts of Nepal provide even more remarkable contrasts, because the change from wet to dry takes place over such a short distance. Pokhara, to the south of Annapurna and open to the monsoon, has an annual average rainfall of 3477 mm. Only 65 km north of Pokhara lies Jomsom, with an annual average of 295 mm. Jomsom lies within the rainshadow of Annapurna and Dhaulagiri.

The existence both in Nepal and in the Western Himalaya of quite large areas of low rainfall to the north of the main ranges enables one to see there elements of a dry flora which is typical of Tibet.

Aspect

In steep mountain country a north-facing slope receives much less sunshine than a south-facing slope. Those who trek during the winter months will be particularly aware that many slopes are in shade for much of the day, and the moment when the sun clears the crest of a ridge and warms one with its rays is always very welcome. East Himalayan species tend to favour slopes with a north or west aspect because they retain far more moisture, and the vegetation on these slopes often merits attention even where adjacent south or east slopes do not hold much of interest. The degree of moisture of a slope also controls the frequency of fire. South-facing slopes with dry combustible undergrowth are often burnt to improve the grazing, whereas north-facing slopes with damp undergrowth do not burn so easily. In consequence a southern slope will frequently be treeless, while a neighbouring northern slope will be forested.

At higher altitudes the snow on a north-facing slope lies longer into the spring. This affects the flora, particularly in the Western Himalaya with its higher snowfall. Often a north slope will be covered by fir trees, with birch in the gulleys where the snow packs and lies long into the spring. An adjacent south-facing slope, and especially a steep rocky one from which the snow clears quickly, may well be covered by *Quercus semecarpifolia*, the spiny-leaved oak.

Geology

Those who know the flora of other lands may be surprised to learn that in the parts of the Himalaya which are subject to the monsoon the chemical composition of the underlying rock is not of prime importance in determining the type of vegetation. It is generally accepted that altitude, rainfall and aspect are far more important factors. One often passes from one rock formation to another without seeing any great change in the forest cover. For example, *Pinus roxburghii*, the long-leaved pine, grows on acid or alkaline rock without any apparent discrimination. However, this is true only of the vegetation as a whole, and certain species have definite preferences. *Cupressus torulosa*, for example, is largely confined to limestone country. Incidentally, despite the reputation of the genus *Rhododendron* for being calcifuge, the authors have seen *Rhododendron arboreum* growing in some quantity on limestone, both in Nepal and in the Western Himalaya, and *Rhododendron lepidotum* has also been seen growing on limestone screes.

The depth of soil and its ability to hold water during the dry season also seem to be of more importance than the composition of the underlying rock. Certain species such as *Quercus floribunda* grow only where the soil is moist and deep.

Man and animals

The Himalaya at lower altitudes are quite densely populated, and the activities of man have had a great effect on the flora found there. Large areas have been cleared of the original forest cover, and are now terraced for cultivation or used for grazing. In most regions cultivation ceases at about 2400 m, but in dry areas along the Tibetan border irrigated crops are grown as high as 4400 m. The zone of cultivation contains many exotic species and weeds with a world-wide distribution. Some forest survives within the zone of cultivation, though for the most part it has been much modified by cutting for firewood, lopping for fodder, and grazing of animals. Most of the best Himalayan forests are confined to the upper temperate zone which lies above the zone of cultivation, but here a distinction must be made between Nepal and the Western Himalaya. The latter have had the benefit of a Forest Service with a policy of conservation and planned timber extraction for over a hundred years. This policy is enforced with a fair degree of success, and in many places fine stands of timber can be seen quite close to villages. Nepal too has a Forest Service, but there does not seem to be the political will, in a country where the pressure of population is very high, to enforce a conservation policy which inevitably would cut across the traditional rights of villagers. In Nepal good stands of trees are almost all confined to higher altitudes and remoter areas. Recently the need for conservation and for reafforestation has become more generally appreciated.

The composition of the vegetation at all altitudes has been much modified by fire. Some fires may be accidental, but many slopes are burnt regularly to improve the grazing. Mature trees of *Pinus roxburghii*, can withstand burning, and if one looks at the bases of their trunks one can see that most of them are blackened by fire. Seedlings of this pine are not fire-resistant, and frequent burnings make regeneration impossible. The blue pine, *Pinus wallichiana*, is not fire-resistant, but its spread is favoured by fire. It regenerates quickly on slopes which have been cleared by fire of other types of forest, and the abundance of blue pine in many of the drier parts of the Himalaya must in part be due to burning. Conversely, the restricted distribution of *Cupressus torulosa*, Himalayan cypress, is undoubtedly due to fire, for this tree is exceptionally susceptible to burning. A typical habitat of this cypress is steep limestone rocks devoid of inflammable undergrowth. The silver-fir and hemlock, *Abies spectabilis* and *Tsuga dumosa*, are also often destroyed by fire. Where this happens the usual succession is to dense thickets of bamboo, amongst which the dead and blackened conifers stand forlornly for a number of years.

xix

Uncultivated land around the villages is almost always heavily grazed, though some slopes may be preserved for hay and in consequence have quite a rich summer flora. Cliffs and steep rocky places are always worth searching for plants, because they escape both fire and grazing. During the summer many animals are moved up from the villages to graze in clearings in the temperate forest and in the alpine meadows. The herdsmen reach the alpine zone in June and leave before the end of September, and during this period alpine camp-sites, which at other times of year are lonely empty places, will be full of life. Grazing eliminates many species from the meadows, and where the grazing is heavy there will not be many flowers. On the other hand some species such as *Iris kemaonensis* or *Anemone rivularis* thrive under these conditions, and display an abundance on grazed ground which they do not achieve elsewhere. Some of the larger *Primula* and *Meconopsis* species are also unpalatable to animals, and often occur gregariously around herdsmen's camps where there is much dung. They form attractive drifts of colour, but they do not improve the pastures. Grazing is particularly heavy in the Western Himalaya, and is a factor which must always be borne in mind when planning an excursion there in search of alpine flowers. In Nepal the problem is not so acute. Grazing is heavy on the flat floors of the big inner valleys such as Khumbu or Langtang, but elsewhere there are many meadows which are only lightly grazed before the animals are moved on.

Many tourists now visit the Himalaya, and in recent years trekking parties in the more popular areas have begun to make an impact on the environment. Disposal of rubbish is now a serious problem on some routes, but this is not primarily the concern of a botanist. Nor is another problem of which any sensitive visitor must surely be aware: the impact of large numbers of tourists upon the way of life of local people, who until very recently were shielded from such intrusion. But the plants also suffer. It must be confessed that neither author is well qualified to lecture others on the need for plant conservation, for in earlier years when the flora of Nepal was little known they both spent much time collecting and pressing many plants in order to make herbarium specimens. However, on the principle of poacher turned gamekeeper they would like to point out that times have changed. Given the number of tourists who now visit the Himalaya, it requires only a few years before the flora in the more popular parts will be much impoverished unless visitors exercise restraint. Epiphytic orchids seem particularly vulnerable. The camera or the seed-packet are to be recommended rather than the trowel.

Most trekkers when in camp rely on wood for fuel. In forested areas this creates no problems, but in the alpine zone the shrubs and small trees of the uppermost forest and alpine shrubberies are already in many places under severe pressure as a source of fuel for local herdsmen. A succession of large trekking parties occupying the same alpine camp-sites may well cause difficulties, and there seems to be a good case for encouraging large parties to bring their own stoves and fuel.

Where and when to visit the Himalaya in search of flowers

Before considering various areas open to trekkers in the Western Himalaya and in Nepal, it may be helpful to offer some brief general advice to those planning a first Himalayan trip.

Choice of an area to visit will depend very much on the time of year. From mid-March to the end of May, go to Nepal. The flowering season there is earlier than in the west, the spring shrub flora much richer, and the weather should be good. From June to the end of September, go to the Western Himalaya. At that time the monsoon rains are falling heavily in Nepal. Between late June and early September monsoon conditions also affect the Western Himalaya, but there are a number of areas in the west which are to a large extent protected from the rain. In the autumn, after the monsoon has finished, trekking conditions are excellent throughout the Himalaya, though the main flowering season is now over. A trekker might well prefer Nepal at this time of year, for there is more in flower here at medium altitudes than in the west. Those whose interests lie in the temperate and alpine flora will find little in flower in any part of the Himalaya between November and late March. At this time there is much to be said in favour of a trek through the 'terai' jungles which lie at the base of the Nepal foothills, but the flora there is tropical and beyond the scope of this book.

The Western Himalaya

For many years after Independence the Indian Government excluded foreigners from all but a few small areas of the Western Himalaya. Recently this policy has changed. The Inner Line still closes the Tibetan border areas, but sufficient country is now open to satisfy the most energetic traveller. No trekking permits are needed, though passports must be carried and may be inspected at police posts.

Apart from Srinagar in Kashmir, which can be reached by air, the usual way of approaching one's chosen area is by road. In recent years much money has been spent in building roads to open up the mountain country. These roads tend to destroy that feeling of remoteness which to most visitors is a major attraction of the hills, but they do at least enable one to reach the high mountains quickly. Most of these roads run through country which is quite heavily populated. Those who are accustomed in other parts of the world to view the flora with the help of a car are warned that the Himalaya are too big for this to be a successful method of transport except in a very few places. To see the flora adequately one must trek with porters or ponies.

In the Western Himalaya there are a number of hill-stations: Dalhousie, Dharamsala, Simla, Mussoorie, Lansdowne and Naini Tal, while far to the east and outside our area lies Darjeeling. These places are a legacy from the British, and Nepal has no similar resorts. They were developed in a pre-motor age and all are sited on the outer edge of the hills, within easy reach of railheads in the plains below. These hill-stations, with the possible exception of Darjeeling, are too far from the snow-ranges to make good centres of operation for anyone interested in flowers. Manali in the Kulu valley lies much deeper into the hills. It is not a hill-station in the old sense, but it has become a considerable holiday centre where arrangements can be made for mountain travel. Further west, Srinagar in Kashmir has for generations been a place where tents and transport can be hired. There is no real trekking centre in Kumaon or Garhwal, though porters are available in some of the villages. Those who intend to attempt any travel other than on a very limited scale here may have to employ experienced men from Manali or Kathmandu.

Another legacy of the British is a number of bungalows scattered throughout the hills for the use of officials on tour. In more accessible parts these bungalows during the summer months tend to be full of officials and their families, but in remoter parts one can often get a room. On the other hand, the custom which has started in recent years on the main trekking routes in Nepal of villagers offering a night's board and lodging in return for a small payment does not prevail in the west. Any serious trekker will have to be equipped with tents.

Winter lingers long in the Western Himalaya, and there is little in flower in March. Much snow still lies at higher altitudes during April and May, and most of the passes remain closed. The fruit trees are then in blossom, and in clearings in the lower conifer forest *Corydalis diphylla*, *Colchicum luteum*, *Gagea elegans*, *Paeonia emodi* and *Tulipa stellata* are in flower. *Parrotiopsis jacquemontiana* and *Prunus jacquemontii* are attractive West Himalayan endemic shrubs, and the western form of *Viburnum grandiflorum*, commonly known as *V. foetens*, is ubiquitous, but the spring shrub flora is not nearly so rich as in Nepal. Irises make a beautiful show in the Kashmir valley during the spring, and perhaps this is the best place to see something of the spring flora. At this time of year much can be achieved in day trips from bungalows around the valley. The trout fishing too is extremely good before melting snow begins to cloud the rivers. Another pleasant medium-altitude district suitable for a spring visit is Saraj to the south of Kulu.

By the end of May the snow is melting fast and the rivers rise in spate. Many herdsmen with their flocks come up from the lower country and enter the forest zone of the hills which surround the Kashmir valley. In consequence the June flora of the forest zone and of the country surrounding resorts such as Gulmarg and Pahalgam is much grazed, and the alpine flora above is not at its best until July. In a normal year the Zoji La pass which leads into Ladakh is open to vehicles by early June, and a visitor based in Srinagar might therefore consider crossing the pass and taking a look at the drier country around Dras and Kargil. It is still much too early for the high-altitude Ladakh flora, but the 'oasis' flora of the irrigated hay meadows is at its best, and the big yellow-flowered umbellifers *Ferula jaeschkeana* and *Prangos pabularia* are prominent. The flora here is far richer in June than one would believe possible when viewing the stony hill-sides in the heat of July. This is one of the few parts of the Himalaya where one can work conveniently from a car.

For a visitor based on Manali there are several June choices. Early in the month one can ascend one of the Kulu side-valleys and see something of the lower alpine flowers before the sheep and goats arrive. *Corydalis cashmeriana, Corydalis govaniana, Fritillaria roylei, Iris kemaonensis, Megacarpaea polyandra* and *Thermopsis barbata* are then at their best, even though one will be regretfully aware that the majority of the summer alpines are still to come. The more ambitious can cross one of the passes into Chamba, where the mountain country is very fine and the flora rather similar to that of Kulu. In June the weather should be excellent for trekking in all this country, though there may be occasional showers.

Another possible choice is to cross from Manali over the Rohtang pass into Lahul. This is not normally open to vehicles until late June or early July, and before then it is necessary to cross the snow on foot. Lahul is much drier than Kulu, and has an irrigated 'oasis' flora rather similar to that of Kargil. In June the pink-flowered *Rosa webbiana* covers the hillsides, and on stony slopes there are white drifts of *Eremurus himalaicus*. By July this lower Lahul flora is almost completely burnt up. There is a route leading from Lahul down the Chenab to Kishtwar and Kashmir running at medium altitudes, which in June is very suitable for exploration in search of flowers.

In July and the first half of August the alpine flora throughout all the Western Himalaya is at its best, and there is an embarrassment of choices. The summer monsoon flora is very rich but it cannot be seen without getting wet. There is much mountain country fully open to the monsoon rains in the states of Himachal and Uttar Pradesh. For convenience it would be as well to choose an area approachable by tarred roads which are kept open during the monsoon, though even these may for a time be closed by landslides after heavy rain. Parts of Kulu district which are within easy reach of Manali might be a suitable choice, or the Valley of Flowers, which is reached from Joshimath in Garhwal. Although trekking conditions will be wet in these places the flora is rich and colourful. However, anyone who selects the Western Himalaya rather than Nepal for a high summer visit probably does so in order to avoid the rain, in which case Kashmir, Ladakh or Lahul are suitable areas.

Kashmir does get some rain during July and August, but the Pir Panjal range keeps off most of the monsoon and the hills escape the almost constant drizzle which at this time shrouds many other parts of the Himalaya. In Kashmir there are almost sure to be long intervals of fine weather. A favourite route for flowers, and rightly so despite the fact that parts of it are quite heavily grazed, is that which leads from Sonamarg via the Vishenshar and Gangabal lakes to Wangat. This route can be extended by starting from Pahlgam and travelling via Amarnath or one of several other passes. Although this is popular trekking country one is not likely to suffer from much overcrowding. The alpine flora of Kashmir is as attractive as anywhere in the Himalaya, and the mountain scenery, although not on the grandest scale, is very beautiful.

Lahul too escapes most of the monsoon rains. By July the Rohtang pass is cleared of snow, and one can travel by road from Manali to the foot of the Bara Lacha pass which leads over into Ladakh, or turn westwards to visit Udaipur and Miyah Nullah. One's first impression of the alpine zone of Lahul may be of stony slopes largely devoid of plants, but a little perseverance will reveal much of interest. Below 4500 m the flora is at its best in July. The high-altitude flora, which is not fully in flower until August, can conveniently be seen on the long flat summit of the Bara Lacha pass (5000 m).

Leh in Ladakh became accessible to foreign tourists in 1974, and since then has become a very popular tourist objective. Much of the country around Leh and the Indus valley is so arid that it is largely devoid of vegetation, and most people go there to see the Buddhist monasteries rather than the plants. Other parts such as Suru and Zanskar, which lie immediately to the north of the Himalayan range, have a higher rainfall and a richer flora. A trek from the Bara Lacha pass in Lahul, through Padam and Zanskar to Suru and Kargil, enables one to cover a wide sweep of this richer country. Much of the route consists of alpine steppe country, and if one is to find much in flower in this type of country it is essential to get one's timing right. July to mid-August is best, varying according to altitude. Either before or after this period there will be very little in flower. The pea family, Leguminosae, is prominent here, with low spiny shrubs of *Caragana versicolor* covering large areas, and there are a number of species of *Astragalus* and *Oxytropis*. Several *Potentilla* species are common, and there is much *Lindelofia stylosa* and *Lonicera spinosa*. On rocks *Biebersteinia odora* and *Parrya nudicaulis* are attractive but much rarer plants. The landscapes are big and sunny, often dominated by mountains of brown and grey rock. Most people will be charmed by the Buddhist villages with their *chötens* and *gompas*, and their irrigated fields bright with geraniums, aquilegias and louseworts. The Suru district is rather wetter than Zanskar, and a number of plants which grow in the hills surrounding the Kashmir valley can be found here also. It is in some ways unfortunate that a motor road has now been built from Suru to Padam, for it spoils what previously was a very good walk.

By late August nights in camp in the alpine zone are getting sharp, the first frosts come, and the flowers are mostly over. Some of the gentians will still be in flower, and there will be plenty of seed to collect. Soon the interest switches to autumn colours, and by October the Kashmir valley displays yet another aspect of its charms, with the poplars and plane-trees turned to gold and the saffron in flower in the fields. The wildfowl begin flighting in from their breeding grounds to the north, the first heavy snowfalls shut the Zoji La, and the valley settles down to the long winter.

Readers may notice that in making the above suggestions for trekking there is little mention made of the districts of Kumaon and Garwhal which lie within the state of Uttar Pradesh. It must be confessed that both authors have travelled much less in the hills of Uttar Pradesh than elsewhere in the Himalaya, and it may be relevant to state why this is so. Within the area covered by this book, the East Himalayan flora is richest in species in the wet eastern parts of Nepal. Conversely, the West Himalayan flora is best seen in drier parts such as Kashmir, Lahul and Ladakh. The authors, having previously worked the country at either end of our area, have not found it necessary to spend much time in this central portion. That is not to say that the central portion does not have an interesting flora, for many of the species which occur to east and west occur here also, and a small part of the flora is endemic. It also contains much very fine mountain country, with peaks such as Kamet and Nanda Devi. However, it remains true that if one wishes to see either the East or West Himalayan flora at its best one should not choose to visit this central portion.

Nepal

Anyone wishing to trek in Nepal must first visit Kathmandu to obtain the necessary trekking permits. The limits of the areas open to foreign tourists vary from time to time, but there is plenty of country to choose from. There are now a number of firms in Kathmandu who specialize in making trekking arrangements. In recent years a road system has been built which extends east and west of Kathmandu, and other roads are under construction. These roads run only through the lower country, but one may well be able to use them to approach one's chosen mountain area. To reach a few places one may be able to use light planes.

For anyone whose primary object is to see the flora, there is not enough in flower to make it worthwhile to take the field until late March or early April. From then until the beginning of June most of the trees and shrubs of the temperate zone come into flower. Without doubt the spring shrub flora is one of the most attractive features of Nepal. By the end of March the shrubberies around the villages resound to the call of cuckoos, and the shrubs of the evergreen oak forests begin to flower. A little later shrubs come into flower in the conifer forests which cover many of the slopes above. The full beauty of the Himalaya has not been experienced until one has seen the rhododendrons in their masses flowering beneath hemlock and fir. *Rhododendron arboreum* and *R. barbatum* are ubiquitous, and in eastern parts many other forest species can also be seen, such as *R. falconeri*, *R. hodgsonii* and *R. grande*. *Magnolia campbellii* also is common in the east. The forest floor is carpeted with mauve primulas, and there are a number of strange-looking *Arisaema* species. By early June, just before the coming of the monsoon, the upper birch forest and alpine shrubberies begin to come to life. Of the sub-alpine rhododendrons *R. campanu-latum* is much the commonest, but in the east there are several others such as *R. campylocarpum*, *R. fulgens* and *R. wightii*.

The shrub flora is richest on the south side of the main ranges at altitudes between 2400–3800 m; slopes which later in the year are fully exposed to the monsoon rains. The flora is particularly rich on the ridges which run down to the Arun and Tamur rivers in extreme east Nepal, but more easily reached country to the north of Kathmandu or Pokhara is also well worth visiting at this time of year. The weather in general should be fine, though places which lie immediately beneath the great wall of snow-mountains tend to get some heavy spring rainstorms. The Arun and Tamur valleys can be especially wet. Spring weather is rather more predictably fine in the high inner valleys, but here it is still too early for most flowers. On the other hand, the low-altitude village country, through which at some stage one will almost inevitably find oneself travelling, has a number of interesting plants in flower in the pre-monsoon period, despite the fact that much of it is under cultivation. Many of the epiphytic orchids flower here in the spring.

By the beginning of June the weather begins to break, and the monsoon rains gradually spread from east to west throughout the country. From now until the end of September the question must be asked whether it would not be preferable to trek in the much drier conditions of the Western Himalaya. However, the fact must be faced that anyone who really wishes to see the Nepal flora will have to carry on regardless of the rain, for it is during the monsoon months that the great majority of plants are in flower. Travelling conditions then are uncomfortable but not impossible. Marches will mostly be made in the rain. Visibility is often limited to less than a hundred yards, and the snow-peaks remain invisible for months on end. There may be diversions due to flooded rivers, and there will be precarious bridges to be crossed. Camp life is difficult, and there are many leeches. But those who persist will be rewarded by seeing many species in flower.

After the rains have been falling for some time there is much in flower at all altitudes, but most people will probably want to concentrate on the alpines. On the south side of the main ranges the open slopes which lie above the rhododendron-conifer forests have a rich alpine flora. By late June the big *Meconopsis* and *Primula* species are in flower, together with the alpine rhododendrons and a great many other plants. The alpine flowers continue in succession into September, when species of *Aconitum, Codonopsis, Cremanthodium, Cyananthus, Leontopodium* and *Saxifraga* are prominent. During the monsoon these southern ridges are extremely wet places, and in the almost constant mist and rain which prevail there even the hardiest person is unlikely to wish to camp for more than a week or so at a time before descending to dry out at lower levels. One of the more accessible areas, such as Gosainkund to the north of Kathmandu or the southern side of Annapurna close to Pokhara, might well be chosen to see something of this type of flora.

At this time of year it might be preferable to visit parts of Nepal which are not so constantly wet. The drier areas which are available can be summarized as (a) the heads of the inner valleys, (b) the country north of Annapurna and Dhaulagiri, and (c) Humla-Jumla. The flora of all these places, though different from that of the rain-soaked outer slopes, is just as interesting.

The heads of the big inner valleys of central and east Nepal are very much drier since the surrounding mountains shelter them from much of the rain. The main valleys from east to west are Kambachen, Barun, Khumbu, Rolwaling, Langtang, Shiar Khola and Buri Gandaki. Recent glaciation has given to these valleys a smooth ice-worn appearance, and walking tends to be much less laborious than in other parts of Nepal. For much of the year there are stupendous views of the great mountains at close range, and one can climb without any difficulty to quite high altitudes. For these reasons the inner valleys are popular with tourists, and some such as Khumbu and Langtang both before and after the monsoon become crowded. Anyone seriously interested in the flora should not aim to visit these places at these times, for the great majority of the alpines will then either not yet be in flower or be over. During the monsoon the inner valleys are full of flowers and almost devoid of tourists. In July and August the cloud cover will be fairly constant, and there will be much drizzle, though probably not much heavy rain. In June or September one may well enjoy rather better weather, and even in July and August there may be the odd fine morning when for a few hours the mountains stand clear and sparkling in the sun before being swallowed up once more in the rainclouds which come drifting up from below. The rainfall does not fall off significantly until the tree-line in the valley-heads is reached. Above this the flora soon shows signs of the drier climate, with much *Ephedra gerardiana, Hippophae tibetana, Myricaria rosea, Rhododendron nivalis* and dwarf black juniper. The flora, the people, the yaks, the *gompas*, the *chötens* and prayer wheels all serve to remind one that only just over the other side of the mountains lies Tibet.

The big Annapurna and Dhaulagiri ranges protect the country to the north from the rain. As a result there is a large area running along the Tibetan frontier in Central and West Nepal which has a flora more Tibetan than Nepali. This dry area falls into three parts; the upper Marsiandi valley; the upper Kali Gandaki valley; and the upper Bheri and Dolpo district. Permission to visit some of these border areas is difficult to get, but at the present the routes up the Marsiandi and up the Kali Gandaki to Jomsom and Muktinath are open. Usually most of June is dry here, and July and August are not unduly wet.

June is still too early for most alpines, but by then the pretty *Stellera chamaejasme* will be scenting the hillsides and the pink *Primula tibetica* in flower in the marshes. In some places the dry steppe-country is covered with wormwood or with low spiny bushes of *Caragana*. In others the soft rock has been eroded into pinnacles, with *Cordydalis moorcroftiana, Dicranostigma lactucoides* or *Incarvillea arguta* growing on ledges. If permission can be obtained for the Phoksumdo lake north of Dhaulagiri, it is well worth a visit in June. A number of Himalayan rarities such as *Ceratostigma ulicinum, Clematis phlebantha, Phaeonychium parryoides* and *Rhododendron lowndesii* can then be seen in flower round the lake, as well as a good display of more common alpines. By July there is much in flower throughout the whole of the dry area, and conditions remain dry enough for one to enjoy the mountain scenery and view the flora in comfort. The higher alpines are not at their best until August.

The short trek from Pokhara up the Kali Gandaki to Jomsom is likely to prove fascinating to anyone interested in plants, despite the fact that the route these days tends to be over-crowded. The change in climate and vegetation over a short distance is more sudden here than anywhere else in the Himalaya. In the course of a few days walk, one passes from a warm wet monsoon flora to the dry windy vegetation of the Tibetan plateau. The contrast can be observed at any time of year, but is at its most striking when all to the south is shrouded in monsoon clouds while to the north the skies are blue.

The Humla-Jumla district in West Nepal is another area which has a lower rainfall. It is protected from much of the rain by a lesser range of mountains up to 4500 m high which cuts it off from the plains. The climate here is not as arid as in the country north of Annapurna and Dhaulagiri, but it is by no means as wet as other parts of Nepal. The country is of an attractive open nature, the houses have flat roofs, and in many respects the district has something of the appearance of a 'little Kashmir'. Many West Himalayan plant species can be found here which do not grow any further east in Nepal, and there are extensive mixed coniferous forests of a type widespread in the West Himalaya. Unfortunately the area is somewhat remote from the main centres of communication. At certain times of the year air transport can be used to get there, but flights cease with the advent of the monsoon. The Rara lake to the north of Jumla is a pretty place, and has become a popular objective for tourists. This is high, cold country and one cannot recommend a visit in search of flowers in the spring. There will be much peach blossom around the villages, but little else. By October the season is too late. Anyone prepared to walk up from the plains or from Pokhara and spend the monsoon months trekking in this district would see many flowers. There will be some rain, but it will not be nearly so heavy or constant as elsewhere.

October and November are very popular months for trekking in all parts of Nepal. The rains are over, the mountain views are marvellously clear, and the winter snows have not yet arrived. This is by no means the best season for flowers, but many people will wish to give priority to the mountain scenery. Those with an interest in plants will not be entirely disappointed. In the alpine zone there will be seed to collect, and the late gentians such as *G. algida, G. depressa* and *G. ornata* will be in flower. The uppermost forest will be brightened by the turning autumn leaves of *Acer, Berberis, Betula* and *Sorbus*. The rhododendron-conifer and evergreen oak forests lower down are rather sombre places in the autumn, but lower down again at village level *Prunus cerasoides* flowers on many of the wayside resting places, and some shrubs of a rather subtropical kind such as *Luculia gratissima* and *Oxyspora paniculata* will also be in flower. The rice is harvested, the villages are decorated for the autumn festivals, and most people other than dedicated plant-hunters will find this the most attractive time of all to travel in Nepal.

Note to the Reprint

The reprinting of this book has enabled the authors to correct a few typographical errors. The only corrections of substance concern Colour Plates 105 and 106.

The photo previously named 1145 *Nepeta govaniana* is now corrected to 1147 *Nepeta erecta*. The two species are very similar, and the foliage identical. The flowers of *Nepeta erecta* are mauve, in whorls (sometimes many-flowered) of which only the lower are stalked, the whorls forming a rather stiffly erect interrupted spike. The flowers of *Nepeta govaniana* are yellow, in few-flowered distant whorls which are all stalked and the lower whorls long-stalked, the whole forming a rather lax raceme.

Likewise, the photo of 1181 *Aconogonum campanulatum* is now corrected to 1178 *Aconogonum molle*.

RANUNCULACEAE   Buttercup Family

A large cosmopolitan family, mostly of herbaceous plants, with its main concentration in the temperate and cool regions of both Hemispheres. Usually perennials, with leaves commonly much divided or lobed, either all basal from a rhizome or swollen root, or alternate on the flowering stem, except in *Clematis*. Perianth segments in 2 or 3 whorls: the outer segments often 5, either green and sepal-like, or coloured and petal-like; the inner whorls of 5 or more coloured segments, or sometimes absent, or sometimes forming nectaries. Stamens many, spirally arranged. Ovary of few or many spirally arranged free carpels (rarely 1), each carpel either with several ovules and splitting in fruit, a *follicle*, or each carpel with one ovule and not splitting in fruit, an *achene*. Fruit a cluster of follicles, or achenes, or a berry as in *Actaea*.

Note: in this family the green perianth segments are referred to as sepals, the coloured segments as petals in the following descriptions.

Fruit of 1 or more follicles, or a berry
  a Flowers regular
    i Flowers large, more than 2 cm across
      x Flowers not spurred   1 CALTHA   2 TROLLIUS
                                                        3 PARAQUILEGIA
      xx Flowers spurred   4 AQUILEGIA
    ii Flowers small, less than 2 cm across
      x Flowers in long racemes   5 ACTAEA   6 CIMICIFUGA
      xx Flowers solitary   7 ISOPYRUM
  b Flowers irregular (symmetrical in one plane only)   8 ACONITUM
                                                        9 DELPHINIUM

Fruit of 1 or many achenes
  a Herbaceous plants; leaves alternate or whorled
    i Flowers with outer green sepals present   10 CALLIANTHEMUM
          11 OXYGRAPHIS 12 RANUNCULUS 15 THALICTRUM
    ii Flowers with the outer whorl of coloured petals   13 ADONIS
                          14 ANEMONE 15 THALICTRUM
  b Usually woody climbers; leaves opposite   16 CLEMATIS

1 **CALTHA** Leaves entire, rounded. Petals 5–8; sepals and nectaries absent. 3 spp.

1 **C. palustris** L. var. **himalensis** (D. Don) Mukerjee MARSH MARIGOLD
Pakistan to Bhutan. 2400–4000 m. Open slopes, damp places on grazing grounds; very common. May–Aug.
    A rather fleshy-looking hairless plant of damp places, distinguished by its large shining rounded or heart-shaped leaves, and its terminal cluster of

few usually yellow or white cup-shaped flowers. Flowers 2.5–4 cm across, with 5–8 ovate petals, and numerous stamens and carpels. Leaves mostly basal, long-stalked, blade 3–15 cm across, finely toothed. Fruit a cluster of beaked follicles each *c.* 1 cm. **Pl.1.**

2 TROLLIUS GLOBE FLOWER Differing from *Caltha* in having much divided leaves; 5–15 yellow petals, and an inner whorl of strap-shaped nectaries. 2 spp.

2 **T. acaulis** Lindley
Pakistan to W. Nepal. 3000–4300 m. Alpine slopes. May–Jun.
Flower solitary, golden-yellow, 5 cm across, flowering before or after the deeply divided basal leaves develop. Petals broadly oval, 5–10; nectaries 12–16, yellow, oblong, shorter than stamens. Leaf-blade rounded in outline, deeply 5-lobed, the lobes further cut into oblong-lanceolate acute, toothed segments; stem leafy above the middle, 8–15 cm in flower, and to 30 cm in fruit. Fruit *c.* 2.5 cm, of several follicles with persistent styles. **Pl. 1.**

3 **T. pumilus** D.Don
Uttar Pradesh to S.W. China. 3600–5500 m. Open slopes. Jun.–Jul.
Very like 2, but flowers smaller 2.5–3 cm across, and petals shallowly notched; nectaries stalked. Flowering stem leafy below the middle.

3 PARAQUILEGIA Flowers solitary; petals stalkless. Follicles 3–7. 1 sp.

4 **P. microphylla** (Royle) J.R. Drumm. & Hutch. (*P. grandiflora* auct. non J.R. Drumm. & Hutch.)
Pakistan to S.W. China. C. and N. Asia. 3400–4900 m. Rock crevices. Jun.–Jul.
A very distinctive small rock-crevice plant, often forming large clumps, with delicate often somewhat pendulous white, blue or lilac flowers and rather glaucous ferny foliage. Flowers solitary, cup-shaped, 2–3.5 cm across; petals ovate, borne on slender leafless stems 4–6 cm; bracts 2, linear or lanceolate. Flowers very variable: W. Himalayan forms, flowers small, white; E. Himalayan forms, flowers large, lilac. Leaves all basal, long-stalked, ternate, with leaflets 1–2 cm across which are further divided into deeply lobed segments; rootstock tufted and covered with the bases of old leaf-stalks. Follicles 5–10 mm, with persistent styles. **Pl. 2.**

4 AQUILEGIA COLUMBINE Distinguished by the 5 backward-projecting spurs to the inner petals. Leaves compound. 4 spp.

Flowers solitary
5 **A. nivalis** Falc. ex Jackson (*A. vulgaris* subsp. *jucunda* Hook.f. & Thoms.)
Pakistan to Himachal Pradesh. 3000–4000 m. Alpine slopes, screes, rocks. Jun.–Aug.
Flowers deep purple with almost blackish-purple inner petals; spurs usually short, incurved. Leaves few, ternate, divided into broad

2

kidney-shaped 3–lobed leaflets; stem unbranched, 10–20 cm, with 1–2 leaves. **Pl. 1.**

Flowers several
a *Spurs long, straight or slightly curved*

6 **A. fragrans** Benth.
Pakistan to Uttar Pradesh. 2400–3600 m. Shrubberies, meadows. Jun.–Aug.
Flowers white or cream-coloured, sweet-scented, large 3–5 cm, with long straight or hooked spurs. Outer petals 2.5–3 cm, blunt or acute; spur, 15–18 mm. Leaves twice-ternate; leaflets 3–lobed, lobes with shallow teeth, more or less glaucous above; stem 40–80 cm, branched above. Follicles 6–9, densely hairy, with long styles. **Pl. 1.**

7 **A. moorcroftiana** Wallich ex Royle
Afghanistan to S.E. Tibet. 2700–4200 m. Shrubberies, open slopes. Jun.–Jul.
Flowers purple, usually several, occasionally solitary, distinguished by their long straight or slightly curved spurs which are longer than the blade of the inner petals. Basal leaves twice-ternate; leaflets glaucous, obovate and further lobed, 1–2 cm across, hairless above and finely hairy beneath; stem 20–40 cm. Follicles 5–6, hairy, styles, 7–10 mm.

b *Spurs short, incurved*

8 **A. pubiflora** Wallich ex Royle
Pakistan to W. Nepal. 2400–3300 m. Shrubberies, open slopes. Jun.–Aug.
Like 7, but spurs shorter than the blade of the inner petals and strongly incurved, and outer petals much longer than the inner and tapering to a narrow point. Flowers purple, 2–4 cm. Leaves finely hairy beneath. Follicles long-hairy, styles hooked. **Pl. 1.**

5 **ACTAEA** Sepals absent; petals 3–5. Fruit a berry. Leaves doubly-compound. 1 sp.

9 **A. spicata** L. var. **acuminata** (Wallich ex Royle) Hara
Afghanistan to S.E. Tibet. 2400–3600 m. Forests, shrubberies. May–Aug.
Distinguished by its terminal cylindrical cluster of numerous small white flowers, and its large ternate twice-cut leaves with lanceolate conspicuously sharp-toothed leaflets. Flower clusters 3–5 cm long; flowers *c.*1 cm across; petals soon falling; stamens conspicuous, white, much longer. Leaves *c.*40 cm, the upper smaller; stem 60–90 cm. Fruit in a terminal cluster of black shining stalked berries each 1 cm.
Whole plant and fruits poisonous. Roots used medicinally. **Pl. 3.**

6 **CIMICIFUGA** Like *Actaea*, but fruit a cluster of 4–8 follicles. 1 sp.

10 **C. foetida** L. (incl. **C. europaea** Schipcz.)
Pakistan to S.E. Tibet. Temperate Eurasia. 2700–4000 m. Shrubberies, forest clearings. Jul.–Sep.

A tall perennial 1–2 m, with compound leaves, and with a pyramidal branched inflorescence of small yellowish flowers crowded into usually slender often drooping clusters. Flowers *c.* 6 mm; sepals and petals similar, ovate, some inner ones 2–lobed; stamens conspicuous, long yellowish. Leaves ternate, 20–40 cm; leaflets ovate to lanceolate, sharply toothed, the terminal leaflet usually 3–lobed. Fruit of several follicles 1–2 cm, with hooked styles. **Pl. 4.**

7 ISOPYRUM Differing from *Paraquilegia* in having petals narrowed to a stalk. Follicles 2, wide-spreading and fused at the base. 1 sp.

11 **I. adiantifolium** Hook.f. & Thoms.
C. Nepal to Arunachal Pradesh. Burma. 2100–3000 m. Forests. May–Jun.
A delicate perennial with solitary white flowers *c.* 1.3 cm across, borne on a slender stem, and with compound leaves. Petals 5–6, soon falling. Leaves ternate, with rounded, toothed leaflets 5–10 mm; stem leaves opposite or whorled, basal leaves long-stalked; stems one or few, 5–12 cm. Follicles distinctive. **Page 445.**

8 ACONITUM ACONITE A difficult genus with many species separated by small botanical characters. Distinguished by their irregular flowers (symmetrical in one plane only), with 5 outer coloured petals, the upper forming a large erect hood, and 2–10 inner petals, the two upper forming nectar-secreting spurs placed inside the hood. Stamens numerous; follicles 2–5. *c.* 33 spp.

Hood 3 times taller than broad

12 **A. laeve** Royle
Pakistan to W. Nepal. 2000–3300 m. Forests, shrubberies. Jun.–Aug.
A tall robust perennial with a branched inflorescence of long spike-like clusters of many small white, pale yellow or dull purple flowers, each with a tall cylindrical hood. Inflorescence to 50 cm or more; flowers 1.5–2.5 cm long. Leaves with rounded-heart-shaped blade 15–30 cm, deeply cut into 5–9 broad lobes, the lobes coarsely toothed; stems 1–2 m. Follicles 3, diverging in fruit. **Pl. 3.**

Hood not taller than broad
a *Upper leaves toothed or shallowly lobed*

13 **A. heterophyllum** Wallich ex Royle
Pakistan to C. Nepal. 2400–4000 m. Forests, shrubberies, open grassy slopes. Aug.–Sep.
Distinguished by its rather large greenish-purple conspicuously darker-veined flowers, and its coarsely toothed but otherwise entire upper leaves. Flowers 2.5–3 cm, usually in lax spike-like cluster with very variable bracts which are either small linear, small ovate, or large ovate and enclosing the

lower part of the flower and fruit. Hood rounded, broader than long. Leaves with ovate-heart-shaped to rounded blades, usually 4–8 cm, the upper clasping the stem, all with large rounded teeth; lowest leaves deeply lobed and long-stalked; stem 30–120 cm. Alpine plants are sometimes much reduced in size and numbers of flowers. Follicles 5, 16–18 mm, shortly hairy, erect.

The tubers are used medicinally as a tonic.

b *Leaves all deeply cut nearly to the base*
  i *Ultimate leaf segments mostly 1–3 mm wide*

14 **A. hookeri** Stapf
C. Nepal to S.W. China. 3600–4800 m. Alpine slopes, rock ledges. Aug.–Sep.

A small perennial with a slender stem usually less than 10 cm, with few rounded deeply cut leaves, and one or several deep blue flowers 2.5–3 cm. Flowers long-stalked, sparingly hairy; hood strongly curved and leaning forward. Leaf-blades 1–1.5 cm across, deeply 3–5-lobed, the lobes further cut into narrow usually blunt segments; stem leaves much reduced. Roots paired, tuberous. Follicles 5, densely hairy. **Pl. 2**.

15 **A. violaceum** Jacquem. ex Stapf
Pakistan to C. Nepal. 3600–4800 m. Shrubberies, open slopes. Jul.–Sep.

A very variable perennial, with a dense spike of many dark or pale blue flowers variegated with white (var. **robustum** Stapf), or a small plant with one to several flowers, and leaf-blades only 2–3 cm across. Flowers 2–2.5 cm; hood broader than long; bracts small, linear. Leaves with rounded blade 2–10 cm across, palmately cut to the base, the lobes much cut into narrow segments 1–3 mm wide; stems 10–30 cm (30–60 cm in var. **robustum**). Follicles 5, densely hairy. **Pl. 3, Pl. 4**.

16 **A. gammiei** Stapf
C. Nepal to S.E. Tibet. 3300–4800 m. Alpine shrubberies, open slopes. Aug.–Sep.

Distinguished by its pale blue, white or greenish-yellow flowers borne in a lax few-flowered inflorescence. Flowers 2–3 cm. Leaves few, 4–5 cm across, deeply divided into narrow acute segments 1–2 mm wide, upper leaves smaller and less divided; stems 20–70 cm.

The roots are much used as a stomach medicine.

  ii *Ultimate leaf segments mostly more than 3 mm wide*

17 **A. ferox** Wallich ex Seringe
C. Nepal to Bhutan. 2100–3600 m. Shrubberies, forest clearings. Aug.–Oct.

Flowers dull-blue, in a terminal spike-like cluster 15–30 cm long, simple or sparingly branched below. Flowers 2–3 cm; hood broader than long, with a short acute beak. Lower bracts of inflorescence pinnately lobed, upper entire; bracteoles linear. Leaves with blade rounded or oval, 8–15 cm across, with 5 ovate-wedge-shaped lobes which are further deeply lobed with ultimate segments at least 3 mm wide; stem 1–2 m. Follicles 5, usually finely hairy.

An extremely poisonous plant. Also used medicinally. **Pl. 3**.

18 **A. spicatum** (Brühl) Stapf
W. Nepal to S.E. Tibet. 3300–4300 m. Forests, shrubberies. Aug.–Sep.

A tall erect perennial 1–2 m, with a dense terminal spike and sometimes with lateral spikes, of purple to greenish-white, veined flowers 2–2.5 cm. Inflorescence 15–40 cm; hood about as broad as long. Leaves often softly hairy; blades 6–12 cm across, deeply cut into ovate lobes which are further cut into toothed and pointed segments. Follicles 5, swollen, *c.* 1 cm, hairy.

The tubers are the largest in the genus; they are used medicinally, and are poisonous. **Pl. 3**.

9 **DELPHINIUM** Distinguished by its irregular flowers (symmetrical in one plane only) with 5 outer coloured petals, the uppermost with a backward-projecting spur, and 4 shorter inner petals, the upper two forming nectar-secreting spurs which are inserted in the outer spur; stamens 8. Follicles usually 3. A very difficult genus with *c.* 24 spp.

Plants very hairy
a *Flowers few, usually in a rounded or flat-topped cluster*

19 **D. brunonianum** Royle
Pakistan to E. Nepal. S.E. Tibet. 4300–5500 m. Stony slopes, screes, in drier areas. Jul.–Sep.

Distinguished by its large woolly-haired often conspicuously-veined blue to purple flowers with broad blunt spurs, borne in a small dense cluster on short stems to *c.* 20 cm. Flowers 3–5 cm including spur, appearing inflated and rather papery; inner petals blackish. Leaf-blade rounded in outline, 3–8 cm across, lobed to two-thirds, the lobes toothed. Some plants with a strong musky odour; also varying in degrees of hairiness. Follicles 4–5.

Juice of leaves used to destroy ticks in animals. **Pl. 4**.

20 **D. cashmerianum** Royle
Pakistan to Uttar Pradesh. 2700–4500 m. Alpine slopes, also on irrigated land in Ladakh. Aug.–Sep.

Like 19, but differs in not being glandular-hairy in the upper parts, and with smaller less inflated flowers mostly 2–3 cm, bluish-purple with a stout spur to 1.5 cm, hairy and veined. Flowers several, long-stalked, clustered into a dense somewhat flat-topped head. Leaves deeply lobed, 3–5 cm across. Follicles 3–7.

Roots used medicinally. **Pl. 2**.

21 **D. viscosum** Hook. f. & Thoms.
C. Nepal to S.E. Tibet. 3600–4500 m. Alpine slopes; not common. Aug.–Sep.

Like 20, but flowers usually greenish-yellow and with a longer and more slender spur often curved at the tip. Flowers less commonly pale blue or mauve, solitary or few, borne on long stalks and clustered into a head. Leaf-blade lobed to about one half, with rounded teeth; stems usually 10–20 cm. Follicles 3. **Pl. 2**.

b *Flowers many, in elongated spikes*

22 **D. drepanocentrum** (Brühl) Munz
Endemic to E. Nepal and Sikkim. 3300–5500 m. Shrubberies, open slopes. Jul.–Sep.

Distinguished by the very long down-curved spurs of the mauve or dull purple hairy flowers. Flowers in long one-sided spikes 10–30 cm; outer petals *c*. 1.5 cm, bristly-haired; spur to 2 cm. Leaf-blades rounded 5–10 cm, 5–lobed to two-thirds, the lobes broad-rhombic, toothed; stem unbranched, 40–60 cm, with very distinctive long bristly brown spreading hairs. **Pl. 4**.

23 **D. vestitum** Wallich ex Royle
Pakistan to E. Nepal. 2700–4300 m. Shrubberies, open slopes. Aug.–Sep.

A very hairy perennial with a long spike of rather small dull purple to blue flowers borne well above the leaves. Flowers 2–2.5 cm including short conical spur, veined and very hairy; bracts lanceolate. Leaf-blade rounded, shallowly or deeply 5–7–lobed, the lobes with acute teeth. Follicles 3. **Pl. 4**.

Plants hairless or with few hairs
a *Leaves dissected into narrow lobes mostly 1–3 mm broad*

24 **D. denudatum** Wallich ex Hook. f. & Thoms.
Pakistan to C. Nepal. 1500–2700 m. Grass banks, margins of fields. Jun.–Aug.

The commonest low altitude species of the W. Himalaya, distinguished by its relatively small blue or violet flowers borne in a widely branched inflorescence with a few terminal spike-like clusters. Flowers *c*. 2.5 cm with the upper outer petals *c*. 12 mm and spur 14–15 mm; upper inner petal white, the others blue. Leaves with rounded blade 5–15 cm across, cut into 3–5 narrow lobes *c*. 2–3 mm wide; stem leaves widely spaced up stem, more deeply cut into narrower lobes. Follicles nearly hairless.

Roots used medicinally.

25 **D. kamaonense** Huth
Uttar Pradesh to C. Nepal. 3000–4300 m. Open slopes; common in Nepal Jul.–Aug.

Like 24, with widely-branched inflorescence, but flowers mostly solitary at the ends of branches. Flowers deep blue, much larger, the upper outer petal to 2 cm; spur slender 15 mm. Leaf-blades smaller, 2–4 cm broad, cut into narrow lobes; stems quite hairless. Follicles with dense spreading brown hairs.

26 **D. roylei** Munz (*D. incanum* Royle)
Kashmir to Himachal Pradesh. 1800–2400 m. Shrubberies, open slopes;

7

common in Kashmir. Jul.–Sep.

Flowers bright blue in a dense elongated spike 10–20 cm (branched only at the base if at all); outer petals *c.* 1.5 cm, spur straight 1–1.5 cm, horizontal; bracts linear. Stem leaves stalkless, widely spaced, cut into many narrow pointed lobes 1–2 mm broad; stem 50–100 cm. Follicles 3, 1–1.5 cm, finely hairy. **Pl. 3**.

b *Leaves with lanceolate or ovate lobes mostly more than 5 mm broad*

27 **D. himalayai** Munz
W. to C. Nepal. 2400–4300 m. Open slopes. Jul.–Aug.

Flowers purplish-blue in a long one-sided spike 10–15 cm. Flowers 2–2.5 cm including spur to 15 mm; outer petals bristly-hairy outside, hairless inside; inner petals blackish; spur ascending; flower-stalks erect. Leaf-blades to 10 cm wide, 5–lobed nearly to base, the lobes broad and further lobed and toothed; stems 40–60 cm. Follicles densely hairy. **Pl. 5**.

28 **D. pyramidale** Royle (*D. ranunculifolium* Wallich)
Pakistan to C. Nepal. 2000–3600 m. Shrubberies, open slopes; common in Kashmir. Jul.–Sep.

Like 27, but distinguished by its usually more open and branched inflorescence with many terminal clusters of blue or purple flowers each *c.* 3 cm. Petals hairy on both sides; outer petals 15–16 mm; inner petals bluish-black. Leaves larger, 6–15 cm across, with more widely-spaced lobes. Fruit with spreading follicles, styles 3 mm; seeds blackish.

29 **D. scabriflorum** D. Don (*D. altissimum* Wallich)
W. Nepal to Arunachal Pradesh. 1500–2400 m. Grassy slopes; common. Aug.–Sep.

Flowers deep blue, densely hairy, with long spurs, borne in a dense central spike-like cluster, and with a few flowering branches below. Outer petals *c.* 15 mm; spur 2–2.2 cm, slightly curved; inner petals bright blue; flower-stalks spreading. Leaf-blade to 15 cm, deeply 5–lobed, the lobes further lobed and toothed.

10 **CALLIANTHEMUM** Petals 8–12, each with a nectary-pit; sepals 5, soon falling. 1 sp.

30 **C. pimpinelloides** (D. Don) Hook. f. & Thoms. (*C. cachemirianum* Cambess.)
Afghanistan to S.E. Tibet. 2700–4000 m. Forests, open slopes. May–Jun.

A small perennial with one or several white flowers borne on short stems arising from a tuft of basal leaves with small rounded, lobed, glaucous leaflets. Flowers 2.5–3 cm across; petals oblong with a rounded shallow-notched apex; sepals 5, ovate, much shorter. Leaves 2–3–times divided; leaflets deeply cut, 5–10 mm; flowering stems 5–10 cm. Achenes few, ovoid, wrinkled. **Pl. 4**.

11 OXYGRAPHIS Like *Ranunculus*, but sepals enlarging in fruit; petals 10–15. 1 sp.

31 **O. polypetala** (Royle) Hook.f. & Thoms. (*O. glacialis* auct. non Bunge) Kashmir to S.E. Tibet. Burma. 2700–4500 m. Open slopes recently cleared of snow; very common. Apr.–Jun.

A tiny early-flowering alpine plant with solitary or several yellow or bronze flowers with many narrow petals, and small somewhat fleshy entire or lobed, rounded to heart-shaped leaves. Flowers 1–2.5 cm across; petals oblong-spathulate; sepals 5, elliptic, blunt. Leaves all basal, stalked, blades 0.5–5 cm; flowering stem 2.5–10 cm. Achenes in a globular head surrounded by enlarged calyx. **Pl. 5.**

12 RANUNCULUS BUTTERCUP Sepals 5, green; petals usually 5 and shining yellow. Achenes numerous in a cylindrical or globular head. *c.* 30 spp.

Hill and valley species found below 3000 m
a *Flowers yellow, 1.5 cm across or more*

32 **R. lingua** L. GREATER SPEARWORT.
Kashmir to Himachal Pradesh. Temperate Eurasia. 1500–1800 m. Riversides, wet places. Jun.–Jul.

A tall erect hairless perennial usually growing in water, with spear-shaped leaves and a lax branched cluster of few large glossy yellow flowers. Flowers 3–5 cm across; petals 5, roundish-ovate; sepals spreading. Leaves to 25 cm, the lower stalked, the upper clasping stem; stem hollow, leafy, 1–1.5 m. Achenes hairless, minutely pitted.

33 **R. laetus** Wallich ex D. Don
Afghanistan to S.W. China. C. Asia. 1500–2700 m. Shrubberies, damp places; common. Apr.–Jun.

A rather robust erect hairy perennial 30–60 cm, with a branched, nearly leafless inflorescence of relatively large conspicuous glossy yellow flowers. Flowers 1.5–2 cm across, long-stalked; sepals spreading. Leaf-blades rounded in outline 8–12 cm, deeply 3–5–lobed, the lobes oblanceolate pointed and further 3–lobed and sharply toothed, densely hairy, the lower leaves long-stalked. Achenes smooth, somewhat flattened, with thickened margin and conical beak. **Page 445.**

b *Flowers yellow, small, usually less than 1.5 cm across*

34 **R. arvensis** L. CORN BUTTERCUP
Afghanistan to Uttar Pradesh. Temperate Eurasia. 600–2400 m. Cornfield weed; common in Kashmir. Apr.–Jun.

An annual weed of cultivation, distinguished by its spiny achenes, its upper leaves which are divided into 3 narrow entire lobes, and its small pale yellow flowers 4–12 mm across. Basal leaves spathulate or wedge-shaped, entire, or apex 3–5–toothed. Stem 15–60 cm.

35 **R. diffusus** DC.
Pakistan to S.W. China. Burma. 1500–4000 m. Forests, shrubberies, open slopes; common. Apr.–Jul.

A softly hairy spreading plant producing runners which root at the nodes, with solitary long-stalked axillary bright yellow flowers 1–1.5 cm across, with hairy sepals. Leaf-blade heart-shaped 2–6 cm, long-stalked, deeply 3–lobed, the lobes with acute or rounded teeth. Achenes 1–2 mm, hairless, flattened with thickened margin.

36 **R. sceleratus** L. CELERY-LEAVED BUTTERCUP
Throughout the Himalaya. India. North Temperate Zone. To 1800 m. Fields, grazing grounds, irrigation channels; common. Mar.–May.

A pale green shining-leaved, hairless, somewhat fleshy annual, with narrow upper leaflets and small pale yellow flowers, with reflexed green sepals. Flowers 5–10 mm across. Stem leaves deeply cut into 3 narrow lobes which are further toothed or lobed; basal leaves rounded in outline, long-stalked, with obovate cut and rounded-toothed lobes; stem hollow, grooved, 15–30 cm. Fruit an oblong head of swollen, hairless achenes.

A poisonous plant.

Mountain and alpine plants usually found above 3000 m (see also 35)
a *Flowers yellow; terrestrial plants*
   i *Leaves mostly entire, much longer than broad*

37 **R. pulchellus** C. Meyer
Pakistan to S. W. China. C. Asia. 3000–5200 m. Open slopes, rock ledges. Jun.–Aug.

A small erect plant usually 5–10 cm, with entire linear to elliptic leaves, or leaves sometimes toothed or sometimes palmately lobed, and with small yellow flowers 8–15 mm across. Flowers often long-stalked; sepals elliptic, often black-tipped. Leaves very variable. Fruit an oblong head of tiny beaked achenes, each achene *c.* 1.5 mm. **Page 445.**

   ii *Leaves deeply or shallowly-lobed, as broad as long*

38 **R. brotherusii** Freyn
Pakistan to S.W. China. 3300–5000 m. Open slopes, drier areas. Jun.–Aug.

A small plant usually 5–12 cm, with finely dissected leaves, and small terminal yellow flowers 5–10 mm across, and reflexed hairy sepals. Petals to 5 mm, ovate. Upper leaves cut into linear acute segments 5–15 mm long, lobes of lower leaves shorter and broader. Achenes round, smooth, less than 1 mm. **Pl. 5.**

39 **R. hirtellus** Royle ex D. Don
Afghanistan to S. W. China. 3000–4800 m. Alpine slopes, pastures, damp places; common. May-Jul.

A small erect or spreading buttercup usually 6–15 cm, with basal leaves rounded and deeply cut into broad lobes contrasting markedly with the narrow-lobed stem leaves, and with bright yellow flowers 1–1.5 cm across. Sepals densely hairy beneath. Basal leaves 1–3 cm across, with oval lobes with rounded teeth, stem leaves with 3–5 narrow lanceolate entire lobes; stems hairy above. Achenes 1–3 mm, with a straight or curved beak. **Pl. 5.**

39a **R. adoxifolius** Hand.–Mazz.
C. Nepal to S. E. Tibet. 3600–4300 m. Meadows, damp places; common. May–Jul.

A small tufted plant with many stems bearing solitary small yellow flowers to 1 cm across, and with tiny 3–lobed leaves. Petals rounded, 5; sepals elliptic, spreading; flowering stem hairy, with few leaves. Leaves with rounded blade 6–10 mm across, the lobes 3–lobed; leaf-stalk very long, to 5 cm; stems erect, 5–10 cm. Achenes *c.* 1 mm. **Pl. 5.**

40 **R. tricuspis** Maxim.
Kashmir to Bhutan. Tibet. N. China. 3300–4400 m. Marshes; Tibetan border areas; prominent along the Indus at Leh. Jun.

A tiny hairless plant often growing in wet places, with long creeping stems and small 3–5–lobed leaves, and erect stems bearing solitary yellow flowers 8–10 mm across, usually shorter than the leaves. Petals 5–8; sepals spreading or reflexed. Leaf-blades 6–12 mm; creeping stems to 30 cm or more, rooting at the nodes.

b *Flowers white; submerged aquatic plants*

41 **R. trichophyllus** Chaix (*R. aquatilis* var.*t.* Hook. f. & Thoms.)
Kashmir to S. E. Tibet. North Temperate Zone. 2700–4500 m. Sluggish streams in drier areas. Apr.–Aug.

A delicate plant with submerged dark green leaves 2–5 cm across cut into thread-like segments. Flowers carried above water, white, 8–15 mm across; petals widely spaced. Achenes 1–2 mm. **Page 445.**

13 ADONIS Differs from *Ranunculus* in having usually 8 or more petals, which have no nectaries at their bases. Leaves compound, much divided into narrow segments. 3 spp.

42 **A. chrysocyathus** Hook. f. & Thoms.
Pakistan to W. Nepal. Tibet. 3300–4300 m. Damp slopes recently cleared of snow; often gregarious. Jun.–Sep.

A handsome perennial with terminal golden-yellow flowers 3–5 cm across, with many petals twice as long as the similar-coloured sepals, and with feathery leaves often only partially developed at flowering. Petals 16–24, narrow-obovate; sepals 7–8; stamens and carpels numerous. Mature leaves 8–16 cm, triangular in outline, much divided into narrow pointed segments 5–10 mm long; stems several, leafy, 15–23 cm when flowering, up to 38 cm in fruit. Fruit a dense globular head 1.5 cm; achenes hairless, beak hooked. **Pl. 5.**

11

43 **A nepalensis** Simonovicz

C. Nepal to Sikkim. 3600–4500 m. Open slopes. Jun.–Jul.

Like 42, but differing in being a smaller plant in all its parts, with flowers 2.5–4 cm across with 12–16 blunt petals and reddish-brown sepals. Leaf-blade elliptic to oblong-ovate in outline. Achenes smaller with a very short straight beak.

44 **A. aestivalis** L.

Pakistan to W. Nepal. Temperate Eurasia. 1200–3000 m. Cultivated ground. Apr.–Jun.

An annual weed of cornfields, with small scarlet or golden-yellow flowers with dark purple centres, and with much divided feathery leaves. Flowers 1–2 cm across; sepals green or slightly coloured; petals a little larger, 5–8. Leaves pinnately divided into thread-like segments; stem leafy, 15–30 cm. Fruiting head oblong; achenes with a network of depressions.

14 ANEMONE Flowers with a single whorl of 5–14 coloured petals, green sepals, and nectaries absent. Often with 3–4 partially united leaves forming an involucre below the flower or flowering cluster. Fruit a cluster of achenes. *c.* 17 spp.

Achenes embedded in dense woolly hairs

45 **A. biflora** DC.

Afghanistan to Kashmir. Iran. 1500–2400 m. Open slopes; in Kashmir one of the earliest species to flower. Feb.–Apr.

A small plant 5–8 cm, with flowers which are white within and bluish outside and which become dull red when old, and with an involucre of lobed leaves below the 1–3 stalked flowers. Flowers 2–3 cm across; stalks hairy. Basal leaves long-stalked; blade 3–lobed, *c.* 2 cm; root tuberous, ovoid. **Page 446**.

46 **A. rupicola** Cambess.

Afghanistan to S.W. China. 2700–4300 m. Rocks, open slopes. Jun.–Aug.

Flowers usually solitary, large showy, white, sometimes pinkish outside; stems several *c.* 20 cm. Petals 2–4 cm long, oval, downy outside. Involucral leaves stalkless, 3–lobed to the middle, the lobes further cut and sharply toothed. Basal leaves long-stalked, blade 3–lobed to the base, the lobes deeply 3–lobed and sharp-toothed. Fruiting head globular, white-woolly. **Pl. 7**.

47 **A. vitifolia** Buch.–Ham. ex DC.

Afghanistan to S.W. China. Burma. 2100–3000 m. Shrubberies. Aug.–Sep.

A rather robust perennial 30–90 cm, branched above, with terminal lax umbels of few large white flowers each 3.5–5 cm across, borne on long silky-haired stalks. Petals broadly ovate, silky-haired beneath; ovaries in a globular woolly-haired cluster. Flower buds white-woolly. Basal leaves with rounded-heart-shaped blade 10–20 cm across, shallowly 5–lobed, the inner lobes conspicuously toothed, pale or white-woolly beneath; involucral leaves much smaller, short-stalked.

Woolly hairs from achenes used as tinder. **Page 446**.

Achenes not imbedded in woolly hairs
   a *Achenes oblong-cylindrical or slightly compressed*

48   **A. obtusiloba** D. Don
Pakistan to S. E. Tibet. Burma. 2100–4300 m. Forests, shrubberies, open slopes, grazing grounds; very common. May–Jul.

A very variable perennial with flowers 2–5 cm across, either white, or blue, or at higher altitudes in W. Himalaya commonly buttercup-yellow, borne on short spreading tufted stems usually 5–15 cm. Petals elliptic, 5–7, silky-haired beneath. Basal leaves many, blades 2–5 cm across, softly hairy on both sides, rounded in outline, deeply 3–lobed, the lobes further toothed or cut; involucral leaves *c.* 2.5 cm, 3–lobed, stalkless. Achenes coarsely hairy, but not imbedded in woolly hairs. **Pl. 7**.

49   **A. rivularis** Buch.–Ham. ex DC.
Kashmir to S.W. China. India. Tibet. 2100–3600 m. Meadows, shrubberies; very common on grazing grounds. May–Jun.

Flowers white, often flushed violet outside, usually in a lax terminal umbel of several long-stalked flowers, borne above a leafy involucre of deeply cut narrow lobes. Flowers 1.5–3 cm across; petals 5–8, silky-haired beneath. Basal leaves with rounded blade 8–15 cm, silky-haired on both sides, deeply 3–lobed, the lobes further cut and shallowly toothed, and usually conspicuously broader than the involucral lobes. Stems 30–90 cm. Achenes hairless; styles hooked. **Page 446**.

   b *Achenes strongly compressed, almost winged*

50   **A. demissa** Hook.f. & Thoms.
W. Nepal to S.W. China. 3300–4500 m. Open slopes. Jun.

Like 48, but fruit differing: achenes *c.* 4 mm, flat, circular in outline, hairless, with reflexed style. Flowers 3–6 in a simple umbel, white, blue or purple, often on spreading stems 15–40 cm. Leaves with rounded blades 1.5–3 cm across, variously cut into obovate stalked segments, hairy on both sides; involucral leaves small, narrow, 3–lobed. **Pl. 5**.

51   **A. tetrasepala** Royle
Afghanistan to Himachal Pradesh. 2100–3600 m. Rocks, open slopes; common in Kashmir. Jun.–Aug.

A robust perennial with an umbel of many white flowers each *c.* 3 cm across, and leathery rounded, long-stalked, lobed basal leaves. Petals often 4, but frequently 5–7; umbels sometimes compound. Leaves with heart- or kidney-shaped blades 8–20 cm across, deeply 5–lobed, the lobes broadly ovate, acute and strongly saw-toothed; involucral leaves much smaller, stalkless, with usually narrower, toothed lobes; flowering stem 30-75 cm. Achenes flattened, large *c.* 1 cm; style hooked. **Pl. 6**.

13

52 **A. polyanthes** D. Don

Pakistan to Bhutan. 2400–4300 m. Forests, open slopes; common in Nepal. May–Jul.

Like 51, but a somewhat smaller densely silky-haired plant to 50 cm with umbels of white, red, or mauve flowers each 2–3 cm across. Leaf-blades 5–10 cm across, with 5–8 oval lobes with usually rounded teeth. Achenes flattened, *c*. 5 mm; style straight. **Pl. 6.**

15 THALICTRUM MEADOW RUE Flowers small, in branched clusters; petals present, or absent; stamens many, conspicuous. Fruit a head of stalked or stalkless achenes, variously ribbed, winged, or angled. *c*. 20 spp.

Petals conspicuous
a *Flowers mauve to purple*

53 **T. chelidonii** DC.

Kashmir to S.E. Tibet. Burma. 2700–3600 m. Forests, shrubberies. Jul.–Sep.

A slender perennial to 3 m, with few relatively large lilac or mauve flowers 1–2 cm across in branched hairless clusters. Petals 4, elliptic blunt. Leaves large compound, with numerous rounded-heart-shaped lobed or toothed leaflets 2–3 cm, pale beneath; stems usually much branched. Achenes to 1 cm, long-stalked, flattened, net-veined.

54 **T. reniforme** Wallich

C. Nepal to Bhutan. 2800–3500 m. Forests, shrubberies. Jul.–Sep.

Differs from 53 in having larger mauve flowers 2–3.5 cm across, and leaflets and flower-clusters glandular or minutely hairy. Leaflets about as broad as long (longer than broad in 53). Achenes *c*. 5 mm, flattened, short-stalked. **Pl. 8.**

b *Flowers white*

55 **T. virgatum** Hook.f. & Thoms.

W. Nepal to S.W. China. 2400–4000 m. Rock ledges, steep banks; often pendulous. Jun.–Jul.

A rather delicate perennial distinguished by its small white flowers 1–2 cm across, and its nearly stalkless simple leaves which are deeply rounded-lobed. Flowers often few, in lateral and terminal branched clusters; petals 4 or 5, stamens little longer. Leaves broader than long, *c*. 2 cm wide, lobed to one third, the lobes with rounded teeth; stem slender, leafless below, 18–50 cm. Achenes spindle-shaped, ribbed, stalked. **Pl. 8.**

Petals inconspicuous or absent
a *Achenes spindle-shaped, not flattened*

56 **T. alpinum** L. ALPINE MEADOW RUE

Pakistan to Bhutan. N. Temperate Zone. 3000–4800 m. Alpine slopes. May–Aug.

A tiny delicate alpine plant, rarely more than 10–15 cm, with a simple leafless spike-like cluster of few greenish-purple flowers with long spreading stamens with purple thread-like filaments. Petals 2–3 mm, much shorter than stamens. Leaves all basal, ternate, further divided into pairs of rounded 3–5–lobed or toothed leaflets 3–8 mm long. Achenes stalkless, ribbed. **Page 445.**

57 **T. foliolosum** DC.
Kashmir to S.E. Tibet. Burma. 1300–3400 m. Forests, shrubberies; common in Nepal. Jul.–Sep.

Flowers small, white to dull greenish-purple, many in much-branched often dense clusters borne on tall leafy stems 1.2–2.6 m. Petals ovate, 3–5 mm, green, soon falling; stamens much longer, filaments white. Leaves 15–45 cm, many times divided into oblong-ovate, rounded-toothed, 3–lobed leaflets 1–1.5 cm. Achenes 2–5, ellipsoid, strongly ribbed, stalkless; beak curved.

The root is used medicinally.

b *Achenes flattened and ribbed, less than twice as long as broad*

58 **T. cultratum** Wallich
Pakistan to S.W. China. 2400–4200 m. Shrubberies, open slopes; common. Jun.–Aug.

Flowers small, greenish-white, usually in large lax much-branched clusters, borne on flexuous leafy stems 60–120 cm. Petals *c.* 5 mm; stamens twice as long, filaments purple thread-like, anthers slender. Leaves several times divided into stalked oval leaflets less than 1.3 cm wide, with rounded or wedge-shaped base and few rounded or acute teeth. Achenes shortstalked, flattened, 3–ribbed on each side, with a slender curved beak. **Pl. 8**.

59 **T. foetidum** L. (*T. minus* L.) LESSER MEADOW RUE
Pakistan to C. Nepal. Temperate Eurasia. 1500–4700 m. Forests, shrubberies, meadows. Jun.–Sep.

A wiry-stemmed, widely branched leafy perennial, with numerous nodding greenish-yellow flowers often tinged purple, and with numerous conspicuous hanging yellow stamens. Flowers small; petals greenish, *c.* 4 mm. Leaves glandular-hairy, 3–5–times divided into rounded, irregularly lobed or toothed leaflets, 3–25 mm wide, but very variable; stem 30–120 cm. Achenes flattened, ribbed, glandular-hairy, stalkless, with an enlarged lanceolate beak.

16 **CLEMATIS** Distinctive in the Ranunculaceae in having usually woody climbing stems and opposite pinnate or trifoliate leaves. Petals 4 or 5; achenes with long persistent often silky-haired styles. *c.* 24 spp.

Leaves with 3 leaflets; flowers solitary, long-stalked, in axillary clusters

60 **C. barbellata** Edgew.
Pakistan to C. Nepal. 2100–4000 m. Forests, shrubberies. Jun.–Jul.

Distinguished by its brown-purple bell-shaped flowers, and its very hairy stamens. Flowers with spreading lanceolate pointed petals 2–2.5 cm, hairy on both sides. Leaves with 3 ovate-lanceolate acute leaflets 4–10 cm, which are often further lobed and sharply toothed, becoming hairless; stems rather thin, hairless or nearly so. Achenes hairless except on margin; style 4–5 cm long. **Pl. 9**.

61 **C. montana** Buch.–Ham. ex DC.
Afghanistan to S.W. China. 1800–4000 m. Forests, shrubberies; very common. Apr.–May.

Flowers white or pinkish, 2–6 cm across, with oblong blunt hairless or silky petals 2.5–5 cm; stamens hairless. Leaves with 3 ovate-lanceolate acute toothed leaflets 2.5–7.5 cm. A vigorous climber to 10 m. Achenes flat, hairless; style more than 2.5 cm. **Pl. 8**.

Leaves pinnate or 2–pinnate (rarely with 3 leaflets); flowers usually borne in branched clusters (often solitary in 63 and 66)
a *Flowers cup-shaped, petals spreading from the base*
  i *Filaments of stamens hairless*

62 **C. grata** Wallich
Afghanistan to C. Nepal. China. 600–2400 m. Shrubberies, cultivated areas. Jul.–Sep.

Distinguished by its clusters of many small cream-coloured fragrant flowers, with spreading petals. Petals 4–8 mm, ovate-oblong, woolly-haired beneath; stamen filaments hairless. Leaves pinnate, with usually 5 leaflets each 3–8 cm long, ovate-lanceolate long-pointed, strongly toothed or lobed, hairy beneath. A vigorous climber with deeply furrowed stems. Achenes hairy; styles 2.5–4 cm long.

63 **C. phlebantha** L.H. Williams
Endemic to W. & C. Nepal. 2700–3500 m. Rocky, stony places; N.W. of Dhaulagiri. May–Jul.

A trailing shrub to 1.5 m, with white flowers with purple veins, usually solitary 3–4 cm across. Petals elliptic, 5–7, silvery-haired beneath, spreading; stamens yellow, filaments hairless. Leaves pinnate, leaflets 5–9, triangular, lobed at apex, 0.5–1.5 cm, green above, white-woolly beneath. **Pl. 9**.

  ii *Filaments of stamens hairy*

64 **C. graveolens** Lindley
Afghanistan to C. Nepal. 900–3000 m. Dry areas, banks, edges of cultivation. Jul.–Sep.

Flowers sweet-scented, pale yellow, with elliptic spreading petals usually notched at apex, and conspicuously hairy with a band of hairs on the margin outside, hairy inside, and with hairy stamen-filaments. Flowers 2–3 cm across, several in branched clusters, or the upper solitary, with pinnate

16

leafy bracts. Leaves distinctive, 2–3–pinnate with small lanceolate nearly entire leaflets 1–2.5 cm; a slender climber. Achenes hairy.

65 **C. tibetana** Kuntze
Pakistan to Uttar Pradesh. 1800–3600 m. Shrubberies, banks, cultivated areas; common in Ladakh. Jul.
 Flowers yellow spotted with brown, nodding, long-stalked, often 3 borne on axillary branches. Petals elliptic-pointed 2–2.5 cm, woolly-haired above and on the margin; stamen-filaments hairy. Leaves 2–pinnate, with long-stalked linear to lanceolate long-pointed entire leaflets, or leaflets coarsely toothed at the base, mostly c. 5 cm. Achenes silky-haired.

66 **C. vernayi** C. Fischer
W. & C. Nepal. C. & S. Tibet. 1800–4000 m. Irrigated land in dry Tibetan border areas. Jun.–Aug.
 Like 65, but differing in its larger dull reddish-yellow to yellowish-green flowers, with broad blunt spreading petals to 3 cm long, which are densely hairy inside and hairless outside, and by its 1–pinnate leaves with more or less ovate leaflets which are deeply 3–lobed, the lobes elliptic. **Pl. 9**.

 b *Flowers funnel- or urn-shaped; petals erect with recurved tips* (see also 65)
  i *Filaments of stamens hairy throughout*

67 **C. buchananiana** DC.
Pakistan to S.W. China. Burma. 1800–3000 m. Forests, shrubberies. Jul.–Oct.
 Flowers cream or yellow in long leafy branched clusters, sweet-scented, with narrow-oblong pointed petals recurved only at the tips. Petals thick, 2–2.5 cm, densely woolly-haired outside, ribbed; stamen-filaments densely hairy. Leaves 1–pinnate, with leaf-bases more or less united round stem, with 5–7 broadly ovate-heart-shaped, coarsely toothed, often lobed hairy leaflets 5–10 cm. A strong woody climber with hairy shoots. Achenes woolly-haired; styles to 5 cm.

68 **C. connata** DC.
Pakistan to S.W. China. 1800–3300 m. Shrubberies. Aug.–Sep.
 Flowers yellowish-white, 1–2 cm long, in branched axillary clusters. Distinguished by its leaf-stalks which are fused together at the node often into a large flat disk. Petals 2–2.5 cm, oblong blunt, finely woolly-haired on both sides, not ribbed; stamen-filaments hairy. Leaves pinnate, with 3–7 hairless, oval-lanceolate to heart-shaped, sharply and irregularly toothed stalked leaflets, 6–12 cm long; stems stout. Achenes silky-haired; styles 4–5 cm long. **Pl. 8**.

69 **C. grewiiflora** DC.
Uttar Pradesh to Arunachal Pradesh. Burma. 1200–2100 m. Shrubberies. Nov.–Dec.
 Readily distinguished by its leaves which are densely covered with

17

woolly rufous hairs. Flowers cream-coloured; petals 1–3 cm, densely woolly-haired, the tips only slightly recurved; stamen-filaments long thread-like, with soft spreading hairs. Leaves pinnate, with 3–5 ovate pointed or broadly heart-shaped, toothed leaflets 8–10 cm; stems woolly-haired.

ii *Filaments of stamens hairy on the lower half only*

70 **C. roylei** Rehder (*C. nutans* Royle)
Himachal Pradesh to C. Nepal. 450–2200 m. Shrubberies, cultivated areas; not common. Oct.–Dec.

Flowers cream-coloured, drooping, in leafy branched clusters; petals oblong with recurved tips, 1.8–2.5 cm, densely silky-haired outside; filaments hairy on the lower half only. Leaves pinnate, with ovate to lanceolate, irregularly toothed, undivided or 3–5–lobed leaflets 2.5–7.5 cm, or leaves rarely entire; stems slender, the branchlets silky-haired. Achenes densely hairy; style 2.5–4 cm long. **Pl. 9**.

PAEONIACEAE   Peony Family

A small family of the North Temperate region, in the past included in Ranunculaceae, but now recognized as distinct and distinguished by small botanical characters. These include persistent sepals; a fleshy disk at the base of the carpels to which the petals and stamens are attached; and seeds with arils.

PAEONIA   1 sp.

71 **P. emodi** Wallich ex Royle   HIMALAYAN PEONY
Afghanistan to W. Nepal. 1800–2500 m. Forests, shrubberies; often gregarious. Apr.–May.

A robust rather shrubby perennial, with large white flowers with numerous orange-yellow stamens and large deeply-cut leaves. Flowers 8–12 cm across, with 5–10 elliptic petals and 5 persistent outer sepals. Leaves 30–60 cm, with lanceolate long-pointed leaflets or lobes to 14 cm long; stem 30–75 cm. Fruit usually a single follicle 3–4 cm, densely hairy or hairless; seeds scarlet then brown-black.

The tubers, seeds and flowers are used medicinally. **Pl. 9**.

MAGNOLIACEAE   Magnolia Family

An ancient family of trees and shrubs showing a number of primitive characters such as indefinite numbers of spirally arranged perianth-segments, stamens and carpels. Leaves alternate, with large stipules which fall off as the leaf expands. Flowers usually large, solitary, with many white petals and no outer green sepals; stamens numerous. Ovary of many or few carpels, arranged spirally and either free or partly fused. Fruit of a number of carpels which split to release the seeds which in *Magnolia* are attached by a silky thread.

MAGNOLIA   Flowers terminal. Fruit a dense spike of more or less fused carpels. 2 native spp.

72 **M. campbellii** Hook. f. & Thoms.
C. Nepal to S.W. China. Burma. 2400–3000 m. Forests. Mar.–Apr.
A large deciduous tree with grey bark, bearing large white or cream scented flowers on bare branches. Flowers 13–20 cm across, with 12–15 large white petals, the outer silvery-haired. Leaves oval 10–30 cm, hairless above, with silky hairs beneath when young. Fruit a cylindrical spike 15–20 cm long; seeds red. A pink-flowered form is cultivated around Darjeeling and occurs wild in E. Himalaya. Only white-flowered trees are found in Nepal. **Page 446**.

73 **M. globosa** Hook. f. & Thoms.
E. Nepal to S. W. China. 3200–3400 m. Forests; rare in Nepal. May.
Distinguished by its rather globular pendent white flowers, with striking deep pink anthers, which are borne with the leaves. Flowers *c.* 8 cm across, scented; petals usually 6, thick. Leaves oval to 18 cm, dark glossy green above, rusty-felted beneath. A tree to 8 m. Fruit cylindrical 5–7 cm, crimson, drooping.
M. grandiflora L. is commonly grown in gardens in Kathmandu.

MICHELIA   Differs from *Magnolia* in its axillary flowers, and stalked ovaries. Fruit a dense or lax elongated spike of many carpels. 4 spp.

74 **M. doltsopa** Buch.–Ham. ex DC. (*M. excelsa* Wallich)
C. Nepal to S.W. China. Burma. 2100–2500 m. Forests. Mar.–Apr.
A large deciduous tree with young buds and flower-stalks covered with red or brown hairs. Flowers pale yellow to white, very fragrant, 8–10 cm across, solitary in the leaf axils. Petals 12–16, obovate to oblanceolate. Leaves oval-oblong 8–18 cm, dark glossy green above, paler and somewhat glaucous beneath with short adpressed rufous hairs when young, becoming hairless. Fruit in a lax spike 10–20 cm; seeds red. **Pl. 10**.

75 **M. kisopa** Buch.–Ham. ex DC.
Uttar Pradesh to Bhutan. 1400–2800 m. Forests. Oct.–Nov.
A tall tree to 30 m, with grey bark, differing from 74 in having much smaller pale yellow flowers with hairy ovaries which appear in the autumn. Flowers 2–3.5 cm across; petals narrowly oblanceolate; flower-stalks with greyish hairs. Leaves lanceolate, greyish-haired beneath, to 15 cm. Young parts of tree clothed with grey hairs but soon becoming hairless. Fruit spike to 8 cm.

SCHISANDRACEAE

A small family of two genera, differing from the Magnoliaceae in being woody climbers, and in having unisexual flowers. Fruit a slender spike of numerous stalkless carpels each with 1–5 flattened seeds.

SCHISANDRA 3 spp.

76 **S. grandiflora** (Wallich) Hook. f. & Thoms.
Himachal Pradesh to S. W. China. 2100–3300 m. Forests, shrubberies. Apr.–Jun; fruiting Jul.–Aug.

A hairless woody climber with drooping globular pinkish-white sweet-scented flowers 2–3 cm across. Petals fleshy, circular, about 9; stamens about 40 in male flowers. Female flowers larger and richer-coloured. Leaves 6–15 cm, ovate to oblong-lanceolate long-pointed, with conspicuous veins beneath. Fruit of numerous scarlet carpels borne on a long cylindrical fleshy axis 10–18 cm. **Pl. 11**.

BERBERIDACEAE Barberry Family

A family of shrubs and herbs, mainly of the North Temperate Zone, with simple, pinnate, or peltate leaves. In *Berberis*, spines take the place of some leaves. Flowers bisexual and regular, with several whorls of usually 4–6 perianth segments, one or all whorls coloured and petal-like; stamens usually 6. Ovary superior, of one carpel with several to many ovules. Fruit usually a fleshy berry.

BERBERIS BARBERRY Distinguished by its undivided spiny-toothed leaves and its spiny stems with yellow wood. Individual species are often difficult to distinguish from each other, being separated by small botanical characters. Many may be locally prominent. The wood of *Berberis* species produces a yellow dye used for dying leather. Leaves and berries of many alpine species turn a vivid red in autumn. *c*. 38 spp.

Leaves evergreen; fruit black or purple

77 **B. asiatica** Roxb. ex DC.
Himachal Pradesh to S.W. China. 1200–2500 m. Shrubberies, grassy and rocky slopes; common in cultivated areas of Nepal. Mar.–May.

A much branched shrub 2–4 m, with pale yellow branches, and thick rigid evergreen leaves with usually 2–5 spiny teeth, shining dark green above and greyish beneath. Flowers pale yellow, in somewhat flat-topped clusters shorter than the leaves, with red hairless stalks; petals obovate, notched, 5–7 mm. Leaves ovate to elliptic, 1.8–7.5 cm; stem spines 1–1.5 cm. Fruit glaucous, dark purple, oblong-ovoid, 8 mm.

Fruit edible. The wood, bark, and plant extracts yield *berberine*, a bitter alkaloid used medicinally.

20

78 **B. lycium** Royle
Pakistan to E. Nepal. 1500–3000 m. Shrubberies; common in cultivated areas of W. Himalaya. Apr.–Jun.

A shrub to 3–4 m, with white stems, and narrow fine-pointed leathery leaves mostly without spiny teeth, glaucous beneath. Flowers dull yellow, 7–8 mm across, borne in axillary clusters 3–5 cm, longer than the leaves. Leaves oblanceolate, 2–4 cm; stems hairless, stem spines 6–20 mm. Fruit black, bloomed, 9–10 mm including the 1 mm style.

Plant used medicinally.

Leaves deciduous; fruit usually red
a *Flowers mostly solitary, or few*

79 **B. angulosa** Wallich ex Hook f. & Thoms.
C. Nepal to S.E. Tibet. 3400–4500 m. Forming thickets in alpine regions. May–Jun.

A stiff stout-stemmed shrub 1–1.3 m, with conspicuously dark brown hairy stems. Flowers drooping, usually solitary or sometimes clustered, large 14–20 mm across; yellow petals of outer whorls 8–10 mm, inner petals shorter and with rounded apices. Leaves 1–2 cm, elliptic to obovate, without spiny teeth, with rounded and mucronate apex, shining yellow-green above and below; stem spines usually slender, 4–16 mm. Fruit shining red, elliptic to globular, large 10–12 mm. **Pl. 11.**

80 **B. erythroclada** Ahrendt
C. Nepal to S.E. Tibet. 3000–4000 m. Shrubberies; forming thickets on grazing grounds. Jun.–Jul.

A densely branched low shrub 30–100 cm, with shining dark red, grooved, hairless branches, and tiny obovate leaves with 5–15 stiff spiny teeth. Flowers yellow, usually solitary, on slender stalks longer than the leaves; petals 5–7 mm. Leaves 1–2 cm; stem spines slender 5–10 mm, orange red. Fruit red, to 1.5 cm.

81 **B. concinna** Hook. f.
C. Nepal to Sikkim. Tibet. 2700–4400 m. Shrubberies, open places; forming dense thickets in the lower alpine zone. Jun.–Jul.

Differs from 80 in having leaves which are conspicuously glaucous beneath (leaves green beneath in 80) and with 2–6 spiny teeth; blade *c.* 1 cm, oval to elliptic; stem spines often as long or longer than the leaves. Flowers yellow, usually solitary. Fruit dull red, oblong, 13–16 mm.

b *Flowers rather numerous, in elongate or flat-topped clusters*

82 **B. aristata** DC. (and var. **floribunda** (G. Don.) Hook. f. & Thoms.)
Himachal Pradesh to C. Nepal. 1800–3500 m. Shrubberies; common and showy. Apr.–Jun.

A shrub usually *c.* 2 m, with arching pale yellow branches, usually elliptic often spineless leaves, and short-stalked clusters of numerous yellow flowers much longer than the leaves. Flowers *c.* 6 mm long. Leaves

2.5–5 cm; stem spines absent, or with 1–3 branches, short 4–10 mm. Fruit dark red, finally blue-purple, 6–8 mm; style very short.

Extract from wood, bark and roots used medicinally. **Pl. 10**.

83 **B. ceratophylla** G. Don

Himachal Pradesh to Bhutan. 1800–4000 m. Open scrub, forests, cultivated areas. Apr.–Jun.

Distinguished from 82 by its conspicuously grooved pale yellow-grey stems, and fairly short stem spines 5–14 mm, and its spiny oblanceolate to wedge-shaped leaves. Flowers clear yellow, in many-flowered clusters which are branched below. Leaves 2.4–5 cm, with numerous marginal spines. Fruit very dark red or purple, oblong-ovoid *c*. 12 mm, including style which is 1 mm or more.

84 **B. wallichiana** DC.

Uttar Pradesh to S.W. China. 2000–3300 m. Forests. Apr.–May.

An erect shrub 2–3 m, with relatively large oblong-lanceolate usually acute leaves with numerous spiny teeth. Flowers 7–10 mm across, borne in axillary clusters much shorter than the leaves; petals obovate entire. Leaves 4–8 cm, paler shining beneath; stem spines slender 1.5–2.5 cm. Fruit black, ovoid 8–9 mm. **Pl. 10**.

85 **B. jaeschkeana** C. K. Schneider

Pakistan to E. Nepal. 2700–4000 m. Alpine meadows, stony slopes, shrubberies; common. May–Jun.

A deciduous shrub 1–3 m, with yellow-brown angled mature stems and oblong-elliptic usually spineless leaves, or with 2–4 spiny teeth. Flowers yellow, 8–10 mm across, 3–8 in a somewhat flat-topped cluster 2–4 cm long; petals entire or notched, *c*. 6 mm, the inner much shorter than the outer. Leaves 2–3 cm; stem spines 1–1.5 cm. Fruit red, oblong-ovoid 9–12 mm; seeds dark purple.

MAHONIA Readily distinguished from *Berberis* by its compound leaves, spineless stems, and inflorescence of several dense spikes. 3 spp.

86 **M. napaulensis** DC.

W. Nepal to Bhutan. 2000–2900 m. Forests. Oct.–Apr.

An erect evergreen shrub to 3 m, with stout little-branched stems and large pinnate leaves borne at the ends of the stems. Flowers yellow, in a terminal cluster of several or many dense many-flowered spikes, 10–25 cm long. Petals *c*. 6 mm, notched. Leaves oblong-lanceolate, to 40 cm, with 4–7 pairs of stiff, slightly overlapping ovate leaflets with spiny marginal teeth and pointed apices; leaflets largest in the middle of the leaf, 6–9 cm long. Fruit in dense cylindrical clusters; berries blue, strongly bloomed, 9 mm long.

Fruit edible. **Pl. 10**.

87 **M. acanthifolia** G. Don

Uttar Pradesh to S.W. China. S.E. Asia. 2100–2800 m. Forests. Oct.–Nov.

Like 86, but style 1 mm, and leaves dull above (style very short; leaves shining above in 86). Flowers yellow, in long spikes 15–23 cm, clustered into terminal bunches of 3–4. Leaves pinnate with 8–11 pairs of thick ovate-oblong leaflets 4.5–6.5 cm long, with 2–4 spiny marginal teeth. Berries blue, bloomed, 10–12 mm.

PODOPHYLLUM Herbaceous plants with a solitary terminal flower, and peltate leaves. 1 sp.

88 **P. hexandrum** Royle (*P. emodi* Wallich ex Hook. f. & Thoms.)
Afghanistan to S.W. China. 2400–4500 m. Forests, open slopes. May–Aug.

A distinctive herbaceous plant, with an erect unbranched stem bearing two large terminal lobed leaves encircling the single large white or pale pink flower. Flower cup-shaped, 2–4 cm across; sepals 3, soon falling; petals 6; stamens 6. Leaves usually 2; blade rounded in outline 10–25 cm, deeply cut into 3 ovate toothed lobes, sometimes further lobed; stem 15–40 cm. Fruit a large scarlet or reddish berry 2.5–5 cm, with many seeds imbedded in pulp.

Quite an important medicinal plant, the rhizome containing *podophyllin*. Recently investigated as a possible drug used in treatment of cancer. **Pl. 11**.

LARDIZABALACEAE

A small family of climbing shrubs largely found in the Himalaya and eastwards to Japan. Leaves alternate, palmate. Flowers unisexual, often on separate plants, borne in clusters from the axils of scale leaves situated at the base of the branches. Sepals often coloured and petal-iike, 3 or 6; petals 6, smaller; stamens 6. Ovary of 3 or more carpels arranged in whorls. Fruit a berry or follicle, coloured, fleshy, many-seeded.

HOLBOELLIA 1 sp.

89 **H. latifolia** Wallich
Pakistan to S.W. China. 1500–4000 m. Forests, shrubberies, shady ravines. Apr.–May; fruiting Sep.–Oct.

A large woody climber with palmate leaves with 3–9 long-stalked leaflets, and with tubular to bell-shaped, very sweet-scented, purplish flowers. Flower in drooping axillary clusters, much shorter than the leaves; petals 1–1.5 cm long. Leaflets oblong-ovate entire, 4–10 cm, stalks of leaflet 1–4 cm (leaflets linear-lanceolate, and flowers light green tinged purple in var. **angustifolia** (Wallich) Hook. f. & Thoms). Fruit an oblong rosy-purple or pinkish fleshy berry 3–6 cm, with numerous seeds imbedded in pulp.

Fruit edible. **Pl. 10**.

NYMPHAEACEAE   Water Lily Family

A cosmopolitan family of aquatic plants with large underwater rhizomes, and peltate or heart-shaped, often floating leaves. Flowers solitary, with 3–6 green or coloured sepals; 3–many petals; stamens numerous. Ovary with 3–35 free or united carpels; fruit a spongy berry splitting apart by the swelling of mucilage within.

EURYALE   Leaf-stalks and stems prickly; leaves floating, corrugated. Fruit a spongy berry crowned with persistent sepals. 1 sp.

90   **E. ferox** Salisb.
   Kashmir. India. China. To 1400 m. Lakes. Jun.
   Leaves floating, rounded or elliptic 30–120 cm across, green and corrugated above, red or purple with thick spiny veins beneath. Flowers partially submerged, bright red within, shining green outside; petals many; stamens many. Fruit 5–10 cm across; seeds 8–20, black, imbedded in pulp.
   Seeds edible; also used medicinally. **Pl. 10**.

NELUMBO   Lotus   Leaf-blades carried above water, smooth and glaucous. Fruit top-shaped, with several carpels deeply imbedded in the enlarged spongy head of the stem. 1 sp.

91   **N. nucifera** Gaertn. (*Nelumbium speciosum* Willd.) INDIAN LOTUS
   India to N. China. 600–1400 m. Lakes; widely cultivated in Kashmir also near Pokhara Nepal. Jul.–Aug.
   Flowers white or pink, very large 10–25 cm across, carried above water, with many petals and stamens. Leaves with a circular glaucous blade 60–90 cm in diameter, attached to the stalk beneath in the middle (peltate); leaf-stalk 1–2 m. Fruit top-shaped, flat-topped, 5–10 cm across, borne well above water, with several deeply imbedded seeds.
   Seeds edible. Many parts of the plant are used medicinally. **Pl. 11**.

NYMPHAEA   WATER LILY   Leaves floating, smooth. Flowers floating. Fruit a spongy berry ripening under water. *c.* 3 spp.

92   **N. tetragona** Georgi (*N. pygmaea* Aiton; *N. alba* var. *minor* DC.)
   Kashmir. N. Asia. To 1400 m. Shallow water in lakes and channels; Jun.–Jul.
   Distinguished by its small floating white flowers *c.* 4 cm across with about 10 blunt petals. Floating leaves with broadly elliptic blades 3–5 cm long with diverging basal lobes. **Pl. 11**.

92a   **N. mexicana** hybrid has larger medium-sized canary-yellow flowers, and floating leaves with roundish-oblong blades 10–15 cm long with slightly diverging lobes. Widely cultivated in Kashmir and elsewhere. **Pl. 11**.

PAPAVERACEAE   Poppy Family

A family mostly of annuals and herbaceous perennials with milky or watery juice (latex), mainly of the North Temperate Zone. Leaves spirally arranged, variously dissected. Flowers usually regular, less commonly irregular and symmetrical in one plane only. Sepals 2, usually falling as bud opens; petals 4–6, often crumpled; stamens 2, 4 or numerous. Ovary superior, of 2 to many fused carpels, with one chamber which may be further partially divided. Fruit a capsule, splitting by pores or valves, rarely not splitting as in *Fumaria*.

Flowers large, regular; stamens numerous
  a Capsule short, opening by pores or valves at the top   1 ARGEMONE
                                             2 MECONOPSIS   3 PAPAVER
  b Capsule long, opening its whole length   4 DICRANOSTIGMA

Flowers small, irregular; stamens 4–6
  a Fruit a 2-valved, many-seeded capsule   5 CORYDALIS
                                             6 DICENTRA
  b Fruit a 1-seeded nutlet   7 FUMARIA

1 ARGEMONE Like *Papaver* but fruit spiny, opening by 4–6 short valves; styles absent. 1 sp.

93   **A. mexicana** L. PRICKLY POPPY
Native of C. America; naturalized world-wide in warm countries. To 1500 m. In Himalaya in cultivated areas, waste lands. Mar.–Apr.
    Distinguished by its prickly stems and leaves, and its yellow poppy-like flowers 5–6 cm across. A glaucous annual to 90 cm, with oblong thistle-like spiny-lobed, clasping leaves. Capsule ellipsoid, spiny.
    The roots, seeds and yellow sap are used medicinally.

2 MECONOPSIS Distinguished from *Papaver* by its fruit which has a short projecting beak or style with 4–6 distinct stigmas, and which opens by 4–6 short valves. Many species occasionally have white flowers. 17 spp.

Flowers yellow (see also 97)

94   **M. paniculata** Prain
Uttar Pradesh to S.E. Tibet. 3000–4100 m. Open slopes, grazing grounds; gregarious around herdsmen's camping grounds. Jun.–Aug.
    A robust plant to 2 m, with a long cylindrical terminal cluster of large yellow or occasionally white nodding flowers, and densely golden-bristly-haired pinnately-lobed lanceolate leaves. Upper flowers stalked, single, the lower borne on branched stalks from the axils of the upper leaves; petals usually 4, rounded, *c.* 5 cm. Basal leaves in a dense rosette, and overwintering in this form; leaves to 90 cm, variously cut into lanceolate, toothed lobes, the upper stem-leaves clasping, all covered with long bristly hairs and short star-shaped hairs. Fruit ellipsoid-oblong, densely bristly-haired. **Pl. 12.**

95 **M. regia** G. Taylor
West to East Nepal. 3600–4300 m. Open slopes, grazing grounds; often gregarious. Jun.–Jul.

Like 94, but with fewer and larger yellow flowers, and entire lanceolate leaves which are regularly and coarsely toothed, not lobed, and densely covered with silvery or golden silky hairs. Petals *c.* 6.5 cm long. Capsule *c.* 5 cm including beak of 2 cm. **Pl. 12.**

These two species and 100 **M. napaulensis** show many intermediate forms in the E. Himalaya: 95b **M. taylorii** L. H. Williams with pink flowers; 95a **M. dhwojii** G. Taylor, **Pl. 13**, and 95c **M. gracilipes** G. Taylor both with yellow flowers, are also part of this complex and are not easy to distinguish in the field.

96 **M. villosa** (Hook. f.) G. Taylor (*Cathcartia v.* Hook. f.)
E. Nepal to Bhutan. 2700–3600 m. Forests, open slopes. Jun.–Jul.

Flowers yellow, drooping, borne singly on long flower-stalks from the upper leaves which are distinctive in being rounded in outline and deeply lobed. Petals 4, broader than long, 2.5 cm by 3 cm broad; stamens numerous. Lower leaves stalked, blade 5–11 cm across, lobes rounded and further cut into blunt rounded segments, hairy, glaucous beneath. Stem with reddish bristles, unbranched, 38–50 cm. Capsule narrow-cylindrical 9 cm, splitting for more than half its length into 4–7 valves. **Page 447.**

Flowers blue, purple, red, or rarely white
a *Leaves all basal* (see also 101, 103)

97 **M. discigera** Prain
W. Nepal to S.E. Tibet. 3300–4500 m. Rocky slopes. Jun.–Jul.

Flowers red, purple, pale blue or pale yellow, 6–10 cm across, 13–20 borne on stout bristly stems to 40 cm. Petals 4–5 cm long; filaments of stamens darker than petals. Leaves to 18 cm, oblanceolate and tapering to a long stalk, mostly basal in a dense rosette, entire or lobed, covered with golden-brown bristles. Unlike most other species in having a flat disk-like extension at the base of the style covering the top of the ovary or fruit. Capsule oblong, conspicuously ribbed. **Pl. 12.**

98. **M. simplicifolia** (D. Don) Walp.
C. Nepal to S.E. Tibet. 3300–4500 m. Alpine shrubberies, open slopes. Jun.–Jul.

Flowers purple to sky-blue, solitary nodding, borne on long stalks arising from the rosette of narrow leaves. Petals 5–8, obovate, 2–4 cm long; flower-stalks several, bristly-haired, to 60 cm. Leaves oblanceolate, variable in size, gradually tapering to a long leaf-stalk, entire, irregularly toothed, or shallow-lobed, with weak bristly hairs. Capsule erect, narrow-oblong to 5 cm, hairless or bristly. **Pl. 13.**

99 **M. bella** Prain
C. Nepal to S.E. Tibet. 3500–5000 m. Rock crevices; a rare but attractive species. Jun.–Sep.

26

Flowers blue, purple, or pink, borne singly on short stems arising from the rosette of leaves, with often many flowers borne from the same rosette. Flowers somewhat nodding; petals with undulate margins, usually 4, rounded, 1.5–3 cm; stamens with dark purple filaments and golden yellow anthers. Leaf-blade 1–5 cm, very variable, entire, or once or twice cut into rounded or oblong lobes, often further 3–lobed, hairless, or with few bristly hairs, leaf-stalk long. Capsule cylindrical 2–3 cm, ribbed. **Page 447**.

b *Leaves borne on flowering stem as well as basal* (see also 97)

　i *Plants with large overwintering rosettes*

100　**M. napaulensis** DC. (*M. wallichii* Hook.)
W. Nepal to S.W. China. 3000–4500 m. Open slopes, grazing grounds; often gregarious. Jun.–Aug.

Flowers very numerous, nodding, red, purple, or occasionally blue or white, borne on robust leafy stems to 2.4 m. Individual flowers long-stalked; petals 4, rounded, *c.* 4.5 cm. Leaves large lanceolate to 50 cm, deeply lobed, the lobes usually further cut into blunt segments, covered in the wintering rosettes with reddish bristly hairs; stem leaves smaller. Capsule ellipsoid-oblong to 3 cm, with beak 2 cm, with very dense adpressed or spreading bristles.

　ii *Plants without large overwintering rosettes*

　　x *Leaves entire, or margin toothed or shallowly lobed*

101　**M. horridula** Hook. f. & Thoms.
W. Nepal to S.W. China. 3500–5500 m. Rocks, stony slopes; common. Jul.–Aug.

The most bristly-haired species of the genus, with usually a spike-like cluster of many light blue to claret-coloured, or occasionally white flowers, and narrow leaves with bristly spiny blades. Petals 4–8, obovate to rounded, minutely toothed at apex, 2–4 cm long. Leaves to 30 cm, the blade elliptic to linear-oblong and narrowed to the leaf-stalk, usually entire, covered with yellow to purple bristly spines with darker bases. Capsule ellipsoid to globular 2.5–3 cm, covered with reflexed, spreading, or adpressed bristly spines. The flowers are often borne on leafless stems arising directly from the rootstock. **Pl. 12**.

102　**M. latifolia** Prain
Endemic to Kashmir. 3500–4000 m. Rocky slopes. Jul.–Sep.

Flowers blue, rarely white, numerous in a spike-like cluster borne on a leafy stem to 1 m. Petals 4, broadly ovate 3.5–4 cm long; stamen filaments deep blue, anthers orange-yellow. Basal leaves long-stalked; blade ovate to oblong to *c.* 20 cm, coarsely and regularly shallowly rounded-lobed, with sparse straw-coloured bristly spines; upper leaves stalkless. Capsule obconic *c.* 3 cm, with style covered with spreading bristly spines. **Pl. 13**.

27

103 **M. grandis** Prain
W. Nepal to S.W. China. Burma. 3300–4500 m. Shrubberies, grazing grounds, open slopes; gregarious. Jun.–Jul.

Flowers usually few, rich blue or purple, borne on long stalks from the uppermost leaves, or sometimes arising directly from the basal rosette of leaves. Flowers large, with commonly 4 rounded petals to 7 cm (petals sometimes as many as 9); stamen-filaments whitish. Leaves stalked, the stalk sheathing at base, with oblanceolate to oblong-elliptic coarsely toothed blade 10–30 cm, covered in rufous bristles; uppermost leaves in a false whorl from which the flowers arise. Capsule ellipsoid-oblong to 5 cm. **Pl. 13**.

104 **M. sinuata** Prain
C. Nepal to S.E. Tibet. 2700–4300 m. Shrubberies, open slopes. Jun.–Jul.

Flowers blue to violet, drooping, long-stalked, in a lax cluster of 3–8, borne from the axils of the upper narrow leaves, and opening one at a time. Petals 4, obovate, *c.* 3 cm long. Basal leaves few, often withered at flowering; stem leaves strap-shaped to oblanceolate blunt, usually shallowly lobed, with bristly hairs, uppermost leaves half-clasping; stem with bristly hairs, 23–76 cm. Capsule cylindrical to narrowly obovoid to 5 cm, with 3–4 ribs, with spreading bristly hairs. **Page 447**.

105 **M. lyrata** (Cummins & Prain) Fedde ex Prain (*Cathcartia* l. Cummins & Prain)
C. Nepal to S.W. China. 3000–4300 m. Open slopes; not common. Jul.–Aug.

A slender rather delicate and sparsely hairy plant with 1–5 pale blue nodding flowers borne on long stalks. Flowers 3–4 cm across; petals usually 4. Leaves small 1–5 cm, the lower long-stalked, usually oval and entire, sometimes with rounded basal lobes, but very variable; stem weak 5–45 cm. Capsule narrow-oblong, to 5 cm, splitting by 3–4 valves to one third its length. **Page 447**.

xx *Leaves deeply lobed*

106 **M. aculeata** Royle
Pakistan to Uttar Pradesh. 3000–4000 m. Rocky slopes, damp rocks. Jun.–Aug.

Like 102 *M. latifolia*, but a more widespread species with leaves deeply lobed and flowers fewer, blue (very rarely purplish-red to red), 5–7 cm across. Petals usually 4, rounded to obovate 2–3 cm; stamen-filaments darker than petals. Leaves deeply and irregularly pinnately-lobed, lobes usually rounded, toothed and widely spaced, sparsely bristly-haired; stem to 60 cm, with bristly hairs. Capsule ovate to oblong c. 1.5 cm, with a beak 1 cm, bristly-haired. **Pl. 12**.

3 PAPAVER Poppy Like *Meconopsis* but differing in its ovary and fruit the latter being a flask-like capsule with a cap-like disk and opening by pores beneath the disk. 5 spp.

Flowers red; cornfield weeds

107 **P. dubium** L. LONG-HEADED POPPY
Afghanistan to C. Nepal. Temperate Eurasia. 1800–2700 m. Cornfield weed, cultivated areas. Apr.–May.

An annual 30–60 cm, with usually red, pinkish or pale scarlet flowers with a dark basal blotch to each petal. Petals 2–4 cm; stamens as long as ovary. Leaves pinnately cut into acute and lobed segments, nearly hairless to bristly-haired. Capsule oblong, narrow to the base, ribbed, hairless. The hairless form, var. **glabrum** Koch is common in cornfields around Jumla, Nepal.

108 **P. macrostomum** Boiss. & Huet ex Boiss.
Afghanistan to Kashmir. N. & W. Asia. 1500–1800 m. A common cornfield weed of the Kashmir valley. Apr.–Jun.

Like 107, but differing in its ovary in which the stigmatic segments are keeled and the stigmatic disk deeply lobed, and in its ellipsoid, smooth, glaucous capsule. Flowers red, often with a prominent black basal cross within; petals 2–3 cm long. Leaves like 107, but ultimate lobes entire and with a few teeth. **Pl. 13**.

Flowers yellow to reddish-orange; mountain plants

109 **P. nudicaule** L.
Afghanistan to Kashmir. C. Asia. 3600–4800 m. Rocky hillsides. Jul.–Aug.

Flowers yellow, or orange to reddish-orange, solitary and borne on long leafless bristly-haired stems 10–40 cm which are much longer than the basal leaves. Flowers usually 2–4 cm across; petals obovate to rounded. Leaves in a rosette, bristly-haired, somewhat glaucous, often 3–lobed with the lobes further cut into oblong segments. Capsule oblong to rounded-oblong 1–1.5 cm, bristly-haired. **Page 447**.

Flowers purplish or white, large; cultivated ground

110 **P. somniferum** L. OPIUM POPPY
W. & C. Asia, Europe. To 1800 m. Sometimes a few plants cultivated in village gardens in the Himalaya. Apr.–May.

Readily distinguished by its large showy purplish flowers usually with dark basal blotches, and its glaucous clasping leaves, borne on a robust stem 30–100 cm. Flowers 5–10 cm across, sometimes white; flower-buds large, sepals soon falling revealing crumpled petals which are rounded, with toothed or cut margin, with or without a darker basal blotch. Leaves oblong-ovate, toothed or shallowly lobed, hairless or nearly so. Capsule rounded, 5–9 cm with a conspicuous deeply-lobed disk, smooth, nearly hairless.

A valuable drug plant; a source of *morphine* and *codeine*, used widely in medicine to relieve pain, and in terminal diseases. A source of opium which, when smoked or chewed, causes intoxication and is an addictive drug. Oil obtained from the seeds used in preparation of food.

4 DICRANOSTIGMA Capsule splitting along its whole length; style erect with 2–4 ascending stigmas. 1 sp.

111 **D. lactucoides** Hook.f. & Thoms. (*Stylophorum 1.* (Hook.f. & Thoms.) Benth. & Hook.f.)
Uttar Pradesh to C. Nepal. Tibet. 2400–4000 m. Stony slopes, river gravels; Tibetan border areas. Jun.–Jul.

Flowers yellow, 3–5 cm across, few, borne on stems 10–30 cm, with pinnately-lobed leaves which are glaucous beneath. Petals 4, circular; sepals 2, hairy with horny tips. Leaves mostly basal, numerous, the lower with a winged, hairy leaf-stalk, the upper stalkless; leaf-lobes broadly ovate, toothed; tap-root thick. Capsule oblong to linear, to 6 cm, splitting by 2–4 valves which remain attached to the stigma. Stems and roots with orange juice. **Pl. 13.**

5 CORYDALIS Flowers irregular, distinctly 2–lipped; the upper petal hooded and spurred, the lower forming a boat-shaped keeled lip, the 2 lateral petals narrow. Fruit a 2–valved capsule. *c.* 48 spp.

Flowers blue, purple or pinkish-white

112 **C. cashmeriana** Royle
Kashmir to S.E. Tibet. 3000–4500 m. Shrubberies, open slopes, screes; quite common. May–Aug.

A small delicate perennial with 3–8 sky-blue flowers with darker blue tips, in a lax terminal cluster, and with delicate, deeply cut glaucous leaves. Flowers 1.2–2 cm, including the spur which is half as long as the flower and slightly down-curved. Leaves mostly basal, *c.* 1 cm across, ternate, the leaflets 3–lobed; stem leaves one or two, smaller with narrower lobes; stem unbranched 3–10 cm. Root-stock tuberous, with bulbils. Capsule linear, borne on deflexed or spreading stalks. **Pl. 14.**

113 **C. crassissima** Cambess. (*C. crassifolia* auct. non Royle)
Pakistan to Himachal Pradesh. 3600–4500 m. Screes. Jul.–Aug.

A striking plant growing in otherwise barren screes, with rounded, deeply lobed, fleshy, conspicuously glaucous leaves, and a dense cluster of relatively large violet to pinkish-white flowers succeeded by large globular inflated capsules. Flowers 2–2.5 cm, with a slightly down-curved spur, subtended by spathulate to lanceolate green bracts. Leaves with a rounded to kidney-shaped 3–lobed blade 5–10 cm across, with rounded-toothed or shallowly lobed margin; stem leaf similar, stalkless. Fruit inflated, balloon-like, 1.5–2.5 cm, easily dispersed by the wind or gravity. **Pl. 14.**

114 **C. rutifolia** (Smith) DC. (*C. diphylla* Wallich)
Afghanistan to W. Nepal. W. Asia. Crete. 1800–3300 m. Shady banks, shrubberies, forests; common in W. Himalaya. Apr.–Jun.

A delicate pink to purple-flowered perennial with a rather dense terminal cluster of few flowers borne above a pair of opposite, dissected, hairless stem leaves. Flowers 5–15, 1.5–2.5 cm long, with a stout, paler, conspicuously up-curved spur with often down-curved apex; bracts entire, acute. Leaves ternate, mostly twice divided into oblong entire leaflets, the middle leaflet largest; basal leaves absent; stem unbranched, 5–15 cm. Capsule oblong-elliptic 1–1.5 cm, with a curved style. **Pl. 14.**

Flowers yellow
a *Spur less than half the length of the flower*

115 **C. flabellata** Edgew.
Pakistan to C. Nepal. Tibet. 2700–4000 m. Stony hillsides of dry Tibetan borderlands; common in Ladakh. Jun.–Aug.

Flowers yellow, in spike-like clusters 10–20 cm long, borne at the ends of the branches, and somewhat fleshy pinnate leaves with broad fan-shaped leaflets. Flowers many, 12–18 mm, with a short spur up to one third as long as the flower, down-curved and somewhat swollen at the apex. Leaves with 3–6 pairs of leaflets and a terminal leaflet, each leaflet 1–2 cm across, often deeply 3–lobed or cut; stem erect, rigid, much branched, glaucous, 20–60 cm. Capsule linear, 12–20 mm, style 4–5 mm, recurved at apex.

116 **C. meifolia** Wallich
Himachal Pradesh to S.E. Tibet. 3900–5200 m. Rocky slopes. Jun.–Aug.

Flowers yellow or sometimes orange with brown-purple or violet tips, borne in a dense almost flat-topped cluster 3–5 cm long. Flowers 12–18 mm including the spur which is much less than half as long as the flower; upper petal with a conspicuous dorsal wing. Leaves glaucous, many times cut into linear or thread-like acute segments; basal leaves large 10–30 cm, long-stalked. A leafy, erect or spreading, usually unbranched perennial 15–30 cm, with a thick twisted rootstock covered with old sheathing leaf-bases. Capsule oblong-elliptic, 7–10 mm, with a curved style. **Pl. 14.**

b *Spur half the length of the flower or more*
i *Stem leaves several*

117 **C. chaerophylla** DC.
Uttar Pradesh to Bhutan. 2100–4000 m. Forests, shrubberies; common. May–Oct.

Flowers golden-yellow, with slender acute spurs longer than the dorsal petal, borne in terminal branched clusters. Flowers 1.5–2 cm; lower bracts leafy, oval to oblong, toothed, the upper entire. Leaves 15–25 cm, long-stalked, three times cut into oblong acute segments, paler beneath; stems 60–120 cm.

118 **C. thyrsiflora** Prain
Pakistan to Kashmir. 3000–4300 m. Open slopes, wet places; common in Kashmir. Jul.–Aug.

Flowers yellow, borne in a widely branched inflorescence with few or many dense terminal clusters carried above the leaves. Distinguished by its short broad capsule 5–7 mm with a curved style almost as long. Flowers 12–15 mm; spur broad, blunt, about half as long as the flower, straight or slightly down-curved. Leaves hairless, somewhat glaucous, three times cut into oblong to linear usually pointed segments 5–10 mm long; stems branched, 15–30 cm. **Pl. 14**.

ii *Stem leaves usually two, deeply lobed*

119 **C. falconeri** Hook. f. & Thoms.
Pakistan to Kashmir. 2700–3300 m. Open slopes. Jul.–Aug.

Flowers yellow with reddish-brown streaks or purple tips, many in a spike-like cluster 3–10 cm long, carried well above the basal leaves. Flowers 12–18 mm; spur about half as long as flower, slightly hooked; bracts usually entire, linear-lanceolate pointed. Basal leaves long-stalked, twice cut into oblong-obovate acute segments 1–2 cm long; stem leaves two, opposite, stalkless, usually placed above the middle of the stem which is 15–40 cm. Capsule oblong, 12–18 mm, with curved style.

120 **C. govaniana** Wallich
Pakistan to E. Nepal. 2400–4800 m. Shrubberies, open slopes; common alpine species. May–Aug.

Like 119 with yellow flowers, usually in a dense cluster 5–10 cm long, but stem leaves, if present, solitary or paired, arising from below the middle of the stem. Flowers larger 2–2.5 cm long; spur straight or somewhat down-curved; outer petals dorsally winged; bracts variable, much dissected to entire. Leaves much dissected, ultimate segments ovate-oblong, blunt or acute, 1–1.5 cm; rootstock slender, covered with bases of old leaves.

Root used medicinally. **Pl. 14**.

iii *Flowering stems leafless or with entire leaves* (see also 120)

121 **C. juncea** Wallich
C. Nepal to S.E. Tibet. 2500–4300 m. Alpine shrubberies, open slopes. Jun.–Jul.

Flowers bright yellow with purple tips, in a many-flowered terminal cluster borne on a slender unbranched stem. Flowers 8–20 mm; spur about half the length of the flower, blunt; upper petal broadly winged dorsally; bracts linear entire. Stem leaves if present bract-like, linear-lanceolate 1–3 cm; basal leaves stalked, ternate, the leaflets 3–5–lobed; stem 30–45 cm. Roots with many spindle-shaped tubers. **Pl. 14**.

6 **DICENTRA** Flowers 2–lipped, distinguished from *Corydalis* by the outer two petals which have swollen sac-like bases. Slender climbing herbaceous perennials. 3 spp.

122 **D. macrocapnos** Prain (*Dactylicapnos m.* (Prain) Hutch.)

Uttar Pradesh to E. Nepal. 1500–3000 m. Rambling on banks, shrubberies. Apr.–Oct.

A slender climbing plant to 2 m, with short-stalked pendulous clusters of yellow urn-shaped flowers, tipped with purple (or flowers orange) and with compound leaves with tendrils. Flowers 2–2.5 cm long; sepals 2, small, triangular; petals 4, the outer 2 with inflated bases, the inner 2 long-stalked, keeled, enlarged and cohering at the tips; stamens 2. Leaves with many leaflets which are stalked, oval to rounded, 5–25 mm, and a terminal branched tendril; stem slender, flexuous, angled. Capsule cylindrical, with papery valves; seeds shining. **Pl. 15**.

7 FUMARIA FUMITORY Distinguished from *Corydalis* by its globular one-seeded, non-splitting fruit. Annuals. 3 spp.

123 **F. indica** (Hausskn.) Pugsley
India. W. Asia. To 2400 m. Cornfield weed; common throughout the Himalaya. Apr.–May.

A delicate much-branched, glaucous, leafy annual 5–30 cm, with clusters of tiny pale pinkish to whitish flowers each 5–6 mm long. Sepals minute; upper petal with a short somewhat down-curved sac-like spur; flower-stalks erect, equalling or slightly shorter than the lanceolate bracts. Leaves 2–3–times cut into narrow pointed segments *c*. 1 mm wide. Fruit globular *c*. 2 mm.

CRUCIFERAE   Mustard Family

A large and important cosmopolitan family of herbaceous plants with a distinctive and uniform floral structure. Leaves usually alternate, without stipules, simple, or less commonly compound. Flowers usually in branched racemes or cymes, usually without bracts. Sepals 4, free; petals 4, free, arranged in a cross; stamens usually 6. Ovary of 2 fused carpels, with 2 chambers each usually containing many ovules. Fruit usually splitting into 2 valves leaving a persistent cross-wall. Fruit either short, less than 3 times as long as broad – a *silicula*, or fruit long, more than 3 times as long as broad – a *siliqua*. Some fruits not splitting when ripe, others break up into 1–seeded sections. Ripe fruits are necessary for the identification of both genera and species. Many genera are differentiated by small and often obscure botanical characters, particularly of the fruit, about which the experts do not always agree; no attempt has been made here to define or describe these characters.

Fruit short, a silicula (less than 3 times as long as broad)

a Leaves entire, or toothed	1 BRAYA 2 CRAMBE 3 DRABA
	4 ISATIS 5 LEPIDIUM
	6 PEGAEOPHYTON 7 THLASPI
b Leaves deeply lobed	8 MEGACARPAEA

CRUCIFERAE

Fruit long, a siliqua (more than 3 times as long as broad)
  a Leaves entire, or toothed
    i Flowers white, violet, purple, or pink   9 ARABIDOPSIS
         10 ARABIS   11 ARCYOSPERMA   12 CHRISTOLEA
    13 ERMANIA   14 MATTHIOLA   15 PHAEONYCHIUM
           16 PYCHNOPLINTHOPSIS (see also 19, 21)
    ii Flowers yellow or orange   17 ERYSIMUM
  b Leaves deeply lobed, or compound
    i Flowers white, pink or violet   18 CARDAMINE
        19 CHORISPORA   20 LIGNARIELLA
    21 PARRYA   22 RAPHANUS   23 RORIPPA
    ii Flowers yellow   24 BARBAREA   25 BRASSICA

1 BRAYA Small tufted alpine perennials with entire leaves. Fruit elliptic, several seeded. 1 sp.

124 **B. oxycarpa** Hook. f. & Thoms. (*Aphragmus o.* (Hook. f. & Thoms.) Jafri)
Afghanistan to S.W. China. Tibet. C. Asia. 3600–5500 m. Screes, damp places in drier regions. Jun.–Jul.
   A small tufted perennial 2–8 cm, with tiny white, pinkish, or pale lilac flowers with yellow centres borne in a dense head, and hairless, somewhat fleshy, entire narrow leaves in a basal rosette. Flowers 3–4 mm across; sepals *c.* 2 mm; stigmas yellow. Rosette leaves to 2.5 cm; stem leaves few or many, smaller; stem branched from the base, sparsely hairy below. Fruit flattened-elliptic, 6–10 mm, hairless. **Page 448.**

2 CRAMBE Fruit two-sectioned, the lower stalk-like and seedless, the upper globular one-seeded. Flowers white. 1 sp.

125 **C. kotschyana** Boiss.
Pakistan to Himachal Pradesh. 2100–4300 m. Rocky slopes; distinctive and abundant in upper Chenab valley, Himachal Pradesh. Jun.
   A stout erect branched perennial 60–180 cm, with a large lax branched cluster of numerous white flowers, and with large heart- or kidney-shaped coarsely toothed leaves. Flowers 7–10 mm across; petals 5–10 mm, ovate-oblong with wedge-shaped base. Basal leaves with blades 20–40 cm broad, with wavy toothed margins, long-stalked, the upper few, smaller. Fruit with stalk-like base, and globular one-seeded upper part *c.* 5 mm.
   Leaves and roots edible; good fodder for goats.

3 DRABA Whitlow Grass Small often tufted perennials, distinguished largely by their fruits which are usually ovate to elliptic, flattened, and with 2 chambers each with two rows of seeds. Flowers white or yellow, rarely mauve. *c.* 19 spp.

34

Flowers mauve, large

126 **D. amoena** O. E. Schulz
Uttar Pradesh to E. Nepal. 3300–4300 m. Open slopes, rocks. May–Jul.

An erect unbranched perennial, with a rosette of greyish leaves, and with a terminal head of mauve flowers which elongates in fruit. Flowers about 1 cm across, but variable; petals obovate, about twice as long as the sepals, or more. Stem leaves clasping, lanceolate, obscurely toothed, 1.5–3 cm; rosette leaves spathulate, densely covered with star-shaped hairs; stems sometimes branched above. Fruit cylindrical, with a conspicuous style, 1–1.5 cm. **Page 448**.

Flowers white, small

127 **D. altaica** (C. Meyer) Bunge
Afghanistan to Bhutan. Tibet. C. Asia. 4000–4800 m. Rocks, screes, open slopes; common. Jul.–Aug.

A small tufted perennial with short stems each bearing a flat-topped cluster *c.* 5 mm across of tiny white flowers, and with rosettes of tiny lanceolate to elliptic leaves. Flowers *c.* 3 mm across; petals *c.* 2 mm long. Leaves all hairy, 5–10 mm, entire or toothed at apex; stem leaves broader, few or none; stems 2–5 cm. Fruit elliptic 3–4 mm, usually hairless. **Page 448**.

128 **D. lasiophylla** Royle
Pakistan to Bhutan. Tibet. 4300–4800 m. Stony slopes. Jun.–Jul.

Like 127, but a taller plant 5–20 cm, with a terminal cluster of white flowers 5–10 mm across, which elongates in fruit. Rosette leaves oblong-lanceolate, 5–15 mm; stem leaves several, similar or broader, stalkless, densely covered with star-shaped hairs. Fruit 5–8 mm, narrow elliptic, twisted, usually hairy.

Flowers yellow
a *Flowering stems leafy*

129 **D. gracillima** Hook.f. & Thoms.
Uttar Pradesh to Bhutan. Tibet. 3600–4500 m. Open slopes. May–Jul.

An erect or spreading perennial with many slender stems bearing lax clusters of tiny yellow flowers each *c.* 3 mm across. Petals 2.5 mm long; sepals shorter. Basal leaves in a lax rosette, obovate-spathulate, entire or toothed, hairy, 1–2 cm; stem leaves few, ovate to elliptic; stems *c.* 10 cm, lengthening in fruit. Fruit linear 6–14 mm, hairless.

130 **D. radicans** Royle
Uttar Pradesh to C. Nepal. 3000–3600 m. Shrubberies, open slopes. May–Jun.

A rather straggling perennial with lax rosettes of lanceolate leaves, and ascending stems bearing a lax terminal rather flat-topped cluster of bright yellow flowers. Flowers about 8 mm across; petals to 8 mm; sepals one half to a third as long, sparsely bristly-haired. Leaves lanceolate, mostly 1–1.5 cm, with shallow widely spaced teeth, the upper stalkless; lower stem

35

leaves somewhat larger and slightly broader. Fruit narrowly elliptic 1–2 cm, with persistent style.

b *Flowering stems leafless; cushion-forming plants*

131 **D. oreades** Schrenk
Pakistan to S.W. China. C. Asia. 3600–5500 m. Screes, rocky slopes. Jun.–Jul.

A tiny often mat-forming perennial, with rosettes of leaves, and short leafless stems bearing a dense rounded cluster *c.* 1 cm across of small bright yellow to yellowish-white flowers. Flowers *c.* 4 mm across; petals *c.* 4 mm; sepals half as long. Leaves variable, mostly 5–10 mm, lanceolate to obovate, with simple or branched bristly hairs; stems 2–5 cm in fruit. Fruit ovate, 4–5 mm, usually hairless. **Page 448**.

132 **D. setosa** Royle
Afghanistan to Himachal Pradesh. 3000–5000 m. Shrubberies, open slopes; common in Ladakh. Jul.–Aug.

A small cushion-forming plant with tiny rosettes about 6–8 mm across of densely hairy leaves, and erect leafless stems bearing a lax cluster of few yellow flowers. Flowers *c.* 3–5 mm across; sepals much shorter than petals, with papery margins, hairless. Leaves linear 1–1.5 cm long with bristly margins; flowering stems slender 3–4 cm. Fruit elliptic 4–6 mm, hairless. **Pl. 15**.

4 ISATIS Fruit flattened, broadly winged, one-seeded. Flowers yellow. 1 sp.

133 **I. costata** C. Meyer
Afghanistan to Himachal Pradesh. 1500–2700 m. Stony slopes. May–Jul.

A robust biennial 30–120 cm, with a branched cluster of numerous small yellow flowers, entire clasping upper leaves, and distinctive flattened winged fruits. Flowers 4–5 mm across; petals 3–5 mm, obovate-wedge-shaped. Basal leaves in a rosette, usually oblanceolate, 5–15 cm, but very variable in size; stem leaves smaller 3–5 cm, narrower and clasping the stem, hairy or hairless. Fruit paddle-shaped, 1–2 cm long by 3–5 mm broad.

Sometimes used as a dye plant, producing blue and green colours.

5 LEPIDIUM Fruit rounded to elliptic, winged and flattened at right angles to the cross-wall, with 1 pendent seed in each cell. Flowers white. 4 spp.

134 **L. latifolium** L. DITTANDER
Afghanistan to Kashmir. W. Asia. Europe. N. Africa. 3000–3600 m. Stony slopes; common in Ladakh. Jul.–Aug.

A tall erect much-branched perennial, often somewhat woody at the base, with a spreading or pyramidal branched inflorescence with terminal clusters of numerous tiny white flowers. Flowers 2–3 mm across; petals 2–3 mm; sepals half as long, with a broad white margin. Leaves mostly

lanceolate, entire or toothed, hairless or sparsely hairy, the lower rather leathery; stem 30–120 cm. Fruit rounded to broadly ovate 1–2 mm.
Plant used medicinally. **Page 448**.

6 PEGAEOPHYTON Leaves entire, all basal. Fruit oblong, not splitting, few-seeded. Flowers white. 2 spp.

135 **P. scapiflorum** (Hook.f. & Thoms.) Marquand & Shaw
Kashmir to S.W. China. 4000–4700 m. Open slopes, damp places, streamsides. May–Jul.
A rosette perennial with a thick rootstock, numerous narrow leaves, and many tiny white flowers with violet veins, borne individually on short stalks directly from the rootstock. Flowers variable in size, 4–7 mm across; petals 4–6 mm long; sepals shorter. Leaves 1.5–4 cm, linear to elliptic, long-stalked, entire or sometimes sparsely toothed, hairless or with few hairs. Fruit 6–8 mm, hairless. **Page 449**.

7 THLASPI PENNY-CRESS Fruit flattened with valves distinctly keeled and usually winged. Upper leaves clasping stem. Flowers white or pinkish. 6 spp.

136 **T. andersonii** (Hook.f. & Thoms.) O.E. Schulz
Pakistan to S.E. Tibet. 3000–4400 m. Open slopes, damp places. May–Aug.
A small spreading or ascending perennial with hairless somewhat fleshy glaucous, ovate to elliptic leaves, and with spike-like clusters of small white or pinkish flowers. Distinctive in having stem leaves with clasping eared bases and contrasting basal leaves which are narrowed to a short stalk. Flowers 6–7 mm across; petals 5–6 mm, often veined; sepals ovate, with papery margins. Basal leaves 1–1.5 cm, in a lax rosette, entire or toothed; stem 5–10 cm. Fruit elliptic, not winged, 6–8 mm by 2 mm broad. **Page 449**.

137 **T. griffithianum** (Boiss.) Boiss. (*T. cochleariforme* subsp. *griffithianum* (Boiss.) Jafri)
Pakistan to Uttar Pradesh. 2700–5500 m. Alpine slopes; common. May–Aug.
Like 136, with white flowers veined with purple, but flowers smaller c. 5 mm across, and fruit slightly notched at apex and usually slightly winged (fruit with entire apex, not winged in 136). Flowers in compact terminal clusters c. 1.5 cm across borne on unbranched stems, with elliptic blunt leaves with clasping ear-shaped basal lobes. Basal leaves spathulate, in a rosette; stems usually several, 10–20 cm. Fruit remaining in a close cluster with spreading fruit-stalks; capsule 5–8 mm. **Page 449**.

8 MEGACARPAEA Fruit distinctive, comprising two broadly winged, one-seeded units which separate when ripe. Flowers white or cream. 1 sp.

138 **M. polyandra** Benth.
Kashmir to C. Nepal. 3000–4300 m. Open slopes, light forests. Jun.–Jul.

A robust perennial 1–2 m, with a stout stem, large deeply pinnately-lobed basal leaves, and a large terminal dense branched cluster of small white or creamy flowers. Flowers *c*. 1 cm across; petals rounded 5–10 mm long, stalked; stamens 8–16. Leaves to 60 cm, with 7–9 pairs of oblong-lanceolate pointed toothed lobes 10–20 cm long; uppermost leaves linear entire. Fruit 3.5–5 cm broad, of paired broadly winged one-seeded units. The young leaves are edible when cooked. **Pl. 15.**

9 ARABIDOPSIS   Fruit long slender with valves one-veined; seeds ovoid, in 1 row. 6 spp.

139 A. himalaica (Edgew.) O. E. Schulz
Afghanistan to Bhutan. Tibet. 2400–4300 m. Grazing grounds, screes; common. Jun.–Aug.

Flowers tiny *c*. 3 mm, lilac, in a crowded terminal cluster, distinguished by the presence of bracts subtending the flowers. Petals 3–5 mm; sepals smaller; stamens 2. Usually a slender erect biennial, but sometimes much branched from the base, with bristly-haired stems 20–50 cm. Basal leaves often in a rosette, oblong-spathulate, coarsely toothed, 2–5 cm; stem leaves smaller, elliptic, clasping. Fruit linear 1.5–3.5 cm by 0.5 mm broad, usually hairless. **Page 449.**

10 ARABIS   Rock-Cress   Fruit long narrow, strongly compressed; seeds in one row. Flowers white or pale yellow. 5 spp.

140 A. amplexicaulis Edgew.
Afghanistan to W. Nepal. 1500–2700 m. Shrubberies; in Kashmir a prominent early spring flower. Apr.–May.

A tall erect perennial with a terminal cluster of small white flowers, and with usually a large basal rosette of bristly-haired leaves. Flowers 8–10 mm across, borne on spreading hairless stalks; petals 7–10 mm long; sepals with white margins. Stem leaves elliptic to lanceolate, sparsely toothed, clasping; rosette leaves oblanceolate, stalked, often toothed, 5–15 cm, bristly-haired; stems to 40 cm. Fruit linear, 3–6 cm, hairless, borne on spreading or horizontal stalks. **Page 449.**

141 A. glabra (L.) Bernh. (*Turritis g.* L.)
Pakistan to W. Nepal. Temperate Eurasia. 1500–3000 m. Stony slopes. May–Jun.

A glaucous annual or biennial to 60 cm, with branched clusters of white or pale yellow flowers, and contrasting basal and upper stem leaves, the latter arrow-shaped and clasping the stem. Flowers *c*. 5 mm across, pale yellow, becoming whitish; petals 4–6 mm. Rosette leaves 5–15 cm, oblong-obovate, stalked or stalkless, entire or toothed, hairy; stem leaves 2–5 cm, hairless, glaucous. Fruit linear, hairless, often pressed to stem, 3–9 cm.

11 ARCYOSPERMA   Like *Arabis*, but rootstock thick and somewhat woody; basal leaves much larger than stem leaves; fruit oblong curved many-seeded; stigma stalkless. 1 sp.

142 **A. primulifolium** (Thoms.) O. E. Schulz (*Eutrema p.* (Thoms.) Hook. f. & Thoms.)
Pakistan to Bhutan. 2100–3600 m. Damp rocks. Apr.–May.

A dwarf perennial with a dense rosette of large ovate leaves, and short unbranched stems bearing a few small white or pinkish fragrant flowers, and much smaller stem leaves. Flowers *c.* 5 mm across, borne on short spreading stalks; petals 5–7 mm. Leaves all more or less fleshy, the basal leaves spathulate, 5–15 cm, narrowed to a long stalk; stem leaves elliptic half-clasping, 1–5 cm. Fruit 15–22 mm, slightly curved upwards.

12 CHRISTOLEA Leaves fleshy. Fruit broadly linear to oblong, flattened; seeds many. 1 sp.

143 **C. crassifolia** Cambess.
Afghanistan to C. Nepal. Tibet. C. Asia. 3300–4200 m. Dry areas, stony slopes. Jun.–Aug.

A spreading or more or less erect perennial, with several stems with terminal spike-like clusters of white to mauve flowers often with purple bases, and thick ovate-oblong, toothed, greyish, often densely hairy leaves. Flowers 5–7 mm across; petals 5–6 mm. Leaves 1–4 cm, variable in shape and hairiness; stems usually 15–30 cm. Fruiting spike up to 8 cm long; fruit broadly linear, 1–3 cm by 3–4 mm broad. **Page 450**.

13 ERMANIA Like *Christolea* but differing in its densely woolly-haired fruits. 2 spp.

144 **E. himalayensis** (Cambess.) O. E. Schulz (*Cheiranthus h.* Cambess.)
Pakistan to Sikkim. Tibet. 4500–5500 m. A high altitude scree plant. Jun.–Aug.

A small densely grey-haired perennial with a compact rosette of fleshy oval, toothed leaves, and several stems with short spike-like clusters of small lilac flowers. Petals strap-shaped *c.* 6 mm; sepals hairy, half as long; flower-stalks ascending, often on one side of stem. Basal leaves 1–1.5 cm; stem leaves smaller, narrower, the uppermost subtending flowers; flowering stems 5–15 cm. Fruits finely hairy, linear-oblong, 2–3 cm by 2–3 mm. **Page 450**.

14 MATTHIOLA STOCK Fruit long narrow, with deeply two-lobed stigma. Flowers brown-purple in our species. 1 sp.

145 **M. flavida** Boiss.
Pakistan to Kashmir. 2700–4500 m. Stony slopes; common in Ladakh. Jun.–Jul.

A rather woody-based perennial with a widely branched inflorescence with whitish woolly-haired stems, narrow leaves, and widely spaced brownish-purple flowers often with inrolled petals. Flowers often fragrant; petals 1.5–2 cm long, linear-oblong; sepals woolly-haired, shorter. Stem leaves 2–8 cm, lanceolate, entire, toothed or shallow-lobed, very variable; basal leaves usually broader; stems branched from the base, 15–100 cm. Fruit 6–10 cm long, often densely hairy. **Page 450.**

15 PHAEONYCHIUM Like *Christolea* but hairs star-shaped; leaves narrow entire. 1 sp.

146 **P. parryoides** (Kurz ex Hook. f. & T. Anderson) O. E. Schulz (*Cheiranthus p.* Kurz ex Hook. f. & T. Anderson)
Pakistan to W. Nepal. Bhutan. Tibet. 3600–4400 m. Cliffs, stony places; drier Tibetan border areas. Jun.–Jul.

A densely tufted rosette-forming perennial with narrow leaves, and erect leafless stems bearing a rather dense cluster of white or mauve flowers with lilac bases. Flowers *c.* 1 cm across; petals rounded 7–8 mm; stamens 3–4. Leaves linear to oblanceolate 4–9 cm, narrowed to the leaf-stalk, usually entire, densely covered with minute star-shaped or branched hairs; flowering stems 10–30 cm. Rootstock stout, branched above and covered with the bases of last year's leaves. Fruit flattened *c.* 2 cm long, covered with star-shaped hairs. **Page 450.**

16 PYCNOPLINTHOPSIS Leaves all basal in a rosette. Flowers solitary. Fruit linear to elliptic-oblong, often with curved apex. 1 sp.

147 **P. bhutanica** Jafri
C. Nepal to Bhutan. 3300–4500 m. Rocks, screes, cliff faces. May–Jun.

A rare but remarkable plant locally abundant on route to Annapurna South base camp, looking like *Primula*, with many short-stalked white flowers surrounded by a rosette of numerous spathulate leaves with conspicuous rounded teeth, but distinguished readily by the 4–petalled flowers. Flowers solitary 1–1.5 cm across, arising directly from the rootstock on stalks to 3 cm; petals rounded. Rosettes 5–15 cm across. Fruit linear to 1.5 cm, hairless. **Pl. 15.**

17 ERYSIMUM Fruit long narrow, four-angled, with valves conspicuously one-veined. Seeds in 1 row; style distinct, stigma usually two-lobed. Flowers yellow. *c.* 4 spp.

148 **E. hieraciifolium** L. (sensu latu)
Pakistan to S.E. Tibet. Temperate Eurasia. 2400–4000 m. Open slopes, cornfield weed. Jun.–Aug.

A very variable biennial or perennial with terminal clusters of numerous bright orange-yellow flowers, and with narrow entire or toothed stem leaves. Flowers *c.* 1 cm across; petals 8–10 mm, with long stalks and rounded blades; sepals finely hairy, 5–7 mm. Leaves variable, narrow elliptic to oblanceolate, stalked, 4–6 cm or more long; stem leaves stalkless; flowering stem 30–60 cm. Fruit 3–5 cm, with adpressed hairs. (Further research is required into the Himayal forms of this plant.)

149 **E. melicentae** Dunn
Pakistan to Kashmir. 1800–3000 m. Rocks, open slopes. May–Aug.
Very like 148, but distinguished by its larger yellow flowers 1–1.5 cm across, with longer and broader petals 12–18 mm by 6–9 mm (petals 3–4 mm broad in 148). Hairs adpressed, mostly with 2 branches (with 2–4 branches in 148). Leaves linear acute, with or without few widely spaced teeth. A tall leafy plant 30–80 cm. Fruit 3.5–7 cm. **Pl. 15**.

18 CARDAMINE Cress Fruit long narrow, strongly compressed, with valves that coil up suddenly from the base ejecting the seeds. Flowers white, purple, or pinkish. 10 spp.

Leaves pinnate

150 **C. loxostemonoides** O. E. Schulz
Pakistan to Bhutan. Tibet. 3000–5000 m. Rocks, screes. May–Jul.
A small delicate but rather showy perennial, with relatively large lilac or pinkish flowers borne on flexuous spreading stems, and with small pinnate leaves. Flowers 1–1.5 cm across; petals strongly veined, 3 times as long as sepals. Leaves with 4–7 pairs of rounded to narrow-oblong leaflets usually 5–15 mm, the terminal leaflet similar, entire or 3–lobed; stems 10–20 cm. Rhizome creeping, with bulbils. Fruit 2.5–3 cm long with a style 2–3 mm, flattened, hairless. **Pl. 15**.

151 **C. macrophylla** Willd.
Pakistan to S.W. China. 2100–3600 m. Forests, streamsides; common in W. Himalaya. May–Aug.
Flowers conspicuous, lilac, pink or sometimes white, 1–1.5 cm across, in a dense terminal cluster which elongates in fruit. Petals 1–1.5 cm long; sepals *c*. 5 mm, with papery margins. Leaves with 3–10 pairs of usually large lanceolate pointed, toothed leaflets 3–8 cm long, but leaflets variable in shape and teeth; stems erect, robust 30–80 cm. Fruit 2.5–5 cm, hairless. **Page 451**.

Leaves entire, toothed

152 **C. violacea** (D. Don) Wallich
W. Nepal to Bhutan. 2700–3600 m. Forests, shrubberies. Apr.–Jun.
Flowers large, deep purple, drooping, in a terminal spike-like cluster, borne on a stout simple erect stem, with clasping lanceolate toothed leaves with arrow-shaped bases. Petals to 1.5 cm long; sepals to 8 mm; style long but not exserted. Stem leaves 8–13 cm, tapering to the apex; basal leaves broader, stalked, withered at flowering. Fruit *c*. 2 cm. **Page 451**.

19 CHORISPORA Fruit long-cylindrical, constricted between seeds, not splitting; stigma erect. Flowers pink, violet or white. 1 sp.

153 **C. sabulosa** Cambess. (*C. elegans* var. *s.* Cambess.)
Pakistan to Himachal Pradesh. 3300–5000 m. Stony slopes; common in dry areas. Jun.–Aug.

A small rosette-forming perennial, with narrow shallowly pinnately-lobed leaves and short stems bearing lax clusters of pink, violet or white flowers. Flowers to 1 cm across; petals 6–10 mm, shallowly notched; sepals half as long. Leaves 2–6 cm, oblanceolate in outline with a rounded terminal lobe and short or long blunt lateral lobes, glandular-hairy. Fruit *c.* 1.5 cm, glandular-hairy, constricted between seeds; style prominent. The fruits are edible. **Pl. 15**.

20 LIGNARIELLA Leaves three-lobed. Fruit oblong, curved upwards; style conspicuous; seeds few. Flowers solitary, axillary, violet. A genus of 2 spp. confined to the Himalaya.

154 **L. hobsonii** (Pearson) Baehni (*Cochlearia h.* Pearson)
Kashmir to Bhutan. Tibet. 2700–4300 m. Open slopes, damp places. Jun.–Jul.

A hairless annual with slender spreading branches forming a loose mat, with small 3–lobed leaves, and with small axillary violet flowers. Petals to 5 mm; sepals ovate; flower-stalks glandular. Leaves with blades 5–10 mm and leaf-stalk as long, lobes narrow-oblong to elliptic. Fruit flattened laterally. Very variable in size: the larger forms with stems 10–20 cm and leaves to 2 cm; other forms much smaller. **Page 451**.

21 PARRYA Fruit linear-oblong, flattened, with distinct style; seeds many, narrowly winged. Flowers white to lilac. 2 spp.

155 **P. nudicaulis** (L.) Boiss.
Afghanistan to Kashmir. Bhutan. Tibet. 4300–5300 m. Rocky slopes; quite common in Ladakh. Jun.–Jul.

Flowers showy, lilac, violet, or white, *c.* 2 cm across, borne on leafless stems, and with a dense basal rosette of long-stalked oblanceolate leaves. Petals 15–22 mm long, the stalks of petals much longer than sepals which are *c.* 10 mm. Leaves 4–15 cm, stalked, glandular, usually entire but some with small lobes; stems to 10 cm in flower, 15–30 cm in fruit. Fruit 3–6.5 cm, glandular, often with undulate margin. **Pl. 15**.

22 RAPHANUS Fruit cylindrical and prolonged into a narrow seedless beak, jointed at the base and more or less constricted between seeds. Flowers white or violet. 1 sp.

156 **R. sativus** L. RADISH
Of unknown origin; widely cultivated for its edible roots in temperate countries, and in the Himalaya to 4400 m. Apr.–Jul.

Flowers white or violet, usually in branched clusters on stems 20–90 cm, and with deeply lobed, long-stalked lower leaves. Flowers 1.5–2 cm across;

petals 1.5–2 cm long, stalks of petals as long as the sepals which are 6–8 mm. Leaves very variable, with 3–5 pairs of lobes and a larger rounded, toothed terminal lobe; uppermost stem leaves linear to oblong, entire or toothed. Tap-root swollen, cylindrical or turnip-shaped. Fruit 2–6 cm, including the long beak which is more than half as long as the lower 1–3–seeded part.

An important vegetable for people living in the Himalaya. The roots are dug in late autumn, sliced and dried for use in the winter and spring.

23 RORIPPA Fruit cylindrical, valves usually without mid-veins; seeds in 2 rows. Flowers white or yellow. 5 spp.

157 **R. nasturtium-aquaticum** (L.) Hayek (*Nasturtium officinale* R. Br.) WATERCRESS
Afghanistan to Bhutan. Temperate Asia. Europe. N. Africa. 1500–4000 m. Marshes, streamsides; very common. Apr.–Jun.

A glossy green, somewhat fleshy-stemmed perennial, found growing in shallow water, with lax terminal clusters of small white flowers. Flowers *c.* 4 mm across; petals 4–5 mm. Lower leaves stalked, with 1–5 leaflets; upper leaves with eared bases and with 5–9 narrower blunt leaflets; stem hollow, rooting from the lower nodes. Fruit oblong, 1–2 cm, on spreading stalks.

Plant edible, and used medicinally.

24 BARBAREA WINTER CRESS Fruit bluntly 4–angled in section; valves with strong mid-vein and netted lateral veins; stigma slightly two-lobed. Flowers yellow. 2 spp.

158 **B. intermedia** Boreau
Pakistan to Bhutan. India. C. Asia. Europe. N. Africa. 3000–4300 m. Open slopes, marshy ground. Jun.–Jul.

An erect, rather pale green biennial, with terminal branched spike-like clusters of small bright yellow flowers, and deeply pinnately-lobed leaves. Flowers 5–6 mm across; petals *c.* 5 mm, twice as long as the often purple-tipped sepals. Lower leaves with 3–5 or more pairs of ovate lateral lobes and a larger rounded terminal lobe; upper leaves with narrower linear lobes and a broadly lanceolate, toothed, terminal lobe; stem angled, 15–60 cm. Fruit linear, 1–3 cm, hairless, erect. **Page 451.**

25 BRASSICA MUSTARD, TURNIP Fruit long narrow, with a conspicuous beak, with 0–3 seeds; valves with prominent mid-vein. Flowers white or yellow. 4 spp.

159 **B. juncea** (L.) Czernov MUSTARD
Probably native of S.E. Asia. Commonly cultivated for oil in the Himalaya; at lower altitudes in winter when fields are yellow with flowers, and in summer at higher altitudes, to 3600 m.

An erect annual to 1 m or more, with lower leaves with 2–3 pairs of toothed lobes and a larger terminal lobe, and upper leaves almost entire or toothed, hairless. Flowers golden-yellow, *c*. 7 mm across; petals 6–9 mm, stalked; sepals 4–6 mm, yellowish, hairless. Fruit linear 2.5–4 cm, somewhat constricted at intervals, narrowed into a beak 5–10 mm long; seeds reddish-brown.

160 **B. rapa** L. (*B. campestris* L.) WILD TURNIP
Possibly native of N. & C. Europe. To 4000 m. Commonly cultivated throughout the Himalaya. May–Aug.

An annual with branched clusters of golden-yellow flowers, with entire oblong, clasping, glaucous hairless upper stem leaves, and contrasting bristly-haired often deeply lobed basal leaves. Flowers 6–10 mm across; petals 8–10 mm; sepals spreading. Stem to 1 m, hairy below and hairless above; basal leaves often absent at flowering. Fruit 3.5–8 cm long including the beak which is 1–2 cm long.

Cultivated for its leaves which are eaten as a vegetable and used for cattle fodder. The seeds yield oil. Seeds and roots used medicinally.

CAPPARIDACEAE   Caper Family

A medium-sized family of the Tropics, Sub-tropics and Mediterranean region, consisting of herbaceous plants, trees and shrubs. Leaves alternate, simple or compound; flowers solitary, or in racemes, usually regular. Sepals 4, free; petals 4, free; stamens numerous. Ovary superior, of 2 fused carpels, 1–chambered with many ovules, long-stalked or stalkless. Fruit usually a capsule splitting by valves, sometimes a berry.

CAPPARIS   Shrubs with entire leaves; flowers large, showy. Fruit a fleshy berry. 4 spp.

161 **C. spinosa** L. CAPER
Afghanistan to E. Nepal. W. Asia. Europe. 2000–3000 m. Stony slopes, in dry valleys. May–Sep.

A straggling spreading shrubby plant with two ranks of rounded rather leathery leaves, with spiny branches, and with large solitary axillary whitish to pinkish flowers with numerous long purplish stamens and a longer stalked ovary. Flowers 3–8 cm across, long-stalked, borne in the axils of the younger leaves; petals 4, obovate 2–5 cm; stamens much longer. Leaves variable, ovate to elliptic, 1–4 cm, mostly spine-tipped; stipules of 2 hooked brown or yellow spines. Fruit fleshy, oblong-ellipsoid 2–5 cm, long-stalked, with red flesh and brown seeds.

The flower buds are edible. The root, bark and leaves are used medicinally. **Pl. 16**.

CLEOME   Annuals with digitate leaves. Flowers in racemes. Fruit a dry capsule. 3 spp.

162 **C. speciosa** Raf.

Native of S. America; commonly cultivated and sometimes semi-naturalized in Nepal. To 1800 m. Aug.–Sep.

Flowers pink, scented, in large terminal clusters to 12 cm across, with 4 spathulate long-stalked petals, and much longer thread-like stamens, and a long-stalked ovary projecting beyond the petals in mature flowers. Flowers long-stalked; bracts leafy; petals *c*. 4 cm. Leaves digitate with 5–7 lanceolate long-pointed lobes to 10 cm; uppermost leaves bract-like entire 1–2 cm; stems to 1 m. Capsules cylindrical to *c*. 6 cm, borne on very long slender stalks, comprising both the flower-stalk and the ovary-stalk and with persistent sepals at the junction one third of the way from the base.

VIOLACEAE   Violet Family

A medium-sized cosmopolitan family mostly of herbaceous perennials with its main concentration in the temperate regions. Flowers usually solitary, regular, or irregular in *Viola*. Sepals 5, free, persistent; petals 5, free; stamens 5, united by their bases in a ring around the ovary. Ovary superior, of 3 fused carpels, with one chamber with many ovules. Fruit a capsule which splits explosively into 3 valves.

VIOLA   VIOLET   Flowers irregular, with 5 unequal petals, the lowest larger with a long or short sac-like spur at its base. Capsule splitting into 3 valves. A difficult genus, with *c*. 20 spp.

Flowers yellow

163   **V. biflora** L.

Pakistan to S.W. China. N. Temperate Zone. 2400–4500 m. Open slopes, shrubberies, forests. May–Jul.

Flowers solitary or paired, bright yellow with dark brown streaks to the centre, and a very short spur. Flowers *c*. 1.5 cm across; sepals oblong, blunt or acute. Leaves 1–3 cm broad, kidney-shaped with rounded-toothed margin, long-stalked; stipules oval; stem erect, to 10 cm or more. **Pl. 16**.

164   **V. wallichiana** Ging. ex DC.

C. Nepal to Bhutan. 2100–3000 m. Mossy rocks in forests. Jun.–Jul.

Differs from 163 in having a long narrow spur to 5 mm, and narrow linear-acute sepals to 5 mm, spreading or reflexed after flowering. Flowers yellow with a dark centre. **Pl. 16**.

Flowers violet, mauve or white
a *Leaves heart-shaped, or little longer than broad*
  i *Leaves mosty with blunt apex*

165   **V. canescens** Wallich

Kashmir to Bhutan. 1500–2400 m. Shrubberies, shady banks. Mar.–May.

Distinguished by its minutely and densely grey-haired leaves and leaf-stalks. Flowers pale violet and often paler at centre, *c.* 1 cm across, with a short blunt spur; sepals hairy. Leaves ovate-heart-shaped to kidney-shaped with a blunt apex; blade rather thick, and covered with grey hairs; leaf-stalks with dense down-curved hairs; stipules lanceolate, fringed, often brown. Runners usually present. Capsule hairy. **Page 452.**

166 **V. rupestris** F. W. Schmidt
Pakistan to Kashmir. 3000–4000 m. Forests, alpine slopes; prominent in the alpine zone in Kashmir. Jun.–Aug.

A small rather tufted plant with comparatively large mauve flowers with pale centres borne above the leaves. Flowers to 1.5 cm or more across; spur stout, short; sepals lanceolate. Leaves with rounded to heart-shaped blunt blade 1–1.5 cm long, finely hairy, long-stalked; stipules broadly lanceolate, fringed.

ii *Leaves mostly with an acute or long-pointed apex*

167 **V. indica** W. Becker
Pakistan to Himachal Pradesh. 1500–2400 m. Shrubberies, shady banks; prominent in spring in Kashmir valley. Apr.–May.

Readily distinguished by its relatively large fragrant violet-blue flowers to 2 cm, with a stout hooked spur to 1 cm. Leaves with a broadly ovate-heart-shaped, acute to long-pointed blade with coarse regular teeth.

168 **V. pilosa** Blume (*V. serpens* Wallich ex Ging.)
Afghanistan to S.W. China. Burma. S.E. Asia. 1200–3000 m. Shrubberies, shady banks. Mar.–May.

Like 165 with liliac flowers, but leaf-blades narrower and longer, ovate-lanceolate, acute to long-pointed, with thin white hairs, or almost hairless above. Flowers 1–1.5 cm; upper petals normally with hairs at the base; stigma 3–lobed, beaked. Stipules entire or toothed, not fringed. Capsule hairy or hairless.

b *Leaves elliptic to lanceolate, much longer than broad*

169 **V. betonicifolia** Smith
Pakistan to S.W. China. India. S.E. Asia. Australia. 1000–3000 m. Shrubberies, shady banks. Apr.–May.

Usually distinguished by its lance-shaped or spear-shaped blunt leaves which have a cut-off base abruptly narrowed to the usually winged upper part of the leaf-stalk. Flowers violet with darker veins, *c.* 1 cm, with a broad spur. Leaves variable, mostly 2–4 cm long, usually with shallow blunt teeth, sometimes with a shallow heart-shaped base; stipules mostly entire. **Page 452.**

170 **V. kunawarensis** Royle
Kashmir to C. Nepal. 3000–4700 m. Rocky hillsides. May–Jul.

A tiny perennial with a basal cluster of tiny elliptic long-stalked leaves, and small violet flowers borne on short leafless stalks direct from the rootstock. Flowers 5–10 mm, with a short blunt spur, and petals often darker veined. Leaves with a blade 1–2 cm, obscurely toothed. Capsule elliptic acute. **Page 452**.

POLYGALACEAE   Milkwort Family

A small cosmopolitan family of herbaceous plants, shrubs, and small trees, with characteristic irregular, rather pea-like flowers. Leaves alternate, always simple, usually without stipules. Flowers generally in spikes or racemes, each flower subtended by a bract and 2 bracteoles. Sepals 5 unequal, two often larger and coloured; petals 3, the lowest, or *keel*, saucer-shaped, and with a fringed crest; stamens 8, fused by their filaments into a split sheath. Ovary of 2 fused carpels, 2-chambered, each chamber one-seeded; fruit a capsule.

POLYGALA MILKWORT   10 spp.

171　**P. arillata** Buch.–Ham. ex D. Don
C. Nepal to S.W. China. India. S.E. Asia. 1500–2700 m. Forests, shrubberies. Jun.
　　An erect shrub 120–240 cm with entire lanceolate to oblong-elliptic pointed leaves, and pea-like flowers in drooping spike-like clusters. Flowers yellow to orange-yellow often with reddish tips to the petals, *c*. 1.5 cm long; keel with a conspicuous crest and partially fused to the 2 lateral petals. Sepals very unequal, 2 large rounded. Leaves 5–15 cm, stalked. Capsule large 1.3–2 cm, turning dark shining red, ribbed, rather fleshy, broadly kidney-shaped. **Page 452**.

172　**P. sibirica** L.
Pakistan to S.E. Tibet. N. Asia. E. Europe. 1600–3600 m. Open slopes. May–Jun.
　　A low tufted herbaceous perennial with woody rootstock and with many slender erect leafy stems bearing a terminal cluster of purple flowers. Distinguished from other species by the shining netted upper surface of the leaves. Flower-cluster slender, 2.5–8 cm long; flowers 6–8 mm; keel conspicuously crested; sepals shorter, broadly lanceolate, greenish-purple. Leaves narrow elliptic usually 8–10 mm; stems 8–15 cm. A very variable plant. **Pl. 16**.

CARYOPHYLLACEAE   Pink or Carnation Family

A large family of the temperate regions, consisting of herbaceous annuals and perennials, and having a distinctive and rather uniform character. Leaves opposite, arising from swollen nodes, always simple and undivided. Flowers commonly in complex branched clusters, rarely solitary. Flowers regular, with 4–5 sepals, either free, or fused as in *Silene* and *Dianthus*; petals 4–5, free, often notched or lobed; stamens twice as many as petals. Ovary superior, of 2–5 fused carpels, with a single chamber and many ovules borne on a central axis; styles 2–5. Fruit usually a capsule splitting by teeth at the apex.

Sepals free
  a Petals bi-lobed   1 CERASTIUM
  b Petals entire or notched   2 STELLARIA 3 ARENARIA
                                                                  4 MINUARTIA
Sepals fused below into a tube with 4–5 terminal lobes or teeth
  a Calyx with overlapping bracts at its base   6 DIANTHUS
  b Calyx without bracts at its base
    i Styles 2 or 3   5 THYLACOSPERMUM 7 GYPSOPHILA
                                                                  8 SILENE
    ii Styles 5   9 LYCHNIS

1 CERASTIUM CHICKWEED Styles usually 5; capsule teeth usually 10. Flowers white; petals usually lobed. *c.* 7 spp.

173 **C. cerastioides** (L.) Britton
Pakistan to Himachal Pradesh. C. Asia. 3000–4800 m. Stony slopes; common in Ladakh and Lahul. Jun.–Aug.
    A lax spreading perennial with erect stems bearing several stalked white flowers 1–1.5 cm across, in a lax branched cluster. Petals conspicuously shallowly bi-lobed; sepals narrow-elliptic *c.* 5–6 mm, with papery margins, hairy; styles usually 3. Leaves linear acute, 1–1.5 cm; flowering stems 5–15 cm. **Pl. 17.**

174 **C. dahuricum** Fischer
Pakistan to Uttar Pradesh. C. Asia. 3000–4000 m. Meadows, a common cornfield weed. Jul.
    A lax slender scrambling plant, with relatively large white flowers *c.* 1.5 cm across, with bi-lobed petals, borne on shining stems in a wide-spreading repeatedly branched lax cluster. Petals 1.5–2 cm; sepals shorter, with narrow white margins. Leaves broadly elliptic to oblong, 4–6 cm, stems scrambling to 4 m. Capsule with recurved teeth. **Page 452.**

2 STELLARIA STITCHWORT Styles 3; capsule with 6 teeth. Flowers white; petals usually bi-lobed. *c.* 22 spp. (Mostly small annuals or perennials not described here).

175 **S. decumbens** Edgew.
Pakistan to S.W. China. Tibet. 3300–6135 m. Rocks, screes. Jun.–Aug.
    Quite a common high alpine plant, forming dense mats or cushion-like masses of closely overlapping tiny narrow elliptic fine-pointed leaves. Flowers 3 mm, solitary or in stalked clusters; sepals 4–5, *c.* 4 mm, narrow-

elliptic acute; petals white, deeply bi-lobed, usually minute. Leaves numerous, shining or dull, 3–8 mm. Capsule shorter than the sepals. **Pl. 18**.

Listed as the highest recorded species in *The Guinness Book of Records*; but see also 178 *Arenaria bryophylla*.

3 ARENARIA SANDWORT Sepals 5, free; styles usually 3, capsule with 6 teeth. Flowers white; petals not lobed or notched. *c.* 25 spp.

Leaves awl-shaped or grass-like
a *Flowers several, borne usually in flat-topped clusters*

176 **A. festucoides** Benth.
Pakistan to Uttar Pradesh. 3600–4500 m. Rocks, stony ground. Jun.–Jul.

A densely tufted plant with bristle-like leaves, and short erect flowering stems with white flowers, and conspicuous white-margined calyx. Petals obovate *c.* 1.5 cm; sepals lanceolate pointed 1 cm; flowering-stems glandular, 2.5–15 cm; bracts with broad papery margins. Leaves shining, usually recurved, spiny-pointed, 6–25 mm.

177 **A. griffithii** Boiss.
Afghanistan to Uttar Pradesh. 2100–3600 m. Rocks, stony hillsides; common in drier areas. May-Aug.

A loosely tufted perennial with slender bristle-like hairy leaves, and with few white flowers *c.* 1 cm across, borne on slender nearly leafless stems. Petals spathulate, *c.* 1 cm; sepals broadly elliptic, sharp-pointed *c.* 5 mm or narrower and 3–veined. Leaves slender-pointed, *c.* 0.5 mm broad, ribbed, densely tufted. **Pl. 17**.

b *Flowers always solitary, stalkless or short-stalked*
  i *Flowers small, 8 mm or less across*

178 **A. bryophylla** Fern. (*A. musciformis* Wallich ex Edgew. & Hook. f.)
Kashmir to Sikkim. Tibet. 4300–6180 m. Rocks, stony slopes; Tibetan borderlands. Jul.–Aug.

A densely tufted plant forming hard mats or tight round cushions with pointed rigid, densely overlapping shiny leaves, pale green when dry, and with small white solitary stalkless flowers. Flowers *c.* 8 mm across; petals linear to lanceolate; sepals lanceolate, little shorter, green with papery margins. Leaves linear, *c.* 5 mm, minutely ciliate, persistent on old stems. A. Wollaston records this as the highest known species on Everest. **Pl. 16**.

179 **A. polytrichoides** Edgew. ex Edgew. & Hook. f.
Kashmir to S.E. Tibet. 4300–5500 m. Rocks, stony slopes of Tibetan borderlands. Jul.–Aug.

Forming compact hemispherical moss-like tufts of minute bright shining leaves 2–3 mm long, and with tiny stalkless white flowers. Flowers minute, 1–2 mm; petals broader and longer than the obovate sepals, which are without papery margins. Leaves densely overlapping, broadly lanceolate, with thick margins and spiny tips; stems densely clothed with old leaves below.

180 **A. densissima** Wallich ex Edgew. & Hook. f.

W. Nepal to Bhutan. 3600–5500 m. Overhanging rock faces; common. May–Aug.

Forming large compact tufts of rigid leafy stems with shiny curved leaves 4–6 mm, and very small solitary stalkless white flowers. Petals *c.* 4 mm, ovate, stalked; sepals ovate, half as long as petals. Leaves ovate to lanceolate with a fine point, densely clustered on stems. **Pl. 16.**

ii *Flowers large, 1 cm across or more*

181 **A. edgeworthiana** Majumdar

C. Nepal to S.E. Tibet. 4500–5000 m. Sandy screes; common in Langtang valley, Nepal. Jul.–Sep.

Forming large green dense mats 15–30 cm across like 178 and 179, but with relatively large flowers 1–1.5 cm across with obovate petals. Petals 8–10 mm, persistent; sepals broadly lanceolate, long-pointed, with papery margins, and with 3 close-set veins. Leaves rigid, linear, spine-tipped, to 8 mm, shining bright green, with a conspicuous mid-vein and bristly margin. Stems many to *c.* 3 cm.

182 **A. globiflora** Edgew. & Hook. f.

W. to E. Nepal. 3600–4500 m. Stony slopes in dry areas. Aug.

Forming small compact hemispherical tufts of many leafy stems with rigid spiny-tipped spreading leaves 8–10 mm, and with relatively large white, short-stalked flowers 1–1.3 cm across. Petals *c.* 8 mm, rounded-ovate, slightly notched, rather longer than the orbicular-oblong palmately-veined green papery sepals; anthers purple; flowering stem 2.5–5 cm.

Leaves ovate to oblong (see also 180)

183 **A. debilis** Hook. f. ex Edgew. & Hook. f.

Uttar Pradesh to S. E. Tibet. 3000–4000 m. Grassy slopes, forest clearings; common. Jun.–Sep.

A leafy erect widely branched perennial with lax clusters of small white flowers, each *c.* 8 mm across, and elliptic pointed leaves. Petals oblong, notched, *c.* 6 mm; calyx with bristly glandular reddish hairs; anthers white. Leaves mostly 1.5–3 cm; internodes of stem long, with reddish hairs; stems 20–50 cm.

184 **A. glanduligera** Edgew. ex Edgew. & Hook. f.

Kashmir to Bhutan. 4000–5500 m. Screes, rocky slopes. Jul.–Sep.

A loosely tufted plant with short-stalked solitary reddish to pinkish flowers 6–12 mm across, and with tiny elliptic glandular-hairy leaves. Petals obovate 4–7 mm, little longer than the glandular-hairy lanceolate sepals. Leaves 4–8 mm; stem to 5 cm. Var. **cernua** F. N. Williams has long-stalked larger flowers *c.* 12 mm across, and is less densely tufted. **Pl. 17.**

185 **A. ciliolata** Edgew.

Uttar Pradesh to S. E. Tibet. 3600–5200 m. Stony slopes, screes. Jul.–Sep.

Like 184, but flowers white. Petals oblanceolate *c.* 1 cm; calyx lanceolate, shorter, bristly-haired. A small, low, tufted plant with elliptic bristly-pointed hairy leaves 6–8 mm.

4 MINUARTIA   Like *Arenaria* but capsule opening by 3 (not 6) teeth. Leaves linear, fine-pointed. 4 spp.

186 **M. kashmirica** (Edgew.) Mattf. (*M. lineata* (Edgew.) R. R. Stewart)
Afghanistan to W. Nepal. Tibet. 1500–5000 m. Rock crevices. May–Aug.
A densely tufted glandular perennial with narrow awl-shaped leaves, and with short slender erect stems bearing rather flat-topped clusters of few white flowers. Petals long-stalked with elliptic entire blade to 8 mm, twice as long as the ovate long-pointed sepals which are white margined and strongly 3–veined. Leaves 1–2 cm, awl-shaped, fine-pointed. Capsule ovoid, rather longer than the sepals. **Pl. 17**.

5 THYLACOSPERMUM   Calyx tubular with 4–5 lobes; petals 4–5, entire. Capsule opening by 4–5 teeth. 1 sp.

187 **T. caespitosum** (Cambess.) Schischkin (*T. rupifragum* (Karelin & Kir.) Schrenk)
Pakistan to Sikkim. Tibet. C. Asia. 4800–5700 m. Screes, stony slopes. Jun.–Jul.
A large hard compact cushion-forming plant found at very high altitudes, with tiny densely overlapping triangular acute leaves, the cushions to 30 cm across or more. Flowers solitary, white, *c.* 2.5 mm across. Leaves 2–5 mm, shining, hairless, with sharp-pointed tips. Capsule globular, shining leathery. **Pl. 16**.

6 DIANTHUS   PINK   Distinguished by its tubular 4–toothed calyx which has 1–3 pairs of overlapping bracts round its base. Petals long-stalked. *c.* 3 spp.

Petals fringed with long teeth

188 **D. angulatus** Royle
Pakistan to Himachal Pradesh. 2700–4000 m. Stony hillsides of drier areas; common in Lahul. Jun.–Jul.
Flowers white, fading to pink, with petals with a deep fringe of narrow teeth cut to about one third the width of the petal. Flowers 1–1.5 cm across; calyx cylindrical slightly enlarged at base, 1.3–1.6 cm long; bracts 4–6, ovate long-pointed, often coloured. Leaves linear-acute, very finely toothed, 1–2 cm; flowering-stems one to many, stiff, 10–20 cm. **Pl. 17**.

Petals shallow-toothed

189 **D. anatolicus** Boiss.
Pakistan to Kashmir. W. Asia. 1800–4500 m. Stony ground; common in Ladakh. Jul.–Aug.
A slender densely tufted perennial, with many ascending stems 15–25

cm, bearing one or more pink flowers *c.* 1 cm across. Petals with a broad toothed blade, without hairs on the upper surface; calyx narrow cylindrical 1–2 cm; bracts usually 6, broad and with a tooth-like tip. Leaves linear 1–2 cm, stiff, with a thick mid-vein and margin. **Pl. 17**.

190 **D. jacquemontii** Edgew.
Pakistan to Kashmir. 1500–2100 m. Rocks, stony ground; common in Kashmir valley. Jun.–Jul.
Flowers white, relatively large, to 2 cm across, usually solitary, with petals with narrow conspicuous teeth *c.* 1 mm deep. Calyx broad cylindrical to 2.5 cm long, with long rigid teeth 3–5 mm; bracts 4, very broad and contracted to a long point. Basal leaves numerous, *c.* 1 cm; stem leaves few, longer, all with minutely toothed margins; stems often many, 10–25 cm.

7 **GYPSOPHILA** Calyx tubular, 5–veined, with papery seams between the veins, 5–lobed; styles 2. 1 sp.

191 **G. cerastioides** D. Don
Pakistan to Bhutan. 2100–4700 m. Banks, rocks, open slopes; common. May–Jul.
A low-growing perennial with spreading stems, with small obovate leaves, and with numerous small white flowers streaked with purple, borne in rounded branched clusters 1–2 cm across. Flowers variable in size, to 1 cm across; petals obovate, shallowly notched; calyx hairy; stamens 10. Leaves hairy, blades mostly 6–15 mm, the lower stalked; stems 8–20 cm. Capsule opening by 4 valves. **Pl. 18**.

8 **SILENE** CAMPION, CATCHFLY Calyx tubular with 5 short lobes; petals 5, long-stalked, often bi-lobed. Styles usually 3; capsule usually opening by 6 teeth. *c.* 24 spp.

Annuals; flowers pink

192 **S. conoidea** L.
Pakistan to C. Nepal. Temperate Eurasia. N. Africa. 1200–3300 m. Cornfield weed. May–Jun.
A stout erect glandular-hairy annual, with few small pink flowers, borne in erect terminal branched clusters. Calyx 15–25 mm, tubular narrowed upwards, lobes long linear; petals small obovate, slightly notched. Leaves narrow-lanceolate 2.5–8 cm; stem 15–45 cm. Capsule hard, shining, enclosed in the globular inflated calyx.

Perennials or biennials; flowers white or purple
a *Calyx much inflated in flower*

193 **S. vulgaris** (Moench) Garcke (*S. cucubalus* auct. non Wibel) BLADDER CAMPION
Pakistan to C. Nepal. W. Asia. Europe. N. Africa. 1800–4000 m. Meadows, rocks. Jun.–Aug.

A hairless perennial, readily distinguished by its few large drooping white or greenish-white flowers with deeply lobed petals, and its ovoid inflated green calyx with a network of green veins. Flowers 1.5–2.5 cm across, usually in lax branched clusters. Calyx 1.5 cm, enlarging to 2 cm in fruit, lobes triangular. Leaves pale green to glaucous, ovate to lanceolate acute, 2.5–8 cm; stems branched, 30–90 cm. Capsule stalked, globular with a conical apex, encircled by the inflated calyx.

194 **S. gonosperma** (Rupr.) Bocquet ssp. **himalayensis** (Rohrb.) Bocquet (*Lychnis apetala* auct. non L.)
Afghanistan to S.W. China. C. Asia. 3300–5000 m. Stony slopes. Jun.–Jul.
The commonest high altitude species, with solitary or few nodding balloon-shaped flowers with conspicuous brown-ribbed inflated calyx, and with slightly longer dull purple to white petals. Calyx *c.* 1 cm, lobes triangular, ribs 10 and conspicuous with brown bristly hairs. Leaves mostly basal, narrow-lanceolate 2–4 cm; stems mostly 5–20 cm.

195 **S. nigrescens** (Edgew.) Majumdar
Kashmir to S.W. China. Burma. 3500–5200 m. Rock ledges, screes, open slopes; common. Jul.–Sep.
Like 194, but calyx larger and more inflated to 1.6 cm in diameter, and with broader papery-edged lobes. Flowers solitary; calyx with broad brown-purple ribs; petals short. Stems glandular-hairy, more leafy than in 194. Stalk of capsule woolly-haired.

196 **S. setisperma** Majumdar (*Lychnis inflata* Wallich)
Pakistan to E. Nepal. 3400–4700 m. Open slopes, rocks. Jul.–Sep.
Flowers usually solitary and pendulous, with distinctive very inflated calyx with dark brown ribs, and rounded papery lobes, and with brownish petals with short rounded recurved lobes each with 2 large scales in the throat. Leaves elliptic to lanceolate 4–6.5 cm, the lower stalked; stem leafy, 25–55 cm, with swollen nodes, glandular-hairy above. Seeds distinctive, kidney-shaped, flattened, black, with rows of soft bristles. **Pl. 18.**

197 **S. edgeworthii** Bocquet
Pakistan to Sikkim. 2000–3500 m. Alpine shrubberies, meadows, rocks; common in Himachal Pradesh and Uttar Pradesh. Jul.–Oct.
Flowers several, at first nodding, borne on branched stems, with white petals with linear lobes. Petals about one third as long again as the calyx which is large inflated, *c.* 1.5 cm long, and with fine brown or blackish ribs. Leaves elliptic to narrow-lanceolate 3–6 cm, margin and veins hairy; stems ascending 40–80 cm. Capsule globular; seeds rounded-kidney-shaped, with large dorsal swellings. **Pl. 18.**

b *Calyx not or slightly inflated in flower*

198 **S. moorcroftiana** Wallich ex Benth.
Afghanistan to C. Nepal. Tibet. 2700–4500 m. Rocky slopes, wastelands; common in dry areas. Jun.–Aug.

A tufted perennial with a woody rootstock and many erect little-branched stems, bearing one or few white, or dull red flowers. Flowers 1.5–2 cm across; petals deeply lobed, with scales in the throat; calyx 2.5–3 cm, cylindrical and somewhat enlarged above, conspicuously 10–veined, calyx-lobes short, blunt, dark. Leaves broadly linear acute; flowering stems 10–30 cm. Capsule with a stalk *c.* 1 cm. **Pl. 18**.

199 **S. tenuis** Willd.
Pakistan to Uttar Pradesh. 2700–4300 m. Rocky slopes; common in Kashmir. Jul.–Aug.
A slender tufted perennial with several erect unbranched stems bearing terminal rather crowded elongate clusters of dull purple, brownish-purple, or yellowish-brown flowers. Flowers *c.* 8 mm across; petals with deep narrow lobes; calyx cylindrical to narrow bell-shaped, 1 cm, conspicuously 10–veined, with short lobes. Leaves mostly basal, linear to narrow-lanceolate pointed, usually 2–5 cm; stems 15–35 cm.

200 **S. viscosa** (L.) Pers.   STICKY CATCHFLY
Pakistan to Uttar Pradesh. W. Asia. Europe. 2100–3600 m. Rocks, stony slopes. Jun.–Jul.
A rather robust glandular-hairy perennial 40–100 cm, with a long narrow inflorescence of whorled rather large white flowers with very deeply bilobed or multi-lobed spreading petals with their stalks conspicuously longer than the calyx. Petals 2.5–3.5 cm; calyx narrow-cylindrical 1.5–2 cm, 10–veined, densely glandular hairy. Leaves narrow-elliptic with margins frequently undulate; stem leaves stalkless to 4 cm, basal leaves stalked. **Pl. 18**.

9 **LYCHNIS**   Differs from *Silene* in having usually 5 styles and the capsule splitting by 5 teeth. 1 sp.

201 **L. coronaria** (L.) Desr.   ROSE CAMPION
Pakistan to Kashmir. W. Asia. Europe. 1500–2100 m. Rocks, open slopes. Jun.–Sep.
A grey or whitish shaggy-haired perennial with an erect stem bearing a few, long-stalked, conspicuous usually bright reddish-purple flowers *c.* 3 cm across. Petals with a broad heart-shaped blade and a stiff scale at the junction of blade and stalk. Calyx conical, conspicuously ribbed, shaggy-haired, with twisted teeth. Basal leaves spathulate to lanceolate, 8–12.5 cm; stem leaves oblong acute. Capsule *c.* 1.5 cm. **Pl. 18**.

TAMARICACEAE   Tamarisk Family

A small family of heath-like trees and shrubs of S. Europe, and of sub-tropical Asia and Africa. Leaves alternate, small, needle- or scale-like, without stipules. Flowers in usually dense spikes or racemes, tiny, regular.

Sepals and petals 4–5, free; both borne on a fleshy nectar-secreting disk; stamens twice as many. Ovary of 2–5 carpels with a single chamber; styles 2–5, but stigmas stalkless in *Myricaria*. Fruit a capsule; seeds usually with long hairs.

MYRICARIA   Differs from *Tamarix* in having the stamen filaments fused below, and the pappus of the fruit borne on a twisted style. *c.* 5 spp.

Leaves 3–5 mm long

202   **M. germanica** (L.) Desv.
Pakistan to Kashmir. W. Asia. Europe. 1500–2100 m. Riversides; common and gregarious in Kashmir by Sind and Lidder rivers. Jun.–Aug.
Flower-spikes 10–20 cm or more long, terminal and lateral, with large reddish often quite papery bracts which are broadly trapezoid. Flowers pinkish-red, *c.* 5 mm. Spikes elongating in fruit. Leaves 2–3 mm, densely clustered, elliptic blunt, gland-dotted. An erect shrub 1–2 m. Fruit 8 mm.

203   **M. squamosa** Desv. (*M. hoffmeisteri* Klotzsch)
Afghanistan to C. Nepal. C. Asia. China. 2400–4000 m. Riversides; common and gregarious in Ladakh and Lahul. Jun.–Jul.
Like 202, but differing in having largely lateral spikes 3–5 cm or more long of pink flowers. Bracts lanceolate. Leaves glaucous green, mostly 3–5 mm long; stems light brown, to 2 m. **Pl. 19**.

Leaves 6 mm long, or more

204   **M. elegans** Royle
Pakistan to Kashmir. 2700–4000 m. Stony slopes; common in Ladakh. Jun.–Jul.
An erect shrub 3–4 m with brown stems, leafy lateral branches, and many long lateral spikes of white or pinkish-tinged flowers. Spikes to 10 cm or more; petals 5 mm, rounded, stalked; sepals triangular, 2 mm. Leaves elliptic, 1–1.5 cm. Fruit *c.* 1 cm, spindle-shaped, glaucous.

205   **M. rosea** W.W.Smith (*M. prostrata* Hook.f. & Thoms.)
C. Nepal to S.W. China. 3000–4400 m. Riverside gravel; common in Nepal inner valleys. May–Jun.
A prostrate shrub with brown woody branches and numerous tiny leaves, and dense terminal spikes of pink flowers 3–8 cm long, which lengthen in fruit to 20 cm. Flowers 5–7 mm long; sepals linear-lanceolate, 5–6 mm; bracts green, lanceolate, longer than flowers. Leaves 5–10 mm, broadly lanceolate, gland-dotted; stems spreading 15–40 cm. Fruit spindle-shaped, glaucous, 1.5 cm. **Pl. 19**.

GUTTIFERAE   St. John's Wort Family

A large cosmopolitan family which is mainly tropical. The genus *Hypericum* is sometimes placed in a separate family the Hypericaceae. Mainly herbaceous perennials and small shrubs, with opposite or whorled entire leaves which are often gland-dotted. Flowers regular, yellow, in terminal branched clusters. Sepals 5, free; petals 5, free; stamens numerous, in 2 whorls, or in 3 or 5 bundles, or irregularly grouped. Ovary superior, with 3–5 chambers, or 1–chambered; styles 3–5; ovules many. Fruit a capsule, or rarely a berry.

HYPERICUM   St John's Wort   15 spp.

> *Annuals or herbaceous perennials with flowers not more than 2.5 cm across*
> a *Sepals without gland-tipped teeth*

206   **H. japonicum** Thunb. ex Murray
Himachal Pradesh to Bhutan. India. China. Japan. S.E. Asia. Australasia. 1500–2600 m. Cultivated areas, grazing grounds; common. Flowering most of the year.

A very small annual with tufted spreading 4–angled stems, with variable entire clasping leaves, and with small orange or yellow flowers *c.* 5 mm across. Petals equalling the ovate sepals which have pellucid glands at the apex. Leaves narrowly to broadly elliptic to 1.3 cm, gland-dotted; stems many 5–10 cm. **Page 453**.

207   **H. perforatum** L. Common St John's Wort
Pakistan to Uttar Pradesh. W. Asia. Europe. 1200–2400 m. Shrubberies, cultivated areas; common. May–Aug.

A hairless erect branched perennial with stems with 2 raised lines, and with many yellow flowers 1.5–2.5 cm across in a flat-topped or pyramidal cluster. Petals oblanceolate 8–15 mm, with marginal black dots; sepals lanceolate to elliptic acute, without marginal glands. Leaves oblong to elliptic 0.5–2 cm, gland-dotted; stem 10–110 cm.

> b *Sepals with gland-tipped teeth*

208   **H. elodeoides** Choisy
Kashmir to S.W. China. Burma. 1500–3000 m. Shrubberies, open slopes, cultivated areas. Jul.–Sep.

An erect perennial with unbranched rounded stem, regular pairs of lanceolate clasping leaves, and branched clusters of yellow flowers with distinctive long-stalked marginal glands to the sepals and bracts. Petals 7–12 mm, streaked with glands, longer than the lanceolate sepals. Leaves 1–3.5 cm, apex blunt, gland-dotted; stems 20–50 cm. **Pl. 19**.

> *Shrubs with flowers more than 3 cm across*
> a *Leaves linear-lanceolate, parallel-sided*

209   **H. podocarpoides** N. Robson (*H. acutum* Wallich)

Uttar Pradesh to C. Nepal. 800–2100 m. Open slopes, banks. Apr.–Jul.; Sept.–Oct.

A small branched shrub to 1.5 m, with numerous linear-lanceolate pointed leaves, and terminal branched clusters of few yellow flowers each 3–4 cm across. Petals obovate, twice as long as the broadly lanceolate sepals; styles shorter than ovary. Leaves clasping, 3–5 cm, blade parallel-sided, narrowed only in the apical third; stems reddish, 4–angled.

b *Leaves elliptic, ovate-lanceolate to ovate-oblong*
  i *Styles much longer than ovary*

210 **H. oblongifolium** Choisy (*H. cernuum* Roxb.)
Pakistan to C. Nepal. 800–2100 m. Shrubberies, banks; common. Feb.–Apr.

A much-branched arching often pendent shrub 1–2 m, with oblong to elliptic blunt leaves and few large terminal flowers 3.5–7 cm across. Petals mostly narrowly obovate; sepals unequal, ovate-elliptic, with entire margins; styles to 2 cm. Leaves stalkless 3–8 cm, with dense network of veins on underside. **Pl. 19**.

  ii *Styles shorter or as long as ovary*
    x *Sepals lanceolate acute*

211 **H. choisianum** Wallich ex N. Robson
Pakistan to S.W. China. Burma. 2400–3600 m. Rocks, shrubberies. Jun.–Sep.

Differing from 210 in having broader petals which are broadly obovate to rounded, and 2–4 times longer than the stamens (twice as long in 210), and styles shorter, to half as long as the ovary. Flowers yellow, 4–7 cm across; sepals leafy, lanceolate acute. Leaves very shortly stalked, ovate to lanceolate acute. A shrub with spreading branches, 30–200 cm. **Pl. 19**.

    xx *Sepals ovate to rounded*

212 **H. uralum** Buch.–Ham. ex D. Don (*H. patulum* auct. non Thunb.)
Kashmir to Bhutan. S. China. S. E. Asia. 1000–3600 m. Shrubberies, open slopes. Jun.–Oct.

A shrub like 211, but flower smaller 1.5–3 cm across, and young branches and leaves rather fern-like in appearance. Sepals ovate with rounded apex, not leafy; styles as long as ovary. Branchlets 4–angled or 4–lined, later 2–lined (becoming rounded in 211). Leaves smaller 1.5–3 cm, numerous, 2–ranked; stems arching, usually 30–100 cm.

213 **H. hookeranum** Wight & Arn.
C. Nepal to Bhutan. S. India. Burma. 1500–3000 m. Shrubberies, cliffs, rocks. Jul.–Oct.

A shrub like 211 with large yellow flowers, but sepals rounded with a blunt apex. Flowers 3–6 cm across; styles much shorter than ovary. Leaves 3–6 cm, most broadly lanceolate to elliptic, blunt or acute, with conspicuous venation beneath; branches red-brown, rounded; stems 2–3.5 m.

THEACEAE   Tea or Camellia Family

A medium-sized family of trees and shrubs mainly of the tropical and sub-tropical regions of the world. Leaves usually alternate, evergreen and leathery. Flowers very often solitary and showy. Sepals and petals 4–7; stamens usually numerous, often united into a tube below, or free, or in bundles. Ovary superior, with usually 3–5 fused carpels and 3–5 chambers with few or many ovules; styles as many as the carpels. Fruit a capsule or berry.

CAMELLIA   Fruit a capsule; seeds not winged. 2 native spp.

214   **C. kissi** Wallich
C. Nepal to S.W. China. S.E. Asia. 1000–2100 m. Forests, shrubberies. Oct.–Nov.

A large shrub or small tree, with shining evergreen elliptic-lanceolate pointed leaves, and erect white almost stalkless flowers 2–4 cm across. Petals 5, elliptic 1.5–2 cm, soon falling; stamens very numerous, free below. Bracts broad, silky. Leaves 5–8 cm, margin finely glandular-toothed. Capsule globular, downy when young, c. 2.5 cm across, splitting by 3 valves.

Leaves sometimes used as a substitute for tea. **Pl. 19.**

215   **C. sinensis** (L.) Kuntze   TEA
Native of China, and cultivated there for centuries. Possibly also native in the E. Himalaya; extensively cultivated in Assam and Darjeeling since the nineteenth century. In our area cultivated near Ilam in E. Nepal, and around Palampur in Kangra in the W. Himalaya. To 2300 m. Dec.–Mar.

When plucked in tea gardens, a low bush to 1.5 m or less. If allowed to grow, an erect shrub to 5 m or more, with solitary white flowers 2.5–4 cm across. Petals 5, obovate. Leaves 5–10 cm, oblong-elliptic, toothed, leathery, hairless above. Capsule 3–angled, with persisting rounded sepals, 3–seeded.

Several other *Camellia* species are grown in Nepalese gardens.

EURYA   Plants one-sexed; flowers small. Fruit a berry. 3 spp.

216   **E. acuminata** DC.
Uttar Pradesh to Arunachal Pradesh. India. Sri Lanka. S.E. Asia. 1200–2400 m. Forests, shrubberies; very common in Nepal. Sep.–Nov.

An evergreen shrub or small tree with glossy leathery oblong-lanceolate long-pointed leaves, and almost stalkless axillary clusters of tiny cream-coloured flowers ranged along the stem. Flowers one-sexed, to 5 mm across; petals 5, rounded blunt, c. 3 mm; sepals rounded, smaller; stamens many. Leaves 5–10 cm, margin finely toothed, mid-vein hairy beneath. Fruit globular 3 mm, blue when ripe. **Page 453.**

SCHIMA   Flowers usually subtended by 2 bracts. Fruit a capsule; seeds flat, kidney-shaped, winged. 1 sp.

217 **S. wallichii** (DC.) Korth.

C. Nepal to S.W. China. 200–2000 m. Forests, cultivated areas; common. May–Jun.

A small or large tree with dark grey rugged bark, leathery evergreen leaves, and showy terminal clusters of few white fragrant flowers. Flowers 3–4 cm across, with 5 broadly ovate petals; flower-stalks 1–3 cm; sepals rounded; stamens many; flower-buds large globular. Leaves elliptic-oblong, 10–18 cm, entire or slightly toothed, hairless and reddish-veined beneath. Capsule globular, c. 1.5 cm, woody, splitting into 5 valves and remaining on tree for many months.

Bark very irritant to the skin; resistant to fire. **Pl. 20**.

ACTINIDIACEAE

A family of one genus consisting of woody vines with white flowers in axillary clusters, and fruit a fleshy berry.

ACTINIDIA 2 spp.

218 **A. callosa** Lindley

Himachal Pradesh to S.W. China. S.E. Asia. 1500–3000 m. Forests and shrubberies. May–Jul.

A large climber with stems to 10 cm in diameter, with large elliptic evergreen leaves, and small lax axillary clusters of several white flowers much shorter than the leaves. Flowers 1.5–2.5 cm across; petals 5, rounded, overlapping; sepals 5, woolly-haired; stamens many, filaments white. Leaves 8–15 cm, long-stalked, pointed, with fine gland-tipped teeth. Fruit a berry, ovoid c. 1.5 cm, edible. **Pl. 19**.

SAURAUIACEAE

A small family of trees and shrubs of only one genus, found in the Tropics and Sub-Tropics of Asia and America. Flowers in axillary clusters; sepals 5, strongly overlapping; petals 5, usually fused at the base; stamens many, anthers dehiscing by pores. Ovary 3–5–celled, styles as many. Fruit a berry. Often placed in the Actinidiaceae.

SAURAUIA 3 spp.

219 **S. napaulensis** DC.

Himachal Pradesh to S.W. China. Burma. S.E. Asia. 600–2100 m. Forests; planted in cultivated areas as a fodder tree. Mar.–Oct.

A large deciduous shrub or small tree, distinguished by its large elliptic leaves which are conspicuously rusty-haired beneath, and its lax axillary clusters of pink flowers shorter than the leaves. Flowers many, *c*. 1.3 cm across; calyx pinkish; petals 5, with fringed margins; styles 4–6, conspicuous. Leaves mostly 18–36 cm, with acute gland-tipped marginal teeth and many pairs of prominent lateral veins; young leaves particularly densely rusty-haired. Bark reddish; wood soft spongy. Fruit green, 4–5-lobed, fleshy with sweet pulp, edible. **Pl. 19**.

STACHYURACEAE

A family of one genus of E. Asia. Leaves with stipules. Stamens 8; ovary of 4 carpels with 4 chambers and numerous ovules in each chamber; style one. Often placed in the Theaceae.

STACHYURUS   1 sp.

220 **S. himalaicus** Hook. f. & Thoms. ex Benth.
C. Nepal to S.W. China. 1500–3000 m. Shrubberies. May–Jun.
A small deciduous shrub or small tree with straggling branches, ovate long-pointed deciduous leaves, and numerous short dense pendulous spikes of small greenish-yellow, cream to pinkish flowers often borne on bare branches. Spikes mostly 5–8 cm long; flowers 6–8 mm across; sepals and petals 4, ovate. Leaves 6–10 cm, finely toothed, with strongly netted veins beneath; stems with easily extruded pith. Fruit a berry *c*. 6 mm. **Pl. 20**.

DIPTEROCARPACEAE

An important but small family of tropical forest-forming trees, usually with a buttressed base, a smooth trunk of great height and a cauliflower-shaped crown. Leaves simple, entire, with stipules which soon fall. Flowers usually borne in large showy, branched clusters, often scented. Sepals and petals 5; stamens 5 to many, anthers with a distinctive sterile tip. Ovary of 3 fused carpels, with 3 chambers. Fruit a single-seeded nut enclosed in the persistent winged calyx.

SHOREA   1 sp.

221 **S. robusta** Gaertn. SAL TREE
Himachal Pradesh to Bhutan. India. To 1500 m. Dominant over much of the sub-Himalayan areas. Apr.
A large deciduous tree with smooth dark brown bark, large shining pale leaves, and with large lax, branched white-haired inflorescences of small very fragrant pale-yellow flowers. Petals lanceolate, *c*. 1.3 cm, silky-haired

outside, much longer than the ovate grey-woolly sepals; stamens numerous. Leaves thin leathery, ovate-oblong narrow-pointed, 10–30 cm, with 12–15 pairs of strong lateral veins, short-stalked. Fruit ovoid, *c.* 8 mm across, white-haired, with 5 oblong wings 5–7.5 cm long, which turn brown when dry. Tree seldom quite leafless.

The most important hardwood tree of the sub-Himalayan hills and valleys. Yields valuable durable timber for building. Tapped for resin which is used in incense and medicinally. Oil is obtained from the fruits, as well as flour for cooking. The leaves are often used as platters.

MALVACEAE   Mallow Family

A widely distributed family of herbaceous plants, trees and shrubs. Leaves alternate, usually simple, with stipules, often with star-shaped hairs. Flowers regular, with 5 sepals fused at the base, and often with an outer set of 3–8 sepal-like segments, the *epicalyx*; petals 5, free; stamens numerous, the filaments united for most of their length to form a tube surrounding the ovary and styles. Ovary superior, of 4 or more carpels and usually a similar number of styles. Fruit a dry capsule, or splitting into one-seeded units.

Carpels separating from axis at maturity
  a Epicalyx-segments 3
    i Epicalyx-segments free   1 MALVA
    ii Epicalyx-segments fused at base   2 LAVATERA
  b Epicalyx-segments 5   3 URENA
Carpels not separating, fruit a capsule   4 ABELMOSCHUS
                                                5 THESPESIA

1 MALVA MALLOW Epicalyx of 3 free segments. Fruit a whorl of nutlets borne on a central boss. 4 spp.

222 **M. verticillata** L. CHINESE MALLOW
Pakistan to Bhutan. India. Temperate Eurasia. Africa. 2100–3300 m. Cultivated areas. Jun.–Sep.

An erect annual or perennial, with whorls of nearly stalkless crowded small pink or mauve flowers borne in the axils of the upper rounded, lobed leaves. Petals *c.* 7 mm, not more than twice as long as sepals, but the latter enlarging considerably in fruit. Leaves usually downy, with heart-shaped base, 5–7–lobed, the lobes triangular blunt and toothed; leaf-stalks much longer than flower clusters; stem 60–120 cm. Fruit of 10–12 carpels encircled by the enlarged calyx.

Plant used medicinally; cultivated as a salad plant in Europe.

223 **M. neglecta** Wallr.
Afghanistan to Uttar Pradesh. N. Temperate Zone. 1500–3600 m. Cultivated areas, wasteland; common. Jun.–Aug.

Like 222, but petals at least twice as long as sepals, and calyx not enlarging in fruit. Flowers pale lilac to whitish; petals 9–13 mm, hairy. An annual with rounded-heart-shaped, lobed leaves, and axillary flower clusters.

2 LAVATERA Like *Malva* but epicalyx of 3 segments fused together at the base. 1 sp.

224 L. kashmiriana Cambess. WILD HOLLYHOCK
Pakistan to Uttar Pradesh. 1800–3600 m. Meadows, forest clearings; sometimes cultivated in Nepal. Jul.–Sep.

A tall branching densely woolly-haired perennial, with 3–5–lobed leaves, and with terminal spikes of large bright pink flowers with darker veins. Flowers 5–8 cm across; petals 5, spreading, obovate and conspicuously bi-lobed, narrowing to a pale stalk; calyx 5–lobed, lobes triangular and longer than the 3 broader epicalyx-segments. Lower leaves rounded-heart-shaped, with 5 shallow and toothed lobes, the upper leaves with 3–5 narrower lobes. Fruit of many black carpels surrounded by the calyx and epicalyx. **Pl. 20.**

3 URENA Epicalyx-segments 5, fused to the calyx-lobes. Styles twice as many as carpels which are spiny. 1 sp.

225 U. lobata L.
Pakistan to S.W. China. India. Pan-tropical. To 1800 m. Cultivated areas, shrubberies, wasteland; common. Flowering throughout the year.

A variable shrubby perennial to 2 m, with short-stalked clusters of pink flowers, and heart-shaped shallow-lobed leaves which are paler and densely hairy beneath. Flowers 2.5 cm across; petals united below; calyx-lobes linear, equalling the epicalyx-segments. Leaves 2.5–5 cm long, with shallow rounded or triangular, toothed lobes; stems densely woolly-haired. Fruit spiny and hairy.

4 ABELMOSCHUS Epicalyx-segments 4–5, enlarging in fruit. Capsule oblong. 2 spp.

226 A. manihot (L.) Medikus
Uttar Pradesh to S.W. China. Burma. S.E. Asia. 600–2400 m. Wasteland, shrubberies. Aug.–Oct.

A tall bristly yellow-haired annual or perennial, with large yellow flowers with purple centres, and deeply palmately-lobed leaves. Flowers *c.* 13 cm across; petals *c.* 9 cm; calyx 5–lobed. Leaves with blade 12–25 cm across, long-stalked, the lower leaves deeply 7–lobed with oblong pointed, coarsely toothed lobes to 15 cm, the upper leaves 3–lobed; stems 1.5–2 m, usually bristly and black dotted. Capsule 4 cm, bristly-haired.

5 THESPESIA Epicalyx-segments 5–8, soon falling. Distinguished from

Abelmoschus and *Hibiscus* by its stigmas which are coherent in a club-like mass. 1 sp.

227   **T. lampas** (Cav.) Dalz. & Gibson
W. & C. Nepal. S. & S.E. Asia. Africa. 200–1500 m. Shrubberies, wasteland. Aug.–Oct.

A shrub to 2 m, with large yellow bell-shaped flowers with dark crimson centres, and with usually bi-lobed leaves. Flowers to 10 cm across, one, or several in terminal clusters; calyx and epicalyx with bristle-like tips. Leaves 8–15 cm broad and long, heart-shaped, usually 3–lobed, the lobes triangular, paler with star-shaped hairs beneath; leaf-stalk downy. Capsule ovoid, 2.5 cm, surrounded by leaf-like calyx with bristle-like lobes.

Yields a strong white fibre. **Pl. 20**.

Alcea rosea L. HOLLYHOCK is grown in gardens to 3500 m. At subtropical altitudes **Hibiscus** spp. are grown: in particular **H. rosa-sinensis** L. with red flowers, and **H. syriacus** L. with lilac or white flowers.

BOMBACACEAE

A small but quite important family of tropical trees, largely of S. America, with a few species occurring in S.E. Asia. It includes the Baobab, Durian and Balsa trees.

Mostly deciduous trees with compound leaves shed in the dry season, and flowers borne on leafless stems. Flowers regular; epicalyx sepal-like; sepals and petals 5; stamens 5 to many, free or fused into a tube. Ovary with 2–5 carpels with as many chambers; style 1. Fruit a capsule.

BOMBAX   1 sp.

228   **B. ceiba** L. (*B. malabaricum* DC.) SILK-COTTON TREE, 'SIMAL'
Kashmir to Bhutan. India. S. China. S.E. Asia. To 1400 m. Common in sub-Himalayan areas and in lower valleys. Feb.–Mar.

A large showy and distinctive tree when in flower in early spring, with large scarlet flowers clustered towards the ends of bare horizontally spreading branches. Flowers with a fleshy cup-shaped calyx, and much longer recurved fleshy elliptic petals, 5–8 cm, which are woolly-haired outside, and with numerous yellow stamens with pink filaments. Leaves palmate, with 5–7 lanceolate long-pointed stalked leaflets 10–20 cm long, and a leaf-stalk as long. Trunk buttressed at the base; stems of young trees covered with conical spines. Capsule hard woody, oblong-ovoid 10–13 cm, splitting into 5 valves; seeds embedded in white cottony hairs.

Wood soft but buoyant, used for dugout boats and planking; the bark gives a fibre used for ropes. Flower-buds eaten as a vegetable. Yields a gum which is used medicinally, as are the roots of young plants. Cottony hairs of seeds used for pillows and quilts.

LINACEAE Flax Family

A small widely distributed family of the Temperate Zone, mainly of herbaceous plants. Leaves small, usually alternate, entire, without stipules. Flowers regular, with usually 5 free sepals and petals, the former persisting in fruit and the petals soon falling. Stamens usually 5, fused at their base to form a glandular ring, and sometimes alternating with 5 tooth-like sterile stamens. Ovary usually with 8–10 chambers, each 1–2–seeded; styles 4–5. Fruit a capsule splitting by 10 valves.

LINUM Styles 5. Capsule 5–celled. 1 sp.

229 **L. usitatissimum** L. Flax
Native origin unknown. Widely cultivated throughout the Himalaya and occasionally occurring as a weed. To 3800 m. Feb. at low altitudes: Jun.–Sep. at high altitudes.

An annual with erect stems, narrow leaves and blue flowers *c.* 2.5 cm across, in a terminal branched rather flat-topped cluster. Petals 2–3 times as long as sepals; the latter are 6 mm, ovate long-pointed, with a conspicuous mid-vein. Leaves linear pointed, mostly 2–3 mm wide, 3–veined; stems 60–120 cm. Capsule as long or longer than sepals.

Cultivated for its fibre which is used to make linen, and for its oil-bearing seeds. Parts of plant used medicinally.

REINWARDTIA Styles usually 3. Capsule usually 3–4–celled. 2 spp.

230 **R. indica** Dumort. (*R. trigyna* (Roxb.) Planchon)
Pakistan to S.W. China. India. Burma. S.E. Asia. To 1800 m. Banks of terraced fields, grazing grounds; common. Nov.–May.

A spreading or erect branched shrub to 1 m, but commonly grazed to prostrate form, with showy axillary bright yellow flowers, and elliptic to oblanceolate leaves. Flowers to 4 cm long; petals 5, obovate, 2–3 times as long as sepals; fertile stamens 5, alternating with sterile stamens. Leaves usually 2.5–8 cm, entire or minutely toothed, hairless, short-stalked. Capsule globular, 8–10 mm, splitting into 6–8 one-seeded units. **Pl. 21.**

231 **R. cicanoba** (Buch.–Ham. ex D. Don) Hara
C. Nepal. S.W. China. Has larger flowers and larger leaves but shows many intermediates with the above species.

ZYGOPHYLLACEAE

A family largely of tropical and srb-tropical shrubs but with some temperate herbaceous species. Leaves usually pinnate, with stipules. Flowers bisexual, mostly regular; usually with 5 sepals and petals, and twice the number of stamens. Ovary superior, commonly angled or winged. Fruit dry or fleshy.

PEGANUM 1 sp.

232 **P. harmala** L.
   Afghanistan to Kashmir. C. & W. Asia. Europe. N. Africa. 300–2400 m.
   Stony ground, wasteland; prominent along the Indus valley in Ladakh,
   and in Kashmir valley. May–Jul.
   A somewhat fleshy-leaved, erect much-branched perennial, with solitary terminal greenish-white flowers 1–2.5 cm across. Petals 4–5, oblong-
elliptic; sepals linear as long as petals or longer; stamens mostly 15. Leaves
pinnately cut into long widely-spaced linear acute lobes 1–2 mm wide;
stems stout, 20–90 cm. Fruit globular, 6–10 mm; seeds dark brown.
   The seeds and other parts of the plant are used medicinally. **Pl. 20**.

GERANIACEAE   Geranium Family

A family of annual and perennial herbaceous plants and a few small shrubs,
mostly of temperate and sub-tropical regions. Leaves usually lobed, with
stipules. Flowers regular, solitary or often in flat-topped clusters. Sepals and
petals 5, free; stamens in 2–3 whorls with 5 in each whorl. Ovary superior, of
5 one-seeded carpels fused round a central axis which with the styles forms a
beak. Each carpel splits off elastically from the beak throwing out the single
seed.

BIEBERSTEINIA   Differs from *Geranium* in having flowers in spikes, and
   stigmas united at their tips; stamens 10. Leaves pinnate. 1 sp.

233 **B. odora** Stephan ex Fischer
   Pakistan to Kashmir. C. Asia. 4300–5000 m. Stony slopes; a high-altitude
   alpine of Ladakh. Jul.–Aug.
   Flowers yellow in a dense terminal spike-like cluster, and leaves pinnate
with many pairs of small irregularly lobed leaflets. Flowers to 1.5 cm
across; petals 10–12 mm, elliptic; sepals elliptic blunt, densely hairy
outside; filaments hairy. Leaves linear in outline 8–10 cm by 1 cm, stiff,
somewhat fern-like, glandular-hairy, leaflets broadly ovate, deeply lobed,
c. 5 mm; flowering stem 10–13 cm. Carpels not splitting but separating
from a persistent 5–lobed axis. Whole plant strongly aromatic. **Pl. 20**.

GERANIUM   Leaves usually palmately divided or lobed. Fruit of 5 one-
seeded units, with a long straight or up-curved beak formed from the outer
part of the style, which separates upwards in a spring-like coil from the
central axis and usually remains attached at the apex. *c.* 18 spp.

Calyx without glandular hairs
a *Flower-stalks with one flower*

234 **G. nakaoanum** Hara
   C. Nepal to Bhutan. 3300–4500 m. Alpine slopes. Jun.–Aug.

65

Flowers pink, 2–3 cm across, distinctive in being solitary on the flowering stems which have short recurved hairs and blunt bracts. Sepals 6–10 mm, almost hairless. Leaves rounded in outline, 1.5–2 cm across, deeply 7–9–lobed, the lobes deeply cut into linear or lanceolate rather blunt segments, becoming hairless beneath. A herbaceous perennial with slender stems, unbranched above, 5–12 cm. Fruit hairless. **Pl. 21**.

b *Flower-stalks with two flowers*

235 **G. donianum** Sweet
W. Nepal to S.E. Tibet. 3300–4500 m. Alpine slopes. Jun.–Aug.

Differs from 234 in having flowers paired, and sepals with adpressed whitish hairs on mid-vein and near margin, and stamen filaments with long hairs. Flowers 1.5–2 cm across, pinkish-purple. Leaves with long stiff hairs on veins beneath, and ultimate leaf-segments more acute than in 234. Plant usually 5–15 cm. **Pl. 21**.

236 **G. wallichianum** D. Don ex Sweet
Afghanistan to Bhutan. 2400–3600 m. Forests, shrubberies, open slopes; common. Jun.–Sep.

Flowers paired, rose to red-purple with a pale centre, 2.5–4 cm across. Plant readily distinguished by its large ovate often coloured stipules. Sepals bristly-haired on veins; stamen-filaments 5–7 mm. Leaves mostly 4–8 cm across, 3–5–lobed, the lobes broad-rhombic often with a fine point and further lobed and toothed; stems much branched 30–120 cm. **Pl. 21**.

Calyx with glandular hairs
a *Sepals, including awn-like tip, not more than 9 mm long*

237 **G. polyanthes** Edgew. & Hook. f.
Uttar Pradesh to S.E. Tibet. 2400–4500 m. Shrubberies, open slopes. Jun.–Aug.

Distinguished by its bright reddish-purple flowers, borne in a few-flowered umbel-like cluster on erect stems. Flowers 1.5–3 cm across; petals 12–13 mm; sepals with long soft spreading glandular hairs. Leaves rounded in outline, mostly 2–4 cm across, deeply 5–9–lobed, the lobes obovate and further cut at the apex into blunt segments; stipules fused; stems 10–30 cm. Fruits coarsely netted. **Pl. 20**.

238 **G. procurrens** P. F. Yeo
Uttar Pradesh to Arunachal Pradesh. 2400–3300 m. Shrubberies. Jul.–Sep.

Flowers dark red-purple with black centres, 3–4 cm across; sepals 6–8 mm, with long soft spreading glandular hairs. Flower-stalks densely hairy with mostly glandular hairs; bracts 3–6 mm. Leaves 3–6 cm across, 5–lobed to three quarters their width, the lobes rhombic, further cut into blunt or acute shallow ultimate segments; stipules 4–9 mm long, usually free; stems sometimes scrambling through bushes, to 1 m.

239 **G. tuberaria** Cambess.
Pakistan to Himachal Pradesh. C. Asia. 1500–2500 m. Open slopes, shrubberies; common in Chenab valley. Apr.–Jun.

Distinguished by its swollen tuberous underground stems, and its long slender almost leafless flowering stem bearing paired long-stalked, pinkish to reddish-purple flowers each to 4 cm across. Petals obovate notched, to 2 cm; sepals and flower-stalks with glandular hairs; bracts 2–3 mm. Leaves long-stalked, 3–5 cm across, deeply 7–lobed, with lobes further 3–lobed and coarsely toothed or shallow-lobed; stipules linear; stems 30–40 cm. **Pl. 20**.

b *Sepals, including awn-like tip, more than 10 mm long*

240 **G. lambertii** Sweet
Uttar Pradesh to S.E. Tibet. 2400–4300 m. Forests, open slopes, shrubberies. Jun.–Jul.

Flowers large, nodding, pale pink or rarely white, with purple veins, 4–6 cm across; and sepals 10–14 mm, with glandular hairs. Stamen-filaments 8–11 mm, conspicuous and densely clothed with long hairs; anthers and ovary blackish; bracts lanceolate, 9–20 mm. Leaves often to 10 cm across; stems 30–69 cm.

241 **G. pratense** L. MEADOW CRANESBILL
Pakistan to C. Nepal. Temperate Eurasia. 3000–4500 m. Shrubberies, open slopes; gregarious along irrigation channels in drier areas. Jun.–Aug.

Flowers in pairs, 2.5–4 cm across, usually bluish-purple, with darker veins, but colour varying from almost red, to blue, to white. Sepals 10–12 mm; individual flower-stalks becoming reflexed after flowering, erect in fruit. Leaf-blade 3–5 cm across, divided almost to base into 5–7 ovate lobes which are further deeply cut into oblong acute toothed segments; stems erect, 15–40 cm, with spreading hairs. Fruit *c.* 3.5 cm including beak.

242 **G. himalayense** Klotzsch (*G. grandiflorum* auct. non L.)
Afghanistan to C. Nepal. 2100–4300 m. Forests, shrubberies, open slopes. May–Jul.

Very like 241 and often difficult to distinguish from it, but with ascending stems, and blue flowers in a lax not dense cluster, with petals to 28 mm long (to 20 mm in 241), and ripe fruit 4–4.7 cm (usually not more than 3.5 cm in 241).

243 **G. refractum** Edgew. & Hook. f.
C. Nepal to S.E. Tibet. Burma. 3600–4500 m. Shrubberies, open slopes. Jun.–Aug.

Readily distinguished by its nodding flowers with reflexed petals which are white with purple veins, and its conspicuous protruding stamens and style. Flowers 2–3 cm across; petals obovate; stamens with reddish filaments 9–11 mm, anthers dark; flower-stalks densely glandular-hairy. Leaves mostly 2–5 cm across, deeply 5–7–lobed, the lobes broadly rhombic and deeply cut; stipules fused, ovate-oblong; stems 10–50 cm. **Page 453**.

67

ERODIUM   Leaves pinnate or pinnately lobed. Individual carpels with beak usually twisted in a spiral at maturity. 2 spp.

244   **E. cicutarium** (L.) L'Hér.   Common Storksbill
Pakistan to Uttar Pradesh. Temperate Eurasia. 1500–2400 m. A field weed; common in Kashmir. Apr.–May.
　　Flowers pinkish-purple, lilac, or white, in umbels of up to 10 flowers, each flower *c.* 1 cm across. Petals 4–11 mm; sepals 5–7 mm, hairy. Leaves 6–8 cm, pinnate, with the leaflets further lobed or pinnate, ultimate segments 0.5–1 mm wide; stems usually present, up to 60 cm. Fruit with beak usually 4 cm. A very variable plant, usually a small annual.

OXALIDACEAE   Wood-Sorrel Family

A rather small family of annual and perennial herbaceous plants, mainly of the tropical and sub-tropical regions. Leaves in our species trifoliate. Flowers regular, usually in umbels. Sepals 5, free; petals 5, free, contorted in bud; stamens 10, fused at base. Ovary of 5 usually fused carpels and 5 stigmas. Fruit a capsule.

OXALIS   4 spp.

245   **O. acetosella** L. Wood-Sorrel
Pakistan to S.W. China. North Temperate Zone. N. Africa. 2100–3000 m. Forests, shady banks. Apr.–May.
　　Flowers solitary, white with delicate lilac veins, or pale pink, borne on slender stems arising direct from the rhizome, and with long-stalked trifoliate leaves. Petals 1–1.5 cm; flower-stalks 4–18 cm, with paired bracts near the middle. Leaflets 3, obcordate, 1–3 cm broad, pale yellowish-green; leaf-stalk long; rootstock creeping, covered with broad scales.

246   **O. corniculata** L.
Throughout Himalaya and most of the world. To 2800 m. Wastelands, cultivated areas; very common. Mar.–Jun.
　　Flowers small yellow, 8–10 mm across, solitary, or 2–5 in stalked axillary umbels. A small spreading hairy annual to perennial rooting at the nodes, with hairy trifoliate leaves with obcordate leaflets 6–15 mm broad, and with ascending stems usually less than 15 cm. Capsule 1–2 cm, cylindrical, hairy. **Page 453**.

247   **O. latifolia** Humb., Bonpl. & Kunth
Native of C. and S. America. 1200–1600 m. Naturalized in the Himalaya as a weed of waste places; common in the Nepal valley. Jul.–Sep.
　　Flowers relatively large, pink, many in an umbel borne on long stems arising directly from the rootstock, and longer than the leaves. Petals 1–1.5 cm; sepals lanceolate, one third as long. Leaves all basal, long-stalked,

with 3 distinctive broadly triangular leaflets with widely indented apex, the leaflets 3–7 cm across and much broader than long. Underground bulbils arise from stolons.

BALSAMINACEAE   Balsam Family

A family of annual and perennial herbs with watery translucent stems, widely distributed in temperate and tropical regions, except in S. America and Australia. Leaves simple, toothed, without stipules. Flowers irregular. Sepals 3 or 5, the lower larger pouched and with a *spur*, the 2 or 4 lateral sepals small and often greenish. Petals 5, unequal, the 4 lateral fused in pairs, the lower of each pair forming the *lip*, the upper forming the *wings*, the uppermost petal often larger and often helmet-like. Stamens 5, more or less united by their anthers to form a cap over the ovary. Ovary of 5 fused carpels, with 5 chambers with many ovules. Fruit an explosive capsule, with the turgid valves coiling up and throwing out the seeds.

IMPATIENS   A difficult genus with many species which are often not easily distinguished in the field. Generally found by streams, wet rocks, and on damp banks in forest areas. *c.* 41 spp.

Flowers pink, purple, or crimson (or whitish)
a *Flowers solitary in leaf axils; fruit hairy*

248 **I. balsamina** L.
Native of S.E. Asia. Widely grown in gardens in Nepal. To 1800 m. Aug.
    A hairy erect annual 15–50 cm, with narrow lanceolate leaves, and with many solitary axillary pink or nearly white flowers ranged up the stem, with slender curved spurs. Flowers 4 cm long including spur; upper petal with a green point. Leaves alternate, sharply toothed, 5–10 cm. Capsule pendulous, ovoid acute and swollen in the middle, woolly-haired. A variable plant.
    At one time used for dyeing; also used medicinally. Widely cultivated, especially 'double' forms.

b *Flowers in racemes or umbel-like clusters*

249 **I. glandulifera** Royle (*I. roylei* Walp.)
Pakistan to Uttar Pradesh. Nepal? Naturalized widely in Europe. 1800–4000 m. Shrubberies, bushy places; common on grazing grounds; often growing gregariously. Jul.–Sep.
    A tall hairless annual 1–2 m, with opposite or whorled lanceolate leaves, and with flat-topped clusters of axillary flowers, borne near the shoot tips, of conspicuous reddish-pink, pale pink or white flowers spotted with yellow within. Flowers 3–4 cm long; the lower sepal broadly bell-shaped and abruptly contracted into a short slender cylindrical incurved spur *c.* 6 mm long; lower petals forming a broad prominent lip. Leaves 6–20 cm,

sharply toothed with gland-tipped teeth. Capsule club-shaped, to 3 cm. **Pl. 22.**

250 **I. sulcata** Wallich

Kashmir to Bhutan. 1800–4000 m. Forests, shrubberies, cultivated areas; common and gregarious in Nepal. Jul.–Sep.

Like 249, but distinguished by its linear, horizontal capsules 2.5–4 cm long and 2–3 mm broad. A tall hairless annual to 3 m, with pink, purple, or dark crimson flowers and with a darker spotted sac-like lower sepal and curved spur *c*. 8 mm. Leaves opposite, mostly with rounded, not acute teeth. **Pl. 21**.

251 **I. bicornuta** Wallich

Uttar Pradesh to Arunachal Pradesh. 2100–3000 m. Forests, open slopes. Jul.–Sep.

Usually a robust plant to 60 cm, with pinkish-mauve flowers with an orange-yellow and purple-dotted throat, and with a broad cylindrical lower sepal abruptly narrowed to a slender curved spur *c*. 8 mm. Flowers to 3 cm long; wings with slender horn-like points and short rounded side-lobes; upper petal with a narrow point; bracts and sepals with long gland-tipped awns. Leaves to 20 cm, narrow elliptic coarsely and regularly rounded-toothed and with a tail-like tip.

Flowers predominantly yellow
a *Flowers large, 2.5 cm or more including spur*

252 **I. falcifer** Hook. f.

C. Nepal to Sikkim. 2500–3600 m. Forests, shrubberies. Jul.–Sep.

Flowers yellow often spotted with red, with a long slender spur to 2 cm, and wings with basal lobes small curved, and distal lobe longer than broad. Leaves alternate, ovate-lanceolate saw-toothed, mostly 4–8 cm; stems slender, to 50 cm. Capsule slender, often pendulous.

253 **I. scabrida** DC.

Kashmir to Bhutan. 1200–3600 m. Forests, shrubberies, damp places; common. May–Sep.

A rather robust often much-branched plant with rather large golden-yellow flowers spotted with brown within. Flowers several in each axil, 3–4 cm long, with a broad funnel-shaped lower sepal suddenly contracted to a slender spur 1.5–3 cm; upper petal rounded and spurred; wings smaller. Leaves 5–15 cm, elliptic to lanceolate long-pointed, with acute glandular teeth, short-stalked; stem finely hairy, 60–120 cm. Capsule linear acute 3–4.5 cm.

254 **I. urticifolia** Wallich

W. Nepal to S.E. Tibet. 2700–3800 m. Shrubberies, forests, damp places. Jun.–Aug.

Flowers yellow with brown spots or streaks within, and with distinctive lip-petals with narrow pointed tail-like tips, and with a short bag-like lower sepal with a short blunt spur. Flowers large to 3.5 cm long. Leaves elliptic long-pointed with rounded gland-tipped teeth, to 18 cm; stem usually slender, to 120 cm. **Pl. 21**.

b *Flowers small, less than 2.5 cm including spur*

255 **I. edgeworthii** Hook. f.
Pakistan to Kashmir. 1500–2700 m. Streamsides, rocks, grazing grounds; common in Kashmir. Jun.–Sep.

Flowers small, many in branched clusters from the upper leaf-axils, pale yellow, with orange throat and rounded crested orange upper petal. Lateral united petals short elliptic with rounded wings. Lower sepal narrow funnel-shaped, gradually narrowed to a slender curved spur; lateral sepals large rounded, with crested midrib. Leaves grouped in a terminal rosette, stalked, elliptic long-pointed, to 18 cm, with gland-tipped teeth. **Pl. 21**.

256 **I. stenantha** Hook. f.
C. Nepal to Bhutan. 1800–2600 m. Forests. Jul.–Sep.

Flowers small, orange-yellow with red spots on the lower lip within, and with a long up-curved funnel-shaped lower sepal with a curved reddish spur tapering to a fine point; lateral sepals elliptic. Upper petal ovate entire; lip with long strap-shaped lobes; wings with broad side-lobe. Leaves elliptic long-pointed, with regular rounded gland-tipped teeth; stem slender 30–40 cm. **Pl. 21**.

256a **I. racemosa** DC.
Kashmir to S.E. Tibet. 1300–3900 m. Forests, damp places. Jul.–Oct.

Flowers yellow, small 0.8–2 cm long, borne in lax usually axillary clusters mostly longer than the leaves and with persistent narrow elliptic bracts. Spur slender curved, as long as the flower-stalk, lateral sepals ovate; upper petal rounded; wings with thread-like lobes; the lip boat-shaped. A slender erect hairless plant usually to 35 cm but often taller, with elliptic to lanceolate long-pointed rounded-toothed stalked leaves mostly 7–15 cm.

RUTACEAE   Citrus Family

Members of this family are usually distinguished by the translucent oil glands in the leaves, which give a strong acrid aroma when crushed. Mostly trees and shrubs, with few herbaceous species, with usually trifoliate or compound leaves. Inflorescence terminal, branched, often flat-topped. Flowers bisexual, usually regular; sepals 4–5, free or fused below; petals 4–5, commonly free; stamens variable in number and with a thick glandular basal disk. Ovary superior, with 4–5 fused carpels. Fruit a berry, drupe, or capsule.

Herbaceous plants
  a Flowers 6–8 mm; leaflets entire   1 BOENNINGHAUSENIA
  b Flowers 2.5–4 cm; leaflets toothed   2 DICTAMNUS

RUTACEAE

Shrubs or small trees, or woody climbers
  a Leaves oblanceolate entire   3 SKIMMIA
  b Leaves pinnate   4 ZANTHOXYLUM

1 BOENNINGHAUSENIA Flowers regular; petals 4–5; stamens 6–8. Ovary stalked. Leaves ternate. 1 sp.

257 **B. albiflora** (Hook.) Reichb. ex Meissner
Pakistan to S.W. China. S.E. Asia. 600–3300 m. Forests, shrubberies. Jun.–Sep.
    A rather delicate fern-like, nearly hairless perennial, with twice or thrice ternate leaves, and with small white flowers in leafy branched terminal clusters. Petals 5, oblong blunt, *c.* 5 mm; stamens and style longer; calyx very small, 4–lobed. Leaflets 5–12 mm, ovate entire, gland-dotted, glaucous beneath; stems 30–60 cm. Fruit *c.* 4 mm, separating into 3–5 several-seeded units. **Page 454**.

2 DICTAMNUS Flowers irregular; petals 4–5; stamens 12–15. Fruit a capsule. Leaves pinnate. 1 sp.

258 **D. albus** L. BURNING BUSH
Pakistan to Uttar Pradesh. Temperate Eurasia. 1800–2700 m. Open slopes, rocks. Jun.–Jul.
    A strongly aromatic bushy perennial with a lax spike of large conspicuous pink or white flowers, and with pinnate leaves. Stems, bracts and sepals densely covered with conspicuous dark glands. Petals lanceolate, unequal, to 2.5 cm, streaked and dotted with violet; stamens longer, with purple filaments and green anthers. Leaves with 4–6 pairs of leathery, elliptic gland-dotted finely toothed leaflets; stems robust, 30–60 cm. Fruit 1.5 cm, deeply divided into 5 segments which split at the apex, bristly-glandular.
    Extracts from the roots are used medicinally. **Pl. 22**.

3 SKIMMIA Flowers regular, usually with 5 perianth-segments. Fruit a drupe. Leaves entire, aromatic, hairless or nearly so. 4 spp.

259 **S. anquetilia** N.P. Taylor & Airy Shaw (*S. laureola* auct. in part)
Afghanistan to W. Nepal. 2400–4000 m. Forests, shrubberies, shady places; gregarious. Apr.–May.
    A strongly aromatic evergreen shrub, creeping or erect to 150 cm, with simple rather thick leathery, oblanceolate to oblong-elliptic, shortly pointed leaves, and very compact terminal clusters of yellow flowers. Flower clusters solitary, 2–3 cm; flowers not opening widely; petals 5, 4–5 mm, longer than the greenish triangular sepals. Leaves to 17 cm long, but often much less, crowded at the ends of the branches, hairless. Fruit red, fleshy ovoid 1–1.6 cm. **Pl. 22**.

260 **S. arborescens** Gamble
C. Nepal to S.W. China. S.E. Asia. 1600–2700 m. Forests. Apr.–May.
Differs from 259 in being a taller shrub, or tree to 9 m (1.5 m in 259), and in having greenish-white, widely opening flowers. Fruit black. Leaves generally larger, oblong-lanceolate, thin, not leathery, with a long tail-like drip-tip.

261 **S. laureola** (DC.) Sieb. & Zucc. ex Walp. (*S. melanocarpa* Rehder & Wilson)
C. Nepal to N. China. 2400–3600 m. Forests. Apr.–May.
Like 259, but with smaller elliptic leaves, with widely opening whitish flowers in looser clusters, and with black fruit. **Pl. 22.**

4 ZANTHOXYLUM Stamens 6–8. Ovary deeply 3–5–lobed; fruit a capsule. Leaves pinnate. 7 spp.

262 **Z. armatum** DC.
Kashmir to S.W. China. S.E. Asia. 1100–2500 m. Shrubberies, cultivated areas. Apr.–May.
A shrub or sometimes a small tree, with corky bark and with numerous long straight spines on branchlets and leaf-stalks, with pinnate leaves, and with small yellow flowers in short branched lateral clusters. Flowers *c.* 1 mm, one-sexed; calyx with 6–8 acute lobes; petals absent; stamens 6–8, much longer than calyx in male flowers. Leaf-stalk narrowly winged; leaflets 2–6 pairs, lanceolate, 8 cm, toothed, sparsely gland-dotted. Ripe capsule 3–4 mm, globular, red, wrinkled, aromatic; seed shining black.
Branchlets aromatic; used for cleaning teeth. Fruit used as a condiment, and as a remedy for toothache. Bark used for poisoning fish, and used medicinally.

263 **Z. oxyphyllum** Edgew.
Uttar Pradesh to Arunachal Pradesh. Burma. 1800–2700 m. Forests, shrubberies. Apr.–May.
Differs from 262 in having larger lilac flowers 6–8 mm across, in branched clusters, with 4 blunt petals *c.* 5 mm; filaments white. Leaf-stalks and rachis not winged, with hooked prickles; leaflets 3–10 pairs; branches with hooked prickles. A shrub or woody climber. **Page 454.**

264 **Z. nepalense** Babu (*Z. simularis* auct. non Hance)
C. to E. Nepal. 2400–3100 m. Shrubberies. May.
Like 262, but flowers in short stalkless flat-topped branched clusters; leaf-stalks and rachis unwinged, and with shorter oval leaflets 1–3 cm long, with prominent lateral veins. **Pl. 22.**

SIMAROUBACEAE   Tree of Heaven or Quassia Family

A family of trees and shrubs, of the Tropics and Sub-Tropics, distinguished by their usually pinnate leaves, and numerous small flowers in spike-like clusters or branched panicles. The family is similar to the Meliaceae, but differs largely in its fruit which is a winged single-seeded nut, or a capsule. Petals and sepals 3–7; a ring or cup-like disk occurs between the petals and stamens which are as many, or twice as many as the petals. Ovary superior, of 2–5 carpels, free below but united above by the styles or stigmas.

AILANTHUS   1 sp.

265   **A. altissima** (Miller) Swingle (*A. glandulosa* Desf.) TREE OF HEAVEN
Native of China. Planted in the Himalaya to 2100 m; a common roadside tree in Kashmir. Apr.–May.

A medium-sized deciduous tree, with smooth grey bark, large pinnate leaves, and in the autumn conspicuous with terminal clusters of young reddish winged fruits. Flowers greenish, 5–8 mm across, in erect branched clusters 10–20 cm. Leaves 45–60 cm, with 13–25 large oval-lanceolate leaflets, each with 2–4 coarse glandular teeth near the base; leaflets 7–12 cm long. Fruit in hanging clusters, each with a single central nut and propeller-like wing 3–4 cm twisted at the top. A freely suckering tree.

Parts of the tree are used medicinally.

MELIACEAE   Mahogany Family

Trees and shrubs of the Tropics and Sub-Tropics with pinnate usually alternate leaves. Flowers small, numerous, bisexual and regular; sepals 3–6, free or fused; petals 3–6, usually free; stamens commonly 5–10, often united into a tube below. Ovary superior, with 2–6 carpels; fruit a capsule, berry, drupe, or rarely a nut.

MELIA   2 spp.

266   **M. azedarach** L. PERSIAN LILAC.
Native of Iran; doubtfully native in India. 700–2700 m. Often planted in villages and along roadsides in the Himalaya. Mar.–May; fruiting in cold season.

A small spreading deciduous tree with dark grey vertically fissured bark, with twice-pinnate leaves and with branched clusters of sweet-scented lilac flowers. Flower clusters axillary, long-stalked; flowers 7–9 mm long; calyx deeply 5–6–lobed, hairy outside; petals much longer, linear to oblanceolate; stamen-tube conspicuous, purple. Leaves 30 cm, with 3–11 pinnae and many ovate-lanceolate long-pointed opposite leaflets 2.5–5 cm long. Fruit a globular drupe *c.* 1.5 cm, yellow when ripe, at first smooth, then wrinkled and remaining on tree long after the leaves have fallen.

The bark, seeds, young fruit, leaves and gum all have medicinal uses. Fruit used as beads for necklaces and rosaries; worn as a charm against disease. Leaves lopped for fodder. **Page 454**.

AQUIFOLIACEAE Holly Family

A widely distributed family of trees and shrubs with leathery evergreen, alternate, usually simple leaves. Flowers small inconspicuous, in stalked or stalkless clusters in the axils of the leaves, usually one-sexed. Sepals and petals usually 4, each whorl commonly fused at the base. Stamens 4. Ovary superior, of 3 or more united carpels each with one or two ovules. Fruit a drupe containing 1 or 2 or more nuts.

ILEX HOLLY 9 spp.

267 **I. dipyrena** Wallich HIMALAYAN HOLLY
Pakistan to S.W. China. 1500–3000 m. Forests, damp ravines; common in oak forests. Apr.–May.

A moderate-sized evergreen tree with light grey smooth bark, and with dark glossy green, very leathery, oval to elliptic pointed leaves usually with strong spiny teeth on the margins. Flowers 6 mm across, greenish-white, in small rounded stalkless axillary clusters *c.* 1 cm; petals 4, obovate. Leaves 5–10 cm long, paler beneath, spineless on older trees. Fruit red when ripe, globular, 6 mm; nuts usually 2, grooved.

Wood used principally for fuel. **Pl. 23**.

268 **I. fragilis** Hook. f.
C. Nepal to S.W. China. Burma. 2000–3300 m. Forests. May–Jun.

A small tree with very brittle, quite hairless branches, with ovate-pointed bright green papery leaves, strongly netted with many raised veins beneath, and with fine marginal teeth (not spiny). Flowers with parts in 5, *c.* 3 mm across; male and female flowers on different trees, both in small unbranched clusters in the axils of the leaves. Leaves 10–13 cm. Fruit red, globular, with 5–8 nuts. **Pl. 23**.

CELASTRACEAE Spindle-Tree Family

A moderate-sized family of trees, shrubs and climbers, widespread, but largely concentrated in the Tropics and Sub-Tropics. Distinguished by its ovary which has a large fleshy disk often fused with the base of the stamens and by its seeds which are covered with a brightly coloured fleshy coat, or *aril*. Leaves simple, often leathery. Flowers small, bisexual or unisexual, usually greenish and commonly in short flat-topped clusters in the leaf-axils. Sepals and petals 3–5; stamens 3–5 and inserted on the fleshy disk. Ovary superior, with 2–5 carpels. Fruit various; a capsule, winged nut, drupe, or berry.

EUONYMUS Spindle-Tree Leaves opposite; stems spineless. 11 spp.

Climbing or prostrate shrubs; fruit spiny

269 **E. echinatus** Wallich
Pakistan to Sikkim. 1500–2700 m. Forests, rocks. Apr.–May.

An evergreen shrub climbing by means of tufts of aerial rootlets on trees and rocks, or a low spreading shrub. Flowers greenish-white, to 6 mm across, in branched axillary clusters; petals 4, rounded. Leaves leathery dark green, elliptic-lanceolate 2–6 cm, with shallow teeth; leaf-stalk short, channelled. Fruit black, globular and covered with dark spines; seeds dark brown, covered with a scarlet aril. **Pl. 23**.

Small trees or erect shrubs; fruit smooth
a *Deciduous shrubs or trees; flowers not purple-veined*

270 **E. fimbriatus** Wallich
Afghanistan to C. Nepal. S.W. China. 1800–3600 m. Forests, shrubberies, riversides. Apr.–May.

A small deciduous tree with reddish-brown compressed branchlets, and with distinctive broadly ovate, finely double-toothed leaves. Flowers cream-coloured, *c*. 4 mm, many in rather lax flat-topped clusters, borne on axillary stalks 2.5–7 cm long. Leaves 3–8 cm. Fruit turning red, *c*. 2 cm across, with 4 conspicuous long tapering wings; seeds enclosed in a red aril.

Leaves and branches cut for fodder; the seeds are used as beads.

271 **E. hamiltonianus** Wallich
Afghanistan to S.E. Tibet. Burma. 700–2700 m. Forests, shrubberies, shady places; common in Kashmir. Apr.–Jul.

Differs from 270 in having pale grey, thick corky bark (smooth dark grey in 270), and leaves with a single set of fine marginal teeth with bristle-like deciduous tips. Flowers larger, greenish-white, *c*. 7 mm across, in axillary branched clusters. Fruit top-shaped, smaller *c*. 1 cm across, deeply lobed, not winged. **Page 454**.

b *Evergreen shrubs or trees; flowers conspicuously purple-veined*

272 **E. tingens** Wallich
Himachal Pradesh to S.W. China. Burma. 2100–3300 m. Forests. Apr.–Jun.

An evergreen shrub or small tree, with dark ash-grey corky bark, and with thick dull green leathery wrinkled elliptic to ovate-lanceolate leaves, which are closely toothed and paler beneath. Flowers in long-stalked axillary clusters, often at base or apex of branches; flowers yellowish-white and strongly veined with purple, *c*. 1.3 cm across; petals usually 5, toothed; anthers red. Leaves 2–5 cm. Fruit top-shaped *c*. 1.5 cm across,

3–5–angled, not winged; seed half encircled in orange aril.
Bark used medicinally; it yields a yellow dye. **Pl. 23**.

MAYTENUS Differs from *Euonymus* in having alternate (not opposite) leaves, and stems with spines. 2 spp.

273 **M. rufa** (Wallich) Hara
Uttar Pradesh to Bhutan. 600–2100 m. Shrubberies. Mar.–Apr.

A shrub or small tree with slender sparingly spiny branches, with rather leathery hairless leaves, and with small axillary clusters of greenish flowers much shorter than the leaves. Flowers in branched clusters 2.5–5 cm long; petals 4–5, spreading, 1–2 mm. Leaves mostly 4–8 cm, alternate, lanceolate long-pointed, toothed, paler beneath; spines few slender 1.5 cm. Fruit red 4–8 mm across, 3–chambered, pale yellow within, each chamber 1–seeded. **Page 455**.

RHAMNACEAE Buckthorn Family

A large family of trees, shrubs, and some climbers, which is more or less cosmopolitan. Distinguished by its *perigynous* flowers in which the ovary is surrounded at the base by a rim or cup from which the sepals, petals and stamens arise. Leaves simple. Flowers regular and bisexual, small, greenish, usually borne in flat-topped clusters. Sepals and petals 4–5; stamens 4–5. Ovary free, or imbedded in the disk, with 2–3 chambers each with one ovule. Fruit mostly a fleshy drupe, or a nut.

RHAMNUS Buckthorn 10 spp.

274 **R. purpureus** Edgew.
Pakistan to C. Nepal. 1500–3000 m. Forests. Mar.–Apr.

A deciduous shrub or tree to 10 m, with purple young branchlets and alternate ovate-lanceolate to elliptic, pointed and toothed leaves, mostly 5–10 cm. Flowers tiny, reddish, in small axillary clusters; petals absent; calyx cup-shaped *c*. 3 mm across, with 5 triangular lobes. Leaves with 6–10 pairs of lateral veins, hairy beneath. Fruit reddish, obovoid, 5–7 mm; seeds black, heart-shaped. **Page 455**.

VITACEAE Vine Family

A family mainly of woody climbers, but with some shrubs, largely of the Tropics and Sub-Tropics. Many have tendrils. used for support, which are either modified shoots or inflorescences, and these may have disk-like suckers at their tips. Leaves alternate, simple, palmate or pinnately-lobed. Flowers unisexual or bisexual, very small, in flat-topped or elongated clusters

77

arising opposite the leaves. Calyx cup-shaped with 4–5 teeth or lobes; petals 4–5, often fused by their tips; stamens 4–5, inserted on a rim or disk. Ovary superior, of two fused carpels with 2–6 chambers. Fruit a fleshy berry. There are *c.* 27 species of vines in our area; many of them are difficult to distinguish in the field.

AMPELOCISSUS Leaves entire or shallow-lobed. Petals 5, free; disk annular. 5 spp.

275　**A. rugosa** (Wallich) Planchon
　　Kashmir to E. Nepal. 1000–2400 m. Rocks, steep slopes. May–Jun.
　　A large woody vine-like tendril-climber, with large ovate-heart-shaped finely toothed leaves which are conspicuously red-woolly beneath. Flowers reddish to yellow-green, in large branched clusters arising opposite the leaves; petals 5. Leaves very variable, blade 7–20 cm, sharply toothed and often angled or shallowly lobed; stems woolly-haired, as also the leaf-stalk and inflorescence, but hairiness very variable; tendrils long, bifid, usually arising below the long-stalked inflorescences. Berries black, *c.* 8 mm across. **Pl. 23**.

PARTHENOCISSUS Leaves trifoliate. Petals usually 5, free; disk inconspicuous. 2 spp.

276　**P. himalayana** (Royle) Planchon
　　Pakistan to Sikkim. S.W. China. Burma. 1800–3300 m. Coniferous forests. Apr.–May.
　　A very large climber to 18 m, with trifoliate leaves with 3 ovate long-pointed, sharply toothed, stalked leaflets, which are shining dark green above and pale beneath, the lateral leaflets asymmetrical. Flowers yellow-green, in spreading flat-topped clusters; petals *c.* 5 mm, petals and stamens 4–5. Leaflets mostly 10 cm, bristly-haired on the veins beneath; tendrils branched. Berry black, *c.* 8 mm. **Page 455**.

277　**P. semicordata** (Wallich) Planchon
　　W. Nepal to Bhutan. 1800–2700 m. Forests, rocks. May.
　　A very large climber like 276, but usually distinguished by the veins on the undersides of the leaves being hairless. Flowers greenish-yellow, with petals spreading and soon reflexed. Leaflets mostly smaller, 5–6 cm. Fruit blackish-purple.

VITIS Leaves entire. Petals 5, cohering at the tips. 4 spp.

278　**V. jacquemontii** R. Parker
　　Pakistan to E. Nepal. 1000–1800 m. Shrubberies, steep slopes. Apr.–May.
　　A stout woody climber with large ovate-heart-shaped leaves, which are densely white or reddish woolly-haired beneath, and with a leaf-opposed dense branched cluster 10–20 cm across of numerous tiny greenish-yellow flowers. Flowers *c.* 5 mm across, with yellow stamens much longer than

the perianth which falls off when the flowers open. Mature leaves with blades 10–12 cm, toothed; tendrils woolly-haired, stout, branched; all young shoots densely woolly-haired. **Pl. 23**.

SAPINDACEAE

Quite an important tropical and sub-tropical family of trees, shrubs and climbers. Leaves alternate, simple or compound. Flowers borne in branched inflorescences. Sepals and petals usually 5, not fused; stamens in two whorls of 5, but often two stamens missing, thus 8. Ovary superior, of three fused carpels, with 1–6 chambers each 1–2 seeded. Fruit a nut, berry, or capsule.

DODONAEA 1 sp.

279 **D. viscosa** (L.) Jacq.
Afghanistan to Himachal Pradesh. India. Pan-tropical. To 1500 m. Open slopes; often gregarious. Aug.–Feb.; fruiting Jan.–Jun.

An evergreen shrub to 5 m, with leathery bright green shining oblanceolate entire leaves, which are commonly sticky with yellow resin. Flowers one-sexed, 3–5 mm across, greenish-yellow, in terminal and lateral clusters *c.* 2.5 cm long; sepals 4–5, blunt; petals absent; stamens usually 8. Leaves alternate, 3–9 cm long. Fruit a compressed capsule *c.* 1.5 cm, with 2–4 valves, each valve with a broad membraneous wing.

Wood very hard; used for walking sticks and tool-handles; excellent firewood. A quick-growing shrub often used for hedges. **Page 455**.

HIPPOCASTANACEAE Horse – Chestnut Family

A small family largely of the north temperate regions, with large winter-buds with resinous scales, and palmate leaves. Flowers in racemes, the upper flowers male, the lower bisexual. Flowers irregular; sepals 5 united at the base; petals unequal, 4–5, large, coloured; stamens 5–8. Ovary superior, of 3 fused carpels, 3–chambered with 1–2 ovules in each chamber. Fruit a leathery capsule, usually with one large shining seed.

AESCULUS Horse Chestnut 1 sp.

280 **A. indica** (Colebr. ex Cambess.) Hook.
Afghanistan to C. Nepal. 1800–3000 m. Forests, shady ravines. Apr.– Jun.; fruiting Jul.–Oct.

A large deciduous tree to 30 m or more, with very distinctive long-stalked palmate leaves, and with large terminal narrow-pyramidal clusters of white flowers. Clusters 25–40 cm long; petals 4, 1.5–2 cm, white and yellow, often streaked with red at base; calyx tubular, glaucous pinkish;

stamens 7, much larger than petals. Leaflets 5–9, elliptic long-pointed, sharply toothed, stalked, 15–25 cm; leaf-stalk 10–15 cm. Bark peeling in long narrow strips. Fruit ovoid 2.5–5 cm, smooth; seed 1–3, dark brown, shining.

Wood used for turned articles; the leaves are lopped as fodder; the bark is used medicinally. **Pl. 22.** ·

ACERACEAE   Maple Family

A widespread family of the North Temperate Zone, of largely deciduous trees, with characteristically lobed leaves. Flowers in axillary or terminal clusters, regular. Sepals 4–5; petals 4–5; stamens 4–10, but often 8, inserted on a disk. Ovary two-lobed, two-chambered, each with one ovule. Fruit very characteristic, comprising two one-seeded winged units which separate and spin as they fall, thus facilitating their dispersal.

ACER MAPLE 14 spp.

Leaves undivided, entire

281 **A. oblongum** Wallich ex DC.
Pakistan to S.W. China. Burma. S.E. Asia. 1000–2100 m. Forests. Feb.–Apr.

A small or medium-sized evergreen tree, distinguished by its oblong to broadly lanceolate, long-pointed entire leaves which are glaucous beneath. Mature leaves with blades 8–12 cm; young leaves at first finely hairy; young shoots hairless. Flowers greenish-white, *c.* 4 mm, in a dense terminal branched hairy cluster. Fruit hairless, with wings narrowed to the base placed either parallel or widely divergent or almost horizontal.

Wood used for agricultural implements, and turned for drinking cups. **Page 456**.

Leaves with 3–7 equal lobes
a *Lobes with entire or blunt-toothed margins*

282 **A. cappadocicum** Gled. (*A. pictum* auct. non Thunb.)
Afghanistan to W. Nepal. Sikkim. Bhutan. C. Asia. 2100–3000 m. Forests. Mar.–May.

A large to medium-sized deciduous tree, with leaves with 5–7 equal triangular long-pointed lobes with entire margins, hairless except in the axils of the veins beneath, and long-stalked. Flowers greenish-yellow, *c.* 8 mm across, appearing with the young leaves, in lax branched hairless clusters. Wings of young fruit pink, widely divergent, with upper margin slightly curved; nuts thin, compressed.

Tibetan drinking cups are made from the knotty burs of this tree. **Page 456**.

283 **A. pentapomicum** J. L. Stewart
Afghanistan to Himachal Pradesh. 1000–2100 m. Dry inner valleys only;
in association with *Pinus gerardiana* and *Quercus baloot*. Mar.–Apr.

A small deciduous tree, distinguished by its small leathery leaves with
usually 3 blunt lobes with entire or bluntly-toothed margins (leaves
sometimes with an additional pair of small lobes), grey-green above, paler
or glaucous beneath and hairless except in the vein-axils. Flowers
greenish-white, in dense flat-topped hairless clusters, appearing with or
before the young leaves. Leaf-blades 3–6 cm long and broad. Fruit with
nearly parallel wings.

b *Lobes with finely acute-toothed margins*

284 **A. campbellii** Hook. f. & Thoms.
W. Nepal to Arunachal Pradesh. Burma. 2100–3600 m. Forests, shrubberies. Apr.–May.

A tall tree with bright light-green foliage, and leaves deeply divided into
5–7 more or less equal lobes with fine teeth, and with a long tail-like tip.
Flowers cream to reddish, *c.* 6 mm across, in narrow elongated clusters to
10 cm. Leaves 10–12 cm wide. Fruit many in each cluster, with globular
nearly smooth nuts, and diverging wings. **Page 456**.

Leaves with 3–5 very unequal lobes, the lower outer two lobes much smaller
(see also 283)
a *Leaves pale or glaucous beneath*

285 **A. caesium** Wallich ex Brandis
Afghanistan to C. Nepal. 2200–3000 m. Forests, open grassy places;
common. Mar.–May.

A fairly large deciduous tree to 25 m, with thin grey bark. Leaves rather
large, 8–15 cm long, broader than long, with a heart-shaped base and with
5 unequal broad long-pointed coarsely saw-toothed lobes, usually pale and
glaucous beneath, red when young. Flowers greenish-yellow 5 mm across,
appearing with the young leaves, in branched flat-topped clusters. Wings
of fruit divergent or sometimes overlapping; nuts dark brown, with a
hump-like swelling. **Page 456**.

b *Leaves green beneath*

286 **A. sterculiaceum** Wallich (*A. villosum* Wallich)
Kashmir to Bhutan. 2100–3900 m. Forests; common. Feb.–Apr.

Like 285, but differing in its leaves which are green beneath and its
young shoots which are covered with brownish hairs (hairless in 285).
Flowers to 1 cm, usually appearing before the leaves, in elongate clusters.
Leaves large to 20 cm. Fruits large, each winged nut to 5 cm long, usually
rusty-haired and borne on leafless side-branches; nuts pale brown.

Leaves cut for fodder. **Page 457**.

287 **A. acuminatum** Wallich ex D. Don (*A. caudatum* Wallich in part)
Kashmir to C. Nepal. 2100–3000 m. Forests, shady ravines. Mar.–Apr.

An elegant tree with hairless young branchlets and with usually 5–lobed leaves (sometimes 3–lobed), the lobes triangular, sharply, coarsely and closely toothed, often doubly so, and prolonged at the apex into a conspicuous slender tail-like point. Flowers greenish, in hairless short flat-topped clusters appearing before or with the young leaves. Petals *c.* 4 mm, stamens much longer; bracts red. Leaves 6–12 cm long. Fruit bright red when young, in a lax long-stalked cluster, with diverging or parallel wings; nut rough, sharply angled. **Page 457.**

288 **A. pectinatum** Wallich ex Pax (*A. caudatum* Wallich in part)
W. Nepal to S.E. Tibet. Burma. 2700–3800 m. Forests. May–Jun.
Like 287, but flowers and fruits in long pendulous spike-like clusters. Leaves 3–5–lobed, the lobes triangular to ovate, finely and sharply saw-toothed, often doubly so, the teeth tipped with a bristle, and tips of lobes long-pointed or tail-like. Young leaves with veins and mid-rib covered with rusty hairs beneath, becoming nearly hairless. Flowers red. Wings of fruit wide-spreading; young fruit red; nuts flat. **Page 457.**

289 **A. caudatum** Wallich (*A. papilio* King)
Uttar Pradesh to S.E. Tibet. 3000–3600 m. Forests, shrubberies. Jun.
Like 288 with flowers and fruits in narrow elongated clusters, but lobes of leaves differing, having teeth without bristles, and the teeth themselves being finely toothed, and lower surface of leaf usually remaining persistently hairy. Flowers yellowish-green, or reddish, in narrow erect clusters. Wings of fruit diverging at an acute angle; nuts recessed. **Page 457.**

STAPHYLEACEAE   Bladder Nut Family

A small family of trees and shrubs of the temperate and tropical regions, with trifoliate or pinnate leaves. Flowers regular, usually bisexual, in branched clusters. Sepals and petals 5, free; stamens 5, free. Ovary superior, of 2–4 fused carpels. Fruit an inflated capsule, or berry-like.

STAPHYLEA   1 sp.

290 **S. emodi** Wallich ex Brandis
Afghanistan to W. Nepal. 1800–2700 m. Forests, shady ravines. Apr.–May; fruiting Jul.–Aug.
A large deciduous shrub, distinguished by its trifoliate leaves, its distinctive whitish brown-streaked bark, and its large white or yellowish inflated fruit. Flowers white, in terminal usually drooping clusters, borne on axillary shoots. Flowers 1–1.5 cm long; petals and sepals similar, spathulate; stamens persistent. Leaflets 3, ovate to elliptic, 5–15 cm, finely toothed, long-pointed, pale and usually finely hairy beneath. Fruit an inflated bladder-like ovoid capsule 5–7 cm, splitting from the apex; seeds shiny greyish-brown.

The stems with their ornamental bark are much sought after to make walking sticks as they are supposed to have the property of keeping off snakes. **Pl. 23.**

SABIACEAE

A small tropical and sub-tropical family of trees, shrubs and some climbers, comprising 4 genera. Leaves alternate, simple, or pinnate. Flowers in branched clusters. Sepals 3–5, unequal, opposite (not alternating with) the 4–5 petals; stamens 3–5, opposite the petals. Ovary of 2 fused carpels, with 2 chambers each with 1 ovule. Fruit a berry.

SABIA 4 spp.

291 **S. campanulata** Wallich ex Roxb.
Kashmir to S.W. China. 1800–3300 m. Forests, shrubberies, shady banks. Apr.–May.
A small deciduous weakly climbing shrub, with elliptic entire pointed leaves, and solitary stalked green to brown-purple flowers borne from the leaf axils. Flowers variable to 1.5 cm across; petals 4–5, obovate, concave; stamens 5, borne on a 5–lobed annular disk. Leaves mostly 2.5–6 cm, short-stalked. Fruit blue, 6–8 mm, rounded, flattened, wrinkled, solitary or paired. **Page 458**.

ANACARDIACEAE   Mango or Cashew Family

A medium-sized family of woody plants mainly of the tropical and sub-tropical regions, but with some temperate species. Most have resinous tissues. Leaves usually pinnate, but entire in mango and *Cotinus*. Flowers small, regular, in clusters. Sepals 5, fused; petals 5, free; stamens usually 5–10. Between the stamens and ovary there is usually a fleshy disk. Ovary usually superior, of commonly 1–5 fused carpels, with 1–3 styles. Fruit usually a drupe with a single seed.

COTINUS   Like *Rhus* but leaves simple and entire. Fruit-stalks long, with spreading hairs. 1 sp.

292 **C. coggygria** Scop.   Wig Tree, Smoke-Tree
Pakistan to C. Nepal. W. Asia. S. Europe. N. China. 1100–2400 m. Forests, shrubberies. Apr.–Jun.
A deciduous shrub or rarely a small tree, with neat broadly elliptic blunt to nearly rounded entire leaves with conspicuous lateral veins, and with terminal much-branched greyish feathery fruiting clusters. Flowers 3 mm, yellowish to pale purple, in lax spreading or drooping, branched hairy clusters 8–15 cm long; after flowering the stalks of the numerous sterile

flowers elongate and are covered with long grey or purple silky hairs. Leaves 5–12 cm, paler beneath, stalked. Fruit ovoid 5 mm long, unequally lobed, hairy.

The bark, leaves and young branches are used for tanning and dyeing. Wood used for inlay work. **Page 458**.

RHUS   Petals 5. Leaves pinnate. Fruit short-stalked. 7 spp.

Leaflets hairless beneath

293   **R. succedanea** L. Wild Varnish Tree

Pakistan to Bhutan. China. Burma. 1200–2400 m. Forests, shrubberies. May–Jun.

A small deciduous tree, with large pinnate leaves, and with small greenish-yellow flowers in slender drooping branched clusters about half as long as the leaves. Flowers 3–4 mm; petals oblong with numerous dark veins. Leaflets 7–15, ovate-oblong to lanceolate with a long and slender apex, 7–16 cm, entire, shining, short-stalked. Fruit light brown, globular, *c*. 6 mm.

The fruit yields a wax which is used in lacquer work, and for candles. Plant used medicinally.

Leaflets woolly-haired beneath
a *Leaflets toothed*

294   **R. javanica** L. (*R. semialata* Murray)

Kashmir to S.W. China. Burma. Japan. 1200–2400 m. Forests, shrubberies. Aug.–Sep.

A shrub or small tree with young parts, leaf-stalks and inflorescence hairy, and with pinnate leaves with 5–13 leaflets and with the upper part of the rachis narrowly winged. Flowers *c*. 3 mm, pale yellowish-green, very numerous, in large branched pyramidal clusters *c*. 30 cm and nearly as long as the leaves. Leaflets variable, lanceolate to ovate, long-pointed, 5–10 cm, coarsely sharp-toothed and woolly-haired beneath; the leaves turn red in autumn. Fruit *c*. 4 mm, woolly, reddish-brown.

The milky juice of this and other species causes irritation of the skin. **Pl. 23**.

b *Leaflets entire*

295   **R. wallichii** Hook. f.

Himachal Pradesh to Bhutan. 1000–2400 m. Forests, shrubberies. May–Jun.

A small deciduous tree to 18 m, with young parts, inflorescence and undersides of leaves covered with rusty hairs. Flowers *c*. 2 mm, yellowish-green with dark veins, in dense axillary branched clusters much shorter than the leaves. Leaflets 7–11, large elliptic to oblong 8–23 cm, entire, often long-pointed, the terminal leaflet long-stalked. Fruit *c*. 8 mm long, in a compact pyramidal cluster, brown-woolly at first but becoming hairless, splitting into 5 segments arranged in a star and exposing the central stone.

A varnish is made from the juice. The wood is used for handles and small implements. Wax is obtained from the fruits. The juice turns black on exposure to air, is corrosive and raises blisters on the skin. **Page 458**.

PISTACIA   Petals absent. Leaves pinnate. 2 spp.

296   **P. chinensis** Bunge subsp. **integerrima** (J. L. Stewart) Rech. f.
Afghanistan to W. Nepal. 600–2400 m. Riversides, cultivated areas. Apr.

A medium-sized tree with short trunk, spreading branches and rough dark grey bark, and with pinnate leaves. Flowers reddish, without petals, appearing with or just before the young leaves. Trees one-sexed; male flowers in compact clusters 5–10 cm long; female flowers in broader, lax branched clusters, 15–25 cm long. Leaves with 4–6 pairs of leathery, hairless, lanceolate long-pointed entire leaflets, 6–13 cm. Fruit grey, wrinkled, globular *c.* 6 mm. Tree prominent in spring with red young leaves.

Wood hard and durable; used for furniture-making and carving. The large crooked dull red galls found on the leaves are used in local medicine. **Page 458**.

CORIARIACEAE   Coriaria Family

A family of only one genus, found in C. & E. Asia, Europe, and America. Distinguished by its fruits in which the carpels are surrounded by the persistent enlarged fleshy petals. Sepals and petals 5; stamens 10; carpels 5–10, each with a single ovule.

CORIARIA   2 spp.

297   **C. napalensis** Wallich
Pakistan to S.W. China. 1000–2700 m. Forests, shrubberies. Mar.–Apr.

A large hairless shrub with arching reddish-brown branches, and elliptic-pointed entire leaves. Flowers *c.* 5 mm, in one or several axillary clusters 2.5–10 cm long; petals greenish, 5, at first smaller than sepals but soon enlarging and becoming fleshy. Stamens and styles red or purplish, conspicuous and protruding. Leaves nearly stalkless, 2.5–5 cm, conspicuously 3–veined. Fruit black, with 5 carpels encircled by 5 larger purple fleshy persistent petals. **Page 458**.

LEGUMINOSAE   Pea Family

A very large and very important cosmopolitan family of herbaceous plants, trees, shrubs, climbers, and some aquatics. It is divided into 3 sub-families with flowers which appear to be very different but which have the same general floral pattern: the Mimosoideae, with regular flowers and numerous stamens; the Caesalpinioideae, with more or less irregular flowers and 10 or fewer free stamens; the Papilionoideae with irregular butterfly-like flowers and 10 stamens usually fused by their filaments. Leaves usually compound, with stipules, and sometimes with tendrils. The inflorescence is usually a raceme with closely clustered flowers. Sepals often fused, 5; petals 5, free; stamens commonly 10. Ovary of a single carpel and style. Fruit quite characteristic: a one-chambered pod, the *legume*, which usually splits into 2 valves. The pod may be dry or fleshy, inflated or compressed, green or brightly coloured, with 1 to many seeds.

Sub-Family **MIMOSOIDEAE** Flowers regular, tiny, numerous in clusters; stamens many, usually free, longer than petals.

1 ALBIZIA 2 MIMOSA 3 ENTADA

Sub-Family **CAESALPINIOIDEAE** Flowers with petals unequal, more or less irregular, medium to large; stamens 10 or fewer, their filaments not fused.

4 BAUHINIA 5 CAESALPINA 6 CASSIA

Sub-Family **PAPILIONOIDEAE** Flowers irregular, with petals unequal comprising a *standard*, 2 *wings* and a *keel*; stamens 10, usually fused by their filaments (or 9 fused).

Trees 7 ERYTHRINA 8 ROBINIA

Shrubs, woody climbers, or herbaceous perennials
  a Leaves with 1–3 leaflets
    i Shrubs or woody climbers 9 BUTEA 10 CAMPYLOTROPIS
      11 CROTALARIA 12 DESMODIUM 13 FLEMINGIA
      14 LESPEDEZA 15 PIPTANTHUS 16 PUERARIA
    ii Herbaceous plants 17 ARGYROLOBIUM 29 LATHYRUS
      18 MEDICAGO 19 MELILOTUS 20 PAROCHETUS
      21 THERMOPSIS 22 TRIFOLIUM 23 TRIGONELLA
  b Leaves with 5 or more leaflets
    i Shrubs or woody climbers 34 ASTRAGALUS
      24 CARAGANA 25 COLUTEA 26 INDIGOFERA
      27 SOPHORA
    ii Herbaceous plants
      x Leaves with terminal tendril or soft awn 28 CICER
        29 LATHYRUS 30 LENS 31 PISUM 32 VICIA
      xx Leaves with terminal leaflet or sharp spine 33 APIOS
        34 ASTRAGALUS 35 OXYTROPIS 36 CHESNEYA
        37 GUELDENSTAEDTIA 38 HEDYSARUM 39 LOTUS

1 **ALBIZIA** Stamens numerous, with filaments united into a short tube below; corolla tubular, 5–lobed. 7 spp.

298 **A. julibrissin** Durazz. (*Acacia mollis* Wallich) Persian Acacia
  Pakistan to Bhutan. Asia. Africa. 1000–2100 m. Forests, riversides,

cultivated areas; often planted as a shade tree. Apr.–May.

A medium-sized deciduous tree, distinguished by its pink flower-heads and its twice-pinnate leaves with very numerous tiny leaflets. Flower-heads globular, stalked, of numerous small fragrant flowers and usually clustered at the ends of branches. Flowers with numerous conspicuous stamens with long pink filaments 2.5–3.2 cm, very much longer than the hairy calyx and corolla; corolla *c.* 8 mm, with triangular lobes, yellow-green. Leaves 10–30 cm long, with 4–15 pairs of pinnae each with 10–30 pairs of unequally oblong acute, adpressed-hairy leaflets 12–18 mm long by 3–4 mm broad. Pod pale brown or yellowish, beaked, 7.5–12.5 cm by 1.5–2.5 cm broad.

Excellent for planting as a roadside tree. **Pl. 24.**

299 **A. chinensis** (Osbeck) Merr. (*A. stipulata* Roxb.)
Pakistan to Sikkim. India. China. S.E. Asia. 300–1300 m. Riversides, cultivated areas. May–Jun.

Distinguished from 298 by its smaller narrower leaflets which are scarcely 3 mm broad, by its conspicuous stipules, and by its yellowish-white flowers with long white filaments sometimes tinged with purple. A large deciduous tree with a broad flat-topped crown and smooth grey bark. Stipules soon falling, 2.5 cm, obliquely heart-shaped. Leaves with 6–20 pairs of pinnae each with 20–40 pairs of leaflets 6–10 mm long. Pod light brown, 10–18 cm.

Foliage lopped for fodder. Planted as a shade tree in tea-gardens. **Pl. 24.**

2 MIMOSA Stamens 8–10, free; petals fused at the base. Pod breaking up into 1–seeded sections. 2 spp.

300 **M. pudica** L. SENSITIVE PLANT
Native of Tropical America. To 1900 m. Very common on rice-field terraces, cultivated areas. Jul.–Aug.

A spreading undershrub with prickly hairy stems, and palmate twice-pinnate leaves with usually 4 pinnae each with many pairs of narrow oblong leaflets. Easily recognized by its very sensitive leaves which collapse downwards when touched, while the leaflets close together in pairs. Flower-heads of many tiny flowers, globular *c.* 1 cm, purplish-pink, usually paired, axillary, borne on a slender bristly-haired stalk. Pods in globular clusters, each pod 1.5–2.5 cm with 3–5 joints, and covered with stiff yellowish bristles.

Leaves and roots used medicinally. **Pl. 25.**

301 **M. rubicaulis** Lam.
Afghanistan to Bhutan. India. 300–1900 m. Wasteland, ravines; common. Jun.–Aug.

A large straggling very prickly shrub with terminal elongate clusters of many tiny pink flowers in globular heads 1–1.5 cm across, and with twice-pinnate spiny leaves. Flowers at length fading to white, numerous in each head, with stamens much longer than the 4–lobed corolla. Leaves

8–15 cm, with a prickly rachis, and with 3–12 pairs of pinnae each with 6–15 pairs of tiny oblong leaflets 4–8 mm. Pods thin, flat, curved, 8–13 cm by 1 cm, breaking into 4–10 rectangular 1–seeded units leaving the remains of the pod attached to the shoot.
Roots and leaves used medicinally. **Pl. 25.**

3 ENTADA Leaves twice-pinnate, with tendrils. Stamens 10, free. Pod very large. 1 sp.

302 **E. phaseoloides** (L.) Merr. (*E. scandens* Benth.)
C. Nepal to Sikkim. Tropical and sub-tropical Asia. 300–1400 m. Climber in forests and shrubberies. Apr.
A very large woody climber with twice-pinnate leaves, with paired woody tendrils, and with tiny pale yellow flowers either borne in long slender axillary spikes or in terminal branched clusters. Flower spikes 25 cm long or more; flowers 2–3 mm across; petals 5, equal; stamens longer. Leaves with 2 pairs of pinnae each with 3–4 pairs of oblong to obovate shining leaflets 3–8 cm long. Pod enormous, 60–120 cm by 8–10 cm broad, with 10–30 rounded or square flattened joints; seeds flat, shining brown.
Seeds, stem and bark poisonous. Used medicinally; the seeds contain saponins. **Pl. 24.**

4 BAUHINIA Leaves 2–lobed (resembling a camel's footprint), palmately veined. Flowers large, with almost equal petals. 6 spp.

Trees

303 **B. purpurea** L.
Pakistan to Bhutan. India. China. To 1600 m. Secondary forests, cultivated areas. Sep.–Oct.; fruiting Jan.–Mar.
A medium-sized deciduous tree with ashy to dark brown bark, and very distinctive leaves which are deeply cleft at the apex into 2 rounded usually blunt lobes. Flowers pink-purple, occurring with the leaves, slightly drooping, few in terminal or axillary clusters, the inflorescence brown-haired. Petals 3–4 cm, oblanceolate narrowed to a long stalk, veined; calyx woolly-haired, splitting into 2 recurved lobes; fertile stamens usually 3, curved upwards. Leaves stalked, blade usually 7–15 cm long, with 7–11 strong veins. Pod flattened, 15–25 cm by 1–2 cm, greenish tinged purple.
Widely planted for its foliage which is lopped for fodder. Bark used for dyeing and tanning; the bast for fibre. Used medicinally. **Pl. 26.**

304 **B. variegata** L.
Pakistan to Bhutan. India. Burma. China. To 1800 m. Forests; planted in villages. Mar.–Apr.
Differs from 303 in that it flowers in the spring on leafless branches; leaves cleft to a third (cleft to half in 303), and calyx-tube as long or longer

than calyx-limb. Flowers fragrant, 5–6 cm; petals all white, or 4 petals pale purple, and the fifth darker with purple veins; fertile stamens 5; calyx spathe-like, not splitting.

Leaves cut for fodder; flowers eaten and pickled; bark used for dyeing and tanning; the wood for building and implements. The bark is used medicinally.

Woody climbers

305 **B. vahlii** Wight & Arn. CAMEL'S FOOT CLIMBER
Kashmir to Sikkim. India. To 1500 m. Forest, rocks. Apr.–Jun.

A very large evergreen climber with rusty-brown-haired leaves and shoots, and with stems not uncommonly 1–1.6 m in diameter. Flowers cream, 3.5–5 cm across, in dense stalked flat-topped terminal clusters; petals obovate 1.5–4 cm, silky-haired beneath; fertile stamens 3. Leaves variable in size, 10–46 cm, rounded with heart-shaped base and cleft to about one third into 2 lobes with rounded apices, dark green above, downy beneath; branchlets with paired tendrils. Pod woody, 22–30 cm by 4–5 cm broad, with rusty velvet hairs; seeds flat, dark brown, polished, 2.5 cm.

Seeds edible; foliage used for fodder, and leaves for plates and wrappings. The bark yields a good fibre used for ropes, and also for tanning. Seeds and leaves used medicinally. One of the greatest enemies of forest trees, consequently eradicated in many forests. **Pl. 26.**

5 CAESALPINIA Leaves twice-pinnate; prickly shrubs. 5 spp.

306 **C. decapetala** (Roth) Alston (*C. sepiaria* Roxb.)
Pakistan to Bhutan. India. China. Japan. S.E. Asia. To 2200 m. Shrubberies, stony open slopes, ravines. Mar.–Apr.

A scrambling shrub with twice-pinnate leaves, with rachis and branches with prickles, and with large erect spike-like clusters of bright yellow flowers. Flower clusters 30–40 cm long; flowers 2–3 cm across, with rounded unequal petals; calyx-lobes ovate, 1 cm or more; stamens 10, free, with filaments which are woolly-haired below. Leaves 23–38 cm, with 5–10 pairs of pinnae, each with 8–12 pairs of oblong blunt leaflets 1.5–2 cm long and glaucous beneath. Pod 6–12.5 cm by 2.5 cm, beaked, woody. **Pl. 25.**

6 CASSIA Leaves once-pinnate. Herbs, shrubs or trees, without prickles. Stamens free, usually 10, but often some infertile or absent. *c.* 7 spp.

Trees

307 **C. fistula** L. INDIAN LABURNUM
India. Burma. In the Himalaya to 1400 m. Forests, shrubberies, open hillsides. Apr.–May.

A small nearly evergreen tree with greenish-grey bark, large pinnate leaves, and with long drooping lax clusters of bright yellow flowers.

Clusters 30-60 cm long; flowers 5 cm across; petals 5, obovate, veined. The 3 lowest stamens longer and curled, with large anthers; 4 or 6 lateral stamens smaller straight; 1 or 3 upper stamens short incurved and sterile. Leaves with 3–8 pairs of opposite ovate to elliptic leaflets 5–15 cm, pale silvery-haired below when young. Pod 40–60 cm, rounded in section, brown to glossy-black, not splitting, many-seeded.

An ornamental tree. Wood very durable, used for implements. Bark used for tanning, the ash as a mordant for dyeing; pod-pulp used to flavour tobacco. Parts of the tree have been used medicinally. **Pl. 24.**

Undershrubs or herbaceous plants

308 **C. occidentalis** L. COFFEE SENNA
Probably native of America; naturalized in India and the Himalaya. To 1400 m. Common in wasteland, cultivated areas; often gregarious. May–Aug.

A shrub to 2 m, or often an annual, with 3–5 pairs of lanceolate to elliptic acute leaflets, and a terminal cluster of yellow to orange-yellow flowers, each flower 2 cm across. Bracts white, tinged with pink, longer than the flower buds, soon falling. Leaflets mostly 3–6 cm long, hairy when young, but becoming hairless, very foetid when crushed. Pod 9–12.5 cm by 6–9 mm broad, erect, slightly curved, flattened.

Leaves, roots and seeds used medicinally.

309 **C. tora** L.
Probably native of S. America. In the Himalaya to 1500 m. Wasteland, cultivated areas. Aug.–Sep.

Commonly a gregarious foetid annual with usually 3 pairs of obovate leaflets, and solitary or paired orange-yellow axillary flowers *c.* 1 cm across borne from the axils of the leaves. Sometimes an undershrub to 120 cm. Petals unequal, obovate, 8–12 mm; stamens 7, unequal. Leaflets usually 2–4 cm long, hairy or hairless. Pod 10–20 cm by 2–6 mm broad, rounded, often curved.

Leaves and seeds used medicinally. **Pl. 24.**

7 **ERYTHRINA** Leaves with 3 leaflets. Flowers scarlet or deep crimson; petals very unequal, the standard much longer than the keel or wings. 3 spp.

310 **E. arborescens** Roxb.
Uttar Pradesh to Bhutan. China. Burma. 1500–3000 m. Shrubberies, cultivated areas. Aug.–Oct.

A low tree with few prickles on stems and leaf-stalks, with long erect spike-like clusters of drooping vivid scarlet flowers appearing in autumn with the large trifoliate leaves. Stalked flower-clusters to 38 cm; flowers with ovate pointed standard 4–5 cm, more than twice as long as the keel, wings shorter than keel. Leaflets 13–18 cm broad and long, brown-velvety when young, becoming hairless. Pod lanceolate, curved, brown-hairy, 15–23 cm long. **Pl. 25.**

311 **E. stricta** Roxb.
W. Nepal to S.W. China. India. Burma. To 1600 m. Forests; common in
cultivated areas of Nepal. Mar.–Apr.

A large tree with trunk and branches covered with sharp conical whitish
prickles, and with scarlet flowers in dense one-sided spike-like clusters
10–13 cm long occurring at the ends of stout leafless branches in spring.
Calyx spathe-like and split to the base (bell-shaped or top-shaped in 310,
312); standard oblong-lanceolate, 4–5 cm, twice as long as keel. Leaflets
to 18 cm, very broadly ovate, broader than long. Pod hairless, lanceolate
5–10 cm.
Flowers and bark used medicinally. **Pl. 25.**

312 **E. suberosa** Roxb. CORAL TREE
Kashmir to Bhutan. India. Burma. 900–1200 m. Forests, cultivated areas.
Mar.–Apr.

A medium-sized tree with spiny, corky, deeply fissured bark, and with
scarlet flowers appearing before or with the young leaves, in small short
crowded clusters at the ends of the branches. Flower clusters 5–10 cm;
standard 4–5 cm, oblong, keel scarcely half as long, wings minute.
Leaflets 3, broadly ovate, the terminal largest, 8–12 cm, all densely hairy
beneath often with reddish-brown hairs. Pod hairless, cylindric long-
pointed, somewhat constricted between seeds, 13–15 cm.

Wood very light, soft, spongy but fibrous and tough, used for making
small articles.

8 ROBINIA Leaves pinnate, with spiny stipules. Flowers white. 1 sp.

313 **R. pseudacacia** L. FALSE ACACIA
Native of C. & N. America. To 2400 m. Commonly planted by roadsides
in the W. Himalaya. Apr.–May.

A small deciduous tree with deeply longitudinally-fissured bark, with
pinnate leaves, and with pendulous clusters of white sweet-scented
flowers. Clusters 10–20 cm long; flowers to 2 cm; the standard rounded,
notched and reflexed, marked with a greenish-yellow spot, and little
longer than the wings and keel; calyx cup-shaped with triangular lobes,
hairy. Leaflets 7–21, oval, paler beneath, 2.5–5 cm; spines present on
branches. Pod 5–10 cm long, flat, brown, hairless.

9 BUTEA Leaves trifoliate. Flowers flame-coloured; standard recurved.
Pod splitting round the single seed, lower portion flat and not splitting. 2
spp.

314 **B. minor** Buch.–Ham. ex Baker
W. to E. Nepal. 300–2000 m. Rocks, steep slopes, river gorges. Apr.–May.

A small shrub to 1.5 m, with very large trifoliate leaves, and with
numerous flame-coloured flowers in long terminal and axillary spike-like
clusters. Flowers c. 2.5 cm; petals and calyx densely silky-haired outside;

standard acute, recurved; keel blunt, little curved, three times as long as the cup-shaped calyx. Leaflets broadly ovate, 15–38 cm, the terminal long-stalked, the two lateral leaflets asymmetrical, leathery, finely silky-hairy beneath; young branches at first silky-haired. Pod densely hairy, 5–8 cm by 2.5–3 cm. **Pl. 27**.

10 CAMPYLOTROPIS Leaves trifoliate. Flowers red; keel acute and much incurved. 2 spp.

315 **C. speciosa** (Royle ex Schindler) Schindler
Kashmir to Arunachal Pradesh. 1800–3000 m. Open slopes, shrubberies. Aug.–Sep.

A much-branched shrub to 1 m or more, with rather leathery trifoliate leaves, and with terminal and axillary spikes 3–15 cm, of many pink to deep red-purple flowers. Flowers *c*. 1.2 cm; calyx with linear teeth, hairy. Leaflets obovate with a wedge-shaped base and a fine bristly tip, mostly 1–2 cm long, paler and densely white-hairy beneath, the middle leaflet stalked. Pod ovate with persistent beak, to 8 mm, densely hairy, often with remains of keel. **Page 461**.

11 CROTALARIA Leaves simple or with 3 leaflets. Flowers yellow; keel much incurved and beaked. *c*. 20 spp., mostly sub-tropical.

316 **C. cytisoides** Roxb. ex DC. (*C. psoralioides* D. Don, *Priotropis c.* (Roxb. ex DC.) Wight & Arn.)
C. Nepal to Bhutan. S.W. China. Burma. 1200–1800 m. Shrubberies. Aug.–Sep.

An erect much branched deciduous shrub 1–2 m, with trifoliate leaves, and with dense spike-like clusters of bright yellow flowers. Petals hairless, *c*. 1 cm, the keel conspicuously up-curved and long beaked, the standard and wings shorter. Calyx tubular, with lanceolate unequal lobes, finely hairy. Leaves with 3 elliptic, hairless leaflets 3–8 cm. Pod flattened, narrowed at both ends, with a persistent style and short stalk, 2.5–4 cm. **Page 459**.

12 DESMODIUM Leaves trifoliate. Pod flattened and conspicuously jointed between the seeds. *c*. 18 spp., mostly sub-tropical.

317 **D. elegans** DC.
Afghanistan to S.W. China. 1200–3000 m. Forests, shrubberies, open slopes; common. Jun.–Aug.

A deciduous shrub with trifoliate leaves with broad leaflets, and with terminal spike-like clusters to 30 cm long of pink to dark purple flowers. Flowers 7–13 mm, borne on slender stalks; standard rounded; calyx 2–3 mm, with blunt lobes, shaggy-haired. Leaflets rounded to obovate mostly 2–4 cm, paler and hairy beneath, the middle leaflet long-stalked; young leaves often densely silvery-white-haired; stems 1.8–3 m. Pod 3–6 cm long by 6 mm broad, the lower margin conspicuously indented between the seeds, the upper nearly straight, hairless or densely silvery-haired. **Pl. 26**.

318 **D. multiflorum** DC.

Kashmir to S.W. China. 600–2400 m. Shrubberies, open slopes, cultivated areas. Jul.–Aug.

Differs from 317 in its smaller lilac flowers 7 mm long, its longer persisting bracts 5 mm (bracts 2–5 mm in 317) and its smaller densely brown-hairy pods 1.8–3 cm long by 4 mm broad. Young branches sharply angular; terminal leaflet often obtuse.

13 FLEMINGIA Leaves simple. Pod small, swollen, 2–seeded, splitting into 2 valves. Bracts large ovate, folded, enclosing flower clusters. 5 spp.

319 **F. strobilifera** (L.) Aiton (*Moghania s.*(L.) J. St. Hil. ex Kuntze)
Tropics and Sub-Tropics of Asia and America. To 2300 m. Sal forests, shrubberies. Sep.–Oct.

A variable erect shrub 1.2–3 m, with simple leaves, and with long terminal and axillary flower-clusters made conspicuous by the 2 ranks of large rounded pale green folded papery bracts which conceal the smaller white flowers. Bracts 1.2–2.5 cm long; flowers 2 or more in each bract, 9–10 mm; calyx hairy, shorter than petals. Leaves broadly lanceolate acute, silky-haired beneath, 5–13 cm. Pod 9–10 mm, densely hairy, concealed by bracts. **Page 459**.

14 LESPEDEZA Leaves usually trifoliate. Pod ovate or rounded, flattened, one-seeded, not splitting. *c.* 7 spp.

320 **L. gerardiana** Graham ex Maxim.
Kashmir to Bhutan. 1200–3300 m. Grassy slopes. Aug.–Oct.

A small undershrub with spreading annual woody shoots to 1 m, and many dense clusters of yellow flowers with purple tips, ranged along the unbranched stems. Clusters almost stalkless, axillary, about as long as the subtending leaf. Flowers several; petals 1–1.5 cm; calyx densely silky-haired, with bristle-like teeth much longer than the calyx-tube. Leaflets oblanceolate to linear-oblong, blunt with a conspicuous spine-like apex, densely silky-haired below, 1–2 cm; branches finely grey-haired. Pod silky-haired, *c.* 3 mm. **Page 459**.

320a **L. elegans** Cambess.
Pakistan to Kashmir. 1200–2400 m. Wasteland, cultivated areas. Aug.–Sep.

Like 320, but leaflets usually more than 5 mm wide (usually less than 5 mm wide in 320) obovate to elliptic 1–2.6 cm. Flowers purple, with a yellow keel *c.* 9 mm long.

15 PIPTANTHUS Leaves trifoliate; stipules fused. Flowers bright yellow; stamens free. Pod linear-lanceolate, not splitting. 1 sp.

321 **P. nepalensis** (Hook.) D. Don

Himachal Pradesh to S.W. China. 2100–3600 m. Forests, shrubberies. Mar.–May.

A distinctive laburnum-like shrub, with dark green shining bark, trifoliate leaves with narrow pointed leaflets, and large bright yellow flowers in short hairy terminal clusters. Flowers 2.5–3 cm; standard with rounded reflexed blade and a long stalk; calyx bell-shaped with large unequal blunt lobes, grey-hairy; bracts elliptic, densely woolly-haired. Leaflets lanceolate, 3–10 cm, dark green above, grey-hairy beneath at first but later hairless and shining; stems 3–4 m. Pod 7–13 cm, flat, long-stalked. **Pl. 27.**

16 PUERARIA Twining shrubs or herbs; leaves trifoliate. Flowers blue or purple. 4 spp.

322 **P. tuberosa** (Roxb. ex Willd.) DC.
Pakistan to C. Nepal. India. 300–2000 m. Shrubberies, streamsides. Mar.–Apr.

A wisteria-like deciduous woody climber, with trifoliate leaves with large leaflets, and dense spikes of numerous bright mauvish-blue flowers appearing when the plant is leafless. Spikes 15–40 cm long, on axillary branches; petals 1–1.8 cm; calyx densely silky. Leaves long-stalked; leaflets 10–20 cm, silky-haired when young and remaining thinly hairy beneath, the terminal broadly ovate long-pointed, the lateral leaflets unequal-sided. Pod flat, constricted between seeds, covered with bristly brown hairs, 5–8 cm.

The large underground tubers are used medicinally, and are fed to ponies. **Page 460.**
Wisteria sinensis (Sims) DC. is commonly grown in gardens in Kathmandu and Srinagar, and occasionally in villages in the hills.

17 ARGYROLOBIUM Herbaceous plants; leaves trifoliate. Flowers red; calyx 2-lipped. Pod linear, flattened. 2 spp.

323 **A. roseum** (Cambess.) Jaub. & Spach
Pakistan to W. Nepal. W. Asia. 1000–3200 m. Open slopes. Apr.–Aug.

A small much-branched prostrate herbaceous plant, with small pink or red flowers borne in axillary stalked few-flowered clusters longer than the trifoliate leaves. Petals to 1 cm; calyx 2-lipped, half to nearly as long as petals. Leaves shortly stalked; leaflets obovate, 4–10 mm. Pod 2–3 cm by 2–3 mm broad, hairy. **Page 460.**

18 MEDICAGO Leaves trifoliate. Flowers yellow or mauve; calyx-teeth equal. Fruit coiled or curved. 3 spp.

324 **M. falcata** L. SICKLE MEDICK
Afghanistan to C. Nepal. Temperate Eurasia. 2700–4000 m. Common fodder plant of irrigated meadows in drier areas. May–Aug.

An erect or spreading herbaceous perennial with trifoliate leaves, and short rounded, stalked, axillary clusters of small yellow flowers each 8–11 mm long. Leaflets mostly narrowly oblong 8–10 mm, toothed; stems much branched 30–80 cm. Pod strongly curved in a semi-circle, 1.3–2 cm long. **Page 460**.

325 **M. sativa** L. LUCERNE
Probably native of W. Asia and the Mediterranean region. 1500–3000 m. Cultivated in irrigated meadows in the W. Himalaya. Jun.–Aug.
Distinguished from 324 by its mauve flowers borne in stalked axillary clusters. Flowers 5–12 mm. Leaflets obovate to nearly linear 5–20 mm, toothed, and with adpressed hairs; stems erect 30–60 cm. Pod sickle-shaped, or in a loose spiral *c*. 5 mm across.

19 MELILOTUS Leaves trifoliate. Flowers white or yellow. Fruit short, straight. 3 spp.

326 **M. alba** Medicus ex Desr. WHITE MELILOT
Temperate Eurasia; introduced Pakistan to E. Nepal. S.E. Asia. Australia. America. 200–3300 m. Meadows. Jun.–Aug.
A rather slender erect annual 30–150 cm, with trifoliate leaves, and with long erect axillary spikes of tiny white flowers. Standard 3–5 mm long, the wings and keel shorter. Leaflets 1.3–2.5 cm, elliptic to obovate, blunt, with acute teeth. Pod ovoid, *c*. 4 mm, with a network of raised veins, hairless.

327 **M. officinalis** (L.) Pallas RIBBED MELILOT
Pakistan to Kashmir. Temperate Eurasia. 2700–4000 m. Common in irrigated meadows in Ladakh. Jun.–Aug.
Differs from 326 in having yellow flowers with the wings longer than the keel. Flowers 5–7 mm. Pod *c*. 4 mm, with transverse ridges.
Plant used medicinally. **Page 460**.

20 PAROCHETUS Flowers violet-blue; standard broad obovate, stalked, much longer than keel or wings. Leaves trifoliate. Pod swollen. 1 sp.

328 **P. communis** Buch.–Ham. ex D.Don
Himachal Pradesh to S.W. China. India. S.E. Asia. 1000–4300 m. Cultivated areas, damp places, open slopes, forests. May–Nov.
A creeping clover-like plant, with trifoliate leaves, and rather conspicuous deep violet-blue usually solitary or paired flowers borne on stalks longer than the leaves. Flowers 1.3–2.5 cm long; standard erect, stalked; calyx bell-shaped, 5–lobed, brown-haired. Leaflets obcordate 8–20 mm; stems slender, rooting at the nodes. Pod 2–2.5 cm, straight, hairless. **Pl. 28**.

21 THERMOPSIS Petals all stalked; stamens free. Leaves trifoliate, with leafy stipules. 3 spp.

329 **T. barbata** Royle
Kashmir to S.W. China. 3000–4600 m. Open slopes. May–Jul.

95

A very distinctive herbaceous perennial with a terminal cluster of dark chocolate-brown flowers, and usually densely rusty-haired, long-stalked trifoliate leaves. Flowers 2–3 cm long; calyx densely hairy, lobes 1 cm long. Leaves with 3 lanceolate to elliptic leaflets usually 1.5–3 cm long, and leafy stipules as long; stems several, tufted, erect 15–45 cm. Pod oblong 2.5–3.5 cm, hairy. **Pl. 26**.

330 **T. inflata** Cambess.
Kashmir to Himachal Pradesh. Tibet. 4000–5000 m. Stony mountain slopes in drier Tibetan areas; Ladakh & Lahul. Jun.

Like 329, but with yellow flowers, 2–3 at each node and forming a terminal cluster. Flowers c. 2 cm long; calyx half as long, hairy; bracts ovate, hairy. Leaflets oval to obovate 1–2.5 cm; stems several, mostly 10–15 cm. Pod inflated, broadly oblong, 4–5 cm.

22 TRIFOLIUM Leaves trifoliate. Flowers numerous in rounded heads. Pod small, more or less enclosed by calyx. 2 spp.

331 **T. repens** L. WHITE or DUTCH CLOVER
Throughout the Himalaya. Temperate Eurasia. N. Africa. 1500–2500 m. Grassy places. Throughout the summer.

Flowers white or pinkish, borne in a globular head, on a slender stalk longer than the leaves. Flower-heads 1.5–2.5 cm across; flowers 8–10 mm; calyx pale. Leaves trifoliate, long-stalked; leaflets obovate 5–20 mm, toothed; stems slender creeping, rooting at the nodes. Pod linear c. 5 mm, with 3–4 seeds.

23 TRIGONELLA Leaves trifoliate. Flowers clustered. Pod straight or curved, flattened; seeds in 2 rows. 6 spp.

332 **T. emodi** Benth.
Afghanistan to Bhutan. 2100–4500 m. Stony slopes, river gravels; quite common. Jun.–Sep.

An erect or spreading perennial, with trifoliate leaves and with small yellow flowers in rounded axillary stalked clusters 1–2 cm across. Flowers 6–10 mm long; petals at least twice as long as calyx. Leaflets obovate 7–23 mm, finely toothed, hairy beneath; stems several 15–60 cm. Pod straight, to 15 mm, flattened, transversely veined, hairless. **Page 461**.

24 CARAGANA Spiny shrubs; leaves pinnate, ending in a hardened spine. Petals much longer than calyx which has 5 equal lobes. Pod oblong, swollen, not constricted between seeds. c. 7 spp.

Leaflets crowded, more or less digitate

333 **C. versicolor** (Wallich) Benth. (*C. pygmaea* auct. non DC.)
Afghanistan to W. Nepal. Tibet. China. 3600–4800 m. Alpine steppe country; dominant over large areas in Nepal north of Dhaulagiri. Jun.–Jul.

A densely branched, spreading or erect, very spiny shrub to 60 cm, with solitary yellow often orange-flushed flowers ranged along the stems, and with tiny leaflets arising from scaly dwarf shoots in the axils of spines. Flowers to 1.5 cm, short-stalked; calyx hairless, with broadly triangular lobes. Leaflets 4, digitately arranged, narrow oblanceolate, mostly 5 mm; stipules of 3 spines, the longest to *c*. 8 mm. Pod *c*. 3 cm, hairless.

Used as firewood in treeless country.

334 C. brevifolia Komarov
Kashmir to Uttar Pradesh. 3300–4500 m. Alpine steppe country. Common in Ladakh. Jun.–Aug.

Very like 333, but flowers larger to 2 cm, yellow streaked with red. Leaflets mostly 1 cm. Pod 3 cm, hairless. **Pl. 27.**

Leaves pinnate
a *Flowers nearly stalkless*

335 C. gerardiana Royle (incl. C. rossii Vassilicz., C. pulcherrima Vassilicz.)
Pakistan to C. Nepal. 3000–4100 m. Dry open slopes in Tibetan border areas. May–Jul.

A densely branched, very spiny shrub 90–120 cm, often forming tight clumps in dry steppe country, with downy pinnate leaves, and with almost stalkless, usually solitary yellow flowers, ranged along the stems which have numerous large spines. Flowers *c*. 2 cm; calyx densely hairy, half as long as the petals. Leaflets 8–12, oblanceolate, 5–7 mm; stipules not spiny but often persisting and encircling the stem. Spines 1–4 cm, formed from the rachis of old leaves. Pod 1.2–2 cm, with dense grey hairs.

Used as firewood in treeless country. **Page 461.**

336 C. sukiensis C. Schneider (*C. hoplites* Dunn, *C. nepalensis* Kitam.)
Uttar Pradesh to Bhutan. 2400–3600 m. Thickets, shrubberies; prominent in Langtang valley, C. Nepal. Jun.–Jul.

Like 335, but differing in the ear-like extensions to the wing petals, and by the stipules which are papery, free and soon fall (stipules leathery, clasping, persistent in 335). A small shrub to 2.5 m, with 1–2 orange-yellow, almost stalkless flowers born in the upper leaf axils. Leaflets elliptic, more than 12. Pod to 4 cm, becoming hairless.

337 C. jubata (Pallas) Poiret
C. Nepal. Bhutan. Tibet. China. Mongolia. Siberia. 3300–4000 m. Drier areas; in Nepal confined to the upper Kali Gandaki, Marsiandi and Shiar Khola (not nearly so common as 335). Jun.–Jul.

Like 335 with solitary yellow flowers often reddish-streaked, but leaves and branches covered with dense woolly hairs. Flowers to 2.5 cm; calyx densely hairy, with bristle-like hairy teeth. Leaflets *c*. 12, narrow-elliptic, densely woolly-haired, 6–10 mm.

b *Flowers long-stalked*

338 **C. brevispina** Royle
Kashmir to C. Nepal. 2400–3200 m. Shrubberies, forests. May–Jun.

An erect spiny shrub 2–3 m, with pinnate leaves, and with stalked clusters of few yellow, often red-flushed flowers. Flowers usually 2 borne on a stalk to 2 cm; bracts linear. Standard 2–2.5 cm long, rounded with reflexed sides, thicker and firmer than the wings and straight keel; calyx 7–9 mm, finely hairy, lobes short with awn-like tips. Leaflets 8–16, ovate to oblong, 8–20 mm, with a fine point, silky-haired beneath. Old leaves hardening into long thick spines to 4 cm with small basal spines; young branches downy. Pod hairless, 4–5 cm.

The young flowers are eaten by children. **Pl. 28**.

25 **COLUTEA** BLADDER SENNA Shrubs with pinnate leaves. Flowers yellow. Fruits inflated, bladder-like and papery. 2 spp.

339 **C. multiflora** Shap. ex Ali
C. & E. Nepal. 2100–2900 m. Shrubberies. Jul.–Aug.

A spineless deciduous shrub to 2 m, with pinnate leaves, lax axillary clusters of yellow to reddish-yellow flowers, and very distinctive large inflated pods. Flowers *c.* 1.5 cm; standard rounded as broad as long; keel blunt up-curved; calyx hairless or very finely hairy, with triangular teeth. Leaves 6–8 cm, with about 7 pairs of widely-spaced, elliptic blunt leaflets each *c.* 1 cm, nearly hairless. Pod ellipsoid, 5–7 cm by 2 cm broad, stalked and beaked. **Pl. 29**.

26 **INDIGOFERA** Spineless shrubs with pinnate leaves, and usually with forked hairs. Flowers in axillary racemes; calyx oblique; stamens with 9 filaments fused and 1 free. Ovary with short style and knob-like stigma. Pod linear, many-seeded. *c.* 9 spp.

Leaflets mostly less than 15 mm long
a *Bracts minute, shorter than flower-buds*

340 **I. cylindracea** Graham ex Baker
Uttar Pradesh to Sikkim. 2100–2700 m. Forests, open slopes. May–Aug.

Usually a spreading shrub with ascending hairless branches, with pinnate leaves, and with lax spike-like clusters of pink or red flowers each *c.* 1.3 cm. Flower-clusters short-stalked, at length 8–10 cm long. Leaflets many, paired, elliptic blunt, 8–13 mm, green above and pale beneath with few short adpressed hairs. Pods swollen 2–4 cm long.

341 **I. heterantha** Wallich ex Brandis (*I. gerardiana* Wallich)
Afghanistan to Bhutan. China. 1500–3000 m. Shrubberies; forms dense thickets in W. Himalaya. May–Jun.

A small or large shrub to 2.5 m, covered with bristly white hairs, and with conspicuous purple or pale pink flowers in erect often almost stalkless spike-like clusters 2.5–5 cm long. Flowers mostly 6–10 mm; calyx bristly-haired, with lobes as long as tube; bracts minute. Leaves and leaflets very variable, leaflets elliptic to oblanceolate, mostly 4–12 mm, with white hairs. Pod 1.3–2.5 cm, straight, hairless. **Pl. 27**.

b *Bracts longer than flower buds*

342 **I. dosua** Buch.–Ham. ex D. Don
Himachal Pradesh to Arunachal Pradesh. 600–3000 m. Open slopes. Apr.–Jul.

Often a low shrub, or sometimes up to 5 m, densely hairy, with deep pink or red flowers in long slender erect spikes 2.5–7 cm long, and densely silky-haired calyx. Flowers 1–1.3 cm long; standard densely hairy on back; bracts narrow-lanceolate long-pointed, silky-haired, much longer than the flower buds. Leaflets opposite, elliptic, usually 6–15 mm long, densely adpressed-hairy. Pod *c.* 2.5 cm, straight, hairy.

Leaflets mostly more than 15 mm long
a *Bracts linear or lanceolate, as long as or longer than flower buds*

343 **I. atropurpurea** Buch.–Ham. ex Hornem.
Kashmir to Bhutan. India. China. 700–3000 m. Shrubberies. Jul.–Sep.

A large nearly hairless shrub 2–3 m, with pinnate leaves, and with usually deep purple flowers in long slender erect spikes 10–30 cm, as long as or longer than the leaves. Flowers very numerous, 6–9 mm long. Leaflets many, oval to elliptic, blunt, 2.5–4 cm, becoming hairless. Pod 2.5–4 cm, 6–10–seeded, hairless. **Page 461**.

b *Bracts boat-shaped, encircling flower buds*

344 **I. hebepetala** Benth. ex Baker
Pakistan to Bhutan. 1500–3000 m. Shrubberies. Jun.–Aug.

A tall shrub like 342, but bracts differing, boat-shaped and encircling the flower buds. Flowers deep red, in long erect lax clusters 10–15 cm. Petals 9–12 mm long; standard hairy outside; calyx finely hairy. Leaves with 7–17 ovate leaflets 2.5–4 cm long; twigs pale brown, sparsely hairy when young. Pod 2.5–4 cm, straight, hairless.

345 **I. pulchella** Roxb.
Kashmir to Bhutan. 300–1500 m. Forests, open slopes; common in *Pinus roxburghii* and *Shorea robusta* forests. Feb.–Apr.

Like 344 and not easily distinguished from it in some forms, but flowers larger, pink fading to violet, often flowering on leafless branches. Flower-clusters 5–15 cm long; petals 1–1.8 cm; calyx densely white-haired; bracts boat-shaped longer than the flower buds. Leaflets broadly elliptic 1.5–2.5 cm, thick, glaucous beneath, and with adpressed hairs. Pod straight, hairless, 2.5–4 cm.

27 SOPHORA Shrubs; leaves pinnate. Flowers cream, yellow, or blue; stamens free. Pod constricted into 1–seeded units. 3 spp.

346 **S. moorcroftiana** (Benth.) Benth. ex Baker

Ladakh. C. Nepal. Tibet. 2800–3900 m. Stony slopes; dry Tibetan border-lands. May–Jun.

A much-branched spiny shrub to 1 m or more, with pinnate leaves, and with stalked axillary clusters of blue or sometimes yellow flowers. Distinguished by its long 5–6–seeded pod, 7–10 cm, which is constricted between the seeds. Flowers *c*. 1.5 cm; standard longer than keel or wings; calyx 6–7 mm, grey silky-haired. Leaves *c*. 5 cm; leaflets many, ovate *c*. 6 mm, silky-haired; lateral stems spine-tipped.

Used for firewood in treeless country.

347 **S. mollis** Graham ex (Royle) Baker
Pakistan to C. Nepal. 1200–2000 m. Rocks in drier valleys. Mar.–Apr.

An erect deciduous shrub to 2 m, differing from 346 in being spineless, with crowded axillary clusters of bright yellow flowers often appearing on bare stems. Flower-clusters 5–10 cm long; petals 1.5–2.6 cm, nearly equal; calyx grey-hairy, with 5 short blunt lobes. Leaflets many, mature leaflets ovate to elliptic blunt, 1.2–2.5 cm long, sparsely hairy; stems at first finely grey-hairy. Pod with 4–10 deep constrictions, 4–winged, 7.5–13 cm long. **Page 463**.

28 CICER Leaves pinnate, with a terminal tendril. Calyx 5–lobed. Fruit ovate or oblong, usually with 2 wrinkled seeds. 2 spp.

348 **C. microphyllum** Benth. (*C. soongaricum* auct. non Stephan)
Afghanistan to W. Nepal. 3300–4800 m. Dry slopes in the Tibetan border areas, and on irrigated ground. Jun.–Jul.

A spreading glandular-hairy perennial, with pinnate leaves ending in a coiled tendril, and with solitary or paired, stalked, axillary blue, purple to white flowers. Flower-stalks ending in a long bristle; flowers 2–2.8 cm long; calyx hairy, about a third as long as the petals, with more or less equal lanceolate lobes. Leaflets many, in widely-spaced pairs, ovate-wedge-shaped with coarsely toothed apex, 5–8 mm long; stipules palmately lobed. Pod 2–3 cm, inflated, explosive, beaked and densely hairy. **Page 462**.

29 LATHYRUS PEA, VETCH Like *Vicia*, but stems usually winged or angled; leaflets usually fewer, parallel-veined. Style flattened, with a tuft of hairs on the upper side only. 5 spp.

Flowers yellow or orange

349 **L. laevigatus** (Waldst. & Kit.) Gren. (*L. luteus* Baker)
Pakistan to W. Nepal. 2100–3300 m. Forests. May–Jun.

An erect perennial with axillary stalked clusters of bright yellow or orange flowers, and with large pinnate leaves. Standard 2.3–3 cm long; wings and keel little shorter; calyx hairless. Leaves pinnate, with 4–10 leaflets, ending in a bristle; leaflets elliptic with a fine point, 5–9 cm long; stipules leafy; stems 1–2 m. Pod oblong 5–7 cm by 7–8 mm broad, hairless. **Page 462**.

350 **L. pratensis** L. MEADOW VETCHLING
Pakistan to C. Nepal. Temperate Eurasia. 1800–3000 m. Forests, meadows, shrubberies. Jun.–Jul.

Flowers yellow, 3–12, borne on stalks much longer than the leaves, and with leaves with 2 narrow leaflets, a terminal branched tendril, and paired arrow-shaped leafy stipules. Flowers 1–1.8 cm long; calyx with unequal narrow lobes. Leaflets linear-lanceolate, 1–4 cm; stipules broader, two-lobed at base, 1–3 cm long; stems angled, 30–90 cm. Pod 3–4 cm, mostly hairless. A variable species.

Flowers usually blue, reddish, purple to white

351 **L. humilis** (Ser.) Fischer ex Sprengel (*L. altaicus* Ledeb.)
Pakistan to Uttar Pradesh. C. Asia. 1800–4000 m. Open slopes, shrubberies. Jun.

A more or less erect perennial with long-stalked clusters of 2–6 reddish- or bluish-purple flowers, and with pinnate leaves with elliptic leaflets, and a terminal tendril. Flowers 1.5–2 cm long; calyx-lobes unequal. Leaflets 6–10, broadly to narrowly elliptic, 2–3 cm, hairless or sparsely hairy; stipules leafy, half-arrow-shaped 1–1.5 cm; tendrils mostly simple; stems not winged, 15–30 cm. Pod 3–5 cm long, hairless. **Pl. 30.**

352 **L. sativus** L. CHICKLING PEA
Worldwide. Cultivated in Himalaya to 3000 m. Feb.–May.

A hairless annual with winged stems 20–100 cm. Leaves with 2 narrow-lanceolate to linear leaflets usually 2–6 cm, and branched tendrils; tendrils of upper leaves usually 3–branched. Flowers solitary axillary stalked, usually blue, 1–1.5 cm. Calyx-lobes nearly equal, longer than the tube. Pod 2–4 cm, with 2 wings on the upper margin, conspicuously beaked.

Grown as an animal fodder crop; seeds not edible.

30 LENS Like *Vicia* but not climbing. Pod with 1–2 flattened disk-like seeds. 1 sp.

353 **L. culinaris** Medicus (*Ervum lens c.* L., *L. esculent a.* Moench) LENTIL
Origin unknown. Widely cultivated at low altitudes in the terai and outer foothills where it flowers in January, and is harvested in March. Also common as a field weed at higher altitudes (to 3300 m) where it flowers in August.

A hairy branched annual with axillary stalked clusters of 1–4 small pale purple flowers 5–6 mm. Calyx-lobes linear, nearly as long as the petals, hairy. Leaves with 4–7 pairs of narrow-oblong leaflets 5–15 mm, and terminating in a bristle or short tendril; stem to 40 cm. Pod smooth, hairless, more or less rectangular 8–14 mm; seeds 2.

An important crop yielding the pulse called 'Masur'. Several other 'dals' are grown in the hills of the Himalaya as winter crops. They include:

Vigna mungo (L.) Hepper (*Phaseolus mungo* L.) 'MUNG'. Native of India; cultivated to 2500 m.

Lablab purpureus (L.) Sweet (*Dolichos lablab* L.) Widely cultivated in the Tropics; and in the Himalaya to 2500 m.

Cajanus cajan (L.) Huth (*C. indicus* Sprengel) 'ARHAR'. Probably native of Africa; widely cultivated in the Himalaya to 2000 m.

Cicer arietinum L. GRAM. CHICKPEA, 'CHANNA'. Origin unknown; cultivated in the Himalaya to 1300 m.

31 PISUM Like *Lathyrus* but stipules leafy, larger than leaflets; tendrils branched; stems not winged. Calyx-lobes leafy, unequal. 1 sp.

354 **P. sativum** L. PEA
Native of S. Europe. W. Asia. To 4400 m. Cultivated in drier Tibetan borderland areas in the Himalaya. Jun.–Jul.

A hairless glaucous climbing annual, with large white axillary flowers, or with standard lilac or red-purple, and with pinnate leaves ending in a branched tendril. Flowers 1–3, borne on a stalk shorter than the subtending leaf; petals 1.6–3 cm long. Leaves with 2–8 oval leaflets and larger and broader oval-oblong toothed stipules usually 4–6 cm; stems to 1 m. Pod 4–6 cm by 1–1.7 cm broad.

Commonly cultivated for its seeds which are used as a vegetable or pulse; also grown for animal fodder.

32 VICIA VETCH Differing from *Lathyrus* in the style which is hairless, or equally hairy all round, or with a tuft of hairs on the lower side only. Stem rounded in section; leaflets often many. *c.* 8 spp.

355 **V. bakeri** Ali (*V. sylvatica* auct. non L.)
Pakistan to C. Nepal. 1800–3000 m. Shrubberies, open slopes. Jul.–Sep.

An annual with stalked axillary erect clusters of pale lilac flowers, and with pinnate leaves with tendrils. Flowers many, drooping, 1–1.5 cm long; calyx-lobes very unequal. Leaflets many, 1.5–2 cm, oblong-lanceolate with a very fine point, hairless or nearly so above; tendrils branched or entire; stipules lanceolate; stem to 1 m, or more. Pod *c.* 4 cm, hairless. **Page 462.**

356 **V. tenuifolia** Roth
Pakistan to C. Nepal. Temperate Eurasia. 1500–2700 m. Field weed. Apr.–Jun.

Like 355 with blue to violet flowers, but a climber with longer and narrower linear-oblong leaflets to 3 cm by 2–4 mm broad (leaflets 4–9 mm broad in 355). Pod to 3 cm.

33 APIOS Climbing herbs; leaves with 3–7 leaflets. Flowers reddish; keel long. Pod linear, flattened. 1 sp.

357 **A. carnea** (Wallich) Benth. ex Baker
Uttar Pradesh to S.W. China. S.E. Asia. 1200–2400 m. Forests, shrubberies. Aug.–Sep.

A tall climbing perennial, with large pinnate leaves, and long-stalked clusters of reddish flowers distinguished by their narrow incurved keels which are longer than the standard and shorter wings. Flower clusters 15–23 cm long; flowers 2–2.5 cm long; standard obovate, streaked; calyx hairless. Leaves usually with 5, or rarely 3, large ovate-acute leaflets 8–12 cm, paler beneath. Pod 10–18 cm by 6 mm wide, hairless.

The seeds are used as beads, but they are poisonous. **Page 463**.

34 ASTRAGALUS Distinguished from related genera by the tubular calyx with 5 short teeth or lobes, by the blunt-tipped keel, and by the pod which is often longitudinally divided into 2 chambers. A very large and important but difficult genus; particularly common in the drier areas of the Himalaya. c. 36 spp.

Small spiny shrubs or shrublets forming low clumps or mats
(species 358-361 are very similar and difficult to distinguish)

358 **A. candolleanus** Royle ex Benth.
Pakistan to C. Nepal. 2700–4500 m. Alpine slopes; common in Chamba and some inner valleys of Nepal. May–Jun.

A low compact shrublet, with yellow flowers in dense stalkless clusters, set amongst pinnate leaves and long spines formed from the rachis of old leaves. Flowers 1.5–2.5 cm long; standard longer than the wings and the much shorter keel; calyx often reddish, densely silky-haired. Leaves 5–10 cm, with many oblong to ovate leaflets 3–8 mm; rachis of old leaves becoming woody and spiny. Pod 1.5–2.5 cm, with spreading silky hairs. **Pl. 28**.

359 **A. grahamianus** Royle ex Benth.
Pakistan to Uttar Pradesh. 1500–3300 m. Open slopes, stony ground; common in Kashmir and Lahul. May.–Aug.

A low very spiny shrub, very like 358 but differing in having a more spreading form with branches spreading to 60 cm, and with a rather different distribution in the W. Himalaya. Flowers yellow, 2–3 cm long; calyx c. 1.5 cm, sparsely hairy, with linear lobes. Leaves small, mostly 2–4 cm, with 8–14 leaflets each 3–5 mm, ovate-blunt with a fine point; spines needle-like slender. Pod c. 1.5 cm, with silky adpressed hairs. **Pl. 29**.

360 **A. zanskarensis** Benth. ex Bunge
Kashmir. 3000–4300 m. Stony slopes; common in Ladakh and Zanskar. Jun.–Aug.

A spiny shrublet very like 358 forming compact rounded clumps but restricted to the dry areas of the N.W. Himalaya. Flowers yellow 1.8–2 cm long; standard longer than wings or keel; calyx densely silky-haired. Leaflets 20–26, oblong-blunt 6–10 mm, with short brown hairs.

361 **A. oplites** Benth. ex R. Parker
Kashmir to C. Nepal. Tibet. 3000–4500 m. Stony slopes in the Tibetan borderlands. Jul.

Similar to 358, but forming large mats to 1 m across with branches radiating from the centre. Flowers yellow, in clusters of 2–5; standards hairless 2.5–2.8 cm long; calyx with spreading black or white hairs. Leaves 7–18 cm long; leaflets 20–40, elliptic-oblong to obovate, blunt or notched, 5–11 mm.

Spineless erect, spreading or prostrate herbaceous plants
a *Flowers mauve, purple, pink or white* (see also 367)

362 **A. donianus** DC.
W. Nepal to Bhutan. 2900–4500 m. Rocks, grassland, open slopes. May–Jul.
  A prostrate perennial with small pinnate leaves, and with one or several purple flowers borne on stalks much longer than the subtending leaves. Flowers 1.5–1.8 cm long; calyx bristly-haired, with linear lobes. Leaves mostly 2–3 cm; leaflets 3–6 mm, elliptic to obovate, hairy beneath, silky-haired when young. Pod inflated, beaked, hairy, 1–2 cm.

363 **A. himalayanus** Klotzsch
Pakistan to E. Nepal. 2400–4500 m. Open slopes; common in Kashmir. Jun.–Sep.
  An erect or spreading perennial, with rather dense rounded clusters of lilac or mauve flowers, borne on stalks usually longer than the subtending leaves. Flowers often many in each cluster, each 1–1.5 cm long; calyx c. 5 mm, with black and white hairs. Leaves usually 3–5 cm; leaflets many, 4–10 mm, oblong, blunt or notched, pale green with white hairs; stems 30–60 cm, with adpressed hairs. Pod 9–13 mm, with black hairs. **Page 463**.

364 **A. strictus** Graham ex Benth.
Pakistan to Bhutan. Tibet. 2100–5000 m. Open slopes, rocks; common. May–Aug.
  Like 363, but flowers smaller and in globular clusters 1–1.5 cm across, borne on stalks shorter than the pinnate leaves. A much smaller rather tufted perennial with spreading branches 10–15 cm. Flowers 8–10 mm long, pale purple to pinkish; calyx with black hairs. Leaves 2–4 cm; leaflets numerous, oblong-blunt 2–8 mm, with adpressed white hairs. Pod c. 8 mm, with adpressed black and white hairs.

365 **A. leucocephalus** Graham ex Benth.
Afghanistan to C. Nepal. 1500–4700 m. Stony slopes. May–Aug.
  Flowers pale pink, or less commonly yellow, in dense globular silky-haired clusters, borne on long stalks longer than the densely woolly-haired upper leaves. Flowers c. 6 mm long; calyx 4–5 mm, with dense long, spreading white hairs. Leaves usually 2.5–5 cm long; leaflets oblong, very densely grey-woolly, 3–7 mm; stems many, densely tufted, spreading, 7–20 cm. Pod 4–5 mm, finely hairy.

b *Flowers yellow* (see also 365)
  i *Prostrate perennials*

366 **A. rhizanthus** Royle ex Benth.
Pakistan to Himachal Pradesh. 3000–3800 m. Stony slopes; common in

Ladakh and Lahul. Jun.–Aug.

Flowers bright yellow, in a dense stalkless cluster at the centre of a rosette of radiating pinnate leaves pressed to the ground. Flowers *c.* 2 cm; standard longer than wings and keel; calyx with unequal lobes, the lower longest to 6 mm. Leaves 10–15 cm; leaflets many, oblong 5–10 mm, bluish-green; stipules obovate, ciliate. Pod 1.2–2 cm, densely silky-haired. **Pl. 28**.

ii *Erect perennials to 50 cm or more*
 x *Leaflets usually less than 2 cm long*

367 **A. chlorostachys** Lindley
Pakistan to Bhutan. 1800–3700 m. Shrubberies, open slopes. Jul.–Aug.

An erect perennial, with stalked axillary spikes of numerous yellow-green, or pinkish to white, drooping flowers borne on stalks longer than the pinnate leaves. Flowers 1–1.5 cm; calyx tubular, hairy, about half as long as the petals, lobes very short. Leaves mostly 5–10 cm; leaflets oblong-blunt, mostly 1–2 cm, finely hairy at least beneath; stem 30–150 cm. Pod *c.* 1.5 cm, narrowed to a stalk and with a curved beak.

368 **A. concretus** Benth.
Kashmir to Bhutan. 2700–3600 m. Shrubberies, open slopes. Jun.–Jul.

Like 367, but with yellow flowers in long-stalked axillary clusters on erect leafy stems. Flowers 1.2–1.5 cm; wings and keel almost equal; calyx 6–7 mm, hairless, with minute hairy lobes. Leaves 7–15 cm; leaflets oblong to lanceolate-blunt, usually 1–1.5 cm, with adpressed hairs beneath; stipules elliptic, fused below, 1.2–2 cm. Pod *c.* 2.5 cm including stalk and beak, hairless.

369 **A. floridus** Benth. ex Bunge
C. Nepal to S.E. Tibet. 4200–4700 m. Meadows, open slopes. Jun.–Aug.

Flowers yellow, in rounded or oblong many-flowered clusters, borne on stalks 2.5–5 cm which are covered with black silky hairs. Flower clusters several, forming a flat-topped inflorescence. Flowers 1–1.8 cm; calyx with distinctive black hairs and narrow curved lobes. Leaves 4–8 cm; leaflets oblong-elliptic, 8–20 mm, glaucous, thinly covered with adpressed silvery bristles beneath; stems erect, moderately stout, bristly-haired, 15–30 cm. Pod c. 1.3 cm, with dense black silky hairs; stalk of pod as long as persistent calyx. **Pl. 28**.

xx *Leaflets mostly 2 cm or more*

370 **A. frigidus** (L.) A. Gray
Pakistan to C. Nepal. Temperate N. Hemisphere. 2700–4300 m. Open slopes. Jul.–Aug.

Flowers yellow, in lax one-sided clusters borne on stalks usually longer than the subtending leaves. Flower clusters lengthening to 5–8 cm; flowers c. 2 cm; standard longer than wings and keel; calyx somewhat inflated, tubular to 1 cm, hairless or hairy. Leaves 10–15 cm; leaflets large, oblong to ovate, 2–5 cm, becoming hairless above and sparsely hairy beneath; stipules free, elliptic 1–2 cm; stems hairless, to 1.3 m. Pod 3 cm including long stalk and beak, densely hairy.

371 **A. stipulatus** D. Don ex Sims
C. Nepal to Bhutan. 2100–3400 m. Open slopes. Jul.–Aug.

A robust herbaceous perennial with stout hollow hairless stems, readily distinguished by its large ovate stipules which encircle the young leaves and flower buds. Flowers bright yellow, 1.5 cm, numerous in a one-sided long-stalked spike 8–15 cm; calyx tubular, covered with fine grey hairs; bracts lanceolate, papery. Leaves to 30 cm; leaflets numerous, oblong-blunt, hairless, 2–4 cm. Pod swollen, curved and narrowed to a beak, hairless, *c.* 2.5 cm. **Page 463**.

35 OXYTROPIS A large and difficult genus, very close to *Astragalus* and only distinguished by a minute botanical character: the presence of a beak at the apex of the keel. Herbaceous perennials. *c.* 18 spp.

Flowers 15 mm long or less
a *Leaves densely white- or silvery-grey-haired*

372 **O. cachemiriana** Cambess.
Pakistan to Kashmir. 2400–4000 m. Stony slopes. Jun.–Aug.

Flowers pinkish-purple, in globular heads *c.* 2 cm across, borne above densely hairy pinnate leaves. Flowers *c.* 12 mm; calyx with dense spreading pale hairs, and bristle-like teeth 4–5 mm long. Leaves 4–5 cm, with many pairs of elliptic entire leaflets 5–10 mm long, with spreading white hairs; stipules lanceolate, encircling the stem; flowering stems several, spreading, to 10 cm, with black and white spreading hairs. Pod 11–13 mm, papery, inflated, densely hairy.

373 **O. williamsii** Vassilicz.
W. & C. Nepal. 2400–4300 m. Riverside gravels, open slopes; common in upper Kali Gandaki valley. May–Aug.

Differs from 372 in being a densely tufted plant, with silvery-grey leaves, stems and calyces, and being restricted to Nepal. Flowers violet, borne in many globular heads 1–2 cm across, on stalks longer than the leaves. Petals 6–8 mm; calyx densely silvery-haired, teeth *c.* 3 mm. Leaves 5–10 cm, with numerous elliptic leaflets 5–10 mm. Pod globular, inflated, *c.* 1 cm, densely woolly-haired. **Pl. 30**.

b *Leaves green, hairy*

374 **O. lapponica** (Wahlenb.) Gay
Pakistan to Kashmir. Temperate Eurasia. 3000–4500 m. Open slopes; common in Ladakh. Jun.–Aug.

Flowers mauve, in ovoid clusters 1–3 cm long, borne on long stalks usually longer than the tuft of basal leaves. Flowers *c.* 1 cm; calyx with dark hairs and bristle-like teeth. Leaves 5–15 cm, with many pairs of narrow-elliptic leaflets 5–15 mm; aerial stems well developed. Pods in dense clusters, each to 1.5 cm long with a persistent beak. **Pl. 30**.

375 **O. mollis** Royle

Pakistan to C. Nepal. 2700–3600 m. Stony slopes; common in Kashmir and Lahul. May–Jul.

Like 374, but with aerial stems very short and flowers larger 9–16 mm, often purple fading to blue, in long-stalked clusters. Leaflets ovate, elliptic or lanceolate, with an acute tip, 7–17 mm, silky-haired. Pod to 2 cm, hairy.

Flowers more than 15 mm long

376 **O. microphylla** (Pallas) DC.

Pakistan to C. Nepal. Tibet. C. Asia. 2700–4700 m. Stony slopes of the Tibetan borderlands. May–Aug.

Distinguished by its tiny leaflets up to 8 mm long by 2–4 mm broad in clusters, and absence of aerial stem. Flowers pink-purple, in stalked clusters; petals *c*. 2 cm; calyx hairy and with brown glandular swellings, calyx teeth 2–3 mm. Leaves to 18 cm; leaf-stalk with white or yellowish hairs; base of stem with persistent leaf-stalks surrounded by dense woolly white hairs. Pod to 2.5 cm, with glandular swellings.

36 CHESNEYA Prostrate herbs; leaves pinnate. Calyx 2–lipped; standard longer than wings and keel. Pod linear-oblong. 3 spp.

377 **C. cuneata** (Benth.) Ali (*Calophaca c.* (Benth.) Komarov)

Pakistan to Kashmir. 2400–4000 m. Stony slopes; common in Ladakh. Jun.–Sep.

A tufted perennial with a thick woody tap-root, with short crowded stems with hairy pinnate leaves, and with short-stalked rather large pink-purple flowers. Flowers 1–4 in each cluster; standard 2–3 cm, hairy outside; calyx cylindrical *c*. 1.8 cm, velvety-haired. Leaves 2–5 cm; leaflets oblong to obovate, 5–16 mm, with blunt or notched apex, velvety-haired; stem to 15 cm. Pod 4–5.7 cm, becoming hairless. **Pl. 29**.

378 **C. nubigena** (D. Don) Ali (*C. crassicaulis* Benth. ex Baker)

Uttar Pradesh to Bhutan. China. Burma. 4000–5200 m. Open slopes in drier areas. Jun.–Jul.

Flowers solitary, yellow to purple, borne on short stalks from lax rosettes of densely hairy pinnate leaves forming small mats. Distinguished from 377 by its smaller leaflets to 5 mm. Petals to 2.5 cm; standard hairy outside; calyx half as long as standard, densely hairy. Leaves mostly 2.5–4 cm; leaflets many, oblong, silvery-haired. Pod 1.4–1.9 cm, finely hairy, stuffed inside with cottony down. **Pl. 29**.

37 GUELDENSTAEDTIA Leaves pinnate. Calyx bell-shaped; keel not more than half as long as other petals. Pod linear, swollen. 1 sp.

379 **G. himalaica** Baker

Uttar Pradesh to Bhutan. Tibet. China. 3300–4500 m. Open slopes; in drier areas and inner valleys. Jun.–Aug.

A mat-forming perennial with small pinnate often densely silky-haired leaves, and with 1–3 deep mauve to red flowers borne on thread-like stalks. Flowers 7–9 mm; standard broader than long; wings broadly obovate; keel very short; calyx 3 mm, hairy, lobes unequal. Leaflets obovate to obcordate, 2–10 mm. Pod *c.* 1.3 cm, hairless. **Pl. 29**.

38 HEDYSARUM Distinguished by its broad flattened pods which are conspicuously constricted between the seeds and break up into one-seeded units. Flowers pink to purplish. 7 spp.

Plants stemless or stems usually less than 30 cm

380 **H. kumaonese** Benth. ex Baker
Uttar Pradesh to C. Nepal. 2400–4600 m. Open stony slopes. Jun.–Jul.
An almost stemless perennial, with red, purplish or pink flowers in rather dense elongated clusters, borne on leafless stalks, and pinnate leaves with leaflets densely silvery-haired on the undersides. Flowers to 1.5 cm, distinctive in having standard and keel almost the same length and wings much shorter; calyx woolly-haired, with bristle-like teeth to 5 mm. Leaflets many, obovate to broadly elliptic, blunt or notched, 5–15 mm; stipules brown, papery. Pod *c.* 1 cm, 1–2 jointed, not winged, with a very short stalk.

381 **H. sikkimense** Benth. ex Baker
E. Nepal to Bhutan. S.W. China. 3600–4700 m. Alpine slopes, riverside gravels. Jun.–Jul.
Differs from 380 in its keel which is distinctly longer than the wings or standard, and in having slender sparsely leafy stems usually less than 20 cm including the flower cluster. Flowers to 1.8 cm, mauve, purple to reddish-purple, spreading or pendulous; calyx usually densely hairy, calyx-lobes triangular, the lower longer than the side-lobes. Leaflets 3–14mm. Pod with narrow wings. **Pl. 30**.

Erect perennials with stems more than 30 cm
a *Wings, keel and standard equal in length*

382 **H. cachemirianum** Benth. ex Baker
Kashmir. 2700–4000 m. Rocky slopes; quite common. Jun.–Jul.
A nearly hairless erect perennial 35–60 cm, with dense clusters of numerous large drooping red-purple flowers borne on nearly leafless stems, and with pinnate leaves with numerous leaflets. Flowers 1.5–2.5 cm long; calyx to 1 cm, with spreading hairs, and long narrow lobes. Leaflets narrow-ovate to narrow-elliptic, 1–2.5 cm long, at first with long hairs, but soon nearly hairless. Pod of 2–3 oval winged joints, each 2 cm by 1 cm. **Pl. 30**.

b *Petals unequal in length*

383 **H. campylocarpon** Ohashi
C. & E. Nepal. 2500–4200 m. Alpine shrubberies, open slopes; prominent in the Langtang valley, C. Nepal. Jun.–Jul.
Like 382, but with rather lax axillary clusters of magenta to purplish

flowers, with the standard conspicuously shorter than the wings; calyx with 4 triangular lobes, the upper 2 with short teeth. Flowers 1.7–2 cm long. Leaflets with sparse to dense silky adpressed hairs; stipules 1–1.5 cm; stem robust, 30–90 cm. Pod pendulous, 2–3–jointed, stalked. **Pl. 30.**

384 **H. microcalyx** Baker
Kashmir to Uttar Pradesh. 2700–4400 m. Streamsides, shrubberies. Jul.–Aug.

Like 382 and 383, but upper leaves whorled, and flowers with wings distinctly shorter than the long keel and slightly shorter standard; stipules conspicuous, elliptic to 4 cm, encircling the young shoots. Flowers c. 1.8 cm, purple, reddish-purple or crimson-purple. Leaves hairless; leaflets to 4 cm; stems erect to 1m. Pod 2–3–jointed, each joint narrowly winged and 1–1.5 cm by 7 mm broad.

39 LOTUS Leaves with 5 leaflets. Fruit elongated, many-seeded, usually with partitions between the seeds. Flowers yellow. 1 sp.

385 **L. corniculatus** L. COMMON BIRD'S-FOOT TREFOIL
Pakistan to C. Nepal. Temperate Eurasia. 1500–4000 m. Cultivated areas, open slopes; common. May–Sep.

A variable prostrate or ascending, hairy or hairless perennial, with usually long-stalked clusters of 3–6 yellow flowers, and leaves with 5 leaflets. Flowers 8–16 mm long; petals more or less equal; calyx with unequal lanceolate lobes, hairy or hairless. Leaflets variable, mostly broadly obovate-wedge-shaped to lanceolate, 5–15 mm long, the lower pair at the base of the rachis and separated from the terminal three. Pod 2–3 cm, cylindrical, hairless.

ROSACEAE Rose Family

A large and important world-wide family of herbaceous plants, shrubs, and trees. Leaves alternate, simple or compound, with stipules; thorns and prickles occur in some genera on leaves and stems. Flowers often large showy; sepals and petals usually 5 (the sepals with an outer *epicalyx* sometimes present); stamens numerous. Ovary of several or many carpels, rarely one as in *Prunus* etc., either superior (*hypogynous*) with the carpels arising from the receptacle above the stamens, or half-inferior (*perigynous*) in which the receptacle is disk- or cup-shaped with the sepals, petals and stamens arising from the rim, or inferior (*epigynous*) where the receptacle encircles the ovary so that the latter arises below the other parts of the flower. Fruit very varied; dry, or fleshy, with one to many carpels which are either free or fused together, or are fused with the receptacle.

Trees and shrubs above 1 m
  a Leaves compound, with 3 or more leaflets

  i Carpels free   1 RUBUS 2 SORBARIA 19 POTENTILLA
  ii Carpels enclosed in a fleshy receptacle 3 ROSA 15 SORBUS
 b Leaves entire, toothed or lobed
  i Carpel 1   4 PRUNUS 5 PRINSEPIA 6 NEILLIA
  ii Carpels several, free 1 RUBUS 6 NEILLIA 7 SPIRAEA
  iii Carpels encircled in a fleshy receptacle in fruit
     8 COTONEASTER 9 CRATAEGUS 10 PYRACANTHA
       11 PHOTINIA 12 STRANVAESIA
       13 MALUS 14 PYRUS 15 SORBUS

Herbaceous perennials or small undershrubs
 a Epicalyx present   16 ALCHEMILLA 17 FRAGARIA
      18 GEUM 19 POTENTILLA 20 SIBBALDIA
 b Epicalyx absent   1 RUBUS 21 AGRIMONIA 22 ARUNCUS
     23 FILIPENDULA 7 SPIRAEA 8 COTONEASTER

1 RUBUS Raspberry Distinguished by its fruit of numerous fleshy carpels in a globular cluster borne on a domed receptacle. Shrubby species,
usually with curved prickles or spines. Epicalyx absent. *c.* 45 spp.

Prostrate creeping plants

386 **R. calycinus** Wallich ex D. Don
Uttar Pradesh to S.W. China. Burma. 2100–3000 m. Shady banks, forests.
Mar.–May.
  An herbaceous perennial with long creeping bristly stems rooting at the
nodes, and rounded entire and toothed or obscurely lobed leaves, and
solitary or paired erect white flowers 2–2.5 cm across. Petals often longer
than the leafy toothed blunt calyx-lobes which are usually densely covered
with long straight bristles. Leaf-blade 2–4 cm broad, stalked; stipules
broadly ovate, toothed; stems creeping 30–90 cm. Fruit scarlet, carpels
few. **Page 464**.

387 **R. nepalensis** (Hook. f.) Kuntze (*R. nutantiflorus* Hara)
Uttar Pradesh to Sikkim. 2100–3000 m. Rocks, banks, shrubberies.
May–Jul.
  Like 386 with creeping stems, but with trifoliate leaves with rounded,
toothed leaflets, and with lanceolate long-pointed usually shaggy-haired
calyx-lobes. Flowers white, nodding, 2–3 cm across. Leaflets broadly
ovate to rhombic, double-toothed, 1–3 cm. Fruit scarlet. **Page 464**.

Large erect or climbing shrubs with prickly stems
 a *Leaves simple or lobed*

388 **R. paniculatus** Smith
Pakistan to Bhutan. 1500–2900 m. Shrubberies; common. Jun.–Aug.
  A rambling climber distinguished by its large simple ovate-heart-shaped
finely-toothed leaves, and its terminal branched clusters of white flowers
with narrow petals which are much shorter than the pale silky-haired

calyx-lobes. Flowers 1–1.5 cm across. Leaves 8–15 cm, including leaf-stalk of 2–3 cm. All parts of plant except the upper sides of the leaves and older stems covered with a dense mat of white or buff hairs; prickles absent or few. Fruit *c*. 1 cm, with many glossy black fleshy carpels, edible. **Pl. 31**.

389 **R. acuminatus** Smith
Utter Pradesh to S.W. China. 1000–2300 m. Forests, shrubberies. Jun.–Aug.

Differs from 388 in being a hairless shrub and having ovate long-pointed, finely-toothed leaves with rounded bases. Flowers white, *c*. 1 cm across, in small terminal and axillary clusters; calyx hairless or nearly so. Leaves 5–12 cm; stem with few spines. Fruit red. **Pl. 31**.

b *Leaves compound, with 3 or more leaflets*
i *Flowers white or yellow*

390 **R. biflorus** Buch.–Ham. ex Smith
Pakistan to S. W. China. 2100–3300 m. Forests, shrubberies, open slopes; common. May–Jun.

Distinguished by its silvery-white waxy-bloomed young stems. A large shrub with numerous recurved prickles, with leaves with 3–5 leaflets, and with solitary or clustered axillary white flowers 1.5–2.5 cm across. Petals obovate, as long or longer than the ovate-acute woolly-haired calyx-lobes. Leaflets toothed, white-woolly beneath, the terminal leaflet often larger and 3–lobed; rachis spiny. Fruit yellow to orange, 1.5–1.8 cm, edible.

391 **R. ellipticus** Smith
Pakistan to S.W. China. S. India. Sri Lanka. S.E. Asia. 600–2300 m. Shrubberies, cultivated areas. Feb.–Apr.

A large shrub with stout stems covered with numerous long rufous bristles, usually trifoliate leaves, and white flowers in dense short branched clusters. Flowers 1–1.5 cm across; calyx-lobes woolly-haired on both sides, shorter than petals. Leaflets leathery, elliptic or obovate, toothed, green above, grey-woolly beneath, the terminal largest 4–10 cm; leaf-stalk with long bristles; stems with recurved spines as well as bristles. Fruit globular, yellow, clustered.

Common around villages; the edible fruit is the best flavoured of the genus. **Pl. 31**.

R. fruticosus L. BLACKBERRY, with white flowers and black fruit is common in hedges in the valley of Kashmir where it has probably been introduced.

ii *Flowers pink or red*

392 **R. foliolosus** D. Don
Himachal Pradesh to C. Nepal. 2100–3600 m. Shrubberies. Jun.–Aug.

Distinguished by its small pink flowers in branched axillary leafy clusters, and its leaves with 3–7 small rounded and toothed leaflets. A scrambling shrub to 2 m, with conspicuously densely hairy undersides of

111

leaflets, and hairy branches. Flowers *c*. 1 cm across; petals equalling the woolly-haired calyx-lobes. Leaflets elliptic-pointed, sharply toothed, mostly 2–4 cm on flowering stems. Fruit small, pink or white, hairy.

393 **R. hoffmeisterianus** Kunth & Bouché
Afghanistan to W. Nepal. 1500–2400 m. Hedges, cultivated areas. May.

A rambling shrub to 2 m, with slender finely-hairy young branches with recurved prickles, with leaves usually with 3 leaflets, and with pink flowers. Flower clusters terminal and axillary, few-flowered, hairy; petals obovate, little shorter than the ovate acute calyx-lobes which are woolly-haired on both sides. Leaflets mostly 3–5 cm, the terminal largest, often 3–lobed, with usually blunt teeth, densely white-woolly beneath; leaf-stalk and rachis with prickles. Fruit red or orange, *c*. 1 cm across, with numerous fleshy carpels.

The fruit is edible. **Pl. 31**.

394 **R. hypargyrus** Edgew. (*R. niveus* Wallich)
Himachal Pradesh to S.W. China. 2200–3300 m. Forests. Jun.–Aug.

Like 393 with pink flowers and with leaves with 3 leaflets which are usually white-hoary beneath, but calyx distinctive with narrow triangular sepals with long slender apices much longer than the petals. Flowers few, usually in short flat-topped finely hairy clusters; calyx to 1.4 cm long. A very variable spiny shrub to 2 m, with variable ovate long-pointed double-toothed leaflets. Fruit orange to red.

2 SORBARIA Like *Spiraea* but distinguished by its pinnate leaves. 1 sp.

395 **S. tomentosa** (Lindley) Rehder (*Spiraea sorbifolia* auct. non (L.) A. Br.)
Afghanistan to C. Nepal. C. Asia. N. China. 1800–2900 m. Open slopes, riversides, cultivated areas; gregarious and common. Jun.–Aug.

A slender graceful spreading deciduous shrub to 3 m or more, with long pinnate leaves, and large terminal, pyramidal branched clusters of tiny white flowers. Clusters 20–45 cm long; flowers 5–7 mm across; petals rounded. Leaves 20–40 cm long; leaflets lanceolate slender-pointed, 5–10 cm, double-toothed, hairy beneath. Carpels 5, 3–4 mm, hairless, many-seeded. **Pl. 31**.

3 ROSA ROSE Distinguished by its fruit, the *hip*, which is fleshy and flask-like with many hard one-seeded carpels borne on its inner wall, and usually with a persistent calyx at the apex. Shrubs with spines or prickles. 4 native spp., and a number of introduced spp.

Climbing shrubs; flowers white

396 **R. brunonii** Lindley Himalayan Musk Rose
Kashmir to S.W. China. Burma. 1200–2400 m. Shrubberies. Apr.–Jun.

A stout climber with hooked prickles, and terminal clusters of many fragrant white flowers each 2.5–4 cm across. Petals usually 5; sepals narrow-lanceolate long-pointed, mostly entire, reflexed and falling in fruit; flower-stalks glandular. Leaves with 5–7 elliptic to oblong-lanceolate, finely toothed leaflets mostly 3–5 cm; leaf-stalk and rachis downy, usually prickly. Fruit to 1.5 cm, globular to ovoid, dark brown, in a cluster; sepals absent. **Pl. 32.**

397 **R. moschata** Miller is very similar and possibly occurs in the extreme west of our area. It is distinguished by its smooth branches and leaf-stalks without prickles.

Erect shrubs
a *Flowers white, or cream*

398 **R. sericea** Lindley
Himachal Pradesh to S.W. China. Burma. 2100–4500 m. Forests, shrubberies, cultivated areas, alpine slopes. May–Aug.

A stiff erect shrub with usually hairless often unarmed twigs, with leaves with small leaflets, and with solitary white or cream, axillary somewhat drooping flowers 4–6 cm across with 4 petals. Petals obcordate; sepals 4, elliptic with a long apex, shorter than the petals, persisting in fruit. Leaflets 5–11, elliptic to narrow-oblong, toothed at apex, terminal leaflet slightly larger, 8–25 mm. Fruit bright red, globular to pear-shaped, 1–1.5 cm, calyx erect.
Fruit edible. **Pl. 32.**

Other white-flowered species introduced to the Himalaya are:-

398a **R. laevigata** Michaux   CHEROKEE ROSE
Native of China. Common in Kathmandu valley. Naturalized in U.S.A. A strong climber with solitary white fragrant flowers. **Pl. 32.**

398b **R. beggeriana** Schrenk
Native of Temperate Asia. Common in Kashmir valley. Usually a bush but sometimes climbing, with few white flowers in flat-topped clusters.

b *Flowers pink or red*

399 **R. macrophylla** Lindley
Pakistan to S.W. China. 2100–3800 m. Forests, shrubberies. Jun.–Jul.

An upright shrub 3–5 m, with dark red or purple stems with few prickles, and with large often solitary bright pink flowers 3–7 cm across. Petals 5, broadly obcordate; sepals long-pointed often with leafy tip, nearly as long as petals or longer, usually with glandular bristles. Leaves 8–20 cm, with 7–11 ovate-elliptic, finely toothed leaflets to 5 cm or more, hairy and glandular beneath. Fruit very large, to 5 cm, red, flask-shaped, bristly and with conspicuous persistent calyx, edible. There are many forms, some hardly distinguishable from 400. **Pl. 32.**

400 **R. webbiana** Wallich ex Royle
Pakistan to W. Nepal. 1500–4100 m. Rocky slopes, villages. Jun.–Aug.
The common pink-flowered rose of the W. Himalaya. A shrub to 2.5 m,

with slender branches with straight prickles, and leaves with small rounded leaflets and prickly leaf-stalks, and with rather dense clusters of 1–3 deep pink flowers fading to pale pink. Flowers 2.5–7 cm across; petals 5; sepals shorter, glandular and sometimes hairy on back, often enlarged at tip. Leaflets 5–9, mostly 0.5–1.5 cm long, coarsely toothed, nearly hairless. Stems and leaves often turning bright pinkish. Fruit red, ovoid to flask-shaped, to 3.5 cm, with persistent sepals. **Pl. 32**.

c *Flowers yellow*

Introduced species include:-

400a **R. foetida** Herrm.
Native of W. Asia. Commonly planted and semi-naturalized in Lahul. A medium-sized shrub with heavy-scented rich yellow flowers. **Pl. 32**.

400b **R. ecae** Aitch.
Native of W. Asia. Common in hedges around Kargil in Ladakh. A small shrub with small solitary golden-yellow flowers.

4 PRUNUS Cherry Fruit a *drupe*, which is fleshy outside and has a stony one-seeded nut. Flowers half-inferior; receptacle cup-shaped. Trees or shrubs. 16 spp.

Fruit hairy, peach- or almond-like

401 **P. davidiana** (Carrière) Franchet
Native of Mongolia and S.W. China. Cultivated in W. & C. Nepal and semi-naturalized at altitudes of 2100–2700 m. Mar.–Apr.
  Distinguished by its stalkless pink flowers, hairy calyx, and crinkled fruit-stone. Flowers *c.* 3.5 cm across, clustered on leafless branches. Leaves elliptic acute, toothed, to 8 cm; a tree to 7 m. Fruit hairy, *c.* 5 cm across.

402 **P. mira** Koehne
Uttar Pradesh to Bhutan. S.W. China. 2700–4000 m. Cultivated areas, open hill sides. Mar.–Apr.
  Like 401, but with shortly stalked flowers, hairless calyx, and a smooth fruit-stone. Flowers pink, *c.* 2 cm across. Leaves lanceolate long-pointed, to 7 cm. Fruit *c.* 2 cm, pale green with brown hairs.
  Widely cultivated in the W. Himalaya and to a lesser extent in Nepal are: **P. dulcis** (Miller) D. A. Webb Almond; **P. persica** (L.) Batsch. Peach; **P. armeniaca** L. Apricot. Their blossom is attractive in spring, and their fruits are sold fresh or dried in the markets.

Fruit hairless, plum- or cherry-like
a *Flowers several in flat-topped clusters, or solitary*
  i *Flowers pink* (see also 406)

403 **P. cerasoides** D. Don
Himachal Pradesh to S.W. China. Burma. 1200–2400 m. Forests; often

114

planted by wayside resting places. Oct.–Nov.

A moderate-sized tree with smooth bark which peels off in thin horizontal strips, with glossy long-pointed toothed leaves, and with usually pink flowers appearing on bare branches or with young leaves in early winter. Flowers long-stalked, often paired or in few-flowered clusters at ends of branches; petals to 1.5 cm, obovate, spreading. Leaves with elliptic blade 5–8 cm, short-stalked. Fruit yellow or red, ovoid 1.3–1.6 cm long, acid; stone rough, furrowed.

Fruit edible. Stones used for necklaces; branches used for walking sticks. **Pl. 33.**

404 P. carmesina Hara

W. Nepal to Bhutan. S.W. China. Burma. 2300–2600 m. Forests, shrubberies; rare in Nepal and possibly introduced there. Mar.–Apr.

Distinguished from 403 by being spring-flowering, and having red-purple, nearly erect (not spreading) petals, and with larger crimson calyx with oblong or ovate-blunt lobes, and bracts which soon fall. Flowers 2.2 cm long. **Pl. 33.**

405 P. jacquemontii Hook. f.

Afghanistan to Uttar Pradesh. 1800–2700 m. Rocky slopes. Apr.–May.

A small straggling shrub, with deeply and sharply toothed elliptic to narrow-oblong leaves, and often solitary nearly stalkless pink flowers. Flowers *c.* 1 cm across; petals 5 mm, obovate; calyx-tube *c.* 7 mm, funnel-shaped. Leaves 1.5 cm, dark green above, hairless. Fruit bright red; stone smooth.

ii *Flowers white*

406 P. rufa Hook. f.

W. Nepal to S.E. Tibet. Burma. 3000–3800 m. Sub-alpine forests, shrubberies; common. May–Jun.

A small tree with hairy young shoots, peeling shiny purplish bark, and solitary or clustered white or rarely pink flowers with small rounded petals little longer than the calyx. Calyx-tube *c.* 12 mm, swollen at base, with triangular lobes. Leaves elliptic-lanceolate, sharply toothed, hairless except on the leaf-stalk and veins beneath, mostly 5–8 cm. Fruit ellipsoid, borne on stout stalks, red to black. **Pl. 33.**

Commonly grown for fruit in the W. Himalaya are: **P. cerasifera** Ehrh. CHERRY PLUM; **P. avium** (L.) L. CHERRY; **P. domestica** L. PLUM.

b *Flowers in long racemes*

407 P. cornuta (Wallich ex Royle) Steud.

Afghanistan to S.W. China. Burma. 2100–3500 m. Forests; common. Apr.–May.

A medium-sized tree with rough grey-brown to brown bark, with oblong to lanceolate leaves, and with long drooping clusters 10–15 cm of numerous small white flowers. Flowers to 10 mm across; petals round; calyx with blunt lobes. Leaves 8–15 cm, long-pointed, closely and finely toothed. Fruits globular *c.* 8 mm, stalked, in a long raceme, red then dark purple to black; fruit often infected by an insect and becoming long and horn-like, hence the plant's name.

The acid fruit is edible; the foliage is cut for fodder. **Pl. 33**.

408 **P. napaulensis** (Seringe) Steud.
Uttar Pradesh to S.W. China. Burma. 1200–2600 m. Forests. Apr.

Distinguished from 407 by its leaves which are glaucous beneath, and remotely saw-toothed; by its dense, usually very hairy inflorescence, and by its short thickened fruit-stalks covered with lenticels. Flowers white, in spikes to 20 cm. It grows at lower altitudes than 407.

5 **PRINSEPIA** Spiny shrubs, with fleshy fruits with persistent calyx. 1 sp.

409 **P. utilis** Royle
Pakistan to S.W. China. 1200–2700 m. Forests, shrubberies, hedges. Apr.–May and Oct.–Nov.

A green-stemmed shrub to 3 m with dark leathery leaves, stout spines, and with short clusters of white flowers generally borne from the axils of the spines. Flowers 7–15 mm across; petals rounded; calyx cup-shaped with 5 unequal rounded lobes, persisting in fruit. Leaves elliptic to narrow-lanceolate, long-pointed, 2.5–7.5 cm, margin entire or minutely toothed; spines usually 2–4 cm. Fruit oblong-cylindrical, 1.3–1.7 cm long, fleshy, dark-purple, bloomed.

Oil is pressed from the seeds. Its spiny branches are cut and used to block gaps in hedges and walls against goats and sheep. **Pl. 31**.

6 **NEILLIA** Spineless shrubs. Leaves entire or lobed. Fruit of 1 or several carpels. 2 spp.

410 **N. rubiflora** D. Don
C. Nepal to S.W. China. 2100–3200 m. Shrubberies. Jul.–Aug.

A large shrub with slender arching branches, with shallowly 3–lobed leaves, and with terminal and lateral clusters of small white flowers. Flowers *c.* 8 mm across; petals obovate, mostly shorter than the brownish woolly-haired, triangular calyx-lobes. Leaves coarsely toothed, 5–10 cm, with a shallow heart-shaped base, usually with 2 shallow lateral lobes and a much longer long-pointed terminal lobe. Fruit of 1 to several hairy carpels, encircled by the persistent calyx. **Pl. 35**.

7 **SPIRAEA** Carpels usually 5, free or fused at the base, each with 2 or more seeds. Spineless shrubs. 11 spp.

Branches softly woolly-haired; flowers white

411 **S. canescens** D. Don
Pakistan to S.W. China. 1500–3000 m. Shrubberies, cultivated areas. May–Jun.

A small stiff deciduous shrub, with grey-hairy arching branches, and numerous dense flat-topped clusters of small white flowers borne on short side branches and forming long terminal clusters. Flowers 4–6 mm across; stamens not longer than the rounded petals. Leaves elliptic to obovate 6–16 mm, entire or toothed towards the apex, hairy, paler beneath. Stems usually *c.* 1 m but often more. Ripe carpels long-haired, partly sunk in calyx-tube. **Pl. 34**.

Branches hairless or nearly so; flowers usually pink

412 **S. arcuata** Hook.f.
Himachal Pradesh to S.W. China. 3300–4500 m. Shrubberies, open slopes; commonly forming low thickets at the heads of inner valleys in Nepal. Jun.–Jul.

A small shrub usually *c.* 1 m or less, with stiff shining dark brown grooved and arched branches, and small dense domed clusters of pale or dark pink or sometimes cream flowers. Flowers 6–8 mm across. Leaves elliptic to obovate 6–15 mm, entire or with blunt teeth, or lobed at the apex. Ripe carpels shining, exposed, not sunk in calyx-tube. **Pl. 33**.

413 **S. bella** Sims
Pakistan to S.W. China. 2100–3600 m. Forests, shrubberies. May–Jul.

Like 412, but a much more laxly branching shrub, with pink flowers in lateral leafy flat-topped clusters. Flowers *c.* 5 mm across. Leaves larger, mostly 2–3 cm, elliptic, coarsely toothed. A shrub usually to 1.5 m.

8 COTONEASTER Like *Crataegus* but leaves entire and stems without spines. Fruit fleshy with 2–5 carpels fused to the receptacle. A very difficult genus, probably with much hybridization. Botanists do not agree on the delimitation of some species. Perhaps 40 spp.

Petals erect and flower not opening widely, commonly pink
a *Leaves hairless or nearly so*

414 **C. roseus** Edgew.
Afghanistan to Uttar Pradesh. 1800–3000 m. Rocky slopes; common in Chenab valley. Jun.–Jul.

Distinguished by its bright pink flowers in flat-topped lax clusters of 3–10. Petals 3 mm; calyx usually quite hairless like the flower-stalks and peduncle. Leaves elliptic, 1–3 cm, with a rounded base and fine apical point, membraneous, almost hairless beneath. An erect straggling deciduous shrub which is very sparsely hairy when young. Fruit bright red.

b *Leaves with adpressed or woolly hairs*

415 **C. acuminatus** Lindley
Kashmir to S.W. China. 2100–4000 m. Forests, shrubberies; common in C. & E. Nepal. May–Jun.

Flowers white or pink, solitary or in flat-topped clusters of up to 5, and with flower-stalks and peduncle with bristly hairs. Flowers 5 mm across.

Leaves ovate to lanceolate 2.5–5 cm, tapering from below the middle to a long point, hairy on the mid-vein beneath, and with scattered adpressed hairs above. A deciduous shrub to 4 m, with young twigs densely clothed with bristly hairs but becoming hairless. Fruit scarlet, with incurved calyx-lobes.

416 **C. falconeri** Klotz (*C. integerrima* Medicus sensu R. Parker, *C. vulgaris* Lindley)
Afghanistan to Himachal Pradesh. 1500–3300 m. Rocky slopes, shrub-beries; common in Kashmir. May–Jun.

Like 415, but with smaller broadly ovate to elliptic leaves commonly 5–10 mm, which are grey-woolly beneath, and usually rounded at both ends. Flowers white, solitary or up to 4 in axillary clusters; flower-stalks hairy. A small erect deciduous shrub to 1.5 m, with young twigs densely hairy but becoming hairless. Fruit red.

Petals spreading and flowers opening widely, usually white
a *Leaves evergreen, thick, margins recurved*

417 **C. microphyllus** Wallich ex Lindley
Afghanistan to S.W. China. 2000–5400 m. Rocks, banks, alpine slopes; common. May–Jun.

A low or prostrate mat-forming evergreen shrub, with rigid much-branched stems, and small axillary white flowers. Flowers *c.* 8 mm across, solitary or 2–3; calyx densely hairy. Leaves 6–13 mm, ovate to elliptic, glossy dark green above and bristly-hairy beneath. Fruit scarlet.

Up to 12 different dwarf prostrate species have been described; they are very similar and are difficult to distinguish from each other. Some have pink flowers in bud. **Pl. 34**.

b *Leaves deciduous, thin*
i *Leaves less than 3 cm; flowers 2–12 in each cluster*

418 **C. nummularia** Fishcher & Meyer
Pakistan to Uttar Pradesh. 600–3000 m. Rocky slopes of dry inner valleys; the commonest low altitude W. Himalayan species. Apr.–May.

A low straggling or prostrate shrub with small leaves to 13 mm, which are rounded to broadly elliptic and with woolly or shaggy hairs beneath, but becoming nearly hairless. Flower clusters very short, with 2–5 flowers; petals 2–3 mm. Fruit black. **Pl. 35**.

What were previously considered to be varieties of this species have now been described as distinct species. They are: **C. afghanica** Klotz, **C. insignis** Pojark, and the less hairy **C. pruinosus** Klotz. They are difficult to distinguish in the field.

419 **C. ludlowii** Klotz
W. Nepal to S.W. China. 2100–3600 m. Shrubberies, drier areas. May–Jun.

Distinguished from 418 in having larger elliptic to narrow elliptic leaves

1.7–2.5 cm, with an acute or wedge-shaped base and blunt or acute apex, shaggy-haired beneath but becoming hairless, and with shaggy-haired leaf-stalk. Flowers more numerous, 5–12, in erect clusters, the flower-stalks softly hairy at first. Fruit red. A shrub to 2m.

ii *Mature leaves 4–12 cm; flower clusters many-flowered*

420 **C. bacillaris** Wallich ex Lindley (including **C. ignotus** Klotz and **C. confusus** Klotz)
Afghanistan to Bhutan. 1800–3000 m. Shrubberies, open slopes; common in W. Himalaya. Apr.–May.

An erect deciduous shrub with branched clusters of many white flowers, and variable lanceolate, ovate or obovate leaves which are hairless above and becoming hairless beneath. Flower clusters 2.5–5 cm across, becoming hairless; calyx minutely hairy. Young shoots brown woolly-haired to nearly hairless; leaves 3–8 cm. Fruit reddish-black.

Wood used for walking sticks. **Pl. 34**.

421 **C. affinis** Lindley
Pakistan to S.W. China. 2000–2800 m. Shrubberies; common in Nepal. Apr.–May.

Like 420, but with shorter and broader ovate or elliptic leaves which are densely white-woolly beneath when young, and with more woolly-haired peduncles, flower-stalks and calyx, and with a denser inflorescence. **Pl. 34**.

422 **C. frigidus** Wallich ex Lindley
Uttar Pradesh to S.W. China. 2200–3400 m. Shrubberies, riversides; common in Nepal. Jun.–Jul.

Like 420, but with larger woolly-haired flower clusters, and with scarlet fruit. Calyx-tube woolly-haired, calyx-lobes acute. A large shrub or small tree with leaves 7–12 cm, oblong to lanceolate acute, narrowed to the base, woolly-haired beneath when young, becoming hairless. **Pl. 34**.

9 CRATAEGUS Hawthorn Leaves lobed; stems spiny. Fruit fleshy, with 1–5 nuts. 2 spp.

425 **C. songarica** G. Koch
Afghanistan to Uttar Pradesh. 1500–2700 m. Cultivated areas; quite common in Kashmir and Chenab valleys. May.

A deciduous tree to 8 m with greyish branches, deeply and sharply lobed leaves, and lax flat-topped clusters of white flowers. Flowers to 1.5 cm across, long-stalked; calyx woolly-haired below. Leaves deeply cut to two-thirds their width into nearly parallel-sided lobes with coarse acute teeth; blade broadly ovate or rhombic in outline, 3–5 cm, leaf-stalk nearly as long. Fruit red, globular c. 1 cm; lenticels few, scattered.

10 PYRACANTHA Evergreen, spiny shrubs. Leaves simple. Fruit fleshy, with 5 nuts. 1 sp.

426 **P. crenulata** (D. Don) M. Roemer
Kashmir to S.W. China. Burma. 1000–2400 m. Shrubberies, open slopes,
cultivated areas. Apr.–May.

A spiny evergreen shrub with usually crowded narrow-oblong blunt
shining leathery leaves, and numerous small white flowers in axillary
clusters ranged along the branches. Flowers to 9 mm across; petals
rounded; calyx fused to ovary, with 5 blunt lobes. Leaves usually 1.5-3 cm,
with obscure usually rounded teeth; spines stout, terminating lateral
branches; shrub usually to 2 m. Fruit globular, *c.* 6 mm, orange-red; nuts
protruding from the persistent calyx.
Wood used for walking sticks. **Pl. 34**.

11 PHOTINIA Evergreen spineless shrubs with entire leaves. Fruit hard,
brittle or leathery, 1–2–seeded. 1 sp.

427 **P. integrifolia** Lindley
Uttar Pradesh to S.W. China. Burma. S.E. Asia. 1500–2700 m. Forests.
Apr.–May.

A quite hairless sometimes epiphytic shrub, with shining oblanceolate
leaves, and large spreading much branched domed clusters 5–15 cm across
of numerous small white flowers. Flowers *c.* 4 mm across; petals 5,
rounded; calyx hairless, lobes short, persistent. Leaves 5–12 cm, quite
entire, narrowed to the short leaf-stalk; shrub or small tree to 7 m. Fruit
reddish to black-purple, globular, *c.* 5 mm. **Page 466**.

12 STRANVAESIA Leaves entire. Fruit fleshy, 5–celled, splitting longitudinally releasing 5 seeds. 1 sp.

428 **S. nussia** (D. Don) Decne.
Uttar Pradesh to C. Nepal. Burma. S.E. Asia. 1000–2100 m. Forests.
Apr.–Jun.

A small leafy evergreen tree with leathery leaves, and numerous small
white flowers in flat-topped clusters 5–10 cm across. Flowers *c.* 1.3 cm
across; petals obovate; calyx densely white-woolly, lobes broadly triangular, persistent. Leaves variable, mostly 8–15 cm, obovate to narrow
lanceolate, entire or finely toothed, stalked. Fruit orange-yellow, *c.* 8 mm,
crowned with incurved calyx-lobes, and with 5 slightly projecting carpels.
Page 466.

13 MALUS APPLE Fruit with 3–5 carpels imbedded in the fleshy receptacle. Petals stalked; styles fused at base. 3 spp.

429 **M. baccata** (L.) Borkh. SIBERIAN CRAB APPLE
Kashmir to Bhutan. N. Temperate Asia. 1800–3600 m. Forests. Apr.–
Jun.

A small deciduous tree with a short trunk and rounded crown, with
elliptic-pointed leaves, and with clusters of white long-stalked flowers.

Flowers 2.5–3.7 cm across; petals elliptic; calyx hairless or hairy outside, lobes lanceolate, hairy within, deciduous. Leaves variable, 4–8 cm, hairless, finely blunt-toothed. Fruit small 7–15 mm, red or scarlet, globular or ovoid.

430 **M. sikkimensis** (Wenzig) Koehne
W. Nepal to S.E. Tibet. 2500–2800 m. Forests. Apr.–May.

Like 429, but calyx woolly-haired; leaf-stalks and leaves mostly woolly-haired beneath, leaf-margin with pointed teeth. Flowers white, *c*. 2.5 cm across. Fruit pear-shaped. **Pl. 35**.

M. domestica Borkh. APPLE is grown for its fruit in the middle hill villages, particularly in the valleys of Kulu and Kashmir.

14 PYRUS PEAR Like *Malus*, but fleshy fruit with gritty cells. Petals stalked; styles free. 1 sp.

431 **P. pashia** Buch.–Ham. ex D. Don
Afghanistan to S.W. China. Burma. 750–2700 m. Shrubberies, cultivated areas. Mar.–Apr.

A small or medium-sized deciduous tree, with ovate to broadly lanceolate long-pointed, toothed, quite hairless shining mature leaves. Flowers 2–2.5 cm across; petals obovate, with darker veins, narrowed to a short stalk; calyx urn-shaped, with spreading white-woolly triangular lobes. Leaves variable, 5–10 cm, those on sucker shoots often 3–5-lobed; young leaves woolly; leaf-stalk slender, often reddish; stems sometimes spiny. Fruit globular, 1.3–2.5 mm, dark brown, with raised lenticels.

Fruits edible when half rotten; leaves cut for fodder. Wood used for fuel and small implements. **Pl. 35**.

P. communis L. PEAR is cultivated for its fruit in the W. Himalaya.

15 SORBUS Flowers half-inferior. Fruit fleshy, of 2–5 carpels each with 1–2 seeds. Leaves either pinnate, lobed, or simple. Distinguished from *Pyrus* by its compound flower clusters. 12 spp.

Leaves simple, toothed or lobed
a *Leaves elliptic-acute, not lobed*

432 **S. cuspidata** (Spach) Hedlund
Uttar Pradesh to Sikkim. 2700–3700 m. Forests; common. May.

A large tree with large elliptic leaves which are white-woolly when young and remain densely woolly beneath when mature. Flower clusters 4–9 cm across, densely white-woolly; flowers white, *c*. 1.3 cm across; petals densely woolly at base within; styles hairless except at base. Leaves 8–20 cm, margin entire or obscurely toothed. Fruit nearly globular, *c*. 1.5 cm, red. **Pl. 35**.

433 **S. rhamnoides** (Decne.) Rehder
C. Nepal to S.W. China. 2700–3500 m. Forests. Apr.–May.

A large tree like 432, but not nearly so common; distinguished by its narrow-elliptic long-pointed hairless leaves, and its terminal flat-topped clusters of small white flowers. Flowers *c.* 8 mm across. Leaves 10–12 cm, margin finely and regularly toothed. Fruit nearly globular, *c.* 5 mm. **Pl. 36**.

b *Leaves blunt, ovate to broadly oblong, shallowly lobed*

434 **S. lanata** (D. Don) Schauer
Afghanistan to C. Nepal. 2500–3400 m. Forests. Apr.–May.

A small tree like 432 with white-woolly flower clusters but leaves usually with shallow acute and toothed lobes, hairless above when mature and with greyish-white wool beneath, or rarely hairless. Flowers white, scented, 1.3 cm across; petals nearly hairless; styles densely woolly in lower half. Fruit yellowish-red, edible. **Page 465**.

Leaves pinnate
a *Leaflets mostly 2 cm or less*

435 **S. microphylla** Wenzig
Himachal Pradesh to Arunachal Pradesh. 3000–4200 m. Forests, alpine shrubberies; common. Jun.

A slender nearly hairless shrub or small tree, with pinnate leaves with 12–18 pairs of small linear-oblong acute, deeply toothed leaflets 1.3–2 cm long. Flowers pink to white, 6–10 mm across, in lax few-flowered flat-topped clusters; petals rounded; calyx hairless or nearly so. Fruit *c.* 8 mm, bluish-white. **Pl. 36**.

b *Leaflets more than 2 cm*

436 **S. foliolosa** (Wallich) Spach
Himachal Pradesh to Arunachal Pradesh. Burma. 2400–4000 m. Forests, alpine shrubberies; common. Apr.–Jun.

A large shrub or small tree with nearly hairless twigs and with bark which peels off in horizontal strips. Inflorescence with white downy hairs; flowers white or greenish-white, in dense flat-topped clusters, smelling unpleasantly; petals rounded, 2–3 mm. Leaves 6–12 cm, with 9–15 pairs of linear-oblong, sharply-toothed, blunt or acute leaflets, 2.3–4 cm long. Fruit ovoid or globular, red or bluish-red.

437 **S. ursina** (Wenzig) Decne.
Uttar Pradesh to S.E. Tibet. Burma. 2900–4300 m. Forests, shrubberies. May–Jun.

Like 436, but inflorescence and midribs of leaves with red hairs. Flowers white to pale pink. Fruit pink to red. A shrub or small tree to 7 m.

Several other species occur in this section; they are not easily distinguished.

16 ALCHEMILLA LADY'S-MANTLE Petals absent; epicalyx present. Fruit a one-seeded carpel. Leaves lobed. 3 spp.

438 **A. trollii** Rothm.

Pakistan to Kashmir. 3000–4100 m. Alpine meadows. Jun.–Aug.

Flowers yellow-green, in a lax branched cluster, borne on sparsely leafy stems to 10 cm. Flowers *c.* 3 mm across, in dense terminal umbels. Basal leaves hairy, with very long leaf-stalks to 10 cm, and with kidney- to heart-shaped blade mostly 3–6 cm across and with usually shallow rounded coarsely toothed lobes; stem-leaves similar but short-stalked and with ovate, toothed leafy stipules. A hairless perennial. **Page 465**.

17 FRAGARIA Strawberry Distinguished in fruit by its fleshy coloured receptacle, with small carpels borne over the surface. Leaves trifoliate. Flowers white; epicalyx conspicuous. 4 spp.

439 **F. nubicola** Lindley ex Lacaita

Pakistan to S.W. China. Burma. 1800–3800 m. Forests, shrubberies, shady banks; very common. Apr.–Jun.

Distinguished by its globular fruit and usually entire calyx-lobes. A small softly silky-haired perennial with trifoliate leaves, with long runners which root at the nodes, and with few white flowers 1.5–2.5 cm across. Petals 5, broadly obovate; calyx with 5 epicalyx = lobes alternating with 5 calyx-lobes which are spreading in fruit. Leaves long-stalked, arising from the rootstock; leaflets ovate 2.5–4 cm, deeply and coarsely toothed; flowering stems 5–10 cm. Fruit *c.* 1 cm, red, insipid. **Pl. 36**.

440 **F. daltoniana** Gay

Uttar Pradesh to Sikkim. Burma. 2000–3600 m. Forests, shrubberies, shady banks; common. Apr.–Jun.

Like 439, but a smaller plant with larger oblong-ovoid fruit to 2.5 cm long, and persistent calyx with irregularly toothed lobes. Flowers white, *c.* 1.5 cm across. Leaflets elliptic, coarsely toothed, 1–2 cm. **Pl. 38**.

18 GEUM Avens Distinguished from *Potentilla* in fruit by the persistent long slender straight or hooked style to each carpel. 3 spp.

441 **G. elatum** Wallich

Pakistan to S.E. Tibet. 2700–4300 m. Shrubberies, open slopes. Jun.–Aug.

A perennial with pinnately-lobed basal leaves, and with few large yellow, or rarely red flowers borne on sparsely leafy erect stems. Flowers 2.5–4 cm across, erect or nodding; petals 5, broadly obovate, much longer than the triangular calyx-lobes; epicalyx-lobes tiny. Basal leaves 5–30 cm, hairy, pinnately cut into larger paired rounded and toothed broad-based lobes alternating with much smaller lobes; stem leaves much smaller; stems 10–30 cm. Fruit a globular head of usually hairy carpels with long nearly straight styles. **Pl. 36, 38**.

442 **G. sikkimense** Prain

C. Nepal to Bhutan. 3000–4100 m. Shrubberies, open slopes. Jun.–Aug.

Differs from 441 in having white to purplish-pink flowers with small petals widely spaced and little longer than the broadly triangular calyx-lobes. Flowers nodding, *c.* 2 cm across. Leaves with the terminal rounded lobe many times larger than the small and unequal lateral lobes which are usually less than 1 cm long. **Pl. 38.**

443 **G. roylei** Bolle
Afghanistan to C. Nepal. 1800–3600 m. Forests, shrubberies. Jun.–Aug.

Like 441 with yellow flowers, but flowers smaller 1.5–2 cm across, and styles jointed in the middle. Upper leaves large 3–lobed, with leafy stipules; basal leaves smaller, unequally pinnate-lobed. Achenes bristly-haired. (Very similar to **G. urbanum** L. WOOD AVENS, of Europe.)

19 POTENTILLA CINQUEFOIL Distinguished by its fruit which is a globular head of dry, 1–seeded carpels with short styles. Epicalyx-lobes 5, alternating with 5 sepals; stamens numerous. Flowers commonly yellow. Leaves trifoliate to pinnate. *c.* 39 spp.

Receptacle with long stiff hairs longer than the achenes
a *Lower leaves pinnate or palmately-lobed*
   i *Shrubs*

444 **P. fruticosa** L. SHRUBBY CINQUEFOIL
Pakistan to Bhutan. Temperate Eurasia. N. America. 2400–5500 m. Shrubberies, open slopes. Jun.–Sep.

An erect or low-spreading shrub, with pinnate leaves with 3–7 small leaflets, and bright yellow solitary flowers 1.5–4 cm across. Petals rounded to broadly obovate, much longer than the calyx; flower-stalks short, silky-haired. Leaflets ovate-lanceolate acute, entire, 3–16 mm; stipules conspicuous, brown papery.
Var. **rigida** (Wallich ex Lehm.) Wolf (*P. arbuscula* D. Don)

Forming thickets in Nepal in the lower alpine monsoon country. A tall shrub to 120 cm. Leaves green. **Page 464.**
Var. **pumila** Hook. f.

Common in dry Tibetan borderlands to 5500 m. A low prostrate shrub with tiny silvery-haired leaves. **Page 464.**
Var. **ochreata** Lindley

Common in drier areas of Nepal around Dhaulagiri and Annapurna to 4500 m. A much branched shrub 60–90 cm, with silvery-haired leaves.

  ii *Prostrate mat-forming woody-based perennials*

445 **P. biflora** Willd. ex Schltr. (*P. inglisii* Royle)
Pakistan to Bhutan. Temperate Asia. 4000–5800 m. Screes, rocks, stony slopes; common in drier areas. Jun.–Aug.

A densely mat- or cushion-forming plant often in large clumps, with tiny palmate leaves with 3–7 narrow leaflets, and with usually solitary, short-stalked, bright yellow flowers 1.5–2 cm across. Petals obovate, notched, longer than the broadly lanceolate calyx-lobes. Leaflets linear-blunt, 4–12 mm, silky-haired, pale beneath.

b *Lower leaves trifoliate*

446 P. cuneata Wallich ex Lehm.
Kashmir to S.W. China. 2400–4500 m. Rocks, stony slopes; quite common. Jun.–Sep.
A small spreading perennial with solitary yellow flowers 1.3–2.5 cm across, borne on short stems, and with leaves with 3 small obovate, coarsely toothed leaflets. Petals broadly obovate, longer than the triangular acute calyx-lobes and blunt elliptic epicalyx-lobes. Leaflets 6–13 mm, 3–toothed or 3–lobed at the apex, hairy or hairless; stems erect 2.5–10 cm. Achenes many, with long silky hairs. **Pl. 39**.

447 P. eriocarpa Wallich ex Lehm.
Pakistan to S.W. China. 3000–5000 m. Rocks. Jul.–Sep.
Like 446, but ends of stout rootstock-branches covered with silky or woolly stipular sheaths, and with slender flowering stems with usually solitary yellow flowers 2–3 cm across. Petals shallowly notched, much longer than the triangular calyx-lobes and smaller epicalyx-lobes. Leaves long-stalked; leaflets wedge-shaped, conspicuously toothed or 3–lobed, bright green; stipules silky, acute; rootstock often much branched. **Pl. 38**.

Achenes not concealed by long hairs
a *Lower leaves at least pinnate*
 i *Leaflets unequal, large and small leaflets mostly alternating*

448 P. polyphylla Wallich ex Lehm. (*P. mooniana* Wight)
Uttar Pradesh to Bhutan. Sri Lanka. 2400–4000 m. Shrubberies, open slopes. Jun.–Jul.
Distinguished by its leaves which have many elliptic toothed leaflets, the larger 1.3–4 cm, alternating with much smaller more rounded leaflets. Flowers yellow, rather small 8–13 mm across, either in a lax erect leafy branched cluster, or in a short flat-topped cluster; calyx shaggy-haired; epicalyx-lobes broad, toothed. Leaves narrow oblong, 15–25 cm; leaflets many, very unequal; stems several, 10–40 cm.

449 P. anserina L. SILVERWEED
Pakistan to Bhutan. N. Temperate Zone. 2400–4800 m. Open slopes; common on irrigated ground in Tibetan borderlands. Jun.–Aug.
Distinguished by its solitary yellow flowers borne on slender stalks from the axils of pinnate leaves with many pairs of alternating leaflets. Flowers 1.3–2.5 cm across; calyx-lobes elliptic, densely hairy. Leaves arising in a tuft from the rootstock; leaflets ovate, deeply toothed or shallow-lobed, 1–1.5 cm, alternating with tiny entire leaflets, silvery beneath. Stems slender creeping, often conspicuously pink, rooting at the nodes. Achenes hairy. **Pl. 38**.

ROSACEAE

ii *Leaflets more or less equal*
x *Ultimate leaf segments .5 mm or less broad; tufted or spreading perennials*

450 P. coriandrifolia D. Don
Uttar Pradesh to S.E. Tibet. 3600–4500 m. Open slopes. Jun.–Aug.
Distinguished by its small white flowers with reddish centres, and its small narrow leaves with leaflets cut into narrow segments. Flowers one or few, 1–1.5 cm across, on slender stems; calyx-lobes half as long as petals, dark, usually bristly-haired. Leaves 5–10 cm, mostly basal, numerous, with 6 to many broadly ovate leaflets deeply cut into linear segments; stems several, spreading 5–15 cm; rootstock stout. Achenes smooth, nearly hairless. **Pl. 38.**

451 P. microphylla D. Don
Uttar Pradesh to S.E. Tibet. 3600–4800 m. Rocks, open slopes. Jun.–Jul.
A dwarf densely tufted or mat-forming perennial, with pinnate leaves with numerous narrowly lobed leaflets, and usually solitary yellow flowers 1–1.3 cm across. Flowering stems leafless, 2–3 cm; petals rounded to obovate, much longer than the woolly-haired calyx. Leaves very variable, 2.5–5 cm, oblong, with densely grouped leaflets spreading forwards and upwards, silky beneath; rootstock stout, often branched at apex, covered with old leaf-bases. **Pl. 39.**

xx *Ultimate leaf segments 1 mm or more wide; usually erect perennials*

452 P. peduncularis D. Don
W. Nepal to S.W. China. 3000–4500 m. Shrubberies, open slopes; gregarious on grazing grounds. Jun.–Aug.
Distinguished by its few dark brown, relatively large achenes c. 2 mm across, and its pinnate leaves with numerous pairs of leaflets of approximately equal size. Flowers yellow, 1.3–2.5 cm across, few, usually on stems as long or longer than the leaves; calyx silky-haired. Leaves to 20 cm, mostly basal, silvery-haired when young, oblong, with many often overlapping oblong and deeply toothed leaflets mostly 1.3–2.5 cm long, with long soft silvery hairs beneath, and silky hairs or hairless above; rootstock stout, covered with old leaf-bases. **Pl. 37.**

453 P. sericea L.
Afghanistan to Kashmir. Tibet. N. Asia. 4000–4800 m. Stony slopes; common in Ladakh. Jun.–Aug.
Flowers yellow, several in dense hairy somewhat flat-topped clusters, and densely silky shining white pinnate leaves with deeply cut leaflets. Flowers 6–15 mm across; petals rounded, shallowly notched. Leaves oblong, crowded from the stout rootstock; leaflets 5–11, oblong, and cut nearly to the base into linear acute ascending lobes with incurved margins; stems to 45 cm, but dwarfed to c. 5 cm at high altitudes.

126

b *Leaves trifoliate or digitate, with 3 or 5 leaflets*
  i *Lower leaves with 5 leaflets*

454 **P. nepalensis** Hook.
Pakistan to C. Nepal. 2100–2700 m. Grazing grounds, cultivated areas. Jun.–Sep.

Distinguished by its dark crimson to bright rose, or orange flowers, borne in branched spreading clusters, and its digitate lower leaves with 5 leaflets. Flowers 1.3–2.5 cm across; petals rounded, shallowly notched, with a darker stalk; calyx shorter, hairy. Lower leaves long-stalked; leaflets ovate 2.5–8 cm long, conspicuously toothed; upper leaves with 3 narrower elliptic leaflets, short-stalked; stipules conspicuous, entire, 1–2 cm; stems leafy erect, 20–90 cm.

  ii *Lower leaves with 3 leaflets*

455 **P. atrosanguinea** Lodd. (*P. argyrophylla* Wallich ex Lehm.)
Afghanistan to Sikkim. 2400–4500 m. Shrubberies, open slopes; abundant on alpine grazing grounds. Jun.–Aug.

A variable perennial with rather large yellow, orange or dark red flowers, and with trifoliate leaves which are either densely silvery-haired, or green with paler undersides. Inflorescence lax, branched, few-flowered; flowers 2–4 cm across; petals obcordate, longer than the silvery-haired nearly equal calyx and epicalyx. Leaflets elliptic to ovate 2–5 cm, coarsely sharp-toothed; basal leaves long-stalked; stems erect or spreading, 10–60 cm. **Pl. 37.**

456 **P. curviseta** Hook.f.
Pakistan to Kashmir. 3000–4500 m. Rocky slopes; often forming large mats; common in Kashmir. Jul.–Aug.

A low mat-forming perennial, with 2–3 yellow flowers borne on short flowering stems, and with leaves with 3 narrow wedge-shaped leaflets which are coarsely 3–toothed at the apex. Flowers c. 1.5 cm across; petals obcordate; calyx silky-haired outside. Leaflets 1.3–2.5 cm, leathery, hairy on both sides; stipules linear forming a narrow wing to the leaf-stalk; old leaves persisting and becoming woody at crown of rootstock; flowering stems 2–8 cm. Achenes 5–6, hairy. **Pl. 37, 39.**

457 **P. gelida** C. Meyer
Afghanistan to Himachal Pradesh. N. Asia. 3000–4500 m. Rocky slopes; common in Kashmir. Jun.–Aug.

A low spreading perennial, with lax clusters of small yellow flowers, and with trifoliate leaves with ovate coarsely toothed leaflets. Flowers 8–15 mm across; petals obcordate; calyx densely hairy. Leaflets mostly 7–15 mm; stipules elliptic, often entire; stems spreading 5–15 cm. Achenes hairless.

20 SIBBALDIA Like *Potentilla*, but stamens usually 5, and achenes 5–12. 7 spp.

ROSACEAE

458 **S. cuneata** Hornem. ex Kuntze
Afghanistan to S.W. China. C. Asia. 3000–4500 m. Open slopes; common in Kashmir. Jun.–Aug.

A spreading mat-forming or tufted perennial, with tiny pale yellow flowers in rather flat-topped clusters, and leaves with 3–5 wedge-shaped leaflets with conspicuously toothed apices. Flowers c. 4 mm across; petals as long as the triangular, bristly-haired calyx-lobes. Leaves mostly basal; leaflets usually 5–12 mm; flowering stems erect, 1–10 cm. Achenes hairless. **Pl. 39.**

459 **S. purpurea** Royle
Kashmir to Bhutan. 3500–5300 m. Open slopes; common in Nepal. Jun.–Aug.

Like 458, but flowers red-purple, with widely spaced elliptic petals longer than the triangular calyx-lobes. Leaflets 5, elliptic to obovate, to 1 cm, silvery-haired. **Pl. 39.**

21 AGRIMONIA AGRIMONY Distinguished by its fruit which has a ring of hooked bristles above. Flowers yellow, in spikes. 1 sp.

460 **A. pilosa** Ledeb.
Pakistan to S.W. China. 1000–3000 m. Shrubberies, cultivated areas. Jun.–Sep.

Readily distinguished by its long slender spike-like cluster of small yellow flowers, and by its top-shaped fruit with numerous hooked spines. Flowers c. 5 mm across; petals oblong-obovate; calyx-tube c. 8 mm, grooved, topped with bristles; lobes triangular. Leaves hairy, pinnate with larger elliptic to lanceolate, coarsely blunt-toothed leaflets 2–5 cm, alternating with much smaller leaflets; stipules large, leafy; stem unbranched 30–150 cm. Fruit ribbed below, c. 5 mm. **Page 466.**

22 ARUNCUS Leaves 2–3–ternate. Fruit of 3–5 free carpels, each several seeded. 1 sp.

461 **A. dioicus** (Walter) Fernald (*Spiraea aruncus* L.)
Kashmir to S.W. China. Burma. Temperate Eurasia. N. America. 3300–4000 m. Forests, shrubberies. Jul.–Aug.

A slender erect herbaceous perennial, with large 2–3–ternate leaves, and with a large terminal pyramidal branched cluster of tiny pinkish-white flowers. Flowers c. 3 mm across, stalkless, usually one-sexed; petals 5; stamens numerous in male flowers. Leaves to 1 m; leaflets ovate acute long-pointed, deeply double-toothed, 2.5–6 cm; stem to 120 cm. Carpels 3, 2–3 mm, shining. **Page 464.**

23 FILIPENDULA MEADOW SWEET Leaves pinnately-cut. Fruit of 5–15 free carpels, each one-seeded. 1 sp.

462 **F. vestita** (Wallich ex G. Don) Maxim.

128

Afghanistan to W. Nepal. S.W. China. 2100–3300 m. Forests, damp places, Jul.–Aug.

An erect leafy perennial, with numerous tiny cream-coloured flowers in an oblong much-branched terminal cluster, and deeply pinnately cut leaves with very unequal, often tiny widely spaced, lateral lobes and a much larger palmate terminal lobe. Flowers 4–6 mm across; petals elliptic, longer than the blunt calyx-lobes. Leaves to 30 cm, the terminal lobe deeply 3–5–lobed, toothed, pale hoary beneath, 3–10 cm long; stems finely hairy above, 30–100 cm. Achenes *c.* 2 mm, with curved styles, hairy. **Page 462**.

SAXIFRAGACEAE Saxifrage Family

A medium-sized family of herbaceous plants mainly of the North Temperate Zone, with usually simple leaves, often forming rosettes. Flowers regular, usually with 5 sepals and petals, but sometimes 4; stamens equalling petals in number or twice as many. Ovary of 2 carpels, rarely more, frequently fused at their bases and tapering into the stigmas, with a flat or cup-shaped receptacle partly encircling the ovary, thus half-inferior; ovules many. Fruit a capsule, splitting along the inner edges of the carpels.

Leaves compound   1 ASTILBE   2 RODGERSIA
Leaves simple, toothed or lobed
  a Leaves large, blade 5 cm or more   3 BERGENIA
  b Leaves small, blade less than 5 cm   4 CHRYSOSPLENIUM
                                                     5 SAXIFRAGA

1 ASTILBE Leaves large, twice- or thrice-pinnate. Fruit of 2 splitting carpels. 1 sp.

463 **A. rivularis** Buch.–Ham. ex D. Don
Pakistan to S.W. China. S.E. Asia. 1800–3300 m. Shrubberies. Jul.–Sep.

An erect hairy herbaceous perennial 1–2 m, with large twice- or thrice-pinnate leaves, and with long terminal branched pyramidal clusters of tiny greenish-yellow flowers. Flower clusters 30–60 cm long; flowers *c.* 2 mm across; petals absent; calyx cup-shaped, deeply 5–lobed; stamens much longer, 5. Leaves 15–45 cm; leaflets ovate long-pointed, 2.5–10 cm, sharply and unevenly toothed, hairless above, bristly along the midrib beneath. Carpels 2, many-seeded, *c.* 4 mm. **Page 466**.

Recalling 461 **Aruncus dioicus** but differing in having no petals, fewer stamens, and 2 carpels.

2 RODGERSIA Leaves pinnate. Petals absent; stamens 10. Carpels fused. 1 sp.

464 **R. nepalensis** Cope ex Cullen
E. Nepal to Sikkim. 2600–3300 m. Forests, shrubberies; not common;

prominent on track up Dudh Kosi to Everest. Jul.

Flowers greenish-white, in a large much-branched domed cluster borne on a stem to 1 m. Flowers *c.* 8 mm across, stalkless; calyx glandular-hairy, the lobes triangular; stamens longer. Leaves with 5–11 large obovate to oblanceolate, long-pointed, sharp-toothed leaflets 15–20 cm. Carpels usually 2. **Page 467**.

3 BERGENIA Distinguished from *Saxifraga* by the large thick leaves with leaf-stalks sheathing at the base, and its stout flowering stem arising from the stout rootstock. 3 spp.

Leaf margins with bristly hairs

465 **B. ciliata** (Haw.) Sternb.
Afghanistan to S.E. Tibet. 1800–4300 m. Forests, rock ledges; common. Mar.–Jul.

Flowers white, pink or purple, in a spreading or dense cluster, borne on a stout leafless stem, and readily distinguished by its large rounded basal leaves with entire bristly margins. Flowers 1.5–2.5 cm long; petals 5; calyx hairless, lobes blunt; styles long. Leaves with ovate or rounded blade 5–15 cm at flowering, but enlarging to 30 cm or more and turning bright red in the autumn, margin fringed with long bristle-like hairs; leaf-stalk short. Flowering stem to 25 cm; rootstock very stout, creeping. **Pl. 40**.

466 **B. stracheyi** (Hook. f. & Thoms.) Engl.
Afghanistan to Uttar Pradesh. 3300–4500 m. Alpine slopes; the common alpine species of the W. Himalaya. Jun.–Aug.

Like 465, but margin of leaves finely toothed and with bristle-like hairs, and leaf-stalks very short with enlarged sheathing bases nearly throughout their length. Flowers pink, 2–2.5 cm long, in a drooping cluster; calyx hairy, with marginal bristles. Leaves obovate, 5–10 cm long, narrowed to the base; stem 10–25 cm. **Pl. 40**.

Leaf margins without teeth or bristles

467 **B. purpurascens** (Hook. f. & Thoms.) Engl.
C. Nepal to S.W. China. 3600–4700 m. Rocks, open slopes; the common alpine species of C. & E. Nepal. May–Jul.

Distinguished by its obovate entire hairless leaves, and dense terminal cluster of nodding purple flowers. Flowers few, 1.5–2.5 cm long; flower-stalks and calyx glandular-hairy; calyx-lobes very blunt. Leaves with blades 3–9 cm, stalked; flowering stem to 30 cm; rootstock stout creeping. Capsule with a long style to 2 cm.

4 CHRYSOSPLENIUM Golden Saxifrage Weak fleshy perennials, with small flowers without petals. Sepals 4; stamens 8. 9 spp.

468 **C. carnosum** Hook. f. & Thoms.
Uttar Pradesh to S.E. Tibet. 3600–5200 m. Stony slopes, screes. Jun.–Aug.

A delicate perennial with several stems bearing glossy green leaves, and a terminal cluster of greenish-yellow to golden flowers encircled by an involucre of leaves. Sepals 4, rounded. Lower stem leaves bract-like, the upper larger; the uppermost involucral leaves with rounded to broadly elliptic blades 5–10 mm across, with conspicuous rounded teeth; stems 4–8 cm. **Pl. 42.**

469 **C. nudicaule** Bunge
W. Nepal to S.W. China. 3600–4500 m. Streamsides, damp places. Jun.–Aug.

Like 468, but flowering stems leafless below, or with one leaf (not bract-like). Flowers greenish-yellow, several in a dense terminal cluster surrounded by golden involucral leaves with rounded to kidney-shaped blades 5–15 mm across with shallow rounded lobes. Basal leaves few, long-stalked.

470 **C. forrestii** Diels
C. Nepal to S.W. China. Burma. 3600–4800 m. Alpine slopes, stony ground; quite common. Jun.–Jul.

Like 469, but larger in all its parts; involucral leaves and basal leaves with blades 3 cm across; stems 12–15 cm. Flowers and involucral leaves shining yellow; flowers larger, *c.* 5 mm across.

5 **SAXIFRAGA** SAXIFRAGE Sepals and petals 5; stamens 10; ovary superior or half-inferior, of 2 carpels. Fruit a capsule. A large genus with *c.* 86 spp. in our area.

Flowers yellow
a *Flowering stems erect, leafy*
  i *Basal rosettes and runners absent*
    x *Upper leaves clasping stem, ovate to triangular-heart-shaped*

471 **S. moorcroftiana** (Seringe) Wallich ex Sternb.
Pakistan to S.W. China. 3600–4500 m. Alpine slopes; common. Aug.–Sep.

An erect perennial 10–40 cm, with terminal clusters of few bright yellow flowers, and with ovate clasping upper stem-leaves and narrower stalked lower leaves. Flowers several, in a terminal cluster; petals 7–10 mm, broadly oblong-ovate; sepals broadly ovate, shorter, glandular-hairy; flower-stalks glandular-hairy. Lower leaves hairless, with elliptic to lanceolate blades 3–5 cm; upper leaves 2–4 cm, progressively smaller above, glandular-hairy.

472 **S. parnassifolia** D. Don
Uttar Pradesh to Bhutan. 1900–4500 m. Shrubberies, open slopes; common in Nepal. Aug.–Oct.

Very like 471, but basal leaves broadly ovate with heart-shaped base (narrowed to the leaf-stalk in 471), and upper leaves triangular-heart-shaped with clasping rounded lobes. Flowers yellow; petals somewhat smaller 5–7 mm; sepals and bracts glandular-hairy. **Pl. 40**.

xx   *Upper leaves not clasping stem, oblanceolate to linear*

473   **S. hirculoides** Decne.
Pakistan to E. Nepal. Tibet. 4300–5000 m. Stony slopes, damp places.
Jun.–Aug.
   A rather delicate plant with solitary or less commonly several yellow
flowers borne on slender stems; upper leaves stalkless, oblanceolate to
linear. Petals elliptic, to 1 cm, longer or shorter than the blunt sepals;
flower-stalks densely brown-hairy above. Basal leaves elliptic to oblanceolate, stalked and with sheathing bases; stem leaves 8–15 mm, usually with
brown hairs at base; flowering stems 1 or several, 5–10 cm. Sometimes
whole plant covered with reddish hairs.

474   **S. brachypoda** D. Don
Uttar Pradesh to S.W. China. Burma. 3600–4800 m. Rocks, stony slopes.
Jul.–Aug.
   Distinctive with its numerous narrow-lanceolate rigid spine-tipped leaves
ranged regularly up the stems, and with solitary terminal golden-yellow
flowers borne on a short glandular flower-stalk. Petals much longer than
the ovate acute glandular sepals; anthers reddish. Leaves 5–9 mm, half-
clasping stem, densely overlapping, often with spiny-toothed margins;
stems many, often forming a moss-like clump, 5–10 cm. **Page 467**.

ii   *Basal rosette of leaves present*
   x   *Runners absent*

475   **S. lychnitis** Hook.f. & Thoms.
Kashmir to S.E. Tibet. 4500–5500 m. Stony slopes; a rare but attractive
species. Jul.–Aug.
   A small plant with basal rosettes of leaves, and erect leafy stems bearing
solitary usually nodding yellow flowers with red glandular-hairy calices.
Petals narrow-elliptic 1–2 cm, 3–veined. Basal leaves *c*. 1 cm, ovate-
spathulate; stem leaves narrower, 5–10 mm, with red glandular hairs.
Capsule hardly as long as calyx. **Pl. 41**.

xx   *Runners present*

476   **S. brunonis** Wallich ex Seringe
Pakistan to S.W. China. 3000–4500 m. Rocks, stony slopes; quite common
in Nepal. Jun.–Sep.
   476, 477, 478 are very similar, all having rosettes of narrow leaves, long
slender often red runners, and terminal cluster of small yellow flowers.
   476 is distinguished by its rigid linear-lanceolate leaves 1–1.5 cm long
and 1–2 mm broad, which have margins with bristle-like hairs and hairless
blades. Flowers yellow, usually 2–4; petals elliptic *c*. 7 mm, twice as long
as the hairless sepals; flower-stalks glandular-hairy. Stems often red, with
few leaves, 5–10 cm. **Pl. 42**.

477 **S. mucronulata** Royle

Uttar Pradesh to S.W. China. 3800–4800 m. Screes in drier areas, stony slopes. Jun.–Aug.

Very like 476, but leaves broader and shorter, the basal mostly 1 cm by 2 mm broad, and with regular stiff bristles ranged along the cartilaginous margin and a conspicuous terminal bristle. Flowering stems with many linear acute overlapping leaves (leaves few scattered, not overlapping in 476). Flowers yellow, with petals twice as long as the glandular-hairy sepals. Stolons glandular-hairy. **Pl. 40**.

478 **S. stenophylla** Royle (*S. flagellaris* auct. non Willd.)

Pakistan to C. Nepal. 3600–5000 m. Open slopes; common in Ladakh and Lahul. Jun.–Aug.

Very like 477, but leaves without marginal bristles but with glandular hairs on surface and margin; stem leaves many, overlapping. Flowers yellow; petals more than twice as long as sepals. **Pl. 41**.

b *Plants forming dense or loose prostrate tufts or cushions*

479 **S. jacquemontiana** Decne.

Pakistan to S.E. Tibet. 4000–5200 m. Stony slopes in drier areas; quite common in Kashmir. Jul.–Sep.

Distinguished by its hard mat-like growth of small densely crowded rosettes of blunt leaves, and its solitary yellow flowers which protrude just beyond the leaves. Flowers almost stalkless; petals 5 mm, elliptic blunt, nearly twice as long as the glandular sepals. Rosettes of leaves 5–8 mm across; leaves oblong *c.* 4 mm, conspicuously glandular-hairy. **Pl. 42**.

480 **S. saginoides** Hook.f. & Thoms.

Uttar Pradesh to S.W. China. 4300–5200 m. Screes, stony slopes. Jul.–Aug.

A moss-like plant forming large hard cushions to 45 cm across, and with tiny yellow flowers *c.* 5 mm across, usually flush with the cushion but sometimes short-stalked. Petals oblong blunt; sepals blunt, hairless. Leaves linear blunt, *c.* 6 mm, hairless. **Pl. 42**.

481 **S. engleriana** Harry Smith

C. Nepal to S.E. Tibet. 3600–4700 m. Screes, stony slopes, damp places. Jun.–Jul.

A delicate mat-forming perennial with tiny solitary yellow flowers with darker often red centres, and very small obovate leaves ranged along the slender spreading stems. Flowers *c.* 6 mm across; petals elliptic, widely spaced; sepals ovate, hairless. Leaves fleshy 1–3 mm; flower-stalk 1–3 cm. **Pl. 43**.

Flowers white, sometimes pink
a *Blade of leaf as broad as long*

482 **S. sibirica** L.

Pakistan to C. Nepal. Temperate Eurasia. 3000–5000 m. Rocks, open slopes, damp places; very common in W. Himalaya. Jun.–Aug.

A small erect leafy perennial, with few relatively large white flowers, and distinctive rounded-kidney-shaped shallow-lobed leaves. Petals 1–1.5 cm, obovate-wedge-shaped; sepals much shorter, blunt, glandular-hairy. Basal leaves with stalk 2.5–4 cm, and blade usually 0.5–2 cm across, with 5–7 shallow rounded lobes or coarse teeth; upper leaves smaller, stalkless, ovate to 3–lobed; stem hairy 7–20 cm; rootstock with bulbils. **Pl. 39**.

483 **S. asarifolia** Sternb. (*S. odontophylla* Wallich ex Hook.f. & Thoms.)
Kashmir to Bhutan. 3000–4300 m. Rocks, open slopes. Jul.–Aug.

Like 482, but a larger plant 15–30 cm, with flowers with white petals which are streaked crimson or red-spotted towards their base. Differing from 482 in the longer and densely hairy leaf-stalks to the basal leaves to 10 cm, and larger blade to 5 cm across, with more numerous rounded teeth or lobes. Flowers to 2 cm long; calyx and flower-stalks glandular-hairy. **Pl. 42**.

b *Blade of leaf longer than broad*
i *Leaves not in a dense rosette; plant not cushion-forming*

484 **S. pallida** Wallich ex Seringe
Pakistan to S.W. China. 3200–4800 m. Open slopes, rocks, moist places; common. Jun.–Aug.

A small rather erect perennial, with a slender branched inflorescence of few small white flowers, and with stalked basal leaves and stalkless upper leaves. Petals to *c.* 5 mm, twice as long as the sepals which are acute and hairless. Basal leaves with ovate toothed blades 1–2 cm long, upper leaves smaller, few; stems mostly 5–15 cm, hairy.

485 **S. pseudo-pallida** Engl. & Irmscher
Kashmir to S.W. China. 3600–5200 m. Screes, stony slopes. Jun.–Aug.

Very like 484, but leaves almost all basal in a lax rosette, with leaf-stalks usually shorter than the blade (longer in 484); stem-leaves if present bract-like, lanceolate. Filaments of stamens linear or narrowed to the apex (club-shaped in 484). Flowers white, 5–10 mm across; petals elliptic. **Pl. 43**.

486 **S. melanocentra** Franchet
C. Nepal to S.W. China. 4100–4900 m. Screes, stony slopes. Jul.–Aug.

Differs from 485 in having larger white flowers with orange spots at the base of the petals, and deep red to chocolate carpels. Flowers to 1.5 cm across; petals *c.* 8 mm, elliptic to ovate, conspicuously stalked. Leaves all basal; blade 1–2 cm, elliptic, narrowed to a long stalk, coarsely toothed; flowering stem often branched above, with 1–5 flowers.

487 **S. strigosa** Wallich ex Seringe
Uttar Pradesh to S.W. China. 2400–4300 m. Forests, mossy rocks, shady banks; common at low altitudes in Nepal. Aug.–Sep.

A small bristly-hairy perennial, with lax rosettes of oblanceolate leaves, and slender stems with distinctive bristly axillary bulbils, and small white flowers with yellow centres. Petals *c*. 5 mm, elliptic; sepals bristly, reflexed. Rosette-leaves 1–3 cm, with few coarse teeth; stem-leaves narrower, to 5 mm long; all leaves with bristly hairs; flowering stems glandular-hairy, often branched above, 10–15 cm. **Page 467**.

ii *Leaves in dense rosettes; plant cushion-forming*
x *Leaves green, not lime-encrusted*

488 **S. androsacea** L.
Pakistan to Kashmir. N. Asia. Europe. 3600–4300 m. Stony slopes; quite common in Kashmir. Jul.–Aug.
A small tufted perennial with rosettes of spathulate leaves, and with usually solitary stalkless or short-stalked white flowers. Petals 5–8 mm; calyx bell-shaped, with blunt glandular-hairy lobes. Leaves oblanceolate to spathulate, 1–1.5 cm long including stalk, entire or 3–lobed, with hairy margin. **Pl. 43**.

xx *Leaves lime-encrusted (Kabschia section)*

489 **S. pulvinaria** Harry Smith (*S. imbricata* Royle)
Kashmir to E. Nepal. 4300–5200 m. Rocks, stony slopes. Jun.–Aug.
A densely tufted plant forming firm cushions to 50 cm across. Leaves blunt, in tiny rosettes; flowers solitary, stalkless, white with yellow bases and violet anthers. Petals obovate 3–4 mm, margins undulate; sepals blunt, glandular-hairy. Rosettes dense, *c*. 5 mm across; leaves oblong 2 mm, densely overlapping, bent back at apex, and with a white terminal pit, lightly lime-encrusted. **Pl. 42**.

490 **S. andersonii** Engl.
W. Nepal to S.E. Tibet. 3600–5200 m. Rocks, stony slopes; common in drier areas of W. Nepal. May–Jul.
A dense cushion-forming plant with small cushions to 10 cm across of tiny lime-encrusted leaf-rosettes, and short compact terminal clusters of white or pinkish flowers. Petals 3–5 mm long; sepals with sparse or very dense glandular hairs; flowering stems 2–4 cm. Rosettes 5–8 mm across; leaves obovate, recurved, 5–7 mm; stem-leaves spathulate. **Pl. 43**.

491 **S. poluniniana** Harry Smith
W. & C. Nepal. 3300–3500 m. On rocks; rare. May–Jun.
A rather lax mat-forming plant, with pale green leaves with white incrustations, and solitary short-stalked white flowers often pink-flushed in the sun. Petals *c*. 10 mm, rounded and narrowed below; sepals ovate-blunt, glandular-hairy; anthers dark red. Leaves glandular-hairy, linear 5–6 mm, spreading and recurved; stems spreading to 12 cm, with dead persistent leaves.

492 **S. roylei** Harry Smith
C. Nepal. 3300–3800 m. Rock ledges; rare. Apr.–Jun.
A rather loose cushion-forming plant, with 1–3 white flowers borne on

short stems to 1.5 cm. Petals 4–6 mm, rounded, abruptly narrowed to a short stalk; sepals ovate, glandular-hairy; anthers reddish. Rosettes *c.* 4 mm across, with ovate hairless leaves 3–4 mm, with a terminal pore and marginal teeth at the base; old leaves hardening and persistent. **Pl. 41**.

PARNASSIACEAE   Grass of Parnassus Family

Often included in the Saxifragaceae but differing in having an additional whorl of 5 sterile stamens bearing nectaries. Ovary superior; carpels 4. Fruit a capsule. Leaves simple. A family of 1 genus of the North Temperate Zone.

PARNASSIA   8 spp.

493   **P. nubicola** Wallich ex Royle
Afghanistan to S.E. Tibet. 3000–4300 m. Open slopes, damp places; quite common. Jul.–Sep.

Readily distinguished by its solitary white flower borne on a slender stem, with a single stalkless clasping ovate leaf arising from below the middle of the stem, and with many basal leaves. Flower 1.5–2.5 cm across, with 5 spreading, often finely toothed, oblong to obovate petals; calyx-lobes blunt, hairless; stamens 5, alternating with the 5 fleshy 3–lobed nectaries. Basal leaves long-stalked, blade broadly elliptic-heart-shaped, acute, 2.5–5 cm; flowering stem 11–30 cm.

494   **P. cabulica** Planchon ex C. B. Clarke (*P. laxmannii* auct. non Pallas)
Pakistan to Kashmir. 2100–4300 m. Damp places; common in Kashmir and Ladakh. Jul.–Aug.

Very like 493, but differing in that the flowering stem is usually leafless, and the spathulate petals are usually 5–veined, 6–12 mm long (petals 7–veined, 12–15 mm long in 493). Nectaries distinctive, deeply 3–lobed, the lobes longer than broad (broader than long in 493). **Pl. 41**.

HYDRANGEACEAE   Hydrangea Family

A rather small family of deciduous shrubs, climbers, or trees, with simple opposite leaves without stipules, mainly of the north temperate and sub-tropical regions. It is often included in the Saxifragaceae. Sepals 4–10; petals 4–10; stamens 4 to numerous. Ovary superior to inferior, with 3–5 chambers and many ovules. Fruit a capsule which splits at the top, or a berry. (This includes the Philadelphaceae which is recognised by many botanists as a distinct family)

Flowers in flat-topped clusters, some outer flowers larger
                                1 HYDRANGEA
Flowers in branched not flat-topped clusters, flowers all equal
  a Stamens 20–40   2 PHILADELPHUS
  b Stamens 10–12   3 DEUTZIA 4 DICHROA

1 HYDRANGEA Sterile outer flowers with 4 large coloured calyx-lobes. Stamens 8 or 10; styles 2–4. 5 spp.

Climbing shrubs

495 **H. anomala** D. Don
Uttar Pradesh to S.W. China. Burma. 1800–2700 m. Climbing on trees and rocks; common in Nepal. Apr.–May.
    A large climbing shrub, with cinnamon-coloured shining bark which peels off in strips, and with large flat-topped clusters of tiny cream-coloured flowers and usually few (to 6) outer very much larger coloured sterile flowers. Clusters 10–15 cm across; fertile flowers *c.* 3 mm across; stamens many; petals falling off as a cap; outer sterile flowers with rounded whitish papery petal-like calyx-lobes 8–20 mm. Leaves ovate long-pointed, coarsely toothed, the blade 8–18 cm, the leaf-stalk 2.5–8 cm. Fruit globular *c.* 3 mm. **Pl. 44.**

Erect shrubs or small trees

496 **H. aspera** Buch.–Ham. ex D. Don
Uttar Pradesh to S.W. China. Burma. 1800–2700 m. Forests. Jul.–Aug.
    A shrub or small tree to 7 m, distinguished by its large lanceolate leaves which are paler and softly hairy beneath. Flower clusters 15–22 cm across, with 10 or more much larger white and green-veined, or pinkish, outer sterile flowers with elliptic usually toothed petal-like calyx-lobes to 2.5 cm; fertile flowers greenish-white. Leaves finely toothed, to 20 cm. Capsule with a cut-off apex, *c.* 5 mm long including styles.

497 **H. heteromalla** D. Don
Uttar Pradesh to S.W. China. Burma. 2400–3300 m. Forests. Jun.–Jul.
    Differs from 496 in having ovate or oblong leaves (not narrowed gradually to the leaf-stalk as in 496), and its capsule which is crowned by the thickened bases of the styles (not cut-off). Calyx-lobes of sterile flowers 2.5 cm long, entire. Leaf-blades 15–25 cm, densely and softly-woolly beneath. **Page 467.**

2 PHILADELPHUS Stamens numerous, filaments not winged. Petals usually 4. 1 sp.

498 **P. tomentosus** Wallich ex G. Don (*P. coronarius* auct. non L.)
Kashmir to S.E. Tibet. 1800–3300 m. Shrubberies, light forests. May–Jun.
    A deciduous shrub with light grey bark, with ovate-lanceolate long-pointed leaves, and with very fragrant white flowers, usually with 4 petals,

in axillary clusters at the ends of branches. Flowers 2–2.5 cm across; petals ovate; calyx hairy within, usually with 4 ovate-triangular pointed lobes. Leaves opposite, 4–10 cm, irregularly toothed, densely hairy, paler and with 5 prominent veins beneath. Capsule top-shaped c. 1 cm, 4–celled. **Pl. 44.**

3 DEUTZIA   Stamens 10, filaments winged. Petals 5. 3 spp.

499 **D. staminea** R. Br. ex Wallich
Pakistan to Bhutan. 1700–3000 m. Shrubberies; common in Uttar Pradesh and Nepal. Apr.–Jun.
A shrub to 2 m, with dark grey bark peeling in long strips, with lanceolate leaves which are grey-haired beneath, and with small axillary clusters of fragrant white flowers. Flowers 8–12 mm across; petals 5, to 1 cm, oblong, hairy beneath; calyx-lobes acute, matted with star-shaped hairs. Leaves 2.5–7 cm, minutely toothed, with grey star-shaped hairs beneath. Capsule rounded 3–4 mm. **Pl. 44.**

500 **D. compacta** Craib (*D. corymbosa* auct. non R. Br. ex G. Don)
Kashmir to S.W. China. Burma. 2100–3300 m. Shrubberies. Jun.–Jul.
Differs from 499 in having larger leaves 5–12 cm which are green beneath, and blunt calyx-lobes. Flowers fragrant, 12–18 mm across, borne on long lateral branches; petals obovate, 5–8 mm, overlapping in bud. A shrub to 3 m. **Pl. 44.**

501 **D. bhutanensis** Zaik.
C. Nepal to Bhutan. 2100–2700 m. Shrubberies. Apr.–Jun.
Distinguished by its larger pinkish-purple flowers with elliptic petals 10–12 mm long, in rather dense axillary clusters flowering with the young leaves or sometimes on bare branches. Leaves broadly lanceolate, to 4 cm, green on both sides, with scattered star-shaped hairs. A shrub to 2 m. **Pl. 43.**

4 DICHROA   Flowers blue or violet. Fruit a berry. 1 sp.

502 **D. febrifuga** Lour.
C. Nepal to China. S.E. Asia. 900–2400 m. Shrubberies, damp places; very common. May–Jul.
A shrub to 3 m, with opposite lanceolate leaves, and with dense terminal branched clusters of pale blue or violet flowers. Petals 5–6, thick, 3–4 mm long; calyx-lobes triangular *c.* 1 mm; stamens blue. Leaves stalked; blade lanceolate 10–20 cm, sharply toothed, almost hairless. Berries 5–6 mm, bright dark blue, topped by the persistent calyx.
The shoots and bark are collected and used as a febrifuge in the Darjeeling region. **Pl. 43.**

GROSSULARIACEAE   Gooseberry Family

A family of only one genus of the North Temperate Zone, often placed in the Saxifragaceae. Shrubs with alternate often lobed leaves. Distinguished by the conspicuous, often coloured, receptacle which is prolonged beyond the ovary, and bears the 4–5 sepals, petals, and stamens. Fruit a berry with persistent calyx at apex.

RIBES CURRANT, GOOSEBERRY   9 spp.

Prickly shrubs; flowers solitary

503   **R. alpestre** Wallich ex Decne. ASIAN GOOSEBERRY
Afghanistan to S.W. China. 2400–3600 m. Shrubberies, open slopes, drier areas. May–Jun.
   A shrub 1–2 m, with lobed leaves, 1–3 stout spines at each node, smooth or prickly stems, and small solitary axillary greenish to pink flowers. Flowers *c*. 1 cm long; calyx bell-shaped with narrow lobes becoming reflexed and larger than the petals; stamens protruding. Leaves rounded in outline, 1.5–3 cm, slightly hairy, long-stalked; lobes 3–5, rounded with blunt teeth. Berries globular or ovoid, 1 cm across, dull red, glandular-hairy. **Page 468**.

Shrubs without spines; flowers in clusters
a *Flower clusters long, pendulous*

504   **R. himalense** Royle ex Decne. (*R. emodense* Rehder)
Pakistan to S.W. China. 2400–3300 m. Forests, shrubberies; in drier areas only. Apr.–Jun.
   An erect shrub with hairless branches, with 3–5–lobed leaves, and with long pendulous clusters of reddish to greenish-yellow flowers. Clusters 5–15 cm long; flowers 5–8 mm long; calyx-tube hairless, the lobes petal-like blunt and much longer than the tiny petals and stamens. Leaves 3–7.5 cm, heart-shaped with acute or long-pointed deeply toothed lobes; leaf-stalk red. Berries red, *c*. 8 mm.
   Fruit edible; used for jams and cakes.

505   **R. griffithii** Hook.f. & Thoms.
W. Nepal to S.E. Tibet. 2700–4000 m. Forests, shrubberies. Apr.–Jun.
   Differs from 504 in having leaves which are hairy on the veins beneath, and flowers in lax pendulous clusters 8–15 cm long, with linear bracts longer than the flower-stalks (mostly shorter in 504). Flowers yellowish-green to reddish, 5–7 mm. Berries red. **Pl. 44**.

b *Flower clusters erect or ascending*

506   **R. orientale** Desf.
Afghanistan to W. Nepal. Bhutan. S.W. China. W. & N. Asia. Greece. 2100–4000 m. Stony slopes, rocks. May–Jun.
   A shrub 1–2 m, with glandular-hairy twigs, with small shallowly 3–5–

lobed leaves, and with erect clusters of small reddish-purple one-sexed flowers. Clusters 3–5 cm long; flowers 3–5 mm long; calyx-lobes rounded, hairy, longer than the tiny petals and stamens. Leaves 1.2–3.5 cm broad, rounded or kidney-shaped, lobes blunt and with rounded teeth, covered with stiff glandular hairs. Berries globular, deep red, sparsely glandular-hairy. **Pl. 44**.

507 **R. glaciale** Wallich
Afghanistan to S.W. China. Tibet. 2600–4400 m. Forests, shrubberies. Jun.–Aug.

An erect shrub 3–5 m, with reddish-brown, blackish-purple to greenish flowers in mostly erect glandular clusters 3.5–5 cm long. Petals minute, much shorter than the nearly hairless petal-like calyx-lobes. Leaves 2.5–5 cm long, heart-shaped and 3–5–lobed, the lobes acute and deeply and sharply toothed, nearly hairless. Berries red, c. 5 mm.

508 **R. takare** D. Don
Uttar Pradesh to S.W. China. Burma. 2200–3300 m. Forests, shrubberies. Apr.–Jun.

An epiphytic shrub with long mostly erect clusters to 9 cm of dark purple to dull red flowers. Flowers 3–4 mm long; petals longer than glandular-hairy calyx-lobes. Leaves triangular-heart-shaped, 6–10 cm including stalk, 3–lobed with the central lobe much larger and longer, the lobes double-toothed; stems to 2.75 m. **Pl. 44**.

CRASSULACEAE   Stonecrop Family

A rather small family of characteristically succulent plants, mainly herbaceous or less commonly small shrubs, widely distributed in the warmer and drier regions of the world. Leaves almost always simple, and often clustered in rosettes, fleshy, waxy, spiny or hairy, and thus reducing dessication. Often reproducing by means of leaves, rosettes or bulbils which root readily. Flowers small, massed, in flat-topped clusters, or branched panicles. Sepals and petals 3–30 but often 5; stamens usually twice as many as petals. Ovary superior, of usually 5 free carpels; ovules numerous. Fruit a cluster of *follicles* each splitting down one side.

Petals fused into a tube at base   1 KALANCHOE   2 ROSULARIA
Petals free to the base
    a Rhizome stout, with persistent scaly leaves; petals 4   3 RHODIOLA
    b Rhizome absent or slender; petals usually 5, or more   4 SEDUM

1 KALANCHOE Corolla-tube flask-shaped, with spreading lobes; stamens 8. 1 sp.

509 **K. spathulata** DC.
Kashmir to Bhutan. China. Burma. Java. 300–1500 m. Rocky slopes.
Sep.–Apr.

Prominent when in flower in winter with its flat-topped clusters of
yellow flowers borne on stout stems, and with large fleshy oblong to spathulate toothed leaves. Flowers numerous, in branched clusters; corolla tubular
2–2.5 cm long, with 4 spreading elliptic lobes; calyx much shorter, divided
nearly to the base into 4 erect lobes. Leaves 10–20 cm, the lower stalked
and crowded, the upper smaller, nearly stalkless, all hairless; stem 30–120
cm. Fruit of 4 carpels enclosed in the persistent calyx and corolla. **Pl. 45.**

2 **ROSULARIA** Perennials with basal rosettes. Corolla-tube about as
long as the erect lobes; stamens 10. 4 spp.

510 **R. alpestris** (Karelin & Kir.) Boriss. (*Sempervivella acuminata* (Decne.)
A. Berger)
Afghanistan to Himachal Pradesh. C. Asia. 2400–4500 m. Rocky slopes;
dry areas only. Jul.–Sep.

Distinguished by its dense rosettes of narrow oblanceolate pointed
leaves (recalling *Sempervivum*), and with one to three erect annual stems
bearing a dense cluster of pink-purple to white flowers. Petals 6–10 mm,
with a red mid-vein; calyx less than half as long; filaments of stamens
white. Leaves hairless, fine-pointed, 2.5–3 cm, the stem leaves smaller,
scattered, narrow oblong, long-pointed; stems 5–10 cm.

511 **R. rosulata** (Edgew.) H. Ohba (*Sedum r.* Edgew.)
Afghanistan to W. Nepal. 1500–3000 m. Forests, damp rocks; common in
the W. Himalaya. May-Jun.

A small plant with a lax terminal cluster of white flowers each about 5
mm across, and lax rosettes of spathulate leaves. Petals *c.* 5 mm, narrow
elliptic, minutely hairy, more than twice as long as the oblong sepals.
Rosette-leaves 6–20 mm, broadly spathulate, numerous; stem-leaves
smaller, 3–5 mm, narrower, distant; stems 4–8 cm. **Pl. 45.**

3 **RHODIOLA** Rootstock thick fleshy, with persistent scales. Flowering
stems with alternate leaves. Flowers in flat-topped clusters. Included in the
past in the genus *Sedum*. Species are difficult to distinguish. *c.* 20 spp.

Plants with bisexual flowers (*ovaries as long as stamens*)
a *Leaves not linear*

512 **R. sinuata** (Royle ex Edgew.) S. H. Fu (*Sedum trifidum* Wallich)
Kashmir to Bhutan. 1200–4300 m. Rocks, open slopes. Aug.–Sep.

A small plant, often epiphytic or growing on boulders, with leafy
clusters of white to pale pinkish flowers, and fleshy oblong to ovate usually
pinnately-lobed leaves. Flowers in terminal branched clusters often surrounded by longer leaves; petals linear-lanceolate 5 mm, twice as long as
sepals. Leaves very variable, 2.5–5 cm, with 3–5 unequal spreading blunt
lobes, or sometimes not lobed; stems several, erect 8–30 cm, hairless.

b *Leaves linear*

513 **R. wallichiana** (Hook.) S. H. Fu (*Sedum asiaticum* auct. non DC., *Sedum crassipes* Wallich ex Hook.f. & Thoms.)
Kashmir to Bhutan. 3000–4800 m. Open slopes, rocks; forming large clumps; common in Kashmir. Jun.–Sep.

Distinguished by its usually pale yellow flowers in a compact terminal cluster, borne on erect stems with numerous overlapping fleshy linear leaves. Flowers bisexual; petals lanceolate, *c.* 8 mm, twice as long as sepals. Leaves 2.5–3 cm, remotely toothed; stems many, stout, 15–30 cm; rootstock stout, unbranched. Fruit red. **Pl. 45**.

Plants one-sexed (ovaries much shorter than stamens in male flowers)
a *Rootstock densely surrounded by tufts of old persistent flower stems*
   i *Leaves 0.5–1.3 mm wide, linear to narrow lanceolate*

514 **R. quadrifida** (Pallas) Fischer & C. Meyer
Pakistan to Bhutan. C. & N. Asia. N. Europe. 3300–4500 m. Rocks, open slopes. Jun.–Aug.

A compact cushion-plant with a much-branched rootstock and many very short aerial stems, with linear leaves, and with terminal few-flowered clusters of yellow to reddish flowers. Petals mostly 4, 2–4 mm, twice as long as the oblong sepals. Leaves *c.* 6 mm long, nearly cylindrical, sometimes fine-pointed. Stems less than 5 cm long, arising from the stout branched rootstock and surrounded by old stems.

515 **R. fastigiata** (Hook. f. & Thoms.) S. H. Fu
Pakistan to S.W. China. 3600–5500 m. Open slopes, rocks. Jun.–Jul.

Like 514, but flowers usually pinkish-white, and stems many, clustered, longer, 6–13 cm. Flowers mostly with 5 petals and sepals; petals 4–6 mm long. Leaves densely arranged throughout, linear to very narrowly ovate, 8–12 mm long; flowering stems 8–15 in number.

   ii *Leaves 2–7 mm wide, oblanceolate to narrowly elliptic*

516 **R. himalensis** (D. Don) S. H. Fu
Kashmir to S.W. China. 3600–4800 m. Streamsides, alpine shrubberies, rocky slopes. Jun.–Aug.

A large showy plant with stout leafy stems 10–35 cm, with dense terminal clusters of dark red, pinkish to yellow flowers, and with oblanceolate to narrow elliptic fleshy leaves. Flower clusters large, 3–5 cm across; flowering stems many, arising from a stout erect rootstock covered with old stems. **Pl. 45**.

b *Rootstock usually lacking tufts of old flower stems*
   i *Stamens longer than petals*

517 **R. heterodonta** (Hook. f. & Thoms.) Boriss.
Afghanistan to C. Nepal. 3900–5000 m. Stony slopes; Tibetan border areas. Jun.–Aug.

Distinguished from 518 by its dense domed clusters usually 2.5–3 cm across, of stalkless yellow flowers without bracts. Sepals often reddish. Leaves flat, lanceolate-triangular, to pentagonal-ovate, toothed; stems several, stout leafy, pale green, 10–25 cm, increasing to 40 cm in fruit. Rootstock very thick. **Pl. 45**.

518 **R. imbricata** Edgew. (*Sedum roseum* auct. non Scop.)
Pakistan to C. Nepal. 4000–5000 m. Rocky slopes; common in drier areas of W. Himalaya. Jun.–Jul.

Flowers pale yellow with green or often reddish sepals, in a very dense compact terminal cluster surrounded by an involucre of leaves. Inflorescence with bracts. Leaves 1.3–3 cm long, oblong to narrow elliptic, nearly entire; flowering stems 3–5, reddish, 10–20 cm. Rootstock massive, 2–2.5 cm across, without suckers.

ii *Stamens nearly as long or shorter than petals*

519 **R. bupleuroides** (Wallich ex Hook. f. & Thoms.) S. H. Fu
W. Nepal to S.W. China. 2000–5000 m. Rocks, stony slopes. Jun.–Jul.

Flowers red to black-purple, in large leafy flat-topped branched often lax clusters, borne on leafy stems with broad ovate to elliptic leaves. Petals lanceolate, twice as long as the narrower sepals. Leaves 1.3–5 cm, obscurely toothed, often with a heart-shaped or auricled base; stems usually few, 7–35 cm. **Pl. 45**.

4 SEDUM STONECROP Succulent annuals or perennials, usually without rosettes. Petals free (or almost so), usually spreading. *c*. 16 spp.

Flowers pink; leaves broadly ovate

520 **S. ewersii** Ledeb.
Afghanistan to Uttar Pradesh. C. Asia. 2700–4500 m. Stony slopes, irrigated areas; prominent in Ladakh and Chenab valley. Jul.–Sep.

Readily distinguished by its flat fleshy, bluish-green rounded leaves borne on spreading stems, and its dense domed clusters of numerous pink flowers. Flowers 6–8 mm across, with 5 petals spreading in a star; stamens 10, filaments rose-purple. Leaves mostly opposite, 1.3–2.5 cm, ovate to rounded in outline with a somewhat wavy margin; stems often many, 8–75 cm. Fruit of 5 erect carpels. **Pl. 45**.

Flowers yellow; leaves linear to lanceolate

521 **S. multicaule** Wallich ex Lindley
Kashmir to S.W. China. 1500–3200 m. Forests, mossy banks. Jun.–Aug.

A small plant with stems usually branched from the base, with terminal clusters of yellow flowers which are at first dense and later spreading and branched. Flowers stalkless, subtended by an involucre of longer leaves; petals *c*. 4 mm, about as long as the sepals; stamens 10. Leaves 1–2.5 cm, fleshy, cylindrical acute, usually numerous; stems 5–12 cm. Carpels 5, spreading in fruit. **Page 468**.

522 **S. oreades** (Decne.) Raym.–Hamet
Kashmir to S.W. China. 3200–5500 m. Open slopes, rocks; a common alpine. Jul.–Aug.
A hairless annual with several short stems arising from the rootstock, with narrow lanceolate fine-pointed leaves, and with 1–3 terminal yellow flowers *c.* 1 cm across. Petals elliptic, longer than the narrow elliptic sepals which are 6–7 mm; stamens yellow, 5. Leaves 3–6 mm, thick fleshy; stems 5–10 cm. **Page 468**.

DROSERACEAE   Sundew Family

A small cosmopolitan family of carnivorous plants, which in the case of *Drosera* have long red glistening gland-tipped hairs on their leaves. These attract insects which are trapped in the sticky gum; the leaf-blade then folds round the prey and slowly digests it. Flowers regular, usually in branched clusters, with commonly 5 sepals, petals and stamens. Ovary superior, of 2–3 or 5 carpels, which are fused. Fruit a capsule.

DROSERA   2 spp.

523 **D. peltata** Smith
Uttar Pradesh to S.W. China. India. S.E. Asia. Australia. 1500–3600 m. Grassy slopes; not common. Jul.–Aug.
Distinguished by its half-moon-shaped leaf-blades *c.* 6 mm across, with the upper side of the blades covered with long sticky glandular hairs. Flowers white, small *c.* 6 mm across, in terminal branched clusters; petals 4–8 (usually 5); stamens many; sepals variable, toothed slightly fringed; ovary with 3 styles. Leaves stalked, borne alternately on an erect slender stem 10–30 cm (in var. **lunata** (Buch,–Ham. ex DC.) C. B. Clarke smaller rosette leaves absent at flowering). **Page 468**.

HAMAMELIDACEAE   Witch Hazel Family

A small family of trees and shrubs largely of the sub-tropical regions, but with a very discontinuous distribution. Flowers in our species bisexual or female, often borne in a spike or head. Calyx of 4–5 fused sepals; petals 4–5, free, or absent; stamens 4–5. Ovary 2–celled, with 2 styles. Fruit a woody capsule.

PARROTIOPSIS   1 sp.

524 **P. jacquemontiana** (Decne.) Rehder

Afghanistan to Himachal Pradesh. 1500–2100 m. Forest undergrowth; gregarious. May.

A shrub or small deciduous tree to 5 m, with grey bark and with rounded, strongly-veined, toothed leaves. Flowers yellow, tiny, in dense globular heads 1.5 cm across. Flowers borne on short lateral shoots usually with young leaves and surrounded by 4–6 large conspicuous creamy petal-like obovate bracts to 2.5 cm. Petals absent; stamens numerous, yellow. Leaves 4–8 cm, with star-shaped hairs on the veins beneath. Capsule woody, surrounded by the calyx-tube, splitting into 4, and with grey star-shaped hairs; seeds brown, shining.

Wood is used for making small articles; the twigs for basket-making. Unpopular with foresters because it may prevent rejuvenation of conifer forests. **Pl. 46**.

MYRTACEAE   Myrtle Family

Callistemon citrinus (Curtis) Skeels   Bottle-Brush Tree
A native of Australia; cultivated in the Himalaya to 1500 m, particularly in the Nepal valley. It has showy cylindrical clusters 7–15 cm long of red flowers with very long red stamens; flowering in Mar. and Apr. A small evergreen tree or shrub with glossy lanceolate leaves.

Eucalyptus spp. Australian Gum Trees
Often planted by the Indian Forest Service in the plains and outer hills. Also planted in the Nepal Valley.

MELASTOMATACEAE

A mainly tropical family of shrubs, small trees, and some climbers and herbs. Distinguished by its leaves which are opposite and decussate (each pair set at right-angles to the next), and leaf-blades with 3–9 nearly parallel veins arising from the base and converging towards the apex and with fine transverse veins between them. Stamens characteristically bent in the middle, and with sterile appendages of various shapes. Inflorescence of large showy solitary flowers, or in flat-topped or long and branched clusters. Sepals 4–5, fused; petals 4–5, large showy; stamens twice as many as petals. Ovary superior or inferior, with 1–14 carpels and 4–5 chambers. Fruit a berry, or capsule.

MELASTOMA   Stamens of two kinds: 5–7 longer, with purple anthers and 2 yellow swellings, and 5–7 shorter, with yellow anthers. Apex of ovary bristly. 2 spp.

525 **M. normale** D. Don
  C. Nepal to Bhutan. China. Burma. S.E. Asia. 1000–2100 m. Open slopes, shrubberies, wasteland. Mar.–Jun.
  A large bristly-haired shrub to 7 m, with terminal clusters of bright pinkish-mauve flowers each 5 cm across. Calyx densely feathery-haired, with long lanceolate lobes; petals 5, obovate, to 3 cm; stamens 5 long curved, 5 short. Leaves broadly lanceolate 8–15 cm, with 3–5 strong equal parallel veins, bristly-haired beneath, stalked. Fruit berry-like, splitting irregularly. **Pl. 46.**

OSBECKIA Stamens all similar, 8–10; apex of ovary hairless. 7 spp.

526 **O. stellata** Buch.–Ham. ex D. Don
  Uttar Pradesh to Arunachal Pradesh. Burma. 1200–2400 m. Shrubberies, banks, cultivated areas. Jul.–Oct.
  A straggling branched shrub, often hanging from rocks and banks, with large pinkish-purple flowers in terminal clusters, each flower 5–7 cm across. Calyx tubular, with erect lobes, conspicuously covered with stiff matted star-shaped bristles; petals usually 4, rounded, c. 2 cm, twisted to the right in bud; stamens curved, all similar. Leaves opposite, 7–15 cm, lanceolate long-pointed entire, with 5 prominent nearly parallel veins. Fruit densely hairy, 1.2–1.8 cm, opening by pores at apex. **Pl. 46.**

527 **O. nepalensis** Hook.
  C. Nepal to S.W. China. S.E. Asia. 600–2300 m. Shrubberies, waste land, cultivated areas. Jul.–Nov.
  Differs from 526 in having the calyx-tube covered with large flat scales fringed with bristles, and anthers with no apical beak (anthers beaked in 526), filaments not curved. Flowers mauve-purple or white, smaller 3–5 cm across, in rather dense branched clusters; calyx-lobes lanceolate, hairless. Shrub to 1 m or more, with lanceolate entire leaves with 3–5 conspicuous parallel veins; stems 4–angled. **Pl. 46.**

OXYSPORA Flowers in branched clusters; stamens 4 long purple, 4 short yellow. 1 sp.

528 **O. paniculata** (D. Don.) DC.
  C. Nepal to S.W. China. Burma. S.E. Asia. 1200–2100 m. Shrubberies, cultivated areas. Aug.–Oct.
  A shrub to 2 m with spreading drooping branches and with long pendulous pyramidal flower clusters, with the calyx, petals and flower-stalks all pink-purple. Flowers c. 2 cm long; petals 1 cm, elliptic, soon falling; stamens 8, 4 longer with curved filaments and purple anthers, 4 shorter with yellow anthers. Leaves elliptic acute, mostly 15–25 cm, with 5–7 very conspicuous parallel veins and numerous fine cross-veins, and with few rusty star-shaped hairs beneath. Fruit spindle-shaped, 8–ribbed, c. 1 cm. **Pl. 47.**

LYTHRACEAE   Loosestrife Family

A small family of herbaceous plants, shrubs, and trees mainly of the Tropics, but also of the temperate regions. Leaves opposite or whorled, simple and entire. Flowers in racemes, panicles, or flat-topped clusters, usually regular and bisexual. Petals and sepals usually 4; stamens typically twice as many, but a considerable variation in numbers occurs; petals and stamens borne on the cylindrical calyx-tube; petals crumpled in bud. Epicalyx often present. Ovary at base of calyx-tube, with 2–6 chambers and a single style which is often either long, short, or intermediate in length (*heterostyle*) in different flowers.

ROTALA   Annuals. Petals obovate, 4, twice as long as calyx-teeth; capsule 4–valved. 6 spp.

529 **R. rotundifolia** (Buch.–Ham. ex Roxb.) Koehne (*Ammania r.* Buch.–Ham. ex Roxb.)
India. China. Japan. Throughout Himalaya to 2300 m. Paddy stubble-fields, damp places; often gregarious. Feb.–Apr.`
A small annual with small paired broadly ovate almost stalkless leaves, and with terminal spikes or branched clusters of tiny rose-coloured flowers. Spikes usually several, 1–3 cm long; flowers closely clustered, *c.* 2 mm across; bracts ovate, purple. Leaves 8–15 mm, often widely spaced up stem, margins somewhat undulate; flowering stem erect, 8–15 cm, usually unbranched; plant creeping and rooting from the nodes below. **Page 468**.

LYTHRUM   Perennials growing in damp places. Flowers showy; stamens 12. 1 sp.

530 **L. salicaria** L. PURPLE LOOSESTRIFE
Kashmir to Himachal Pradesh. W. Asia. Europe. 1200–1800 m. Marshes, ditches, lakesides; common in W. Himalaya. Jun.–Sep.
A rather slender waterside plant, with dense terminal sometimes branched spikes 9–45 cm long of bright rosy-purple flowers. Flowers 1–1.5 cm across; petals oblong; calyx tubular with alternating short triangular lobes and longer awl-shaped lobes, ribbed, hairy; stamens 12. Leaves lanceolate, 1–6 cm, stalkess, opposite or in whorls of 3; stems to 1 m or more.

WOODFORDIA   Shrubs with red flowers; stamens 12. 1 sp.

531 **W. fruticosa** (L.) Kurz
Pakistan to S.W. China. India. Burma. W. Asia. Africa. 300–1800 m. Light forests, open slopes; common. Feb.–Apr.
An evergreen shrub 1.5 m with spreading branches, with opposite lanceolate pointed leaves, conspicuous when in flower with short dense axillary clusters of bright red tubular flowers often borne on lower leafless branches. Calyx tubular, bright red, 9–13 mm long, with 6 short triangular

lobes; petals small, red, scarcely as long as calyx-lobes; stamens with red filaments, much longer than the calyx-tube. Leaves 5–10 cm, usually two-ranked, sometimes in whorls of 3, gland-dotted, paler beneath. Capsule ovate or elliptic, 6–10 mm.

Leaves and twigs give a yellow dye; the flowers a red dye. Bark and flowers used medicinally. **Pl. 47.**

PUNICACEAE   Pomegranate Family

A family of only two species, distinguished by the crown of persistent calyx on the fruit, and the unusual arrangement of the carpels which are in two layers one above the other in ovary and fruit.

PUNICA   1 sp.

532   **P. granatum** L. Pomegranate

W. Asia. S. Europe. Probably native in Pakistan and W. Himalaya. 700–2700 m. Gregarious on dry limestone soils; naturalized elsewhere, and commonly cultivated in gardens. Apr.–May; fruiting Jul.–Sep.

Usually a deciduous shrub but sometimes a small tree, with thin smooth grey bark, with large crinkle-petaled scarlet flowers 4–5 cm across, and with large scarlet or brownish orange-like fruits. Calyx scarlet, 2–3 cm long, tubular, with 5–7 triangular fleshy lobes; petals 5–7, scarlet, broadly obovate, 1.5–3 cm; stamens very numerous. Leaves 2–8 cm, entire, lanceolate to broadly oblanceolate, opposite, shining. Fruit globular 4–8 cm, crowned with calyx, and with a thick leathery rind; flesh pink juicy, with red or pink seeds.

Fruit edible. Most parts of the plant are used medicinally.

ONAGRACEAE   Willow-Herb Family

A cosmopolitan family of annuals or perennials, rarely shrubs; with simple leaves. Flowers in *Epilobium* borne in racemes. Flowers regular or weakly asymmetrical; sepals usually 4, free; petals 4, free; stamens 8. Ovary inferior, of 4 fused carpels. Fruit a capsule splitting longitudinally, with many seeds each with a plume of hairs.

EPILOBIUM   Willow-Herb   *c.* 23 spp.

Stigma deeply 4-lobed
  a *Leaves all alternate; flowers weakly asymmetrical, large*
    i *Inflorescence spike-like, leafless*

148

ONAGRACEAE

533 **E. angustifolium** L. (*Chamaenerion* a. (L.) Scop.) ROSEBAY WILLOW-HERB
Afghanistan to S.W. China. North Temperate Zone. 3000–4300 m. Open
slopes. Jul.–Sep.

An erect leafy perennial to 2 m, with a long nearly leafless terminal
spike-like cluster of conspicuous pink flowers, and numerous spirally-
arranged narrow lanceolate leaves on the stem below. Flowers held
horizontally, 2–2.5 cm across, with 4 spreading obovate stalked petals and
shorter narrower dark purple sepals. Stamens conspicuous, curved outwards; style 1–2 cm, curved downwards, with 4 stigmas. Leaves 2.5–20
cm, nearly hairless. Capsule 2.5–8 cm, finely hairy.

ii *Inflorescence leafy*

534 **E. conspersum** Hausskn. (*E. reticulatum* C. B. Clarke)
C. Nepal to S.W. China. 3600–4500 m. Screes, stony slopes. Jul.–Sep.

Like 533, but leaves hairy beneath, with lateral veins diverging at an
acute angle from the mid-vein (at right-angles in 533), and with larger,
long-stalked flowers c. 3 cm across borne in short leafy clusters.

535 **E. latifolium** L.
Afghanistan to C. Nepal. N. Asia. N. America. 3600–4500 m. Screes,
damp places. Jul.–Aug.

Like 534 with flowers borne in the axils of leafy bracts, but flowers
larger, to 5 cm across, with rounded short-stalked petals; styles short, 4–8
mm, and deflexed throughout flowering (becoming erect in 534). Leaves
narrow-elliptic, 4–8 cm, narrowed above and below, hairy beneath; stems
usually to 30 cm, often spreading. **Pl. 47.**

b *Leaves opposite, at least below*

536 **E. hirsutum** L. GREAT WILLOW-HERB
Afghanistan to C. Nepal. Temperate Eurasia. 1000–3300 m. Cultivated
areas, grazing grounds. Aug.–Oct.

An erect robust perennial usually densely clothed with white hairs, and
with large rosy-purple flowers in a terminal leafy spike-like cluster. Petals
notched, 8–12 mm long; style with 4 spreading then recurved stigmas.
Leaves lanceolate, 2.5–8 cm, clasping stem, margin finely toothed, softly
and densely hairy; stems often branched below, 80–150 cm. Capsule 4–6
cm, hairy.

Stigma entire or nearly so
a *Stem nearly hairless, and with hairless raised lines*

537 **E. cylindricum** D. Don
Afghanistan to S.W. China. 2100–4300 m. Damp places. Jul.–Sep.

An erect perennial 30–90 cm, with linear-lanceolate short-stalked leaves,
and numerous axillary clusters of small pale pink flowers each 6 mm long.
Stigmas club-like. Leaves 2.5–7.5 cm, tapering to a fine point with fine
acute teeth, nearly hairless; stems nearly hairless with hairless raised lines.
Capsule 4–7.5 cm.

b *Stems hairy, or with hairy raised lines*

538 **E. wallichianum** Hausskn.
C. Nepal to S.W. China. 1700–3000 m. Shrubberies, open slopes. Jul.–Sep.

   Distinguished from 537 by its stems which are triangular in section, and by its broader elliptic often blunt leaves with conspicuous fine marginal teeth. Flowers pink; petals 8 mm, notched; stigma rounded. **Pl. 48**.

539 **E. laxum** Royle
Afghanistan to Uttar Pradesh. Temperate Eurasia. 2100–4200 m. Open slopes, damp places. Jul.–Aug.

   Like 538, but flowers larger, petals 6–16 mm, and stems rounded with 2 hairy raised lines. Leaves elliptic to narrow-ovate with a long slender apex, conspicuously toothed; stem often branched. **Pl. 48**.

TRAPACEAE   Water-Chestnut Family

A family of floating aquatic plants consisting of one genus of one or several species, depending upon the differing botanical interpretation of the concept of a species.

TRAPA   1 sp.

540 **T. quadrispinosa** Roxb. (*T. natans* auct. non L.)   WATER-CHESTNUT
Afghanistan to C. Nepal. India. To 1700 m. Lakes and still waters; common in Kashmir valley. Jul.–Sep.

   An aquatic plant with lax rosettes of triangular, coarsely toothed leaves floating on the surface, and with submerged leaf-stalks often with large inflated elliptic air-bladders. Flowers white; petals 4, *c.* 8 mm; calyx hairy, fused to the ovary and becoming hard and spiny in fruit. Leaf-blades 2–4 cm by 3–6 cm broad, sometimes shallowly 3–lobed, shining dark green above, reddish-purple and very hairy beneath. Submerged stems slender, flexible, with numerous comb-like lateral roots. Fruit top-shaped, with 2–4 sharp horns.

   Often cultivated for its edible fruit, which is eaten raw or cooked. **Page 469**.

CUCURBITACEAE   Gourd or Cucumber Family

A medium-sized family of mainly herbaceous climbing plants, which is widespread in tropical and sub-tropical regions, but is poorly represented in temperate regions. Usually with tendrils, and palmately veined, simple or lobed leaves. Flowers one-sexed, usually with yellow petals, and with an inferior ovary. Sepals and petals usually 5, borne on a cup-shaped or tube-like extension of the receptacle; petals usually fused below; stamens commonly 3, fused by different parts in different genera. Ovary inferior, with one chamber. Fruits either firm-walled berries, or dry capsules.

HERPETOSPERMUM Petals 5, fused only at the base, margins entire. 1 sp.

541 **H. pedunculosum** (Seringe) Baillon

Himachal Pradesh to S.W. China. 1800–3600 m. Climber on shrubs, and banks; common. Jul.–Oct.

A large herbaceous climber with long branched tendrils, ovate-heart-shaped, acute or long-pointed leaves, and large bright yellow flowers. Corolla with a slender funnel-shaped tube and elliptic lobes spreading to 4–5 cm. Plants one-sexed; male flowers clustered; female flower solitary. Leaves rough-hairy, 5–15 cm, long-stalked, often coarsely and shallowly lobed. Fruit ellipsoid 5–8 cm long, finely hairy. **Pl. 49.**

TRICHOSANTHES Petals 5, fused only at the base, margins fringed. 6 spp.

542 **T. tricuspidata** Lour. (*T. bracteata* (Lam.) Voigt)

W. Nepal to Bhutan. India. China. S.E. Asia. 600–2100 m. Climber in shrubberies; common. May–Aug.

A herbaceous climber with branched tendrils like 541, but leaves deeply 3–7-lobed, and flowers white, larger, 4–8 cm across, with petals and bracts very conspicuously long-fringed. Leaves heart-shaped, 5–20 cm. Fruit globular 3–5 cm, red striped with orange; seeds in dark green pulp. Fruit used medicinally. **Pl. 48.**

THLADIANTHA Flowers bell-shaped. Leaves entire; tendrils simple. 1 sp.

543 **T. cordifolia** (Blume) Cogn. (*T. calcarata* Hook. f.)

C. Nepal to Bhutan. China. Burma. S.E. Asia. 600–2400 m. Rambler in shrubberies; not common. Apr.–Jun.

A large climber with usually simple tendrils, with ovate deeply heart-shaped pointed leaves, and with yellow bell-shaped flowers *c.* 2 cm with recurved lobes. Male flowers borne in the axils of broad fringed bracts, in stalked axillary clusters 5–8 cm long. Leaves stalked, blade 10–15 cm, with marginal tooth-like projections, rough and bristly-haired. Fruit ellipsoid *c.* 3.5 cm. **Pl. 48.**

The following are often cultivated in village gardens in Nepal:

Cucurbita pepo DC. PUMPKIN

Cucurbita maxima Duchesne GOURD

Cucumis sativus L. CUCUMBER.

Their yellow flowers are prominent during the monsoon months, and so too in the autumn are the pumpkins stored beneath the eves of cottage roofs.

BEGONIACEAE Begonia Family

A small family largely of herbaceous plants and some shrubs, of the Tropics and Sub-Tropics. Most have fleshy jointed stems with pale papery deciduous stipules, and thick rhizomes or tubers. Leaves alternate, often asymmetrical. Flowers in branched clusters, one-sexed, the male and female flowers differing very considerably. The male flowers usually have 4 petal-like segments, the 2 outer larger, and many stamens; the female flowers have 2–5 petal-like segments, and an inferior, usually winged ovary. Fruit a tough or fleshy winged capsule. A botanically difficult family.

BEGONIA   18 spp.

Leaves peltate

544 **B. josephii** A. DC.
C. Nepal to Bhutan. 1800–2700 m. Shady banks, rock ledges. Jul.–Sep.
Recognized by its peltate leaves (leaf-stalk attached to the underside of the leaf-blade). Flowers rather small 1–1.5 cm across, rose-coloured, several borne in branched clusters on leafless stems to 30 cm. Male flowers with the two larger outer petals soon falling; female flowers with 1–6 petals. Leaf-blades ovate-pointed or rounded, to 15 cm, 3–5–lobed or with numerous acute toothed lobes, nearly hairless; leaf-stalk long. Capsule hairless, with 3 wings, one wing much longer (to 2 cm) and with upper margin horizontal. Plant very variable in size and habit.

Leaves not peltate
a *Leaves, stems, flowers and fruit hairy*

545 **B. picta** Smith
Pakistan to Bhutan. 600–2800 m. Shady banks, rock ledges, banks surrounding rice-fields. Jul.–Sep.
The commonest *Begonia* species in wetter parts of our area, with blotched purple or variegated broadly ovate-heart-shaped acute, coarsely and irregularly double-toothed leaves, and with few pink or white flowers borne on hairy stems little longer than the leaves. Flowers 2–3 cm across; the male with two broader pink outer petals and two much smaller inner petals; the female with 5 equal petals, and 3 styles with branched stigmas which are persistent in fruit. Leaf-blades 4–14 cm, bristly-haired, with long hairy leaf-stalk and large stipules; stem 10–18 cm. Capsule 7–20 mm, hairy, with 3 very unequal wings pointing upwards. **Pl. 47**.

b *Plants hairless or nearly so*

546 **B. dioica** Buch.–Ham. ex D. Don (*B. amoena* Wallich ex A. DC.)
Pakistan to Sikkim. 2100–2700 m. Shady banks, rock ledges; sometimes a forest epiphyte. Jul.–Aug.
Differs from 545 in its hairless leaves, its hairless capsule which has

nearly equal wings, and its female flowers with 3 petals. A delicate much smaller plant than 545, with flowers pink and white, *c.* 1.5 cm across.

547 **B. rubella** Buch.–Ham. ex D. Don
C. Nepal to Sikkim. 600–1800 m. Shady banks; locally common in and around the Nepal valley. Aug.–Sep.

A rather delicate plant to 40 cm, with triangular long-pointed leaves with often irregular heart-shaped bases and usually without marginal teeth. Flowers pink, *c.* 2 cm across, borne in a leafless branched cluster. Capsule hairless, with unequal wings, the largest to 1 cm long. **Pl. 48.**

CACTACEAE Cactus Family

A large family of succulent trees, shrubs and shrublets, native of N. & S. America, adapted to dry hot regions with erratic rainfall. Most have spines and tufts of barbed hairs, borne on sunken cushions on the green fleshy photosynthetic stems; leaves mostly absent. Flowers solitary, stalkless, red, purple, yellow to white (never blue), with numerous sepals, petals, stamens and bracts, arranged spirally and merging into each other. Ovary inferior, 1–chambered; ovules numerous. Fruit a berry, splitting to release the seeds.

OPUNTIA PRICKLY PEAR

548 **O. monacantha** (Willd.) Haw. (*O. vulgaris* auct. non Miller)
Native of S. America; commonly planted as a hedge plant and naturalized in cultivated areas and waste ground. To 1800 m. Throughout the summer.

Flowers golden-yellow, *c.* 8 cm across, with outer ovate acute green sepals and inner spathulate acute yellow petals. Stems branched from the base, 2–3 m, usually with thick oblong joints *c.* 30 cm long, but variable, bright grass-green and with long straight needle-like spines to 5 cm. Leaves dark brown, slightly recurved. Fruit pear-shaped, deeply depressed at apex, reddish when ripe, covered with tufts of barbed hairs.

The stem-joints and fruit fall off and can take root; care must be taken in handling this spiny plant. **Pl. 48.**

UMBELLIFERAE Umbellifer Family

A large and important, almost world-wide family, of herbaceous plants with very distinctive inflorescences and fruits. Stems usually hollow; leaves mostly compound, sheathing at the base. Flowers typically in flat-topped *umbels*. Umbels of two kinds: either *simple*, usually with *bracts* at the base, and with several stalks (*primary rays*) each terminating in a flower; or *compound*, consisting of few or many primary rays often subtended by bracts, each ray ending in a *secondary* umbel of flowers. Each secondary umbel usually has *bracteoles* at its base and is composed of several or many *secondary rays* each terminating in a flower. The secondary umbels collectively form a flat-topped inflorescence. Flowers sometimes in compact heads with bracts at their base. Sepals 5, tiny, or absent; petals 5, often notched; stamens 5. Ovary inferior, 2–celled. Fruit of 2 one-seeded units, or *mericarps,* which separate when ripe. Each mericarp may have 5 or 9 ribs, which are often winged, with resin canals usually lying between the ribs. Ripe fruit is essential for the identification of many species. In compound leaves the first or second divisions are described as *lobes*, or leaflets, and the ultimate divisions as *segments*.

Flowers yellow
 a Leaves entire   1 BUPLEURUM
 b Leaves compound   2 FERULA 3 PRANGOS

Flowers blue   4 ERYNGIUM
Flowers white or pinkish

 a Mericarps flattened; lateral ribs winged
  i Ribs of mericarp all winged 5 PLEUROSPERMUM 6 SELINUM
           9 CORTIA 10 CORTIELLA
  ii Dorsal and intermediate ribs of mericarp not winged
        7 ANGELICA 8 HERACLEUM,
 b Mericarps not flattened, nearly rounded in section, not winged
          11 CHAEROPHYLLUM

1 BUPLEURUM Leaves entire. Flowers yellow. Fruit ovoid or oblong, usually with prominent ribs. *c.* 11 spp.

549 **B. candollii** Wallich ex DC.
Pakistan to Bhutan. Burma. Tibet. 2400–4000 m. Open slopes, shrubberies. Jun.–Aug.

 An erect perennial to 1 m, with oblong to oblong-ovate usually fine-pointed more or less stem-clasping upper leaves, with longer and shorter leaves mixed. Flowers yellow, with entire petals, in compound umbels with 2–4 unequal ovate leaf-like bracts subtending the 5–12 primary rays; secondary umbels *c.* 8 mm across and closely surrounded by 4–5 ovate bracteoles. Leaves variable, to 9 cm, often nearly parallel-sided. Fruit elliptic 4–5 mm long, with prominent ribs. **Page 469.**

550 **B. falcatum** L. Sickle-Leaved Hare's-ear
Pakistan to Bhutan. Temperate Eurasia. 1500–4000 m. Forests, open slopes, cultivated areas. May–Aug.

 Distinguished from 549 by the middle and upper stem leaves which are linear to lanceolate and usually curved, and secondary umbels with bracteoles shorter than the flowers. Inflorescence yellow, 1–4 cm across, usually without bracts; primary rays 5–8; bracteoles lanceolate, 4–5. Leaves less than 8 mm broad (usually more than 8 mm broad in 549). Fruit oblong 4–6 mm, ribs distinct.

551 **B. longicaule** Wallich ex DC.
Pakistan to Bhutan. 3000–4400 m. Open slopes, shrubberies. Jun.–Sep.
   A perennial often with several stems 15–40 cm, with narrow lanceolate
long-pointed upper leaves with clasping bases. Flowers blackish-brown, in
2–8 secondary umbels each c. 1 cm across, and subtended by an involucre
of 6–8 oval green bracteoles much longer than the flowers. Leaves 8–10
cm. Fruit oblong c. 4 mm, ribs slightly winged.

2 **FERULA** Flowers yellow. Fruit dorsally compressed with very slender
dorsal ribs and somewhat swollen winged lateral ribs: resin-canals numerous. 1 sp.

552 **F. jaeschkeana** Vatke
Pakistan to Himachal Pradesh. C. Asia. Open slopes; common around
Dras in Ladakh and in the Chenab valley. 2400–3600 m. May–Jun.
   A tall perennial to 2 m, with large 2–3-pinnate leaves with broad
oblong, toothed segments, and with large compound umbels of yellow
flowers. Secondary umbels globular c. 1 cm across; bracts and bracteoles
absent. Leaves to 40 cm, ultimate segments to 10 cm by 1–2 cm broad with
margins finely toothed; upper leaves reduced to large sheaths often with
umbels in their axils. Fruiting umbel to 20 cm across; fruit 1–1.5 cm long
by 1 cm broad, flattened, with lateral wings. **Pl. 50.**

3 **PRANGOS** Flowers yellow. Fruit with all ribs with thick undulate
wings. 1 sp.

553 **P. pabularia** Lindley
Afghanistan to Kashmir. 2100–3300 m. Stony slopes; common around
Dras in Ladakh and the Banihal pass in Kashmir. May–Jun.
   An erect perennial to 1 m, with large umbels of yellow flowers like 552,
but leaves quite different, being many times cut into linear ultimate
segments 1–3 cm long, and having fruit with spongy undulate wings.
Primary umbels with 5–6 linear bracts and 10–15 stout primary rays;
secondary umbels each with 5–6 linear bracteoles. Fruiting umbels 10–15
cm across; fruit oblong 1–1.5 cm by 1 cm broad.
   Plant used medicinally; also cut and dried for cattle fodder. **Pl. 51.**

4 **ERYNGIUM** ERYNGO Spiny perennials. Flowers in globular heads
with spiny involucral bracts. 2 spp.

554 **E. biebersteinianum** Nevski ex Bobrov (*E. coeruleum* M. Bieb.)
Pakistan to Kashmir. W. Asia. 1500–1800 m. Wasteland; common in
valley of Kashmir. Jul.
   Distinguished by its bluish stems and leaves, and its small blue globular
flower-heads with 5–8 much longer spine-tipped involucral bracts. Flower-
head c. 1.5 cm; involucral bracts to 2.5 cm or more; calyx with lanceolate
scales and spine-tipped lobes; petals notched; bracteoles longer. Basal
leaves stalked, oblong to heart-shaped, coarsely toothed; stem leaves

stalkless, deeply spiny-lobed. Fruit ovoid, flattened, scaly. **Page 469**.

5 PLEUROSPERMUM Distinguished by its conspicuous usually white-margined bracteoles, and its spongy-walled, winged fruit. 9 spp.

Tall perennials usually more than 30 cm

555 **P. benthamii** (DC.) C. B. Clarke
C. Nepal to S.W. China. 3500–4300 m. Open slopes. Jun.–Jul.
A stout plant 60–170 cm, with pinnately-cut leaves with lanceolate lobes which are further deeply three-lobed and toothed. Flowers usually pink, in a large inflorescence to 20 cm across, with many large coarsely lobed pale bracts; primary rays 10–25, lengthening to 15 cm in fruit; bracteoles large, oblong, leafy, toothed, with narrow white margin. Fruit ellipsoid, 5–6 mm, with winged ribs and broader lateral wings. **Pl. 50**.

Low-growing perennials usually not more than 20 cm
a *Bracteoles pinnately lobed*

556 **P. govanianum** (DC.) C. B. Clarke
Pakistan to Uttar Pradesh. 4000–4800 m. Alpine slopes and rocks. Aug.
A low-growing plant which is often stemless but with several long primary rays bearing secondary umbels of white flowers *c.* 1.5–2 cm across, with conspicuous deeply lobed white-margined bracteoles spreading beyond the flowers. Leaves usually shorter than the inflorescence, 2–3–pinnate with ultimate toothed segments *c.* 5 mm long. Fruiting stems 1–5 cm; bracts of primary umbel leaf-like. **Pl. 48**.

b *Bracteoles entire*

557 **P. brunonis** (DC.) C. B. Clarke
Pakistan to W. Nepal. 3300–4500 m. Rock ledges, open slopes, screes. Jul.–Sep.
A small plant to 20 cm, but sometimes more, with one or more stems arising from the base, with 2–4–pinnate leaves with lanceolate deeply toothed lobes. Flowers white; primary umbels with 4–6 broadly ovate white-margined bracts, entire or lobed at apex; primary rays 5–10; bracteoles 5–8, ovate, conspicuous, white with green veins, surrounding the dense secondary umbels which are *c.* 1.5 cm across in flower. Fruit oblong 4–6 mm, with broad wings.

558 **P. candollei** (DC.) C. B. Clarke
Pakistan to Uttar Pradesh. 3600–4800 m. Alpine slopes; common. Aug.
Distinguished by its short hollow stout stem 10–20 cm, bearing a solitary terminal compound umbel, and its once-pinnate leaves which are oblong in outline. Umbel with white flowers; bracts leaf-like, or absent; primary rays 5–20; bracteoles 10–15, oblong with broad white margin, longer than the flowers; secondary umbels 1–2 cm across. Leaves with paired, broad wedge-shaped to broadly ovate leaflets, which are usually 3–lobed and

toothed; leaf-bases sheathing; base of stem covered with persistent old leaf-bases. Fruit oblong, 5 mm, with narrow wings. **Pl. 48.**

559 **P. hookeri** C. B. Clarke
Pakistan to S.W. China. 4000–5200 m. Open slopes. Jul.–Sep.

Distinguished by the 3–5 lanceolate white-margined bracts 1–1.5 cm long to each primary umbel, and the pinkish or white flowers. Primary rays 5–10; bracteoles linear to lanceolate long-pointed, white-margined; secondary umbels 5–10 mm across; calyx-lobes prominent. Leaves mostly 2–pinnate, leaflets oval, toothed or lobed, to 5 mm; upper leaves with inflated sheaths; stems several, usually to 10 cm.

6 SELINUM Fruit oblong to broadly ovoid, compressed dorsally, with winged ribs. 6 spp.

560 **S. tenuifolium** Wallich ex C. B. Clarke
Kashmir to Bhutan. 2700–4000 m. Shrubberies, open slopes; common. Jul.–Sep.

A hairless perennial with stems 50–150 cm, and leaves 3–5 times finely divided into very numerous elliptic segments which are deeply toothed or lobed. Flowers white, in compound hairy umbels 5–8 cm across; bracts linear or absent; primary rays 15–30; bracteoles 5–10, linear to lanceolate, white-margined, as long as the flowers. Lower leaves to 20 cm, long-stalked, sheathing at base, the upper smaller, the uppermost reduced to a sheath. Fruit c. 4 mm, with broad lateral wings and dorsal and intermediate ribs narrowly winged. **Pl. 51.**

7 ANGELICA Fruit strongly compressed, with broadly winged lateral ribs and conspicuous dorsal ribs. 1 sp.

561 **A. cyclocarpa** (Norman) Cannon
C. Nepal & S.E. Tibet. 3000–3600 m. Forest clearings, grazings. Jul.–Aug.

A rare but striking plant with a stout stem (to 10 cm in diameter at base) to 3 m or more, with large umbels of greenish flowers to 30 cm across. Primary and secondary rays numerous, bracts and bracteoles absent. Upper leaves with very large boat-shaped sheaths in place of the leaf-stalk and encircling the young umbels; lower leaves large to 60 cm across, 2–3–pinnate with large lanceolate leaflets to 10 cm which are coarsely sharp-toothed. Fruit c. 1 cm, with wide lateral wings and conspicuous ribs. **Pl. 51.**

8 HERACLEUM Fruit strongly compressed, with lateral ribs with thick broad wings, and dorsal slender ribs; resin-canals club-shaped. 7 spp.

Robust perennials 60 cm or more

562 **H. candicans** Wallich ex DC.
Pakistan to S.W. China. 1800–4300 m. Open slopes, drier areas. Jun.–Jul.

A tall plant to 2 m, distinguished by the usually white-felted undersides of the large pinnately lobed leaves. Flowers white, the outer petals of the outer flowers larger, bi-lobed; bracts usually absent; primary rays numerous, hairy; bracteoles 5–8, linear to lanceolate; secondary umbels c. 2.5 cm across. Leaves 20–60 cm, with large elliptic to ovate toothed lobes or leaflets 7–10 cm; upper leaves with large boat-shaped sheaths. Fruit flattened obconic, 7–12 mm long, minutely hairy, with broad lateral wings and slender dorsal ribs.

562a **H. lallii** Norman
Endemic to W. & C. Nepal. 3000–4200 m. Shrubberies, open slopes, edges of fields. May–Jul.
Very like 562, but differing in having larger white outer flowers to 6 mm across, and leaves which are not white-woolly beneath. Leaf-blade to 17 cm long; leaflets to 6 cm. Fruit oblong, 7 mm long, margin narrowly winged. **Pl. 50**.

563 **H. nepalense** D. Don
Himachal Pradesh to Bhutan. 1800–3600 m. Shrubberies. Jun.–Jul.
Like 562, but leaves usually nearly hairless on undersides except for bristly hairs on the veins beneath. Flowers white; bracts 3–5, linear inconspicuous, or absent. Lower leaves pinnate with toothed, deeply lobed, or pinnately-cut large leaflets, the upper leaves ternate with leaflets often 3–lobed; stem 60–160 cm, sparingly hairy. Fruit obovate, to 1 cm, with broad lateral wings and weak ribs.

Slender perennials to 60 cm

564 **H. pinnatum** C. B. Clarke
Pakistan to Himachal Pradesh. 3000–4500 m. Rocky slopes, drier areas; common on field verges in Ladakh. Jun.–Aug.
A rather slender plant to 40 cm, with terminal umbels of white flowers 5–7 cm across. Primary rays long, 6–20, bracts absent or few; bracteoles lanceolate, bristly-haired; secondary rays numerous, with umbels about 1 cm across in flower. Leaves to c. 25 cm, pinnate with 1–3 pairs of widely spaced, small, broadly ovate toothed leaflets 2–3 cm, the lower leaflets stalked, the stalkless upper leaflets 3–lobed. Fruit elliptic, c. 8 mm, with winged lateral ribs and 3 dorsal ribs.
Cut and dried for fodder. **Pl. 49**.

565 **H. wallichii** DC.
C. Nepal to Bhutan. 3600–4100 m. Shrubberies, open slopes. Jul.–Sep.
A rather slender perennial with stems 30–60 cm, which are hairy beneath the nodes, and with upper leaves usually with 3–6 lanceolate long-pointed, toothed leaflets, to 5 cm. Flowers white or pinkish, outer petals much larger, to 8 mm, bi-lobed; bract 1, linear; primary rays 6–10, hairless; bracteoles 0–5, linear; calyx-lobes prominent, linear. Leaf-stalks often with an enlarged sheath below. Fruit c. 6 mm, brown, hairless, with broad lateral wings and weak ribs.

9 CORTIA Fruit with 3 dorsal ribs slightly winged, the lateral broadly so. 2 spp.

566 **C. depressa** (D. Don) Norman

Pakistan to Bhutan. 3600–4800 m. Open slopes in drier areas. Jun.–Aug.

A low-growing usually stemless alpine plant, with many spreading radiating long reddish primary rays arising directly from the rootstock, bearing small secondary umbels *c.* 1.5 cm across of white to dark red flowers. Bracts twice-pinnate; primary rays very unequal; bracteoles 1–2–pinnate, as long as the flowers. Leaves 5–8 cm, twice-pinnate with alternate ovate lobes deeply cut into linear bristle-like segments *c.* 2 mm, and with enlarged pale sheaths surrounding the base of the plant, hairless or nearly so. Rootstock stout, covered with old leaf-sheaths; stem to 10 cm. Fruit elliptic 3–4 mm. **Pl. 49.**

10 CORTIELLA Fruit white, with dorsal and lateral ribs broadly and unequally winged. 1 sp.

567 **C. hookeri** (C. B. Clarke) Norman (*C. glacialis* Bonner)

C. Nepal to S.E. Tibet. 4300–5500 m. Rocks, stony slopes. Jul.–Sep.

Usually a small cushion-like plant with a dense stalkless central domed mass of tiny umbels with white flowers, and with a radiating rosette of pinnate leaves surrounding the inflorescence, the whole plant sometimes only 6 cm across. Bracts leaf-like; bracteoles pinnately-lobed. Leaves 3–5 cm, 1–2–pinnate, with ultimate segments linear to elliptic, 2–4 mm. Fruit *c.* 5 mm, with broad wavy-winged ribs. Larger plants to 20 cm across. **Pl. 49.**

11 CHAEROPHYLLUM CHERVIL Fruit cylindrical, slightly compressed, very shortly beaked, with broad and rounded ribs. 3 spp.

568 **C. reflexum** Lindley

Pakistan to S.W. China. 2100–3500 m. Forests, shrubberies, grazing grounds. Jun.–Aug.

A slender erect little-branched hairless or sparsely hairy perennial, with small umbels of white flowers to 5 cm across, and compound fern-like leaves. Bracts absent; primary rays 3–12; bracteoles 5–8, lanceolate, papery, hairy; secondary umbels *c.* 8 mm across. Leaves 2–3–pinnate with lanceolate lobes further divided into ovate lobed segments; leaf-sheath inflated; stems hairy, 20–80 cm. Fruit 7–12 mm long, smooth, shortly beaked. **Pl. 49.**

569 **C. villosum** Wallich ex DC.

Afghanistan to Bhutan. 2100–3600 m. Forests, shrubberies, cultivated areas. May–Aug.

Differs from 568 in having distinctive long stiff downward-pointing white hairs on the lower part of the stem. Flowers white or pale pink; bracts absent; primary rays 6–10; bracteoles 5–6, linear to lanceolate with

159

white margins, entire or ciliate. Leaves like 568, but leaves larger and often longer than the nodes (mostly shorter than nodes in 568). Fruit 5–9 mm, cylindrical.

In early summer bears dig up the hillsides in search of the carrot-like roots which they eat. **Page 469.**

Cultivated plants include:

Daucus carota L. CARROT
Grown as a vegetable in Kashmir and other regions in the hills.

Coriandrum sativum L. CORIANDER or 'DHANIA'
Grown in the hills up to 3000 m as a herb; that exported from Lahul is especially highly regarded.

ARALIACEAE   Ivy Family

A medium-sized family of trees, shrubs, climbers and herbs, mainly of tropical regions but spreading to temperate regions. Leaves alternate, often large and compound, with small stipules, and often with star-shaped hairs. Flowers small, often greenish or whitish, in small umbels, heads or spikes. Calyx 4–5, small, or absent; petals usually 5; stamens 5, borne on a disk which is situated above the ovary. Ovary of 5 fused carpels, with 5 chambers. Fruit a drupe, containing 5 seeds.

Leaves simple, or shallow-lobed   1 HEDERA   2 HELWINGIA
Leaves deeply lobed, or compound and digitate or pinnate
  a Herbaceous plants   3 PANAX   4 ARALIA
  b Woody climbers, shrubs, or small trees   5 ACANTHOPANAX
                                    6 PENTAPANAX 7 TREVESIA

1 **HEDERA** IVY Leaves simple or shallow-lobed; flowers in branched umbels. Woody climbers. 1 sp.

570 **H. nepalensis** K. Koch
Afghanistan to S.W. China. Burma. 1800–3000 m. Forests. Sep.–Oct.
    A woody climber to 30 m, climbing by aerial roots, with simple lanceolate to ovate, entire or variously lobed, leathery, dark green glossy leaves. Flowers tiny, many, yellowish-green, in stalked globular umbels 1.5–2 cm across which are arranged in domed clusters to 8 cm across. Petals 5, elliptic, 3 mm, placed edge to edge in bud. Leaves long-stalked, very variable, entire or 3–5–lobed, 5–15 cm. Fruit globular c. 6 mm, shining yellow, then black.

2 **HELWINGIA** Leaves simple; umbels tiny, stalkless, borne on leaf-blade. Undershrubs. Often placed in the Cornaceae, or in a separate family. 1 sp.

571 **H. himalaica** Hook. f. & Thoms. ex C. B. Clarke
C. Nepal to S.W. China. 2100–2700 m. An undershrub of dense dark forests; not uncommon in extreme E. Nepal. Apr.–May.

A hairless shrub to 2 m, with lanceolate long-pointed leaves with bristly-toothed margins, easily recognised by its small stalkless umbels arising from the middle of the leaf-blade. Flowers one-sexed; male flowers to 20, green or dark purplish, 3–4–lobed, *c.* 3 mm across; female flowers greenish, 1–4. Leaves stalked, blade 5–10 cm, with a long tail-like point. Fruit 1–3 per leaf, red and fleshy when mature, ovoid, 6–9 mm long. **Page 471**.

3 **PANAX** Herbaceous plants; leaves digitate; styles 2–3. Fruit of usually 2 carpels. 1 sp.

572 **P. pseudo-ginseng** Wallich
C. Nepal to S.W. China. Burma. 2100–4300 m. Forests, shrubberies. Jun.–Jul.

An erect perennial with a stem 15–38 cm, topped by a whorl of 4–6 digitate leaves each with usually 4–6 leaflets, and usually one long-stalked terminal globular umbel of greenish-yellow flowers, 1.5–3 cm. Petals 5, ovate, overlapping in bud; ovary with 2–5 styles. Leaflets lanceolate long-pointed, saw-toothed, 5–15 cm, with scattered bristles on upper surface. Rhizome thick. Fruit globular *c.* 8 mm, shining red, or half red and black. Closely related to the Korean GINSENG. **Page 470**.

4 **ARALIA** Herbaceous plants; leaves pinnate; styles 5. Fruit of 5 carpels. 1 sp.

573 **A. cachemirica** Decne.
Afghanistan to S.E. Tibet. 2100–4300 m. Forests, shrubberies. Jul.–Sep.

A large perennial herb 1.5–3 m, with 2–3–pinnate leaves with usually large ovate long-pointed, toothed leaflets, and with many yellowish-green umbels in a branched cylindrical cluster to 30 cm. Umbels globular 2–3 cm across; calyx-lobes triangular; petals elliptic; stamen-filaments longer than petals; styles 5. Leaves stalked; leaflets stalked, 4–15 cm, hairless. Fruit black, 5–angled, *c.* 3 mm. **Page 470**.

5 **ACANTHOPANAX** Leaves digitate. Petals 4–5, placed edge to edge in bud. Carpels 4–5. 2 spp.

574 **A. cissifolius** (Griffith ex Seemann) Harms
Himachal Pradesh to S.W. China. 2400–3300 m. Forests, rocks. May–Jun.

A shrub, small tree or woody climber, or sometimes an epiphyte, with branches covered with reflexed prickles, and with long-stalked leaves with usually 3–5 lanceolate toothed leaflets. Umbels yellowish-green, solitary, globular 1.5–3 cm, borne on slender stalks 5–13 cm. Leaves with a tuft of fulvous hairs at the end of the leaf-stalk, or with spines; leaflets 4–8 cm, with erect bristly hairs above. Fruit 6 mm long, of 4–5 carpels which are ridged on the back and hairless. **Page 470**.

6 PENTAPANAX Leaves pinnate. Petals 5, overlapping in bud. Styles 4–5. 2 spp.

575 **P. leschenaultii** (DC.) Seemann
Uttar Pradesh to Bhutan. S.W. China. India. Burma. 1600–3700 m. Forests, steep banks. May–Jul.

A woody climber or small tree, sometimes epiphytic, with compound leaves with 3–5 very large elliptic-pointed bristly-toothed leaflets. Flowers *c.* 5 mm across, in branded inflorescence up to 40 cm across, which is formed of long-stalked clusters of 3–6 umbels each 2–4 cm across; petals pinkish to greenish, reflexed. Leaflets 8–15 cm, with a rounded or shallow heart-shaped base, short-stalked. Fruit *c.* 3 mm, ribs not prominent. **Pl. 49**.

7 TREVESIA Leaves very large, palmate or digitate. Petals 8–12, placed edge to edge in bud. 1 sp.

576 **T. palmata** (Roxb.) Vis.
C. Nepal to S.W. China. Burma. S.E. Asia. 250–2500 m. Forests. Mar.–Jun.

A shrub to 3–5 m, with branches covered with reddish hairs and many prickles, and with large palmately lobed leaves often with prickly leaf-stalks. Flowers relatively large, in a branched inflorescence to 45 cm long, of several to many long-stalked rounded umbels each *c.* 3 cm across subtended by small bracts. Petals 8–12, thick, triangular; stamens as many as petals. Leaves variable, 30–60 cm across, many-lobed, with scattered star-shaped hairs below. Fruit in globular umbels 6–8 cm, broad ovoid *c.* 1.5 cm long, crowned by a stout style. **Page 470**.

CORNACEAE   Dogwood Family

A small family of trees and shrubs, mostly of the North Temperate Zone. Leaves usually opposite and simple. Flowers usually in flat-topped clusters sometimes surrounded by large showy bracts. Flowers small, regular, unisexual or bisexual; calyx tubular with 4–5 lobes; petals and stamens 4–5. Ovary inferior, usually of 2 fused carpels, with 1–4 chambers with a single ovule in each. Fruit a drupe, or a berry.

BENTHAMIDIA Flowers in dense heads subtended by 4 large coloured bracts. Carpels coalescing in fruit. 1 sp.

577 **B. capitata** (Wallich) Hara (*Cornus c.* Wallich)
Himachal Pradesh to S.W. China. 1200–3400 m. Forests, shrubberies.
May–Jun.

A small deciduous tree with smooth greyish-brown bark, and with leathery entire oblong or elliptic pointed leaves 4–7.5 cm, which are paler glaucous and minutely hairy beneath. Flowers greenish-yellow, very small, closely packed into hemispherical heads 1–1.5 cm across which are surrounded by 4 or rarely 5 conspicuous white or cream, ovate petal-like bracts 2.5–4.5 cm long. Fruits yellow to red, coalesced into a succulent globular strawberry-like head 2.5–5 cm across.
Fruit edible, used for preserves. **Pl. 51**.

SWIDA Flowers in compound flat-topped clusters, without bracts. Carpels not coalescing in fruit. Often placed in the genus *Cornus*. 3 spp.

578 **S. macrophylla** (Wallich) Soják
Afghanistan to S.W. China. Burma. Japan. 1500–3000 m. Forests, shrubberies. Apr.–Jun.

Usually a medium-sized tree with rough brown bark, and rather large broad ovate-pointed leaves which are glaucous above and finely hairy and paler beneath. Flowers yellowish-white, in terminal flat-topped clusters 5–10 cm across. Flowers *c.* 1 cm across; calyx-tube urn-shaped, white, with fine adpressed hairs; petals strap-shaped, minutely hairy outside; stamens 4, much longer than petals. Leaves quite entire, 7–18 cm, stalked. Fruit black, globular 5 mm, finely hairy. **Pl. 54**.

579 **S. oblonga** (Wallich) Soják
Kashmir to S.W. China. Burma. S.E. Asia. 1000–2400 m. Forests, shrubberies. Sep.–Oct.

Distinguished from 578 by its narrow oblong leaves 1.5–4 cm broad (leaves at least 4 cm broad in 578) which are green beneath, and by its petals which are hairless outside. A shrub or small tree with reddish-brown bark. Flowers yellowish-white, *c.* 8 mm across, in a compound cluster 2.5–6.5 cm across. Fruit ovoid, almost hairless. **Page 471**.

TORICELLIACEAE A family of one genus, usually placed in the *Cornaceae*, but differing in its alternate leaves, its large panicles of flowers, and its 3–5–celled ovary.

TORICELLIA 1 sp.

580 **T. tiliifolia** DC.
W. Nepal to Bhutan. S.W. China. 1600–2500 m. Forests, shrubberies. Apr.

A small tree with thick branches, with broadly heart-shaped coarsely toothed leaves, and with long pendulous branched clusters to 30 cm of small white flowers. Trees unisexual; male flowers with a short 5-lobed calyx-tube, and 5 petals and stamens. Female flowers with ovoid minutely-lobed calyx-tube and no petals or stamens; ovary 3–5–celled; stigmas 3–5. Leaves 10–15 cm broad, long-stalked, nearly hairless. Fruit ovoid c. 5 mm, dark purple, hairless, crowned by the short stigmas. **Pl. 52**.

CAPRIFOLIACEAE   Honeysuckle Family

A rather small family, mainly of trees and shrubs but with some climbers, which has a nearly cosmopolitan distribution. Leaves opposite, simple. Flowers mostly in flat-topped clusters, or in pairs and with the ovaries partly fused, as in *Lonicera*. Calyx fused to ovary, with 4–5 small sepals; corolla with a long or short tube and with 5 spreading lobes which are often two-lipped; stamens 5, fused to the corolla-tube. Ovary inferior, of 3–5 fused carpels with a single ovule in each chamber; style simple. Fruit usually a berry.

Herbaceous plants; leaves fused in pairs by their bases   1 TRIOSTEUM

Shrubs, small trees, or woody climbers
  a Stamens 4; fruit 1–seeded   2 ABELIA
  b Stamens 5; fruit several-seeded
    i Stems hollow   3 LEYCESTERIA
    ii Stems not hollow   4 LONICERA

1 TRIOSTEUM Leaves stalkless, fused in pairs by their bases. Fruit ovoid, 3–lobed. 1 sp.

581 **T. himalayanum** Wallich
Uttar Pradesh to S.W. China. 3000–3800 m. Forests. Jun.–Jul.
    A hairy herbaceous perennial with opposite paired fused leaves, and whorls of funnel-shaped flowers which are green without and purple to reddish within. Whorls of flowers forming short terminal spikes c. 3 cm long. Corolla hairy, c. 1.5 cm, the tube 8 mm long with 5 unequal blunt lobes. Leaves broadly elliptic blunt, with coarse wavy margins, bristly-haired, 10–15 cm, fused below into a broad cup-shaped base encircling the stem. Berries red, c. 1 cm, 3–seeded. **Page 471**.

2 ABELIA Leaves opposite, stalked. Fruit leathery, elongate, 1–seeded. 1 sp.

582 **A. triflora** R. Br. ex Wallich
Afghanistan to C. Nepal. 1500–4200 m. Open slopes, shrubberies, in drier areas. May–Jul.

A large shrub 1–3 m or more, with grey fibrous bark, with lanceolate leaves, and with fragrant pinkish-white flowers borne in small 3–flowered clusters from the axils of leafy bracts, the clusters crowded at the ends of branches. Flowers 8 mm across; corolla tubular-funnel-shaped, the tube reddish *c.* 9 mm long with 5 rounded spreading lobes; calyx with 5 linear, reddish lobes feathered with silky hairs; bracts lanceolate, shining. Leaves mostly 4–5 cm, short-stalked, entire, or with a few teeth or lobes on sterile shoots, sparsely hairy. Fruit linear 5–6 mm, topped with long feathery calyx-lobes. **Pl. 54**.

3 LEYCESTERIA Flowers in clusters in axils of large bracts. Fruit a many-seeded berry. 3 spp.

583 **L. formosa** Wallich
Pakistan to S.W. China. Burma. 2100–3000 m. Forests, shrubberies. May–Jul.

A deciduous half-woody shrub-like plant with hollow glaucous stems to 2 m, with opposite leaves, and with drooping axillary spikes of red or purple-tinged flowers borne in the axils of conspicuous purple leafy bracts. Flowers 1–2 cm long, funnel-shaped, 5–lobed; stamens and style exerted; calyx-tube ovoid with awl-shaped glandular-hairy lobes. Leaves variable, ovate to broadly lanceolate, long-pointed, entire or with small irregular teeth, 5–15 cm. Fruit a dark purple fleshy berry *c.* 8 mm, crowned with the persistent calyx. **Pl. 52**.

4 LONICERA HONEYSUCKLE Flowers usually in stalked pairs, with ovaries often partly fused. Corolla-tube short or long, sometimes swollen at base, and with 5 more or less equal lobes, or two-lipped. Fruit of 2 often fused berries. 25 spp.

Climbers with yellow or white flowers
a *Corolla-tube hairless outside*

584 **L. glabrata** Wallich
C. Nepal to S.W. China. Burma. 1200–2400 m. Forests, shrubberies. Jul.–Sep.

A woody climber with oblong-heart-shaped acute leathery leaves, and short often branched pyramidal leafy spikes to 10 cm or more, of pale yellow flowers turning to buff. Flowers 2–3 cm long, conspicuously two-lipped; corolla-tube hairy within, the lips about as long as the tube or longer; calyx-tube somewhat glaucous, hairless, lobes triangular. Leaves *c.* 8 cm, stalked. Fruit black. **Pl. 53**.

b *Corolla-tube hairy outside*

585 **L. acuminata** Wallich
C. Nepal to S.W. China. S.E. Asia. 2100–3600 m. Forests, shrubberies. Jun.–Jul.

A climber like 584, but differing in its smaller funnel-shaped corolla-tube *c.* 8 mm long which is hairy outside at least in bud. Flowers cream or white, *c.* 2 cm; calyx-lobes with usually few short hairs.

586 **L. macrantha** (D. Don) Sprengel
C. Nepal to Bhutan. China. S.E. Asia. 1500–2400 m. Shrubberies. May.
Distinguished by its hairy very long corolla-tube to 5 cm, with 2–lipped lobes 1.5 cm. Flowers white or yellow, generally in terminal many-flowered clusters at the ends of lateral branches; calyx-tube hairless but with linear conspicuously hairy bristly lobes. A large climbing plant with large elliptic-pointed leaves with a heart-shaped base, *c.* 10 cm, hairy beneath.

Erect shrubs
a *Flowers pink to purple, or orange-red*
  i *Corolla not distinctly 2–lipped* (see also 593)
    x *Flowers several or many from each node*

587 **L. rupicola** Hook. f. & Thoms.
Uttar Pradesh to S.W. China. 3600–4800 m. Shrubberies, open slopes, drier areas. Jun.–Jul.
A dense low shrub with stiff interlacing branches which become somewhat spiny, with tiny oblong leaves, and with almost stalkless clusters of pale pink to purple flowers. Flowers often 6 from each node; corolla-tube 8 mm long, hairy, the lobes rounded, equal, spreading, 5 mm; calyx-lobes hairy. Leaves often in threes, mostly 8–15 mm, hairless above and pale and downy beneath; branchlets ribbed, older stems with greyish bark peeling off in strips. A spreading shrub 60–120 cm or more. **Pl. 53.**

588 **L. spinosa** (Jacquem. ex Decne.) Walp.
Afghanistan to S.E. Tibet. Alpine slopes, drier areas of Tibetan borderlands; dominant with *Caragana* species in parts of Ladakh and in the Dolpo area of Nepal. Jun.–Jul.
A low somewhat spiny shrub forming dense mats like 587, but leaves hairless and margins incurved. Flowers *c.* 1 cm long, pale pink, several to many in almost stalkless clusters; corolla-tube slender, hairless, lobes 4 mm, equal, elliptic; stamens and style longer; calyx hairless. Leaves narrow-oblong 5–13 mm; branchlets spiny, often glaucous (hairy in 587). **Pl. 53.**

    xx *Flowers paired from each node*

589 **L. purpurascens** Walp.
Pakistan to C. Nepal. 2400–3600 m. Forests, alpine shrubberies. Jun.–Jul.
A shrub to 3 m, with hairy branchlets, with hairy oblong to oblanceolate leaves, and with paired dull-purple funnel-shaped flowers borne on rather long drooping stalks. Flowers 1–1.5 cm long; corolla-tube hairy, swollen at the base, lobes short; stamens and style longer; flower-stalks 8–25 mm; bracts narrow lanceolate. Leaves 2–5 cm, slender-stalked. Fruit black, the paired berries more or less fused.

590 **L. cyanocarpa** Franchet var. **porphyrantha** Marquand & Shaw
E. Nepal to S.E. Tibet. 3600–4000 m. Alpine shrubberies. Jun.

A rare but attractive shrub with large distinctive dark red funnel-shaped flowers to 2.5 cm, borne from the axils of large leafy broadly ovate bracts to 2 cm. Stamens and style included in corolla. Leaves oblanceolate 3–5 cm, with bristly-haired margin and underside; stems 1 m or more. Fruit black, paired, not fused together. **Pl. 52**.

ii *Corolla distinctly 2–lipped*

591 **L. webbiana** Wallich ex DC.
Afghanistan to S.W. China. 2100–4300 m. Forests, shrubberies. May–Jun.

An erect shrub with large elliptic glandular-hairy leaves, and with stalked axillary pairs of usually pinkish 2–lipped flowers from each node, but flowers sometimes yellowish or greenish. Corolla-tube short *c.* 5 mm, very glandular-hairy, conspicuously swollen towards base, lips *c.* 8 mm; flower-stalks long, to 4 cm. Leaf-blade 5–10 cm, hairy or becoming hairless above, leaf-stalk short; shrub 2–4 m. Fruit red, not fused. **Page 472**.

592 **L. lanceolata** Wallich
W. Nepal to S.W. China. 2700–3800 m. Forests, shrubberies; common in Khumbu, Nepal. May–Jun.

Very like 591, but with smaller two-lipped red-purple flowers *c.* 1.2 cm, and flower-stalks shorter *c.* 1 cm. Leaves lanceolate long-pointed, blade 5–7 cm. Fruit red-purple.

b *Flowers white, cream, yellow or orange-yellow, pinkish (see also 591)*
  i *Corolla not distinctly 2–lipped*
    x *Flower-stalk more than 1 cm; leaves usually more than 2.5 cm*

593 **L. angustifolia** Wallich ex DC.
Kashmir to S.E. Tibet. 2700–3600 m. Forests, shrubberies. May–Jun.

An erect shrub to 4 m, with lanceolate leaves which are very pale beneath, and with paired white flowers tinged with pink, borne on a slender stalk from the leaf-axils. Flowers scented; corolla to 12 mm, hairy or not, lobes short rounded; calyx-lobes triangular; flower-stalk usually more than 1 cm. Leaves hairy beneath, 2.5–4 cm, short-stalked. Fruit red, globular, fused, edible.

594 **L. hispida** Pallas ex Willd.
Pakistan to S.W. China. C. Asia. Forests, shrubberies, open slopes. May–Jul.

An erect shrub to 2m, with variable elliptic to lanceolate-oblong leaves; distinguished by its long-stalked paired nodding white or yellowish flowers encircled by two large leafy ovate-boat-shaped, bristly-haired bracts 1.5–2 cm. Corolla variable, broadly or narrowly funnel-shaped, 1.5–3 cm long, hairy, the lobes rounded; stamens and style slightly exserted. Flower-buds and fruit enclosed in the bracts; flower-stalk hairy, 1–3 cm. Leaves usually

1–7 cm, dark green above; leaf-stalk very short. Fruit red, paired, fused or separate, with whitish bracts. **Pl. 53**.

xx *Flower-stalk less than 1 cm; leaves usually less than 2.5 cm*

595 **L. asperifolia** (Decne.) Hook. f. & Thoms.
Afghanistan to Uttar Pradesh. 2700–4000 m. Stony slopes, riversides, dry inner valleys; common in Ladakh. May–Jul.
A small shrub, distinguished by its rough-hairy twigs and by its leaves which are bristly-hairy, dark green above and paler beneath, with bristly ciliate margins. Flowers yellow, paired; corolla-tube 1–1.5 cm long, the lobes elliptic, half as long as the tube; bracts ovate, glandular rough-hairy, 8 mm. Leaves ovate to elliptic with rounded or shallow heart-shaped bases, 1.3–3 cm; shrub to 1 m. Fruit red, 6–8 mm, edible but very bitter.

596 **L. myrtillus** Hook. f. & Thoms.
Pakistan to S.W. China. 2400–4000 m. Forests, shrubberies, open slopes; common. Jun.–Jul.
A small nearly hairless much-branched shrub to 1.5 m, often with prostrate branches, with small oblong leaves, and with paired white or cream-coloured flowers sometimes tinged pink. Corolla-tube cylindrical, 5 mm, the lobes blunt c. 2 mm; bracts lanceolate; flower-stalk short, usually to 5 mm. Leaves blunt, pale beneath, 6–12 mm. Fruit red, not fused. **Page 472**.

597 **L. obovata** Royle ex Hook. f. & Thoms.
Afghanistan to S.E. Tibet. 3000–4400 m. Shrubberies, open slopes; one of the commonest alpine shrubs in Nepal. Jun.–Jul.
A straggling sometimes prostrate, hairless shrub to 1 m, with grey twigs, with small obovate to elliptic hairless leaves, and with paired pale yellow flowers with bell-shaped corolla-tube and short rounded lobes. Flowers often appearing with the young leaves, 6–8 mm long; corolla-tube widening upwards, swollen at base, the lobes not spreading; stamens and style exserted; flower-stalk short. Leaves 8–20 mm, pale, netted beneath, hairless. Fruit dark bluish-purple, spherical, fused. **Page 472**.

ii *Corolla distinctly 2–lipped* (see also 591)

598 **L. hypoleuca** Decne.
Afghanistan to C. Nepal. Tibet. 2700–4200 m. Stony slopes, rocks; drier areas of Tibetan borderlands only. Jun.–Jul.
An erect shrub to 2 m, with small broadly ovate blunt leaves which are pale glaucous beneath, and with paired pale yellow 2–lipped flowers 1–1.5 cm, sometimes in spike-like clusters borne at ends of branches. Corolla-tube with a swollen base, the lobes narrow-oblong and as long as the tube; bracts lanceolate, leafy; bracteoles fused into a very hairy cup. Leaves 8–25 mm, rather thick, glandular-hairy on both sides; branchlets often purplish,

hairy. Fruit ellipsoid 5 mm, red, glandular-hairy, half encircled by the bracteoles.

599 **L. quinquelocularis** Hardw.
Afghanistan to S.W. China. 1800–3000 m. Forests, cultivated areas; common. May–Jul.

A large hairy erect shrub to 5 m, or a small tree, with ovate to broadly lanceolate leaves, and with pairs of cream-coloured flowers turning yellow, in axillary stalkless clusters. Corolla 13 mm, tube densely finely hairy outside, the lobes 2–lipped, longer than tube; stamens and stout style exserted; bracts linear; bracteoles small, fused into a cup. Leaves 3–6.5 cm, becoming nearly hairless above, densely hairy beneath; bark grey; young stems purplish. Fruit ovoid c. 6 mm, green or white translucent, not fused. **Pl. 53.**

SAMBUCACEAE Elder family

Sambucus and *Viburnum* are distinguished from the Caprifoliaceae by small botanical characters which include the rotate or broadly campanulate corolla; the single ovule in each cell of the ovary, and the short style with a 2–lobed stigma.

SAMBUCUS ELDER Leaves pinnate. 4 spp.

Herbaceous perennials

600 **S. wightiana** Wallich ex Wight & Arn.
Afghanistan to Himachal Pradesh. 1500–3600 m. Heavily grazed slopes; common in W. Himalaya. Jul.–Aug.

Usually forming large gregarious clumps 1–1.5 m high, with green stems, large pinnate leaves, and with terminal flat-topped clusters of very numerous tiny white or cream flowers. Inflorescence 5–10 cm across, enlarging to 25 cm in fruit, with 3–5 primary branches which are further irregularly branched. Flowers c. 3 mm across; corolla with 5 spreading blunt lobes and a very short tube; stamens exserted. Leaves 15–30 cm; leaflets 5–9, lanceolate often long-pointed, 5–20 cm, sharply toothed; stipules leaf-like, toothed. Fruit globular 4–5 mm, orange or red, then turning black.

Fruit, roots and leaves used medicinally.

601 **S. adnata** Wallich ex DC.
W. Nepal to S.W. China. 1500–3700 m. Grazing grounds, wasteland. May–Sep.

Like 600, but leaves usually with 9 leaflets and with the upper leaflets with decurrent blade, and stipules small or absent. Inflorescence hairy; flowers white c. 4 mm, usually in flat-topped clusters to 23 cm across.

Leaflets lanceolate, saw-toothed, 8–15 cm. Herbaceous perennial 1–2 m. Fruit red. **Page 472.**

Stems woody

602 **S. canadensis** L.
Native of N. America; very common in cultivated areas in the Himalaya. To 1500 m. May–Sep.

A shrub to 4 m, with smooth young branches, and with leaves usually with 7 oval slender-pointed, sharply-toothed, bright green leaflets which are slightly downy beneath. Leaflets 5–15 cm, the lowest leaflets often lobed. Flowers white, in a domed cluster 10–20 cm across. Fruit black-purple.

VIBURNUM Leaves simple, toothed or lobed; stipules absent or minute. *c.* 8 spp.

Corolla tubular or funnel-shaped, the tube twice as long as the lobes or longer

a *Leaves evergreen, leathery, without toothed margins*

603 **V. cylindricum** Buch.–Ham. ex D. Don (*V. coriaceum* Blume)
Himachal Pradesh to S.W. China. India. Burma. S.E. Asia. 1200–2400 m. Forests, shrubberies; common at village level in Nepal. Jul.–Sep.

A shrub or small tree to 12 m, with grey bark, with entire leathery glossy hairless leaves, and with stalked flat-topped clusters of numerous white tubular flowers with projecting purple anthers. Flower clusters erect, 5–10 cm across; corolla 4 mm long, the lobes, blunt erect very short. Leaves with oblong-lanceolate long-pointed blade 6–14 cm, leaf-stalk to 2.5 cm; leaves showing curious white markings when crushed in the hand. Fruit ovoid, 5 mm long, succulent, black. **Page 473.**

b *Leaves deciduous, thin, with toothed margins*

604 **V. erubescens** Wallich ex DC.
Uttar Pradesh to S.W. China. Burma. 1500–2700 m. Forests, shrubberies; common in Nepal. Apr.–May.

A small deciduous shrub to 3 m, distinguished by its lax drooping long-stalked branched clusters of white, cream or pink tubular flowers, borne with the leaves at the ends of short branchlets. Flowers with slender corolla-tube 5 mm, and rounded spreading lobes *c.* 2 mm; anthers dark purple. Leaves with ovate to elliptic blade, toothed in the upper part, usually 3–6 cm, shortly stalked. Fruit red, ellipsoid, *c.* 8 mm long. **Pl. 52.**

605 **V. grandiflorum** Wallich ex DC.
Himachal Pradesh to S.E. Tibet. 2700–3600 m. Forests, shrubberies; common in Nepal. Apr.–May, and intermittently throughout the winter.

Distinguished by its dense nearly stalkless clusters of relatively large pink fragrant flowers, which appear on bare stems before the leaves. Flower clusters 3–4 cm across; corolla-tube slender, to 1.3 cm, the lobes

3–4 mm, spreading, elliptic. Leaves elliptic 8–10 cm or more, toothed, hairy on the veins beneath. Fruit red then black, ellipsoid to 2 cm, edible. A large shrub to 3 m, or a small gnarled tree. **Pl. 54**.

The West Himalayan form, previously known as **V. foetens** Decne. but now regarded by most botanists as a form of 605, is the commonest undershrub of the coniferous forests of Kashmir.

Corolla with a short tube; spreading lobes as long or longer

606 **V. cotinifolium** D. Don
Afghanistan to C. Nepal. Bhutan. 1800–3600 m. Forests, shrubberies. Apr.–May.

A large deciduous shrub, with young shoots and undersides of leaves densely covered with white-woolly star-shaped hairs, and with terminal domed clusters of numerous white flowers tinged pink, especially in bud. Flower clusters dense 5–10 cm across; flowers *c*. 8 mm across; corolla funnel-shaped with spreading rounded lobes about as long as the tube. Leaves 5–12 cm, ovate to rounded with rounded or heart-shaped base, blunt, toothed or nearly entire, impressed with veins above, white-woolly beneath, stalked; stems 2–3 m. Fruit red, turning black when ripe, 8–13 mm, edible. **Pl. 54**.

607 **V. nervosum** D. Don (*V. cordifolium* Wallich ex DC.)
Uttar Pradesh to S.W. China. Burma. 2700–3300 m. Forests; common in the wetter parts of Nepal. Apr.–May.

Flowers fragrant, white tinged with pink, in compact or lax compound umbels 2.5–10 cm across, appearing with or before the leaves. Flowers 1–1.5 cm across; corolla-lobes large ovate to 6 mm long, much longer than the very short tube. Leaves 5–13 cm, ovate to obovate or elliptic, long-pointed, margin sharply toothed, impressed with veins above and with prominent veins and sparsely hairy beneath. Fruit ellipsoid, red, to 1.3 cm. A shrub to 6 m (taller than 606). **Pl. 52**.

608 **V. mullaha** Buch.–Ham. ex D. Don (*V. stellulatum* Wallich)
Pakistan to S.E. Tibet. Burma. 1500–3000 m. Forests, shrubberies; common in Nepal at village level. Jun.–Aug.

A large deciduous shrub, with large rounded compound umbels of tiny white flowers each 3–4 mm across. Inflorescence hairy; corolla-tube *c*. 1 mm long, the lobes hairy without, slightly longer; calyx and inflorescence covered with star-shaped hairs. Leaves 7–15 cm, ovate to ovate-lanceolate long-pointed, with few remote teeth, hairless above, hairless or with star-shaped hairs beneath; bark dark grey; shrub to 5 m. Fruit bright shining red, to 8 mm. **Page 473**.

RUBIACEAE   Madder Family

One of the largest families, mostly of tropical trees and shrubs, but with some temperate herbaceous species. Leaves opposite or whorled, simple and usually entire; stipules present and characteristic. Flowers usually regular, in branched clusters, or heads. Sepals 4–5, usually free; corolla tubular, with 4–5 spreading lobes; stamens 4–5, attached to the corolla; ovary inferior, usually 2–celled, with one to many ovules. Fruit a capsule, berry, drupe, or separating into 1–seeded nutlets.

Trees or shrubs
  a Fruit a splitting capsule
    i Ovary 2–celled, cells many-seeded   1 HYMENOPOGON
                  2 LUCULIA 3 WENDLANDIA
    ii Ovary 5–celled, cells one-seeded   4 LEPTODERMIS
                        5 SPERMADICTYON
  b Fruit fleshy or dry, not splitting
    i Fruit with a pair of 1–seeded nutlets 6 PAVETTA 7 COFFEA
    ii Fruit with more than 1 seed in each chamber 8 MUSSAENDA
                  9 RANDIA 10 XEROMPHIS
Herbaceous plants   11 GALIUM 12 RUBIA

1 HYMENOPOGON Corolla-lobes edge to edge in bud. Flowers in flat-topped clusters; bracts leafy. 1 sp.

609 **H. parasiticus** Wallich
Uttar Pradesh to S.W. China. S.E. Asia. 1600–2800 m. Forest epiphyte. Jun.–Jul.
    A small epiphytic shrub, with oblanceolate leaves crowded at the ends of the branches, and with creamy-white tubular flowers in terminal branched flat-topped clusters with longer narrow often white bracts. Corolla-tube very slender 3–4 cm, with 5 hairy spreading lobes; calyx-lobes linear *c.* 1 cm; bracts narrow-lanceolate, stalked. Leaves entire, 8–30 cm, hairy on the veins beneath; stems to 1 m. Capsule cylindrical 1–1.5 cm, with persistent calyx, splitting into 4 valves at apex, 2–celled, many-seeded. **Pl. 52**.

2 LUCULIA Corolla-lobes overlapping in bud. Flowers in domed clusters; style 2–lobed. 1 sp.

610 **L. gratissima** (Wallich) Sweet
  C. Nepal to S.W. China. 1200–1800 m. Shrubberies. Sep.–Oct.
    One of the most attractive autumn-flowering evergreen shrubs with large pink fragrant flowers in large terminal domed clusters 8–15 cm across. Flowers 2–2.5 cm across; corolla-tube slender 2.5–4 cm long, with

5 rounded spreading lobes; calyx hairy, the lobes spathulate, soon falling. Leaves variable, entire, lanceolate long-pointed, 13–18 cm, stalked. A tall spreading shrub 3–5 m. Capsule 2–celled. **Pl. 55.**

3 WENDLANDIA Corolla-lobes twisted in bud. Flowers in branched clusters. 7 spp.

611 **W. puberula** DC.
Kashmir to Arunachal Pradesh. 700–2000 m. Light forests. May–Jun.

A small evergreen tree with rough brownish-grey bark, with entire oblong-elliptic leaves, and with very small scented white flowers in much-branched terminal and axillary clusters 7–15 cm. Flowers *c.* 5 mm across; corolla funnel-shaped with oblong lobes shorter than the tube; stamens protruding; calyx hairy, with blunt triangular lobes *c.* 1 mm. Leaves with 6–10 pairs of lateral veins, 10–20 cm, hairless above, with distinctive brown hairs on the veins beneath and on leaf-stalks. Capsule globular, 2 mm, nearly hairless. **Page 473.**

4 LEPTODERMIS Corolla-lobes edge to edge in bud. Flowers in short-stalked clusters; styles 5. Fruit 5–valved; with 5 nuts. *c.* 5 spp., but Asian species of the genus require revision.

612 **L. kumaonensis** R. Parker
Uttar Pradesh to Bhutan. 2000–3800 m. Forests, shrubberies, rocky slopes. Jun.–Jul.

A small shrub to 1.5 m with grey stems, with small narrow-elliptic leaves, and with nearly stalkless axillary clusters of small white or pink flowers which turn purple as they mature. Flowers with short corolla-tube to 1 cm, and elliptic blunt lobes spreading to *c.* 1 cm, finely hairy; calyx-lobes with finely fringed margins, densely hairy. Leaves 2–10 cm, entire, short-stalked. Fruit a capsule *c.* 5 mm, shining, foetid when crushed. **Pl. 56.**

5 SPERMADICTYON Corolla-lobes edge to edge in bud; style 5–lobed. Capsule 5–valved at the top. 1 sp.

613 **S. suaveolens** Roxb. (*Hamiltonia s.* Roxb.)
Pakistan to S.E. Tibet. India. 700–2300 m. Shrubberies, steep banks. Oct.–Mar.

A shrub to 2 m, with large tough leathery entire elliptic to ovate leaves, and with usually blue-lilac, or sometimes white, sweet-scented flowers in a much branched terminal dense or spreading inflorescence to 30 cm long. Flowers several, densely clustered at ends of branchlets; corolla with a slender funnel-shaped tube to 1.5 cm, with 4–5 short lobes spreading to 8 mm, hairy; calyx-lobes linear, hairy. Leaves usually 8–18 cm, opposite, acute or long-pointed, with 10–15 pairs of prominent veins beneath. Capsule surrounded by the ellipsoid calyx. **Page 473.**

6 PAVETTA Corolla-lobes twisted in bud. Flowers in flat-topped or elongate terminal clusters; bracts papery, sheathing. Fruit 2–4–celled, a berry or drupe. 2 spp.

614 **P. indica** L.
C. Nepal to Bhutan. India. China. S.E. Asia. To 1500 m. Forests, shady ravines. Apr.–May.

A shrub to 3 m or a small tree, with smooth brownish-grey bark, with opposite rather leathery elliptic acute leaves, and with a terminal domed cluster of many fragrant white flowers with long slender corolla-tubes and long projecting slender styles. Flower clusters 5–10 cm across; corolla-tube to *c.* 2 cm, with 4 linear-oblong recurved lobes spreading to 13 mm; calyx with short triangular lobes, usually hairy; bracts broader than long. Leaves 8–23 cm, with usually 5–6 pairs of lateral veins, stalked; stipules united into a sheath. Fruit black, globular, *c.* 7 mm, with 2 nutlets.

Root and leaves used medicinally. **Pl. 55.**

7 COFFEA Differs from *Pavetta* in having axillary flower clusters, and a cylindrical corolla-tube with 4–9 lobes. 1 sp.

614a **C. benghalensis** Heyne ex Roemer & Schultes
Uttar Pradesh to Bhutan. India. Burma. To 900 m. Sal forests, shrubberies; common in lower valleys in Nepal. Mar.–May.

A deciduous shrub with pure white fragrant axillary solitary or paired flowers, appearing with or before the leaves. Flowers 2.5–3.7 cm across; corolla-tube funnel-shaped 1.3–1.8 cm, lobes ovate-oblong, spreading; calyx with 5 broad indistinct teeth divided into numerous linear segments. Leaves opposite, ovate-elliptic long-pointed, papery, hairless, 5–12.5 cm, suddenly contracted into short leaf-stalks; stipules awl-shaped. Fruit a black drupe *c.* 1.3 cm. **Pl. 55.**

8 MUSSAENDA Corolla-lobes edge to edge in bud. Flowers in flat-topped or elongate clusters. Calyx of some outer flowers with one lobe much enlarged and forming a large stalked white or coloured 'leaf'. Fruit a berry. 6 spp.

615 **M. treutleri** Stapf
Uttar Pradesh to Bhutan. 400–1700 m. Dense damp forests. Jun.–Jul.

A large shrub to 3 m, or a small tree, readily distinguished by the several large white papery 'leaves' surrounding and spreading beyond the flat-topped cluster of orange flowers with yellow hairs in the throat. Flowers with a slender hairy tube 2–2.5 cm and with narrow-triangular lobes spreading to 1 cm. Leaves stalked, with an entire elliptic long-pointed blade 10–20 cm, hairy beneath. Berry globular, 1 cm, many-seeded. **Page 471.**

9 RANDIA Corolla-lobes twisted in bud. Flower clusters axillary. Ovary 2–celled; fruit a berry. 4 spp.

616 **R. tetrasperma** (Roxb.) Benth. & Hook.f. ex Brandis
Pakistan to S.E. Tibet. 1200–2400 m. Shrubberies, open slopes, cultivated areas; common. Apr.–Jul.

A small rigid much-branched hairless shrub to 1.5 m or more, with short branches, and with small densely clustered ovate to oblanceolate leaves. Flowers white green-tinged, scented, solitary, stalkless. Corolla-tube *c.* 1 cm long, with 5 narrow-elliptic long-pointed reflexed lobes; stamens and style protruding; calyx-tube ovoid, with 5 long narrow acute lobes. Leaves very variable, usually 1–3 cm, narrowed to a short stalk, crowded at ends of branches, in 4 ranks. Berry blue, globular *c.* 8 mm, with persistent calyx. **Pl. 55**.

10 XEROMPHIS Differs from *Randia* in its fruit which is leathery outside and fleshy within, with seeds imbedded in gelatinous pulp. 2 spp.

616a **X. spinosa** (Thunb.) Keay (*Randia dumetorum* (Retz.) Lam.)
Himachal Pradesh to Sikkim. India. S. China. S.E. Asia. To 1200 m. *Sal* forests, shrubberies; common in hot lower valleys of Nepal. Mar.–May.

A large spiny deciduous shrub or small tree, with few or solitary fragrant white flowers fading to yellow, borne on short leafy branchlets. Corolla with a short hairy tube and with rounded lobes spreading to 1.3–2 cm, with adpressed hairs; calyx with sharp stiff adpressed hairs, and with large leafy broadly ovate lobes, often with minute intermediate lobes. Leaves obovate 2.5–5 cm; spines long, axillary, 2.5–4 cm. Fruit 2.5–4 cm, globose or ovoid, often crowned with persistent calyx.

Ripe fruit edible; unripe fruit used to poison fish. **Pl. 55**.

11 GALIUM Fruit of 2 one-seeded nutlets. Corolla usually 4–lobed. *c.* 15 spp.

617 **G. boreale** L. NORTHERN BEDSTRAW
Pakistan to Himachal Pradesh. Temperate Eurasia. N. America. 2100–3600 m. Meadows, irrigated land. Jun.–Aug.

Flowers white, 3–4 mm across, numerous in a dense oblong-pyramidal terminal cluster. Leaves usually 3–4 in each whorl, 1.5–4 cm, elongate-lanceolate blunt, 3–veined, somewhat leathery, usually hairless; stems herbaceous, 4–angled, erect, 30–65 cm. Fruit dry, 1–2 mm, usually with adpressed hooked hairs, or hairless.

618 **G. verum** L. LADY'S BEDSTRAW
Pakistan to Himachal Pradesh. Temperate Eurasia. 1800–3600 m. Stony slopes. Jun.–Aug.

Flowers yellow, 2–4 mm across, numerous in a rather dense branched terminal cluster. Corolla hairless, lobes acute. Leaves 6–8 in each whorl, 1.5–2 cm, linear acute, shining and usually hairy above, densely finely hairy beneath, margins usually recurved; stem herbaceous rounded, but with 4 raised lines, 20–40 cm. Fruit usually hairless.

Plant used medicinally.

12 RUBIA Like *Galium* but fruit a fleshy berry, not dry. Corolla 5–lobed. 5 spp.

619 **R. manjith** Roxb. ex Fleming (*R. cordifolia* auct. non L.)
Pakistan to S.E. Tibet. 1200–2700 m. Climber on shrubs. Jun.–Nov.

A herbaceous climbing perennial, with 4–angled stems and branches, with unequal stalked leaves in whorls of 4, and with usually reddish-brown flowers in small clusters aggregated together into a large branched cluster with small leafy bracts. Flowers *c.* 3 mm across, with 5 spreading incurved lobes. Leaves with ovate-heart-shaped long-pointed blade 3–5 cm, with hooked prickles on the veins beneath; leaf-stalk as long as blade, with hooked prickles; stems with hooked prickles on angles. Fruit *c.* 5 mm, globular, black, fleshy with red juice.

A valuable dye, *manjith*, is obtained from roots and stems. Roots used medicinally. **Page 474.**

VALERIANACEAE   Valerian Family

A medium-sized family mostly of herbaceous perennials, characteristic of the Northern Hemisphere. Leaves opposite, often pinnately lobed, often strong-smelling when dry; roots also aromatic. Flowers many, small, in branched flat-topped clusters. Calyx situated on top of a rim of the ovary, minute in flower but variously developed in fruit; corolla tubular, with 5 lobes; stamens attached to the corolla-tube, usually 3–4 in our species; ovary inferior, of 3 fused carpels, with 1 chamber and 1 ovule. Fruit dry, 1–seeded, with persistent calyx variously developed.

NARDOSTACHYS Stamens 4; calyx equally 5–lobed in fruit. 1 sp.

620 **N. grandiflora** DC. (*N. jatamansi* DC.)
Uttar Pradesh to S.W. China. 3600–4800 m. Rocks, ledges, open slopes. Jun.–Aug.

One of the most attractive alpine plants, but not common, with rose-purple to whitish flowers in dense heads borne in terminal often branched clusters. The whole plant has a very distinctive and lingering smell. Corolla-tube variable 6–20 mm, with 5 rounded lobes spreading to 12 mm or more; calyx coloured, 5–lobed, the lobes enlarging in fruit and becoming papery. Leaves elliptic-lanceolate or spathulate, 5–20 cm, mostly basal and arising from the spindle-shaped rootstock which is covered with the dark fibres of old leaves; flowering stem 5–30 cm. Fruit obovate, flattened, 1–seeded.

The SPIKENARD of the ancients; a favourite perfume. The root is an excellent substitute for valerian, and is used in many medicinal treatments, while the oil obtained from the root is used in many medicinal preparations and is well known as a hair tonic. **Page 474.**

VALERIANA   Stamens 3; calyx feathery in fruit. *c.* 9 spp.

621   **V. jatamansii** Jones (*V. wallichii* DC.)
Afghanistan to S.W. China. Burma. 1500–3600 m. Forests, shrubberies, open slopes; common. Feb.–Aug.

A forest perennial, with a terminal flat-topped cluster of small white or pink-tinged flowers, borne on erect nearly leafless stems. Flower clusters 2.5–5 cm across, enlarging in fruit; corolla funnel-shaped, with 5 blunt lobes spreading to *c.* 5 mm. Basal leaves with ovate-heart-shaped acute blade, 2.5–8 cm, toothed or wavy-margined, long-stalked; stem leaves few, small, entire or pinnate-lobed; stems several, 15–45 cm; rootstock thick, horizontal. Fruit crowned with a persistent pappus-like calyx.

Roots used medicinally as a substitute for valerian. **Page 474.**

622   **V. pyrolifolia** Decne.
Pakistan to Himachal Pradesh. 2100–3600 m. Forests, open slopes; common in Kashmir. May–Jun.

Like 621, but differing in having 1–3 pairs of stalkless broadly ovate entire stem leaves. Flowers white flushed pink; clusters 1–4 cm across. Basal leaves with rounded to heart-shaped blade to 2.5 cm; stem 5–12 cm. Fruit hairless, with pappus-like calyx. **Pl. 56.**

623   **V. hardwickii** Wallich
Pakistan to S.W. China. Burma. S.E. Asia. 1500–4000 m. Shrubberies, open slopes, Jun.–Sep.

Distinguished by its 1–3 pairs of stem-leaves which are large, pinnate with 3–7 leaflets, and its usually white or pale pink flowers. Flowers 2–3 mm across, in dense terminal domed clusters forming a branched pyramidal inflorescence. Basal leaves ovate, long-stalked, usually shrivelled at flowering time; stem 30–180 cm. Inflorescence becoming lax in fruit; fruit hairy.

DIPSACACEAE   Scabious Family

A small family, mostly of herbaceous plants, with its main centre of distribution in the Mediterranean region and Asia. Leaves opposite or whorled. Flowers in dense heads, subtended by involucral bracts, less commonly in whorls forming a spike, as in *Morina*. Calyx cup-shaped, or divided into five to many bristles or lobes; corolla usually tubular with 2–4 equal lobes, or 2–lipped; stamens 2 or 4, attached to the corolla-tube. Ovary inferior, of 2 fused carpels, 2–celled, each cell with 1 ovule. Fruit enclosed in an epicalyx, and often topped by the persistent calyx.

Flowers in whorls, forming a spike or dense head　　　　　1 MORINA

Flowers not whorled, in dense heads surrounded by involucral bracts
　　a Corolla 5–lobed　　　　　　2 PTEROCEPHALUS　3 SCABIOSA
　　b Corolla 4–lobed　　　　　　　　　　　　　　4 DIPSACUS

1 MORINA Corolla with long tube, 2–lipped; fertile stamens 2 or 4. 4 spp.

Flowers yellow

624　**M. coulteriana** Royle
Pakistan to Uttar Pradesh. Tibet. C. Asia. 2400–3600 m. Open slopes in drier areas. Jun.–Aug.
　　Readily distinguished by its long interrupted spikes of pale yellow flowers borne on erect unbranched stems 30–90 cm, with a lax basal cluster of spiny-lobed, thistle-like leaves. Flowers in whorls, with long curved corolla-tubes to 2 cm, with 2–lipped lobes *c.* 8 mm; calyx 2–lobed, the lobes with or without spiny teeth; bracteoles with long spines; bracts subtending whorls of flowers, free or almost so at base, linear, leaf-like, spreading, longer than the whorls. Leaves linear-oblong *c.* 15 cm by 2 cm broad, stalkless, usually with spiny margins. Fruit 5 mm, with an oblique apex. **Pl. 57.**

Flowers white or pink
　a *Leaves lobed, with many stiff marginal spines*

625　**M. longifolia** Wallich ex DC.
Kashmir to Bhutan. 3000–4000 m. Open slopes, alpine shrubberies. Jun.–Sep.
　　Like 624 with spiny-margined leaves and a long interrupted spike of flowers, but flowers white to rose-pink, and paired bracts subtending whorls with an enlarged ovate base and fused below. Flowers with long slender corolla-tube to 2.5 cm, somewhat 2–lipped, the lips to 6 mm, hairy; calyx with acute or blunt lobes. Leaves strap-shaped with shallow 3–spined lobes and a long-pointed spiny apex; stem leaves fused in a sheath at the base; bracts 3–4 cm, spiny. Flowering stems to 1 m, hairy above. **Pl. 56.**

626　**M. polyphylla** Wallich ex DC.
Uttar Pradesh to Bhutan. 3600–4300 m. Open slopes. Jun.–Aug.
　　Distinguished from other species by its dense terminal spike-like flower-cluster with large linear-lanceolate spreading bracts which are fused at the bases and largely hide the small reddish or sometimes white flowers. Corolla-tube short, lobes rounded, equal; calyx conspicuous with pale blunt lobes. Leaves to 30 cm, many in whorls, linear acute, fused in a

sheath at the base, with lobed double-toothed spiny margins; stem 20–30 cm. Roots unpleasant-smelling when bruised. **Pl. 56**.

b *Leaves entire, with or without marginal bristles*

627 **M. nepalensis** D. Don
W. Nepal to S.W. China. 3000–4400 m. Open slopes. Jun.–Aug.
A much smaller plant than the preceding species, with flowering stems 8–15 cm, and without spines on the leaf margins. Flowers pinkish-purple, few in closely grouped whorls, forming a small dense terminal head with large leaf-like bracts at its base with conspicuous long bristles (not spines). Corolla-tube 1–1.5 cm, curved, lobes nearly equal to *c*. 3 mm; calyx with 1–2 teeth; fertile stamens hairy, curved, 4. Basal leaves many, strap-shaped entire, 5–15 cm, with pale sheathing bases. **Pl. 57**.

2 **PTEROCEPHALUS** Like *Scabiosa*, but calyx with 20–24 plume-like bristles. 1 sp.

628 **P. hookeri** (C. B. Clarke) Diels
W. Nepal to S.W. China. 3300–4000 m. Open slopes, drier areas. Jul.–Sep.
A small distinctive scabious-like plant with globular heads of small lilac or cream-coloured flowers borne on hairy leafless stems arising direct from the woody rootstock. Flower-heads 2.5–4 cm; flowers all equal, with plume-like calyx bristles which are nearly as long as the corolla; involucral bracts lanceolate. Leaves all basal, linear-spathulate, entire or lobed, to 15 cm; flowering stems 10–35 cm. Fruit crowned with persistent papery limb and the calyx bristles. **Pl. 57**.

3 **SCABIOSA** SCABIOUS Flower-head with outer florets larger and spreading; calyx with 5 bristles. 1 sp.

629 **S. speciosa** Royle
Pakistan to Uttar Pradesh. 2400–3500 m. Open slopes. Jul.–Sep.
A handsome perennial with large mauve flower-heads 2–3 cm across, sometimes to 5 cm, with the outer spreading florets much larger than the inner erect florets. Involucral bracts lanceolate, densely hairy; outer florets 3–lobed, 2–3 cm, the inner florets shorter, nearly as long as the black calyx-bristles. Leaves broadly lanceolate to oblanceolate, toothed or entire, 3–7 cm, some with lanceolate lobes at the base; stems tufted, leafy at base, 30–50 cm. Fruit crowned with the persistent papery limb, and longer calyx bristles. **Page 474**.

4 **DIPSACUS** TEASEL Receptacle scales with bristly tips; flowers funnel-shaped, 4–lobed. 2 spp.

630 **D. inermis** Wallich
Afghanistan to S.W. China. Burma. 1400–4100 m. Shrubberies, open slopes, cultivated areas. Jun.–Sep.

COMPOSITAE

A robust perennial with opposite entire or pinnately lobed leaves, and globular heads of numerous cream-coloured flowers with black anthers, borne on long leafless stalks. Heads 2–3 cm across, with 6–8 spreading linear to ovate green involucral bracts; corolla funnel-shaped, with 4 equal lobes; calyx 4–angled, without bristles; receptacle scales ovate, longer than the flower buds. Leaves 8–30 cm, elliptic, toothed or with a few distant coarsely toothed lobes, the terminal lobe largest; stems angled, smooth or with scattered prickles, usually 60–200 cm. Fruit 8–ribbed, crowned with the cup-shaped calyx. **Pl. 57**.

COMPOSITAE   Daisy Family

One of the largest families, worldwide in distribution, particularly abundant in the Mediterranean, semi-arid, temperate, and montane regions of the world. Largely herbaceous perennials, biennials, or annuals, and some evergreen shrubs. Leaves opposite or alternate, mostly simple and toothed or lobed. The family is characterised by its very distinctive inflorescence, the *capitulum* or flower-head, which is composed of numerous small *florets*, borne on a *receptacle* and surrounded by an *involucre* of bracts, the whole appearing and functioning as a flower.

Individual florets are composed of a *pappus* (in place of a calyx) of hairs, scales, bristles or awns, or pappus sometimes absent; a corolla of 5 fused petals; 5 stamens usually fused by their anthers into a tube surrounding the 2–branched style; an inferior 1–celled ovary with 1 ovule. The florets may be subtended by receptacular-scales, -hairs or -bristles, but commonly the receptacle is naked. Florets of 2 basic types: *disk-florets* which have a tubular corolla with usually 5 short lobes, usually bisexual, and if present occupying the centre or whole of the receptacle; *ray-florets* which have a strap-shaped corolla with 3-5 terminal teeth. The latter may occur on the periphery of the receptacle, or occupy the whole of the receptacle; ray-florets are bisexual, female, or sometimes sterile. Fruit a dry, 1–seeded nutlet, often furnished with an apical pappus which aids in dispersal.

Flower-heads with tubular disk-florets only
 a Fruit with a pappus of hairs, scales, or a crown
  i Florets red, purple, blue or violet
   x Involucral-bracts not spine-tipped   1 AGERATUM
          18 PETASITES 27 SAUSSUREA 28 JURINEA
                    32 SERRATULA 35 AINSLIAEA
   xx Involucral-bracts mostly spiny   25 ECHINOPS
          26 COUSINIA 29 CARDUUS 30 CIRSIUM
                31 ONOPORDUM 33 CENTAUREA
  ii Florets yellow or white   1 AGERATUM 2 EUPATORIUM
       7 GNAPHALIUM 8 LEONTOPODIUM 9 ANAPHALIS

20 GYNURA   22 NANNOGLOTTIS   34 CARTHAMUS
35 AINSLIAEA
b Fruit without a pappus   15 TANACETUM
17 ARTEMISIA   18 PETASITES

Flower-heads with both tubular disk-florets and strap-shaped ray-florets
a Pappus of hairs or scales
   i Flower-heads with purple or sometimes white ray-florets
      4 ASTER   5 ERIGERON   6 PSYCHROGETON
      13 WALDHEIMIA   18 PETASITES   36 GERBERA
   ii Flower-heads with yellow ray-florets   3 SOLIDAGO
      10 INULA   19 CREMANTHODIUM   21 DORONICUM
      23 SENECIO   24 LIGULARIA   36 GERBERA
b Pappus absent (or a crown)   11 GUIZOTIA   12 ANTHEMIS
   14 ACHILLEA   15 TANACETUM   16 CHRYSANTHEMUM

Flower-heads with strap-shaped ray-florets only
a Fruit with a distinct beak   38 PICRIS   40 TRAGOPOGON
   41 LACTUCA   43 DUBYAEA   46 CREPIS
b Fruit not beaked
   i Florets blue, purple or white   37 CICHORIUM
      42 PRENANTHES   45 CICERBITA
   ii Florets yellow   39 SCORZONERA   44 SOROSERIS
46 CREPIS

1 AGERATUM Disk-florets only present, whitish to pale blue; pappus a
ring of scales. 1 sp.

631 **A. conyzoides** L.
Pan-tropical. In Himalaya to 2000 m. Cultivated areas; common. Flowering most of the year.
   An erect softly hairy annual, with numerous small pale blue or whitish
flower-heads, often in dense domed to flat-topped terminal and axillary
clusters, and with opposite, lanceolate to ovate, coarsely rounded-toothed
leaves. Flower-heads *c.* 6 mm across; florets all tubular, dilated upwards;
involucral bracts lanceolate, ribbed, in 2 ranks. Leaves *c.* 7 cm, stalked;
stem 30–90 cm. Fruit black, angled, crowned with a ring of 3–5 minutely
barbed slender scales. **Page 475.**

2 EUPATORIUM Differs from *Ageratum* in having fruit with a pappus of
slender hairs. Florets white. *c.* 5 spp.

632 **E. adenophorum** Sprengel (*E. glandulosum* Kunth)
Native of Mexico; introduced to Himalaya about 100 years ago. To 1800
m. Wasteland, damp places; very common and often gregarious over large
areas in C. & E. Nepal.
   Easily mistaken for 631, but stem and leaves glandular-hairy beneath
(softly-hairy beneath in 631). Flower-heads white, slightly larger, to 8

mm long, usually in dense domed or flat-topped clusters. Leaves with coarse acute teeth (teeth rounded in 631). A shrubby perennial to 1 m or more. **Page 475**.

3 SOLIDAGO Disk- and ray-florets both present, yellow. Fruit with pappus of hairs. 1 sp.

633 **S. virga-aurea** L. GOLDENROD
Pakistan to C. Nepal. Temperate Eurasia. 1800–3800 m. Light forests, open slopes. Jul.–Sep.

An erect usually unbranched perennial, with a terminal leafy spike-like cluster of numerous crowded yellow short-stalked flower-heads *c.* 1.5 cm long, and with lanceolate leaves. Ray-florets yellow, usually 10 or 12, *c.* 8 mm long; involucre cylindrical, bracts narrow-lanceolate, unequal, with papery margins. Lower leaves stalked, 8–13 cm, toothed, the upper stalkless, smaller, narrower, entire or toothed; stem 15–60 cm. Fruit ribbed; pappus long, rough. **Pl. 59**.

4 ASTER Ray-florets lilac, purple, rarely white; disk-florets yellow. Fruit flattened, with pappus of hairs. *c.* 13 spp.

Flower-heads many or several, in flat-topped or spike-like clusters
a Shrubs

634 **A. albescens** (DC.) Hand.–Mazz. (*Microglossa a.* (DC.) Benth. ex C. B. Clarke)
Kashmir to S.W. China. Burma. 2100–3600 m. Light forests, shrubberies, open slopes. Jun.–Sep.

A rambling shrub with rather large lanceolate long-pointed leaves, and with numerous small flower-heads with conspicuous lilac ray-florets borne in a flat-topped terminal branched cluster to 10 cm across. Flower-heads *c.* 8 mm across; ray-florets slender; disk-florets yellow; involucre bell-shaped, with many rows of narrow bracts. Leaves 4–12 cm, narrowed to a short stalk, entire or weakly toothed, hoary beneath; stems usually 1–2 m. Pappus red. **Pl. 58**.

b Herbaceous perennials

635 **A. indamellus** Grierson (*A. pseudamellus* Hook.f.)
Kashmir to C. Nepal. 2100–4000 m. Open slopes. Aug.–Sep.

Flower-heads usually several, long-stalked in a lax flat-topped cluster 3–4 cm across, with many narrow mauve ray-florets 8–13 mm long. Outer involucral bracts green, blunt, larger than inner papery bracts. Leaves stalkless, oblong acute or blunt, entire or toothed, 2–5 cm, regularly arranged on stem; stems many, arising from a woody stock, often branched above, 20–50 cm. Fruit finely hairy, with slender white pappus hairs.

636 **A. trinervius** Roxb. ex D. Don
W. Nepal to S.W. China. 1500–3000 m. Shrubberies, forests, cultivated areas. Jul.–Oct.

Flower-heads numerous, in a lax much-branched flat-topped cluster 20–30 cm across. Flower-heads to 2 cm across, with usually short white ray-florets to 1 cm; involucral bracts oblong blunt. An erect perennial to 1 m or more, with variable lanceolate to elliptic leaves 2.5–10 cm, mostly coarsely and irregularly toothed, stalked or not. Fruit hairy, with reddish pappus.

Flower-heads solitary, or 2–3 at ends of branches (see also 635)
  a *Alpine plants usually found above 3300 m*
    i *Flower-heads 5 cm or more across*

637 **A. diplostephioides** (DC.) C. B. Clarke
Kashmir to S.W. China. 3300–4800 m. Open slopes. Jul.–Sep.
A robust handsome shaggy-haired perennial with large solitary flower-heads 5–7 cm across, distinguished by its many very long narrow somewhat reflexed lilac ray-florets. Disk-florets at first blackish then orange; ray-florets obscurely 3–toothed; involucral bracts leafy, lanceolate, the outer often very long. Leaves oblanceolate to linear-lanceolate, entire and narrowed to the base, 5–8 cm; stem simple, leafy, shaggy-haired, usually 15–45 cm, but sometimes shorter. Fruit with silky reddish pappus. **Pl. 60**.

638 **A. falconeri** (C. B. Clarke) Hutch.
Pakistan to W. Nepal. 3000–4200 m. Alpine slopes. Jul.–Aug.
Rather like 637 with large solitary flower-heads to 8 cm across, but with very numerous spreading bluish ray-florets with white bases and trifid apices, to 3.5 cm long. Involucral bracts green, linear-lanceolate *c*. 1.5 cm. Basal leaves mostly oblong-lanceolate, distantly toothed, narrowed to the base, to 15 cm, mostly shrivelled at flowering, the upper smaller, somewhat clasping at the base; stem densely leafy, 15-35 cm. **Pl. 58**.

    ii *Flower-heads usually less than 4 cm across*

639 **A. flaccidus** Bunge
Pakistan to S.W. China. C. Asia. 3600–5200 m. Open slopes, screes. Jul.–Sep.
Flower-heads solitary, 3–4 cm across, with very numerous (to 60) mauve, linear, spreading ray-florets 12–15 mm long, and oblong long-pointed, woolly-haired involucral bracts. Leaves oblanceolate, narrowed to a stalk, with undulate margins, in a cluster arising directly from the rootstock; stem short 3–12 cm, with lanceolate clasping leaves usually similar to the basal leaves, but sometimes much smaller, or absent. Fruit bristly hairy. **Pl. 60**.

640 **A. himalaicus** C. B. Clarke
C. Nepal to S.W. China. 3600–4500 m. Rocks, open slopes. Aug.–Oct.
A low-growing spreading leafy perennial with solitary flower-heads *c*. 3.5 cm across, with numerous very narrow lilac ray-florets, and elliptic leafy, recurved, hairy involucral bracts. Disk-florets brownish-yellow.

Upper stem-leaves narrow-elliptic, half-clasping, the lower long-stalked, obovate-spathulate, all mostly 2.5–8 cm; stems several, leafy, 8–15 cm, spreading from a thick branched rootstock. Fruit silky; pappus white. **Pl. 60**.

641 **A. stracheyi** Hook. f.
Himachal Pradesh to Bhutan. 3600–4500 m. Rocks, open slopes. Jul.–Sep.

A dwarf perennial, distinguished by its creeping stems with short almost leafless flowering stems bearing solitary lilac flower-heads 1.3–4 cm across. Ray-florets numerous, short slender, often recurved; involucral bracts narrow-lanceolate long-pointed. Leaves obovate-spathulate, toothed, mostly basal, 2–5 cm; flowering stems usually 4–10 cm; plant often with long leafy runners to 20 cm. Fruit silky-haired; pappus reddish. **Pl. 60**.

b *Hill plants usually found below 3300 m*

642 **A. molliusculus** (DC.) C. B. Clarke
Pakistan to Uttar Pradesh. 1800–3300 m. Rocks. Jun.–Jul.

Flower-heads 1.5–3 cm across, long-stalked, usually solitary, with many short slender lilac ray-florets which are often recurved, borne on slender leafy stems 10–60 cm. Involucral bracts narrow-lanceolate long-pointed. Leaves lanceolate to oblong, entire or toothed, acute or blunt, 1.3–4 cm; stems many, slender, arising from a woody rootstock, leafy, often red tinged. Fruit silky-haired; pappus reddish. **Page 475**.

643 **A. thomsonii** C. B. Clarke
Pakistan to Uttar Pradesh. 2100–3000 m. Forests, shrubberies. Jul.–Sep.

Flower-heads usually solitary at ends of branches, 3.5–5 cm across, with many purple spreading ray-florets 1.3–2 cm, and with linear-lanceolate long-pointed, hairy, leafy involucral bracts. Leaves ovate to elliptic, acute or long-pointed, coarsely toothed, narrowed below and half-clasping, 5–10 cm; stems shaggy-haired, erect, branched, 30–90 cm. Fruit hairy, much longer than the reddish pappus.

5 **ERIGERON** Very like *Aster*, but ray-florets usually in 2 or more rows, thread-like; pappus-hairs in 1 row (ray-florets in 1 row, pappus-hairs in 2 or more rows in *Aster*). *c.* 14 spp.

644 **E. multiradiatus** (Lindley ex DC.) C. B. Clarke
Pakistan to Bhutan. 2400–4500 m. Shrubberies, open slopes; common. Jun.–Sep.

Flower-heads very variable in size 1.5–5 cm, long-stalked, solitary or clustered, with long spreading thread-like, dark purple ray-florets. Involucral bracts linear acute, hairy, with or without papery margins. Leaves obovate to lanceolate, acute or blunt, entire or coarsely toothed, the upper stalkless, *c.* 3.5 cm; stem hairy, 12–30 cm. Fruit nearly hairless; pappus red, short. **Pl. 58**.

645 **E. bellidioides** (Buch.-Ham. ex D.Don) Benth. ex C. B. Clarke
Pakistan to Bhutan. 1300–3600 m. Open slopes, meadows, cultivated
areas; common. Apr.–Oct.

Like 644, with solitary or few long-stalked flower-heads, but heads
smaller, 1–2 cm across, with many slender purple or white ray-florets
about twice as long as the disk-florets. Involucral bracts linear long-
pointed, with papery margins. Basal leaves lanceolate, entire or coarsely
toothed; stem leaves stalkless, oblong, usually entire; stem slender,
grooved, usually 10–30 cm. **Pl. 60**.

6 **PSYCHROGETON** Like *Erigeron* but distinguished by its flower-heads
with disk- and ray-florets almost similar in colour, and by the narrow linear
disk-achenes which are sterile. 1 sp.

646 **P. andryaloides** (DC.) Novopokr. ex Kraschen. (*Erigeron a.* (DC.)
Benth. ex C. B. Clarke)
Pakistan to Himachal Pradesh. 3000–4500 m. Dry stony slopes; common
in Lahul and Ladakh. Jun.–Aug.

A grey or white densely softly woolly-haired tufted perennial, with a
very stout woody rootstock and small solitary flower-heads *c.* 2 cm across,
borne on short nearly leafless stems. Ray-florets cream, short, broad, and
recurved; disk-florets cream to yellow; involucre broad-cylindrical, bracts
linear, woolly-haired. Leaves densely crowded at base, spathulate, mostly
entire, but variable, 2.5–5 cm; stems usually 4–8 cm, with small scale-
leaves. Fruit flattened, silky-haired; pappus white. **Pl. 60**.

7 **GNAPHALIUM** Tubular disk-florets only present, the outer usually
female, the inner bisexual; involucral bracts numerous, papery, shining
yellow in our species. Annuals. 4 spp.

647 **G. affine** D. Don (*G. luteo-album* var. *multiceps* (Wallich ex DC.)
Hook.f.)
Pakistan to Bhutan. Sub-tropical Asia. 1200–3000 m. Cultivated areas; a
very common weed. Feb.–Oct.

Flower-heads globular, bright glistening yellow, *c.* 3 mm across, in
dense rounded solitary or branched clusters, borne on erect woolly stems
with woolly-haired leaves. Outer florets female, thread-like; involucral
bracts shining, oblong-blunt *c.* 2 mm. Leaves oblong-spathulate 2.5–5 cm,
grey- or white-felted, the lower blunt, the upper narrower, half-clasping;
stems several and often tufted, 10–30 cm. **Page 475**.

648 **G. hypoleucum** DC.
Pakistan to Bhutan. China. Japan. S.E. Asia. 1500–2400 m. Grazing
grounds, open slopes; common. May–Sep.

Flower-heads many, in dense rounded clusters usually massed into a
branched flat-topped inflorescence, with glistening golden or yellowish
involucral bracts. Distinguished from 647 by its linear pointed stalkless
leaves, which are green and hairless on the upper surface and woolly-

haired beneath. Leaves 3.5–7.5 cm; stem 30–60 cm. **Pl. 61**.

8 LEONTOPODIUM EDELWEISS Like *Gnaphalium*, but the uppermost leaves clustered round the flower-heads and spreading in a star-like involucre much exceeding the flower-heads. A difficult genus with *c*. 9 spp., many of which are difficult to distinguish.

649 **L. himalayanum** DC.
Kashmir to S.W. China. 3000–4500 m. Open slopes; quite common. Jul.–Oct.

A grey woolly-haired tufted perennial, with dense terminal flat-topped clusters of several or many small globular flower-heads surrounded by an involucre of much longer white-woolly leaves spreading to 5 cm. Flower-heads white, turning brown, *c*. 6 mm long, with papery purple-tipped involucral bracts, the outer densely woolly-haired. Leaves linear acute, 1.5–3 cm, basal leaves broader, in a lax rosette; stems several, leafy, 5–15 cm, densely woolly-haired; rootstock branched above, covered with old leaf-bases. **Pl. 61**.

650 **L. jacotianum** Beauverd
Kashmir to S.W. China. 3300–4500 m. Open slopes; quite common. Jul.–Sep.

Like 649, but a smaller more tufted plant with flower clusters *c*. 1 cm across and involucral leaves spreading to 2.5 cm (clusters *c*. 2 cm across, and involucral leaves spreading to 3–5 cm in 649). Stems short, often densely clustered, with small linear to narrow-elliptic grey-woolly leaves *c*. 5 mm, but sometimes 1–2 cm; flowering stem often short, or up to 10 cm. **Pl. 61**.

9 ANAPHALIS Like *Gnaphalium* with globular flower-heads, but with shiny papery involucral bracts which are usually white in our species, and with yellow disk-florets only. Woolly-haired perennials. A very difficult genus. *c*. 20 spp. (The genus can easily be mistaken for *Helichrysum* which does not occur in the Himalaya).

Involucral bracts acute, spreading in flower

651 **A. triplinervis** (Sims) C. B. Clarke
Afghanistan to S.W. China. 1800–3300 m. Forest clearings, grazing grounds; very common. Jul.–Oct.

A robust plant with white-woolly elliptic upper leaves which are usually green above, and with dense domed branched clusters of small white flower-heads with shining white-papery spreading involucral bracts. Flower-heads *c*. 1 cm across; florets yellow; involucral bracts *c*. 5 mm. Leaves 5–10 cm, blunt or acute, with 3–5 conspicuous nearly parallel veins, paler and densely white-woolly beneath, the upper clasping, the lower narrowed to a stalk. Stems leafy, 30–60 cm. **Pl. 61**.

Var. **intermedia** (DC.) Airy Shaw (*A. nepalensis* (Sprengel) Hand.– Mazz.) is found in forests and grazing grounds from 2900–4100 m; flowering Jul.–Oct. A smaller plant usually less than 30 cm, with lanceolate woolly-haired leaves, and with a dense terminal cluster *c.* 1 cm across, of usually few (3–6) small flower-heads. Stem leaves mostly *c.* 3 cm.

Var. **monocephala** (DC.) Airy Shaw (*A. nubigena* auct. p.p. non DC.) is a dwarf usually tufted common alpine plant growing from 3400–5000 m; flowering Jun.–Sep. It has solitary or few flower-heads 1–1.5 cm across, borne on a stem 4–10 cm, with narrow-lanceolate white-woolly leaves 1–1.5 cm, but leaves very variable. **Pl. 61**.

Involucral bracts blunt, erect in flower
a *Leaves lanceolate, not decurrent*

652 **A. margaritacea** (L.) Benth. (*A. cinnamomea* C. B. Clarke)
Pakistan to S.W. China. Japan. N. America. 1800–3000 m. Open slopes, grazing grounds, cultivated areas; common. Aug.–Nov.

A perennial, with erect leafy stems, and with terminal dense domed clusters 5–15 cm across, of white flower each *c.* 8 mm across. Involucral bracts white-papery, broad, blunt, erect in flower and spreading in fruit. Leaves lanceolate 5–12 cm, with 1–3 conspicuous parallel veins, dark green above, grey- or reddish-woolly beneath. Stem woolly-haired, mostly unbranched below, 30–60 cm. **Pl. 61**.

b *Leaves linear to narrow-lanceolate, decurrent*

653 **A. busua** (Buch.–Ham. ex D. Don) DC. (*A. araneosa* DC.)
Pakistan to Bhutan. 1800–3600 m. Shrubberies, grazed ground; common. Jul.–Oct.

An erect perennial with linear to narrow lanceolate pointed leaves, and with broad branched domed clusters 7–15 cm across, of numerous tiny globular flower-heads each *c.* 4 mm across. Flower-heads with blunt white involucral bracts which are erect in flower and spreading in fruit. Leaves numerous, 3.5–6 cm, acute, with inrolled margins, and with decurrent basal lobes, green above, white-woolly beneath; basal leaves oblanceolate; stem 30–120 cm.

654 **A. contorta** (D. Don) Hook.f.
Afghanistan to S.W. China. 1500–4500 m. Shrubberies, grazing grounds. Aug.–Nov.

A small often tufted plant, with slender woolly stems with many overlapping linear leaves which are contrastingly green above and white-woolly beneath. Flower-clusters terminal, compact *c.* 1.5 cm across; flower-heads *c.* 3 mm across, with shining white involucral bracts 1–2 mm. Leaves with inrolled margins, mostly 1–2 cm by 1–3 mm wide; basal leaves densely clustered, obovate, white-woolly; stems usually branched from the base, often spreading, 15–45 cm.

10 INULA Flower-heads usually with both yellow ray-florets and yellow disk-florets; involucral bracts in many rows, green, leafy. Fruit angled; pappus hairs simple. *c.* 13 spp.

Shrubs to 2.5 m or more

655 **I. cappa** (Buch.-Ham. ex D.Don) DC.
Himachal Pradesh to S.W. China. 1000–2400 m. Shrubberies, open slopes; often gregarious. Aug.–Feb.

An aromatic shrub with woolly or silky-haired branches, leaves and inflorescence, with oblong-lanceolate leaves, and with numerous small yellow flower-heads in large terminal domed clusters 8 cm or more across. Flower-heads 8 mm across, without or with few very short ray-florets; involucral bracts lanceolate, rigid, silky-haired. Leaves thick, leathery, weakly toothed, sharp-pointed, green above and softly and very densely white-felted beneath, 8–15 cm; stems robust, to 2.5 m or more. **Pl. 64.**

Herbaceous plants
a *Flower-heads large, more than 3.5 cm across*
  i *Flower-heads solitary, terminal* (see also 658)

656 **I. grandiflora** Willd. (*I. barbata* Wallich ex DC.)
Pakistan to C. Nepal. W. Asia. 2000–3300 m. Shrubberies; quite common in Kashmir. Jul.–Sep.

Flower-heads golden-yellow, terminal and solitary, 4–6 cm across, borne on erect leafy stems. Ray-florets narrow, very numerous, to 2 cm long, 3–lobed at tip; involucral bracts with bristly-haired teeth, the outer leafy and as long as the florets, the inner shorter linear, all bristly-haired. Leaves oblong-lanceolate acute, 5–7.5 cm, lobed at base and with glandular teeth and bristly hairs on margin; stem unbranched, leafy throughout, 30–45 cm. **Pl. 59.**

657 **I. royleana** C. B. Clarke
Pakistan to Kashmir. 2100–4000 m. Forests, shrubberies; common and prominent in Kashmir. Jul.–Sep.

A rather stout erect plant with a very large handsome terminal golden-yellow flower-head 10–12.5 cm across, and with few large elliptic-lanceolate leaves, the upper clasping the stem. Ray-florets very numerous, to 5 cm long; outer involucral bracts ovate, with recurved triangular tips, the inner slender-pointed. Lower leaves elliptic blunt, blade 15–25 cm with a winged leaf-stalk as long; upper stem leaves elliptic with enlarged clasping base, to 20 cm; stem unbranched, hairy and glandular, 30–60 cm. **Pl. 59.**

  ii *Flower-heads usually more than one, in branched clusters*

658 **I. hookeri** C. B. Clarke
C. Nepal to S.W. China. Burma. 2400–3600 m. Shrubberies. Aug.–Oct.

Flower-heads golden-yellow, 3.5–6 cm across, usually 2–3 but sometimes solitary, with an involucre covered with very long shaggy hairs, borne on very leafy stems. Ray-florets thread-like, to 2.5 cm; involucre hemispherical, bracts very slender, recurved. Leaves elliptic-lanceolate, narrowed at both ends, 8–15 cm, hairy and with glandular-hairy teeth; stem usually very shaggy-haired above, 30–60 cm. **Pl. 62**.

659 **I. racemosa** Hook.f.
Afghanistan to C. Nepal. 2000–3200 m. Cultivated areas; quite common in the W. Himalaya where it is cultivated. Jul.–Sep.

A tall stout perennial to 1.75 m, with large short-stalked flower-heads 4–8 cm across, borne in a spike-like cluster, or inflorescence branched with one or two flower-heads at the ends of the branches. Ray-florets slender, to 2.5 cm long; outer involucral bracts broad-triangular somewhat leafy, woolly-haired, the inner oblong blunt, the innermost linear, longer. Lower leaves narrowed to a winged leaf-stalk, blade elliptic-lanceolate 18–45 cm; the upper leaves lanceolate, half-clasping stem with large basal lobes, all leaves with rounded teeth and densely hairy beneath; stem rough, grooved.

Roots used medicinally.

Flower-heads less than 3.5 cm across

660 **I. obtusifolia** Kerner
Pakistan to Uttar Pradesh. C. Asia. 2700–4000 m. Rocks; common in Ladakh and forming large clumps. Jun.–Aug.

A perennial with many leafy stems arising from a stout woody stock, each stem with one or few terminal yellow flower-heads, very variable in size, 1.3–3.5 cm across. Ray-florets to *c.* 8 mm, shorter than the width of the disk; outer involucral bracts with one or two considerably enlarged and leafy bracts, the inner linear long-pointed, rigid, erect. Leaves rather rigid, rough, stalkless, ovate to oblong with rounded base, obscurely toothed, 3.5–5 cm; stems 15–30 cm. Fruit slender, silky. **Pl. 59**.

661 **I. rhizocephala** Schrenk
Pakistan to Kashmir. 2100–3600 m. Open slopes, drier areas. Jul.–Aug.

Quite unlike any other member of the genus, with a rosette of spathulate leaves and a domed stalkless cluster of yellow flower-heads densely grouped at the centre of the rosette. Flower-heads 1.9–2.5 cm across; ray-florets yellow, *c.* 8 mm; involucral bracts linear pointed, the outer green with recurved tips, the inner purplish. Leaves pressed to the ground, blunt, narrowed to a broad stalk, bristly-haired, usually 3–5 cm; plant usually 7–12 cm in diameter.

Var. **rhizocephaloides** (C. B. Clarke) Kitam. differs in having no ray-florets and hairless achenes. **Pl. 64**.

11 GUIZOTIA Disk-florets and ray-florets both present, the latter with a 3–toothed apex. Pappus absent. 1 sp.

662 **G. abyssinica** (L.f.) Cass.
Native of Africa; cultivated in Nepal and also a field weed. 1000–2000 m. Sep.–Oct.

189

A stout leafy annual to 2 m, with clasping lanceolate to oblong leaves, and with yellow flower-heads in a lax branched terminal cluster. Flower-heads 3–4 cm across, with ray-florets few, broadly oblong, with 3–toothed apex; disk-florets tubular, anthers longer, protruding. Outer involucral bracts broadly ovate, blunt, green; receptacle with papery scales. Leaves opposite, 7.5–12.5 cm, toothed, with a fine apical point; stem stout, to 1.5 cm in diameter, branched above. Fruit black, flattened.

Cultivated for oil in Nepal; in some parts fields are golden with it in autumn. Seeds used medicinally. **Pl. 62.**

12 ANTHEMIS Leaves once to thrice cut into linear segments. Flower-heads solitary; receptacle with scales. Fruit ribbed; pappus absent. 1 sp.

663 **A. cotula** L. Stinking Mayweed
Pakistan to Kashmir. W. Asia. Europe. To 1750 m. Waste ground; common in the Kashmir valley. Jun.–Jul.

A foetid much-branched annual, with finely cut leaves with narrow segments, and with solitary long-stalked flower-heads 2–3 cm across, with spreading white ray-florets and yellow disk-florets. Involucral bracts hairy, with narrow papery margins; receptacle-scales linear. Leaves 2–3 cm, almost hairless, 2–3–pinnately cut into linear acute segments; stems to 50 cm. Fruit ribbed, with swellings.

Plant acrid, blistering the skin. Used medicinally.

13 WALDHEIMIA Spreading or tufted alpine plants. Ray-florets present; receptacle naked. Fruit 5–angled, pappus of rigid brown or reddish, flattened wavy bristles. 3 spp.

664 **W. glabra** (Decne.) Regel (*Allardia g.* Decne.)
Pakistan to S.E. Tibet. 4000–5500 m. Screes, stony slopes, drier areas; common. Jul.–Sep.

A low spreading or tufted mat-forming perennial, with usually small rosettes of deeply lobed leaves, and with solitary short-stemmed flower-heads with usually pink to purple or sometimes white ray-florets. Ray-florets 1–1.5 cm, broad- or narrow-elliptic, blunt; disk broader than rays, disk-florets yellow; involucral bracts with purple papery margins, blunt. Leaves 1.3–2 cm, aromatic, wedge-shaped and deeply 3-5–lobed, the lobes blunt, usually hairless; stems often spreading to 10 cm or more. **Pl. 62.**

665 **W. tomentosa** (Decne.) Regel (*Allardia t.* Decne.)
Pakistan to W. Nepal. C. Asia. 3600–5000 m. Stony slopes, drier areas. Jul.–Sep.

A loosely tufted perennial with solitary terminal flower-heads 3–6 cm across, with many pink or white broadly linear ray-florets *c.* 2.5 cm long. Involucral bracts broadly lanceolate, densely woolly-haired. Leaves once to twice cut into narrow segments, usually covered with dense soft white-woolly hairs; stems ascending 10–20 cm. **Pl. 64**.

665a **W. stoliczkai** (C. B. Clarke) Ostenf.
Pakistan to Kashmir. 3000–4500 m. Rocks, stony slopes. Jul.–Aug.
Very like 665 and considered by some to be a form of 665, but differing in appearance in being hairless, and having usually hairless involucral bracts with black papery margins. Leaves to 4–5 cm, completely hairless; stems 10–15 cm; rhizomes creeping, weakly branched. **Pl. 63**.

14 ACHILLEA Flower-heads many, in flat-topped clusters; ray-florets short, broader than long; involucral bracts with papery margins. Fruit strongly compressed; pappus absent. 1 sp.

666 **A. millefolium** L. YARROW
Pakistan to Uttar Pradesh. Temperate Eurasia. 1800–3600 m. Meadows, cultivated areas; common. Jul.–Sep.
An erect leafy perennial, with oblong-lanceolate much-dissected leaves, and with a dense terminal flat-topped cluster of tiny white or pale pinkish flower-heads. Flower-heads *c.* 6 mm across, with usually 5 rounded, reflexed, 3–toothed ray-florets; involucral bracts oblong blunt with brown or blackish papery margins. Leaves 2–3–times cut into linear-lanceolate pointed segments, the lower leaves stalked, the upper stalkless; stems grooved, 15–45 cm. Fruit shining.
Plant used medicinally.

15 TANACETUM TANSY Leaves pinnately cut. Flower-heads with numerous yellow disk-florets; ray-florets usually absent; involucral bracts with coloured margins. Fruit 5–ribbed; pappus absent, or a crown. Many species are now included in the genus *Dendranthema* by some botanists. *c.* 14 spp.

Ray-florets absent
a *Involucral bracts hairless*

667 **T. gracile** Hook. f. & Thoms.
Pakistan to Kashmir. 2800–3600 m. Stony slopes; common in Ladakh. Jul.–Sep.
A slender grey-hairy perennial with many stems arising from a woody base, with pinnately cut leaves, and with rather lax branched terminal clusters of tiny yellow flower-heads. Flower-heads 2–3 mm across; involucral bracts papery, hairless, broadly oblong. Leaves 1.3–2.5 cm, sparsely arranged up stem, cut into linear blunt segments; stems 30–60 cm, branched above, densely grey-hairy like the leaves. Fruit obovate, with a cup-like crown. **Pl. 63**.

b *Involucral bracts hairy*
    i *Flower-heads 8 mm or more across*

668 **T. dolichophyllum** (Kitam.) Kitam. (*T. longifolium* Wallich ex DC.)
Pakistan to W. Nepal. S.W. China. 3000–4300 m. Open slopes; common in Kashmir. Jul.–Sep.

An erect perennial with much-dissected leaves and a dense rounded terminal cluster of relatively large yellow flower-heads. Inflorescence sometimes shortly branched; flower-heads several, 8–15 mm across, with hairy involucral bracts with brown-purple margins. Leaves 12–25 cm, usually grey-green, aromatic, oblong in outline, twice cut into linear pointed segments, the basal leaves long-stalked; stem robust, 15–30 cm; rootstock stout, crowned with old leaf bases. **Pl. 64**.

ii *Flower-heads less than 8 mm across*
x *Dwarf alpine plants*

669 **T. gossypinum** Hook. f. & Thoms. ex C. B. Clarke
C. Nepal to Bhutan. 4300–5500 m. Rocks, screes, sandy ground. Aug.–Oct.

A very distinctive plant of high altitudes forming dense white-woolly cushions of rosettes of leaves. Flower-heads yellow, many in a dense rounded terminal woolly cluster, borne on a short stout stem covered with tiny wedge-shaped 3–6–lobed white-woolly leaves. Flower-heads *c.* 6 mm across; involucre globular, bracts linear-oblong, densely woolly-haired, with purple papery margins. Leaves densely overlapping, in 4 ranks, *c.* 3 mm long, toothed, the lower apparently entire; flowering stem 2.5–3 cm. **Pl. 64**.

xx *Taller plants, 10 cm or more*

670 **T. mutellinum** Hand.–Mazz.
Kashmir to S.W. China. 3600–4800 m. Stony slopes, sandy soils; common in drier areas. Jul.–Sep.

A white-woolly or -felted semi-prostrate perennial, with many stems arising from a woody stock, with small dissected leaves, and with yellow flower-heads 6–8 mm across, in a dense terminal rounded cluster. Flower-heads stalked or stalkless; involucral bracts broadly oblong, woolly-haired, with brown-papery margins. Leaves 6–20 mm, cut into short linear segments; stems usually 10–20 cm.

670a **T. tibeticum** Hook. f. & Thoms. ex C. B. Clarke
Pakistan to Himachal Pradesh. 4000–5400 m. Stony slopes in Tibetan borderlands. Jul.–Sep.

Very similar to 670, and perhaps a dry-country form of 670. Flower-heads larger; corolla glandular. **Pl. 63**.

671 **T. nubigenum** Wallich ex DC.
Uttar Pradesh to Bhutan. 3600–4800 m. Stony slopes, sandy ground. Jul.–Sep.

Very like 670, but differing in its taller stems 18–33 cm, smaller flower-heads 3–5 mm across, and leaves which are 3–times cut into linear acute lobes (leaves palmately or twice-cut in 670). A silvery-grey tufted plant with usually many stems arising from the rootstock.

672 **T. tomentosum** DC. (*T. senecionis* Gay)
Pakistan to Kashmir. 3300–4500 m. Stony slopes; common in Ladakh.
Jul.–Aug.

Like 670 with a dense terminal cluster of yellow flower-heads, but heads more numerous, smaller, 5–6 mm across, and leaves larger, 2.5–7.5 cm. Involucral bracts with woolly centre and broad brown papery margins. Leaves ovate in outline with linear blunt segments *c*. 1 mm broad (*c*. 0.5 mm broad in 670); stems with few leaves; a very variable plant, usually 15–25 cm.

Ray-florets present

673 **T. atkinsonii** (C. B. Clarke) Kitam.
C. Nepal to Bhutan. 3600–4500 m. Open slopes; quite common in E. Nepal. Jul.–Oct.

Distinguished by its large solitary flower-head 3–4 cm across, with many spreading large oblong ray-florets, and by its strap-shaped blunt involucral bracts with brown papery margins. Ray-florets yellow *c*. 1 cm. Leaves oblong to oval in outline, 3–times cut into short linear acute lobes; stems several 20–30 cm, arising from a stout rootstock. A strong-smelling plant. **Page 475**.

16 CHRYSANTHEMUM Like *Tanacetum* but in our species ray-florets present, and flower-heads large terminal. Pappus absent. 1 sp.

674 **C. pyrethroides** (Karelin & Kir.) B. Fedtsch. (*C. richteria* Benth.)
Pakistan to Kashmir. 3300–4800 m. Stony slopes; common in Ladakh.
Jul.–Aug.

A grey woolly-haired plant with a woody base, with few erect stems with solitary terminal flower-head 2.5–3.5 cm across, with many oblong blunt white or pinkish ray-florets. Involucral bracts usually oblong blunt, with purple papery margins. Basal leaves stalked, 5–8 cm, oblong in outline, 2–3–times cut into short rounded to lanceolate toothed lobes, densely woolly-haired; stems with few stalkless leaves below, leafless above, to 25 cm. **Page 475**.

17 ARTEMISIA WORMWOOD Leaves much dissected. Flower-heads small, usually in branched leafy clusters; ray-florets absent; involucral bracts with papery margins. Fruit cylindrical, without pappus. *c*. 25 spp.

Flower-heads 5 mm across, or more

675 **A. sieversiana** Willd.
Pakistan to C. Nepal. N. Asia. 1500–4100 m. Stony ground; common in Ladakh, also in dry areas in Nepal. Jul.–Sep.

An erect annual or perennial, with grey-hoary stalked leaves, and with terminal branched spike-like clusters of many stalked drooping yellowish flower-heads ranged along the branches. Flower-heads 6–8 mm; outer involucral bracts green hoary, the inner papery; receptacle with long straight hairs. Leaves broadly ovate in outline, twice-cut into oblong blunt segments 1–2 mm broad, grey-hoary on both sides, uppermost leaves entire lanceolate; stem branched above, grooved, to 60 cm.

Flower-heads less than 5 mm across
a *Ultimate segments of leaves more than 1.5 mm broad*

676 **A. absinthium** L. WORMWOOD
Pakistan to Kashmir. W. Asia. Europe. 1500–2700 m. Cultivated areas, forest clearings; common in Kashmir valley. Jul.–Sep.

A much-branched silvery-haired leafy perennial, with long rather one-sided spike-like lateral clusters of many tiny drooping yellowish flower-heads collectively forming a narrow pyramidal inflorescence. Flower-heads globular, 3–4 mm; outer involucral bracts green, with woolly hairs, the inner papery. Leaves approximately ovate in outline, unevenly 2–3–times cut into oblong to lanceolate segments *c.* 2 mm broad, grey- or white-hoary on both sides; stems to 1 m.

Plant used medicinally; yields an oil.

'It exercises a powerful influence over the nervous system, and its tendency to produce headache and other nervous disorders is well known by travellers in Kashmir and Ladakh, who suffer severely when marching through the extensive tracts of country covered with this plant.' (Watt, *Dictionary of Economic Plants of India*, 1889)

677 **A. dracunculus** L. TARRAGON
Afghanistan to Uttar Pradesh. W. Asia. Europe. 2700–4700 m. Stony places; common in Ladakh and Lahul. Jul.–Aug.

An erect very aromatic perennial, distinguished from others by its entire linear-oblong acute, green upper leaves. Flower-heads globular *c.* 3 mm across, short-stalked, forming long axillary spikes which collectively form a narrow pyramidal inflorescence. Involucral bracts hairless, ovate, with broad papery margins. Leaves 2.5–3.7 cm, entire or toothed; stem grooved, hairless, 30–60 cm.

Used medicinally, and yields an aromatic oil; also used for flavouring food.

678 **A. dubia** Wallich ex Besser
Pakistan to Bhutan. China. Japan. 1200–3400 m. Cultivated areas, waste-land; widespread. Jul.–Oct.

A tall aromatic shrub-like perennial, with broad leaf-segments often contrastingly pale hoary on the undersides and green and nearly hairless above. Flower-heads often reddish, 2–4 mm across, clustered, or ranged along the side branches in a lax or erect inflorescence. Involucral bracts woolly to nearly hairless, the outer green, the inner papery. Leaves to 8 cm long, usually once-cut into lanceolate pointed sparsely toothed segments often 5 mm broad; stems to 2 m. **Page 476.**

Plant used medicinally. Leeches are sensitive to the crushed leaves of this plant.

679 A. wallichiana Besser (*A. moorcroftiana* Wallich ex DC.).
Pakistan to E. Nepal. 2800–5500 m. Open slopes; common at high altitudes in the inner dry valleys of Nepal. Aug.–Oct.

Flower-heads often reddish, 4–5 mm across, stalkless and drooping, widely spaced in long slender lateral spikes to 20 cm, or in short dense lateral clusters. Involucral bracts ovate, papery, with cob-web hairs. Leaves 2–5 cm, usually pale hairy beneath, 1–2–times cut into lanceolate lobes usually 2–3 mm broad; stem erect, 30–60 cm.

680 A. roxburghiana Besser
Pakistan to C. Nepal. Tibet. 1000–4300 m. Open slopes, wasteland; common in Nepal in dry country north of the Dhaulagiri-Annapurna ranges, and in Kashmir. Jun.–Sep.

Distinguished by its creeping rootstock, its simple stems with deeply dissected woolly or nearly hairless leaves, and its tiny purplish flower-heads 3–4 mm across, in terminal and lateral spikes or spike-like clusters. Leaves twice-cut with linear pointed segments usually more than 2 mm broad.

Ultimate leaf-segments less than 1.5 mm broad

681 A. brevifolia Wallich (*A. maritima* L. sensu Hook. f.)
Pakistan to W. Nepal. Tibet. 2100–4200 m. Dry stony slopes; common in Ladakh and Lahul. Jul.–Sep.

A much-branched stiff, strongly aromatic, shrubby perennial, with very dissected pale grey to almost white leaves, and with narrow branched spikes of axillary clusters of tiny yellowish to reddish flower-heads 2–3 mm across. Heads ovoid, with 3–8 florets; involucral bracts narrow oblong, woolly-haired. Leaves ovate in outline, stalked, 1.5–5 cm, several times cut into linear blunt segments less than 1 mm broad; stems 15–40 cm.

Plant used medicinally, and for the production of santonin used internally against worms.

682 A. gmelinii Weber ex Steckm. (*A. sacrorum* Ledeb.)
Afghanistan to C. Nepal. N. Asia. 2100–4200 m. Dry stony slopes; common in Ladakh and Lahul. Jul.–Sep.

Differs from 681 in having larger nodding flower-heads 3–4 mm across with 15–20 florets; flower-heads short-stalked, green, in rather slender spike-like clusters. Involucral bracts hoary, with broad transparent papery margins. Leaves 2–3–times cut into very fine narrow segments *c*. 1 mm broad, grey- to whitish-hoary, or green, very aromatic; stems erect, very finely hoary, shrubby below, 30–120 cm. **Page 476**.

18 PETASITES BUTTERBUR Flower-heads in spike-like clusters; ray-florets absent or inconspicuous; leaves all basal. Fruit cylindrical, with pappus of simple hairs. 1 sp.

683 **P. tricholobus** Franchet (*P. himalaicus* Kitam.)
Pakistan to C. Nepal. 2700–3600 m. Forests. Mar.–May.
Flower-heads pale purple, in a dense cylindrical terminal cluster, borne on an erect stem with large green ovate to lanceolate pointed bracts in place of leaves, and with heart-shaped basal leaves. Flower-heads *c*. 1 cm across; involucral bracts narrow-lanceolate, dark green and flushed with purple, *c*. 1 cm. Basal leaves with a broadly heart-shaped blade 6–12 cm across, with conspicuously toothed margin, leaf-stalk to 25 cm; flowering stem to 30 cm, elongating to 60 cm in fruit. Often flowering before the leaves on woolly stems only 5–10 cm high.

19 CREMANTHODIUM Flower-heads usually nodding, and with spreading yellow ray-florets; involucral bracts in 1 rank; receptacle naked. Fruit 5–10–ribbed; pappus-hairs many, white or reddish. Leaves entire or toothed. 13 spp.
A genus of attractive species, of some importance in the high-summer Himalayan alpine flora, which is unfortunately not amenable to cultivation in Europe.

Flower-heads in a terminal spike-like cluster
a *Leaves kidney-shaped to ovate with a heart-shaped base*

684 **C. retusum** (Wallich ex Hook. f.) R. Good
C. Nepal to S.W. China. 3000–4500 m. Open slopes. Jul.–Sep.
Flower-heads several, nodding, usually in a spike-like cluster, with many short broad yellow ray-florets, and borne on a short stem with kidney- to heart-shaped leaves. Flower-heads 2.5–3.5 cm across; involucral bracts 10–12, oblong, acute or blunt, usually hairless. Leaves 5–20 cm across, coarsely toothed, with a rounded or notched apex and a deeply or shallowly lobed base: the lower leaves with a stout leaf-stalk, the middle leaves with large ovate basal sheath, the uppermost stalkless, bract-like, oblong-lanceolate; flowering stem 30–45 cm. **Pl. 66**.

685 **C. hookeri** C. B. Clarke (*Ligularia h.* (C. B. Clarke) Hand.–Mazz.)
C. Nepal to S.W. China. 3900–4300 m. Shrubberies, rocky slopes; quite common in Nepal. Jul.–Aug.
Flower-heads several, bell-shaped, with narrow ray-florets *c*. 1 cm long, in a spike-like cluster, borne on a slender nearly hairless stem, and with broad very coarsely toothed basal leaves. Flower-heads usually erect, *c*. 2.5 cm across, short-stalked; involucral bracts oblong to linear, acute, hairless or sparsely hairy. Leaves with broadly ovate to kidney-shaped blades, with triangular teeth, long-stalked; stem slender 20–35 cm, with usually one leaf with an enlarged clasping base to the leaf-stalk, and often upper small bract-like leaves.

b *Leaves oblong to elliptic, narrowed to the base*

686 **C. arnicoides** (DC. ex Royle) R. Good (*Senecio a.* Wallich ex C. B. Clarke)
Pakistan to S.W. China. 3300–4800 m. Shrubberies, open slopes. Jul.–Sep.

Flower-heads several, nodding, in a spike-like terminal cluster 10–30 cm
long, and with oblong to elliptic large blunt leaves. Flower-heads to 5 cm
across, with many broad ray-florets to 1.5 cm; involucral bracts elliptic
pointed, fused by their bases. Basal leaves with a long winged leaf-stalk
and with blade to 30 cm, coarsely toothed; stem leaves progressively
smaller more pointed and clasping stem; flowering stem to 60 cm. **Pl. 66**.

Flower-heads solitary
a *Veins of leaves radiating from the base of the blade*

687 **C. reniforme** (DC.) Benth.
Uttar Pradesh to Bhutan. 3600–4500 m. Shrubberies, open slopes; quite
common in Nepal. Jul.–Sep.

Flower-heads solitary, nodding, 5–7 cm across, with spreading yellow
ray-florets and brown disk-florets. Ray-florets conspicuously 3–toothed,
to 2.5 cm; involucral bracts lanceolate, with black hairs. Leaves kidney-
shaped to rounded with blade 3–10 cm broad, with blunt coarse triangular
teeth, the basal leaves long-stalked. Flowering stem slender 20–45 cm,
leafless or with 1 or more bracts, sparsely black-hairy towards the apex.
Pl. 65.

688 **C. decaisnei** C. B. Clarke
Pakistan to S.W. China. 3600–4800 m. Open slopes; a quite common
alpine species. Jul.–Sep.

Like 687, but a much smaller plant usually 10–20 cm, with smaller
leaves, shorter and stouter leaf-stalks, and broader oblong, brown-hairy,
involucral bracts. Flower-heads solitary, 3.5–6 cm across; ray-florets
yellow, 3–toothed. **Pl. 45**.

b *Veins of leaves branching pinnately from the mid-vein*
i *Leaves with blade purple beneath, rounded to ovate*

689 **C. purpureifolium** Kitam.
W. Nepal to C. Nepal. 3600–4900 m. Screes, open slopes. Jul.–Sep.

Distinguished by the reddish- to purplish-green undersides of the leaves,
which have broadly ovate blades. Flower-heads solitary, nodding, 3.5–5
cm across, borne on a leafy stem 10–15 cm. Ray-florets 16–17 mm long;
involucral bracts broad or narrow elliptic, with few dark woolly hairs.
Basal leaves long-stalked, with a finely toothed blade 5–10 cm long, often
purplish above as well as below; lower stem-leaves similar but smaller and
shorter-stalked; uppermost leaves bract-like.

ii *Leaves with blade not purple beneath, variable in shape*

690 **C. ellisii** (Hook.f.) Kitam. (*C. plantagineum* Maxim.)
Kashmir to S.E. Tibet. 3600–4800 m. Screes, open slopes; common.
Jul.–Sep.

Flower-heads solitary, nodding, 4–7 cm across, with dark disk-florets

197

and narrow elliptic ray-florets 2.5 cm long, borne on stout stems with entire oblong clasping leaves. Disk-florets often greenish; involucre with narrow lanceolate bracts with thick dark woolly hairs. Basal leaves with an oblong blunt blade 5–12 cm, with obscure widely-spaced teeth, and with a long winged leaf-stalk; stem mostly 15–30 cm. **Pl. 63**.

691   **C. oblongatum** C. B. Clarke (incl. *C. nakaoi* Kitam.)
Uttar Pradesh to Bhutan. 4300–5200 m. Open slopes; a common alpine species in drier areas. Jul.–Aug.

A slender plant with a solitary nodding flower-head with yellow ray-florets 1.5 cm long. Disk-florets often greenish to brownish; involucral bracts linear to linear-oblong, hairy or hairless. Leaves small, all basal, blade mostly 3–6 cm, broadly oblong to ovate with rounded or heart-shaped base, minutely or coarsely toothed, conspicuously paler beneath, and with short stout leaf-stalk; flowering stem slender, mostly 8–15 cm, with often many oblong green bracts. **Pl. 65**.

692   **C. nepalense** Kitam.
C. & E. Nepal. 2800–4300 m. Rock ledges, open slopes. Jul.

Like 691, but differing in having ovate, toothed leaves, with veins deeply impressed above and hairy. Flower-heads solitary, nodding, with yellow ray-florets conspicuously 3–lobed at the apex; involucral bracts narrow-lanceolate, hairless. **Pl. 65**.

693   **C. nanum** (Decne.) W. W. Smith
Kashmir to C. Nepal. Tibet. 4500–5500 m. Screes, drier areas; a rare but interesting high-altitude plant. Jul.–Aug.

A white-woolly plant, with an erect solitary flower-head 2–4 cm across, with yellow elliptic ray-florets *c.* 8 mm, with a broad yellow disk with numerous disk-florets, and with very woolly-haired, oblong blunt involucral bracts to 1.5 cm. Leaves few, with elliptic blades 2–3 cm, the upper stem-leaves with a short stout leaf-stalk, the lower leaves with longer blades and leaf-stalks; basal leaves absent. Stem 8–10 cm, with bracts at base; tuberous roots long, slender.

20   GYNURA   Flower-heads with disk-florets only; involucral bracts narrow and with papery margins; style-tips long hairy. Fruit narrow, many-ribbed; pappus hairs many, white. 4 spp.

694   **G. cusimbua** (D. Don) S. Moore (*G. angulosa* DC.)
Uttar Pradesh to S.W. China. 1200–2400 m. Rocks, open slopes. Flowering most of the year round.

A robust hairless perennial 1–2 m, with large obovate to oblong long-pointed leaves, and with large clusters of orange to orange-yellow flower-heads. Inflorescence widely branched; flower-heads 1.6–2 cm long, long-stalked, several or many; involucre cylindrical, bracts 10–12, equal, in one row, hairless and with papery margins, and with several smaller linear outer bracts. Leaves stalkless, 15–30 cm or more, irregularly toothed, the upper narrower, clasping. **Page 476**.

G. nepalensis DC. is very similar to 694 but differs in being a hairy plant, with hoary involucral bracts. It has a similar Himalayan distribution.

21 DORONICUM Flower-heads with both ray- and disk-florets; involucre with 2–3 rows of leafy bracts; receptacle convex, without scales. Leaves alternate. Fruit 10–ribbed, the inner fruits with a pappus. 2 spp.

695 **D. falconeri** C. B. Clarke
Pakistan to Himachal Pradesh. 3300–4000 m. Alpine slopes, rocks. Jul.–Aug.

A rather stout erect perennial, bearing a large yellow usually solitary flower-head 5–7.5 cm across, with very numerous long ray-florets. Involucral bracts glandular-hairy, linear-lanceolate, to 1.5 cm. Leaves elliptic to spathulate, obscurely toothed, nearly hairless, 8–15 cm, the lower short-stalked, the upper stalkless, narrower; stem 30–45 cm, nearly leafless above.

Roots used medicinally for nervous depression.

696 **D. roylei** DC.
Pakistan to Bhutan. 3000–3600 m. Forests, shrubberies; quite common in W. Himalaya. Jun.–Aug.

Like 695, but flower-heads several, smaller 1.5–4 cm across, in a lax cluster borne on conspicuously glandular-hairy stalks, which are swollen at the apex. Ray-florets yellow, 5–12 mm; involucral bracts lanceolate, glandular-hairy. Leaf-blade broadly ovate acute 8–12 cm, shallowly toothed, upper leaves clasping, with enlarged basal lobes, the lower leaves long-stalked; stem 60–120 cm.

Used medicinally; said to prevent giddiness caused on ascending heights! **Pl. 65**.

22 NANNOGLOTTIS
Distinguished by having florets of 3 forms in each flower-head: 2–3 outer rows of female florets, the outer with short strap-shaped corolla, the inner with tubular corolla; inner florets all hermaphrodite with tubular corolla. Fruit oblong, 8–10–ribbed; pappus of one deciduous row. 1 sp.

697 **N. hookeri** (C. B. Clarke ex Hook.f.) Kitam. (*Doronicum h.* C. B. Clarke)
W. Nepal to S. E. Tibet. 3200–4100 m. Alpine shrubberies, open slopes; common in Khumbu and Langtang, Nepal. Jul.

A rather robust perennial with an unbranched or little-branched stem bearing a few long-stalked yellow flower-heads 3–4.5 cm across. Ray-florets to 1.3 cm, about twice as long as the lanceolate long-pointed, glandular-hairy involucral bracts. Leaves elliptic to oblanceolate, blunt or

199

acute, entire or irregularly toothed, the upper with blades continuing in a wing down stem, the lower stalked, 10–15 cm; stem 30–80 cm. **Page 477**.

23 SENECIO RAGWORT Flower-heads with both ray- and disk-florets yellow. Like *Doronicum,* but involucre usually with few shorter additional bracts outside the single row of involucral bracts. Fruit ribbed, without a beak, and with a pappus of rough hairs. *c.* 22 spp.

Climbing, scrambling or shrubby-based plants; leaves much longer than broad

698 **S. scandens** Buch.-Ham. ex D. Don
Uttar Pradesh to Bhutan. Burma. 1800–3600 m. Forests. Sep.–Dec.
A climber with zig-zag, grooved branches, with arrow-shaped long-pointed leaves, and with many lax terminal and lateral domed branched clusters of bright yellow flower-heads. Flower-heads *c.* 8 mm long, with few ray-florets *c.* 5 mm; involucral bracts linear-oblong acute, nearly hairless. Leaves entire, or coarsely toothed, stalked, 7.5–10 cm; stems to 4 m. **Page 477**.

699 **S. triligulatus** Buch.–Ham. ex D. Don
C. Nepal to S.W. China. 1800–2400 m. Forests; not so common as 698 but quite prominent around the Nepal valley. Nov.–Mar.
A rambling shrubby plant with elliptic-lanceolate toothed leaves, and axillary clusters of few yellow flower-heads *c.* 1 cm long. Ray-florets usually 3–4, short *c.* 4 mm; disk-florets few; involucral bracts 8, narrow oblong, hairless. Leaves short-stalked, coarsely toothed, 8–10 cm; stems stout 1–2.75 m.

700 **S. cappa** Buch.–Ham. ex D. Don (*S. densiflorus* Wallich ex DC.)
Uttar Pradesh to S.W. China. Burma. 2100–3300 m. Forests, shrubberies; common. Sep.–Nov.
A shrubby perennial with large narrowly to broadly elliptic leaves, with their undersides with adpressed white or grey cottony hairs like the inflorescence and the involucral bracts. Flower-heads small *c.* 6 mm long, in terminal branched clusters; ray-florets 8–10, short; woolly involucral bracts 8–12, linear acute. Leaves stalked, 12.5–22 cm, hairless or cottony above, margin often with hooked teeth; stems stout, somewhat woody, to 60 cm, or sometimes up to 2 m.

Herbaceous perennials
a *Leaves entire or shallowly lobed, little longer than broad*

701 **S. wallichii** DC.
C. Nepal to Arunachal Pradesh. Burma. 2700–3300 m. Forests; common. Aug.–Oct.
Distinguished by its leafless flowering stem, with a dense spike-like, or branched cluster of yellow flower-heads, and with broadly ovate-heart-shaped basal leaves. Flower-heads *c.* 8 mm long, with a slender involucre

c. 2 mm broad, and with often 2 small ray-florets and few disk-florets; involucral bracts 5, oblong blunt. Leaves long-stalked, with blade 6–15 cm long, with wavy and toothed margin, arising from a short rather woody basal stem; flowering stem 20–40 cm, hairy and woolly above.

702 **S. jacquemontianus** (Decne.) Benth. ex Hook.f.
Pakistan to Kashmir. 3000–4000 m. Alpine meadows; gregarious and common in Kashmir. Jul.
   A stout perennial to 1.5 m, with broadly heart-shaped basal leaves, and a spike-like or conical terminal cluster of many spreading or drooping flower-heads. Inflorescence 10–20 cm, lengthening in fruit, lower flower-heads long-stalked with elliptic leafy bracts. Flower-heads to 4 cm across, with 12–15 spreading ray-florets to 1.5 cm; involucral bracts 8–12, oblong acute, hairless, *c.* 8 mm. Lower leaves with a long winged leaf-stalk and blade to 20 cm long and broad, stem leaves smaller, with sheathing stalk. Not easily distinguished from 707. **Pl. 66.**

b *Leaves at least some deeply lobed, much longer than broad*

   i *Flower-heads 1.3 cm or more across*

703 **S. chrysanthemoides** DC.
Pakistan to S.W. China. 2400–4000 m. Shrubberies, open slopes; common, often gregarious. Aug.–Sep.
   An erect perennial, usually branched above, with numerous yellow flower-heads in terminal branched flat-topped clusters, and with deeply lobed and toothed leaves. Flower-heads usually 1–2 cm across, but sometimes more, with 10–12 spreading strap-shaped ray-florets, and with variable involucral bracts lanceolate to oblong acute and often brown-tipped. Lower leaves mostly with a large toothed terminal lobe and few smaller irregularly toothed lateral lobes, the upper leaves with broad eared and toothed basal lobes and an oval, toothed or lobed terminal lobe; stem robust, 60–200 cm. **Pl. 64.**

704 **S. diversifolius** Wallich ex DC.
W. Nepal to S.E. Tibet. Burma. 2700–4300 m. Shrubberies, open slopes. Jul.–Oct.
   Very like 703, but differing in its larger yellow flower-heads 2–2.5 cm across, with distinctive red pappus, but size and pappus variable. Ray-florets to 1 cm. A robust hairless perennial to 60 cm, with deeply and irregularly lobed and toothed leaves with broad eared bases.

   ii *Flower-heads less than 1.3 cm across*

705 **S. graciliflorus** DC.
Kashmir to S.E. Tibet. 2400–4000 m. Shrubberies, forest clearings. Aug.–Oct.
   An erect perennial with large pinnately lobed leaves, and with much-branched flat-topped clusters of many slender yellow flower-heads each with only 3–5 ray-florets *c.* 4 mm. Flower-heads *c.* 8 mm long, erect or

drooping; involucre narrow-cylindrical c. 2 mm broad, with 5–7 linear blunt, hairless involucral bracts. Leaves 10–20 cm, with 6–8 pairs of rather regular, usually lanceolate long-pointed, coarsely and unequally toothed lobes, and with a leaf-stalk which is not eared at base; stem 60–200 cm. Fruit with white pappus. **Pl. 65.**

24 LIGULARIA Like *Senecio,* but leaves not lobed, the upper with large sheaths. Involucral bracts usually with additional smaller outer narrower bracts. 4 spp.

706 **L. amplexicaulis** DC.
Kashmir to Bhutan. 3000–4300 m. Shrubberies, forest clearings. Jul.–Sep.

A robust, nearly hairless perennial, with large rounded-heart-shaped stalked lower leaves, and with a dense branched flat-topped cluster of many flower-heads each with 5–6 very long slender ray-florets. Flower-heads *c.* 4 cm across; ray-florets *c.* 2 cm; involucral bracts *c.* 8 mm, oblong acute, fused below, hairless. Lower leaves with blade 15–20 cm broad, and interruptedly-winged leaf-stalk, or leaf-stalk not winged; upper leaves with very broad sheathing basal boat-shaped lobes; the uppermost leaves boat-shaped. Fruit with a rufous pappus. **Pl. 66.**

707 **L. fischeri** (Ledeb.) Turcz. (*L. sibirica* var. *racemosa* (DC.) Kitam., *Senecio ligularia* Hook. f.)
Pakistan to Bhutan. China. Japan. 2100–3600 m. Forests, shrubberies, damp places; quite common in Nepal. Jul.

Like 706, but flower-heads numerous, spreading or drooping, in a long spike-like cluster to 30 cm, but often shorter. Flower-heads often subtended by conspicuous green bracts as long as the flower-heads. Ray-florets 1.3–2 cm; involucral bracts to 1.5 cm, the outer narrow acute, the inner broader with broad papery margins. A stout erect but variable perennial, with large heart-shaped to kidney-shaped, coarsely toothed leaves to 30 cm across, the lower long-stalked, the upper with winged and sheathing stalks; stem erect to 80 cm. Easily confused with 702. **Page 477.**

25 ECHINOPS GLOBE-THISTLE Leaves thistle-like, spiny. Flower-heads grouped together in a spherical cluster; each flower-head with one floret surrounded by spiny involucral bracts. 2 spp.

708 **E. cornigerus** DC.
Afghanistan to C. Nepal. 2400–3300 m. Grazing grounds, stony slopes; prominent in Ladakh. Jul.–Sep.

Readily distinguished by its spherical inflorescence of numerous densely packed pale blue flower-heads with long spine-tipped involucral bracts, and by its spiny, deeply-lobed leaves which are white-cottony beneath. Inflorescence 7–8 cm across, solitary or several; florets *c.* 1.5 cm long, with 4 narrow recurved corolla-lobes; involucral bracts many, in several ranks, often one with a long spine, the outer silvery-haired, the inner papery. Leaves 10–20 cm, lanceolate or elliptic in outline, with broadly triangular spiny-toothed lobes; stem very leafy, to 1.5 m. **Pl. 70**.

26 COUSINIA Spiny thistle-like perennials. Flower-heads with disk-florets only, and with an involucre of spine-tipped bracts; receptacle with spiny scales. Fruit with pappus. 3 spp.

709 **C. thomsonii** C. B. Clarke
Afghanistan to W. Nepal. Tibet. 3000–4200 m. Stony ground; common in Ladakh and Lahul. Jul.–Aug.

An erect thistle-like perennial, with a cottony stem and deeply pinnately-lobed spiny leaves which are white-cottony beneath, and with large erect pink or purple flower-heads with globular, spiny, woolly-haired involucre. Flower-heads 5–6 cm across; involucral bracts rigid, linear with spreading or recurved apical spine, the inner shining papery. Basal leaves 12–25 cm, stalked, linear in outline, with numerous unequal linear lobes ending in a long rigid spine; stem-leaves stalkless, broader; stem to 45 cm. **Pl. 67**.

27 SAUSSUREA Leaves without spines. Flower-heads usually purple, bluish, or pink, with disk-florets only; involucre oblong to globular, of many rows of adpressed bracts without spiny tips; receptacle with bristly scales. Fruit with pappus of two rows, the outer usually of rigid bristles, the inner longer feathery, with their bases fused in a ring. *c.* 31 spp.

Densely white-woolly-haired alpine plants
a *Flower-head solitary*

710 **S. graminifolia** Wallich ex DC.
Kashmir to S.E. Tibet. 4000–5600 m. Screes, open slopes. Jul.–Sep.

A small plant with one or several stout woolly stems, with distinctive long narrow-linear leaves with dilated papery bases, and each stem with a solitary globular flower-head with many purple florets embedded in dense white-woolly hairs, and surrounded by an involucre of spreading or reflexed white-woolly leaves. Flower-head 2.5–3.5 cm across; involucral bracts papery. Leaves slender grass-like, usually 1–2 mm broad but sometimes more, the lower glossy, nearly hairless; stems variable, 6–15 cm; rootstock covered with old leaf-bases. **Pl. 68**.

b *Flower-heads many in a dense cluster*

711 **S. gossypiphora** D. Don
Kashmir to S.W. China. 4300–5600 m. Screes, stony places; not uncommon at high altitudes. Jul.–Sep.

A remarkable-looking plant like a snowball, the whole plant more or less globular and densely covered with long white- or grey-woolly hairs. Flower-heads deeply embedded in woolly hairs, cylindrical 1.3–2 cm long, and with many purple florets, densely clustered at the apex of the stout stem, and surrounded by woolly linear leaves. Leaves obscurely or coarsely toothed or lobed, embedded in very dense woolly hairs. Stem 10–20 cm, hollow, enlarged and club-shaped and densely leafy above, the base covered with the black shining remains of old leaf-bases. **Pl. 68**.

712 **S. simpsoniana** (Field. & Gardn.) Lipsch. (*S. sacra* Edgew.)
Kashmir to Sikkim. Tibet. 4400–5600 m. Screes, stony slopes; drier areas. Jul.–Oct.
Like 711, but distinguished when in flower by the purple flower-heads exposed outside the wool. A very variable plant, often much smaller than 711, with stem 5–8 cm, and leaves more or less concealed in wool, but sometimes to 15 cm with leaves free. Inflorescence domed; leaves linear-lanceolate, to 10 cm in large plants, coarsely toothed or shallow-lobed, with woolly hairs. **Pl. 68**.

713 **S. tridactyla** Sch. Bip. ex Hook. f.
W. Nepal to S.E. Tibet. 4500–5500 m. Screes; drier Tibetan border areas. Aug.–Oct.
Like 712 with a domed inflorescence with densely matted woolly hairs, but deep purple flower-heads half-exposed, and plant with a woody rootstock. Flower-heads with linear-oblong acute shining involucral bracts. Leaves very densely woolly-haired, linear to narrow-obovate, or spathulate, entire or 3–6–lobed towards the apex, 1–3 cm; stem enlarged and hollow above, 7–15 cm. The whole plant often looks like a white ball of cotton wool.

714 **S. gnaphalodes** (Royle ex DC.) Sch. Bip. (*S. sorocephala* (Shrenk) Sch. Bip.)
Pakistan to S.W. China. C. Asia. 4000–5500 m. High altitude screes; common in Ladakh. Jul.–Aug.
An almost stemless tufted woolly-haired perennial, with a dense domed cluster of many reddish-purple flower-heads, surrounded but not enclosed by spreading narrow leaves. Flower-heads cylindrical *c*. 1.3 cm long; florets often embedded in dense woolly hairs; involucral bracts broadly lanceolate, papery. Leaves variable, obovate to oblong, entire or obscurely toothed 1.3–2.5 cm, covered with dense or loose woolly hairs. A much smaller plant than 713, with a much-branched rootstock and many flowering rosettes 2.5–4.1 cm across. **Pl. 62. Page 477**.

Plants not densely woolly (except on undersides of leaves)
a *Flower-heads solitary, or rarely 2–3* (see also 722)
    i *Leaves pinnately lobed* (see also 717)

715 **S. nepalensis** Sprengel (*S. eriostemon* Wallich ex C. B. Clarke)
Kashmir to S.E. Tibet. 3200–4900 m. Open slopes; quite common in Nepal. Jul.–Sep.
Flower-heads solitary 2–3.5 cm across, borne on a short rather slender nearly leafless stem, and with many narrow deeply lobed basal leaves

which are conspicuously white-woolly beneath. Florets dark purple; involucral bracts lanceolate, papery, purple-tinged. Leaves mostly basal, 5–15 cm, linear, with many triangular backward-pointing lobes; stem absent or 2.5–15 cm, sometimes leafy. **Pl. 69**.

716 **S. auriculata** (DC.) Sch. Bip. (*S. hypoleuca* Sprengel ex C. B. Clarke)
Kashmir to Bhutan. 3000–4000 m. Shrubberies, open slopes; quite common in Nepal. Jul.–Aug.
   An erect perennial with pinnately lobed leaves, and with usually solitary globular nodding long-stalked purple flower-heads 3–4 cm across. Involucral bracts numerous, linear to narrow lanceolate, papery, dark purplish, spreading or recurved. Lower leaves 15–25 cm, with a much larger triangular terminal lobe and backward-pointing triangular lateral lobes, upper stem-leaves similar, smaller, all leaves white-woolly beneath and rough above; stem leafy, simple or sparsely branched above, 60–160 cm. **Page 478**.

   ii *Leaves entire or toothed*

717 **S. roylei** (DC.) Sch. Bip.
Kashmir to C. Nepal. 3000–4300 m. Shrubberies, open slopes; quite common in W. Himalaya. Jul.–Sep.
   A rather slender plant with a solitary dark purple flower-head, and with long narrow entire leaves conspicuously white-woolly beneath and over-topping the flower-head. Flower-head erect 3–4.5 cm across; involucral bracts narrow lanceolate, woolly-haired. Leaves very variable, linear to linear-lanceolate, long-pointed, sometimes with triangular lobes (in Nepal), mostly 8–15 cm; stem woolly-haired, usually 15–40 cm, with few clasping stem-leaves. **Pl. 69**.

718 **S. uniflora**. Wallich ex Sch. Bip.
C. Nepal to S.W. China. 3600–4500 m. Shrubberies, open slopes; quite common in Nepal. Jul.–Oct.
   Flower-head large, solitary, erect, enclosed when young by large purple papery boat-shaped silky-hairy bracts. Flower-head purple, variable in size, 2.5–4 cm across; involucral bracts ovate-lanceolate with long narrow tips, loosely woolly-haired. Leaves with narrow oblong to elliptic, toothed, hairless blade 10–18 cm; the lower long-stalked, the upper half-clasping, acute, and with blade continued in a wing down stem, the uppermost often purple like the bracts; stem simple 30–60 cm; rootstock stout, covered with old leaf-bases.

   b *Flower-heads several or many, in a dense or branched cluster*
     i *All leaves entire or toothed* (see also 727)
       x *Flower clusters not surrounded by large coloured bracts*

719 **S. albescens** (DC.) Sch. Bip.
Pakistan to C. Nepal. 2400–3000 m. Dry wasteland; common in Kashmir. Jun.–Sep.

Easily recognised by its pink, cylindrical flower-heads which are numerous and borne in branched flat-topped clusters on a simple erect leafy stem. Flower-heads 1.3–2 cm long; involucre cylindrical to 4 mm broad, bracts erect, hairless, the outer purplish, ovate acute, sharp-pointed, the inner paler, papery, narrow lanceolate, 2–3 times longer than the outer. Leaves very variable, the upper stalkless linear to broadly lanceolate, the lower stalked entire to shallow-lobed usually white-woolly beneath; stem slender to 120 cm. **Pl. 69**.

720 **S. candolleana** (DC.) Sch. Bip.
Pakistan to Bhutan. 2700–3300 m. Shrubberies, open slopes; quite common in Kashmir. Jul.–Aug.
An erect leafy perennial with large oblong to ovate-lanceolate toothed leaves, and with a dense terminal head-like cluster of many purple flower-heads. Flower-heads cylindrical 12–15 mm long by *c*. 5 mm broad; involucral bracts ovate blunt, hairless or with long silky hairs, the outer smaller and broader than the inner. Leaves 7–15 cm, stalkless, with blade forming a wing down stem, hairless above, usually with cottony white hairs beneath; stem grooved and winged, 30–90 cm.

721 **S. jacea** (Klotzsch) C. B. Clarke
Afghanistan to Kashmir. 2700–3600 m. Shrubberies, irrigated ground. Jul.–Aug.
An erect perennial with numerous overlapping ovate half-clasping leaves, branched above, and with a lax cluster of few purple flower-heads with conspicuous dark-margined involucral bracts. Flower-heads 1.5–2.3 cm long; involucre bell-shaped, bracts rigid, adpressed, the outer ovate, blunt or acute, the inner longer, lanceolate. Leaves 4–7 cm, margin entire; stem 30–100 cm, woody below. **Pl. 69**.

722 **S. fastuosa** (Decne.) Sch. Bip.
Himachal Pradesh to S.W. China. 3000–3600 m. Shrubberies, open slopes. Jul.–Sep.
Flower-heads solitary, or several in lax cluster, distinguished by their conspicuous involucres with triangular bracts with dark brown margins, the inner progressively longer and narrower. Florets yellow to light brown. Leaves many, elliptic, finely and regularly toothed, usually paler beneath, mostly 8–15 cm; stem erect, unbranched, or branched in the inflorescence above, to 1.25 m. **Pl. 69**.

xx *Flower clusters surrounded by large coloured bracts*

723 **S. obvallata** (DC.) Edgew.
Pakistan to S.W. China. 3600–4500 m. Rocky slopes, streamsides; rare in Nepal except in the extreme east around Topke Gola; prominent in 'The Valley of Flowers', Garhwal. Jul.–Sep.
A very striking perennial with large pale yellow boat-shaped papery bracts surrounding the dense cluster of purple flower-heads. Flower-heads

several to many, in a dense umbel-like cluster, each 1.5–2.5 cm long and with involucral bracts with black margins and tips, hairless or nearly so. Encircling boat-shaped bracts several, overlapping, ovate, to 12 cm long, bristly-margined, translucent and conspicuously veined. Leaves oblong-lanceolate blunt, toothed, the lower stalked, the upper half-clasping with blade continuing in a wing down stem; stem stout, 15–45 cm. **Pl. 68.**

724 **S. bracteata** Decne. (*S. schultzii* Hook.f.)
Pakistan to Kashmir. C. Asia. 3600–5600 m. Alpine slopes. Jul.–Aug.

Distinguished by its often pinkish pale papery boat-shaped bracts, which partly surround the densely crowded purple flower-heads. Flower-heads 1.5–2.5 cm long, borne on short densely woolly stalks; involucral bracts blackish, densely woolly-haired, narrow-lanceolate with very slender tips; boat-shaped bracts broadly lanceolate 2.5–4 cm. Basal leaves linear-oblong acute, obscurely toothed, with a very thick mid-vein narrowed to a broad leaf-stalk, the upper leaves half-clasping; stem often coloured, 15–40 cm; rootstock very stout covered with old leaf-bases.

ii *Basal leaves at least pinnately lobed* (see also 719)

725 **S. costus** (Falc.) Lipsch. (*S. lappa* (Decne.) Sch. Bip.)
Pakistan to Himachal Pradesh. 2000–3300 m. Cultivated as a field crop; also found as a casual in irrigated areas. Jul.–Aug.

A tall robust perennial, with large triangular long-stalked basal leaves and large clasping upper leaves, and with usually a dense rounded terminal cluster of few purple flower-heads with purple involucres. Flower-heads 2.5–3.5 cm across; involucral bracts numerous, ovate-lanceolate slender-pointed, rigid, with twisted and recurved tips, hairless or nearly so. Lower leaves pinnate, 30–40 cm, with an irregularly winged leaf-stalk and often triangular terminal lobe *c.* 30 cm across; stem-leaves smaller to 30 cm, all irregularly toothed ; stem 2 m or more, unbranched.

An important medicinal plant known locally as 'kuth'; cultivated in the Himalaya for this purpose. In Lahul now largely replaced by potatoes which are a more profitable crop. **Page 478.**

726 **S. deltoidea** (DC.) Sch. Bip.
Himachal Pradesh to S.W. China. S.E. Asia. 2100–3300 m. Dense herbage, forest clearings; quite common in Nepal. Aug.–Sep.

An erect perennial with leaves conspicuously white-cottony beneath, the upper entire often tooth, the lower pinnately lobed, and with nodding purple flower-heads. Inflorescence lax, branched; flower-heads 1.5–2.5 cm long; involucral bracts acute, hoary, often with purple margins and tips. Leaves very variable, the upper broadly arrow-shaped with blade mostly to 30 cm, the lower with a broad triangular terminal lobe and much smaller irregular triangular lateral lobes; stem 120–240 cm. **Page 478.**

727 **S. heteromalla** (D. Don) Hand.–Mazz. (*S. candicans* (DC.) Sch. Bip.)
Afghanistan to Bhutan. 3000–4000 m. Dry slopes, cultivated areas, waste-

land; common in W. Himalaya. Mar.–Aug.

Flower-heads erect, pinkish-purple, borne on long cottony stalks in open clusters on tall stems 60–160 cm, and with upper leaves oblong, and the lower entire or deeply pinnately lobed. Flower-heads 1.5–2 cm long; involucral bracts finely hairy, lanceolate long-pointed. Leaves variable,. the uppermost often entire toothed or not, the lower mostly deeply lobed, white cottony beneath and usually hairy above, commonly 6–20 cm long; stem branched above, with few leaves. **Page 478**.

28 JURINEA Flower-heads lilac or purplish, with disk-florets only; involucre with many overlapping bracts; receptacle with bristles. Fruit more or less four-sided, with usually a papery crown round base of pappus; pappus-hairs in several rows, rough or feathery, the innermost longest. 2 spp.

728 **J. dolomiaea** Boiss. (*J. macrocephala* (Royle) C. B. Clarke)
Pakistan to E. Nepal. 3000–4300 m. Open slopes. Jul.–Sep.

A quite prostrate perennial with a dense central domed cluster to 10 cm across, of rather large purple flower-heads, and a rosette of longer spreading lobed leaves often with purple mid-veins. Flower-heads to 4 cm long, very shortly stalked, in an umbel-like head; involucre with outer bracts lanceolate, hairy, the inner narrower long-pointed, dark red, papery. Leaves oblong blunt in outline, pinnately lobed, the lobes toothed or shallowly lobed, white woolly beneath, stalked, arising from a stout taproot and radiating to 30 cm.

Plant used medicinally. **Pl. 67**.

729 **J. ceratocarpa** (Decne.) Benth.
Pakistan to Kashmir. 2100–4800 m. Rocky slopes; common in Kashmir. Jul.–Aug.

An erect branched leafy perennial, with terminal pinkish flower-heads 2.5–3.7 cm across, each with an involucre of usually longer spreading narrow upper leaves. Flower-heads erect, long-stalked; florets *c*. 1.5 cm long; outer involucral bracts green, narrow, often recurved, the inner longer narrower and purple-tipped. Leaves 5–10 cm, narrow elliptic, sometimes entire, or more commonly shallowly lobed or deeply pinnately-lobed; the lower leaves stalked, the upper stalkless with lobed bases; stem usually 20–60 cm. Fruit with a feathery pappus. **Page 479**.

29 CARDUUS THISTLE Like *Cirsium*, but fruit with simple pappus-hairs. Involucre of many overlapping spine-tipped bracts. 2 spp.

730 **C. edelbergii** Rech.f. (*C. nutans* L. var. *lucidus* DC.)
Afghanistan to S.W. China. 1500–4000 m. Cultivated areas, grazing grounds; common. Jun.–Aug.

An erect perennial, with pinnately lobed very spiny-margined leaves and spiny stems, and with large globular or ovoid, solitary or clustered, crimson or rarely white flower-heads, 2.5–4 cm across. Flower-heads usually slender-stalked, in a lax inflorescence, or short-stalked and densely clustered; involucre with spiny spreading or recurved bracts with reddish tips. Stem-leaves alternate, mostly 4–10 cm, lanceolate in outline with triangular long-spined lobes, and blade prolonged into spiny wings down stem; stem 30–120 cm. **Page 479**.

30 CIRSIUM THISTLE Fruit with pappus of several rows of feathery hairs; receptacle-scales bristle-like. Involucral bracts usually with a simple spine. 7 spp.

Leaves with small spines on the upper surface

731 **C. falconeri** (Hook.f.) Petrak (*Cnicus f.* Hook.f.)
Pakistan to S.E. Tibet. 2700–4300 m. Forest clearings, grazing grounds. Aug.–Sep.
There are two distinct forms of this species, which may in the future be considered as separate species. The West Himalayan form is very common in Kashmir. It is a tall conspicuous perennial, the whole plant covered with whitish spines giving a silvery appearance. Flower-heads globular, cream-coloured, nodding, 7–8.5 cm across; involucral bracts many, densely woolly with very long pale spines to 2.5 cm. Leaves linear acute, with triangular toothed or lobed margins with pale spines, leaf-surface covered with spines; stems robust, winged, spiny. **Pl. 70**. The East Himalayan form has smaller purple flower-heads. **Pl. 71**.

Leaves without spines on the upper surface
a *Inner bracts of involucre swollen and with toothed tips*

732 **C. wallichii** DC. (*Cnicus w.* (DC.) C. B. Clarke)
Afghanistan to S.W. China. 1200–3300 m. Cultivated areas, forest clearings; common. May–Aug.
Flower-heads solitary or clustered, borne on leafless stalks or stalkless, varying from white to purplish-white. Flower-heads 2–3.8 cm across; florets *c.* 1.6 cm; involucre with lanceolate bracts terminating in erect or recurved spines, the inner bracts with lanceolate to ovate more or less dilated papery toothed tips. Leaves stalkless, pinnately lobed, with margins with very long strong spines, hairless above, cottony or woolly beneath; stems hairy, leafy, to 3 m. A very variable plant.

b *All bracts of involucre ending in a spine 6 mm or more*

733 **C. verutum** (D. Don) Sprengel (*C. involucratum* DC., *Cnicus argyracanthus* (DC.) C. B. Clarke)
Afghanistan to Bhutan. Burma. 740–2200 m. Cultivated areas, forest clearings. Mar.–Jul.
Flower-heads purple, mauve or pink, in dense clusters, surrounded by the upper leaves; involucral bracts with woolly margins and strong erect or spreading terminal spines. Flower-heads 2–2.5 cm across, stalkless.

Leaves hairless above, nearly hairless to cottony-haired beneath, the stem-leaves with a heart-shaped clasping base, and with spiny lobes and teeth, the basal leaves linear, stalked, with many triangular lobes, each lobe with a large terminal spine and shorter lateral spines; stems densely cottony-hairy, to *c.* 1.5 m. **Pl. 70**.

c *Bracts of involucre with very short spine to 2 mm; roots creeping*

734 **C. arvense** (L.) Scop. Creeping Thistle
Pakistan to C. Nepal. Temperate Eurasia. 250–3600 m. Cultivated areas; common in Kashmir. Mar.–Aug.

Flower-heads pale purple, 1–5, shortly stalked and borne at the ends of the branches. Involucre mostly 1.2–1.7 cm long, with adpressed bracts with short terminal spines, the outer bracts blunt, the inner longer with recurved tips; florets unisexual, 1.3–1.8 cm. Leaves lanceolate to oblong in outline, entire to pinnately lobed, with triangular lobes with weak to stout marginal spines, hairless or sparsely cob-web-hairy; stems much-branched, leafy, mostly 50–100 cm. Underground roots far-creeping and producing many erect stems. A very variable plant.

31 **ONOPORDUM** Flower-heads with disk-florets only, and with spine-tipped involucral bracts; receptacle without scales; filaments of stamens not fused. Fruit with feathery or rough pappus. 1 sp.

736 **O. acanthium** L. Scotch Thistle
Pakistan and Kashmir. W. Asia. Europe. To 1750 m. Wasteland; common in the Kashmir valley. Jul.

A robust white-felted thistle-like plant, with spiny lobed leaves and widely winged spiny stem, and with a lax cluster of 2–7 large globular rosy-purple flower-heads with very spiny spreading often white-haired involucral bracts. Flower-heads 3–5 cm across; involucral bracts *c.* 1.3 cm, narrowed to a stiff yellow-tipped spine. Basal leaves elliptic; stem leaves linear-lanceolate with shallow triangular teeth or lobes, each with a terminal spine. A biennial with stem branched above, 30–150 cm. **Pl. 71**.

32 **SERRATULA** Like *Centaurea*, with flower-heads with disk-florets only, but involucral bracts entire, without papery appendages or spines. Leaves without spines. 1 sp.

737 **S. pallida** DC.
Pakistan to Uttar Pradesh. 1400–3000 m. Open slopes, coniferous forests; common in W. Himalaya. Apr.–Jun.

An erect perennial, with pinnately lobed basal leaves, and with an almost leafless stem bearing a large solitary pinkish-purple flower-head, with many overlapping involucral bracts. Flower-head 2.5–4 cm across; florets with deeply lobed corolla, the lobes linear; outer involucral bracts much shorter than the inner. Leaves 8–30 cm, stalked, variable, broadly ovate to oblong, entire and toothed or deeply pinnately-lobed, often both forms occurring on the same plant; stem usually 30–50 cm. Fruit flattened, smooth; pappus with several rows of minutely barbed, rigid hairs. **Page 479**.

33 CENTAUREA Flower-heads purple, with involucre of spiny bracts in our species. Fruit with pappus. 2 spp.

738 **C. iberica** Trevir ex Sprengel
Pakistan and Kashmir. W. Asia. S.E. Europe. 300–2000 m. Wasteland, roadsides; common in Kashmir valley. Jun.–Jul.

A rather rigid, widely branching perennial, with lower leaves pinnately lobed, and with purple flower-heads with an ovoid involucre with conspicuous long stout spreading spines. Flower-heads 1.3–2 cm across, excluding spines; disk-florets only present, corolla with long narrow pointed lobes; involucral bracts broadly ovate, hairless, with narrow papery margin and a stout terminal spine 1–2.5 cm and with small spines at its base; inner bracts longer, spineless, with papery apex. Lower leaves with widely spaced lanceolate toothed lobes, the middle and upper leaves with few lobes, or entire; stems 20–80 cm. **Page 480**.

34 CARTHAMUS Flower-heads with yellow disk-florets only; involucral bracts green with spiny margin and apex, similar to the upper leaves; receptacle with scales. 2 spp.

739 **C. lanatus** L.
Pakistan and Kashmir. W. Asia. Europe. 1500–1800 m. Wasteland, roadsides; common in Kashmir Valley. Jul.–Sep.

A thistle-like annual with brownish stems, with many spiny elliptic clasping leaves, and with few golden-yellow spiny flower-heads in a lax branched, often flat-topped cluster. Flower-heads globular 2.5–3.5 cm, with outer leafy spiny-margined involucral bracts spreading beyond the florets. Stem leaves 3–6 cm, glandular-hairy, the margins with narrow spiny lobes; stem erect, branched above, 15–75 cm. Pappus greyish. A strong-smelling annual with reddish juice. **Page 480**.

C. tinctorius L. SAFFLOWER Sometimes cultivated in drier areas as an oil-yielding and medicinal plant. It has orange-red flowers and leaves with bristly-toothed margins.

35 AINSLIAEA Flower-heads with 1–4 white or pink disk-florets, each floret slightly two-lipped with 5 long narrow unequal lobes; ray-florets absent; involucral bracts many, erect, unequal. Fruit hairy; pappus feathery. 2 spp.

740 **A. aptera** DC.
Pakistan to Bhutan. 1400–3000 m. Forests, shrubberies. Mar.–Jun.

In spring leafless stems appear bearing slender spikes of drooping white or pinkish flower-heads; later in the year long-stalked broadly triangular to rounded heart-shaped leaves appear from the rootstock. In the rainy season and early autumn a leafy stem is produced with numerous flower-heads with florets that do not open but which later produce seed. Flower-heads cylindrical 1.3–2 cm long, with lanceolate involucral bracts, with papery often purplish margins, hairless, the outer much smaller than the inner. Leaves hairless, or densely silky-haired, stalked, blade 5–10 cm, toothed or not; flowering stems 30–90 cm. **Pl. 70**.

741 **A. latifolia** (D. Don) Sch. Bip.
Pakistan to S.W. China. Burma. S.E. Asia. 1500–3500 m. Shrubberies. Mar.–Jun.

Differs from 740 in having leaves with broadly winged leaf-stalks, and flowers which appear with the leaves in spring. Flowering spikes often with stalked and branched lateral clusters of flower-heads, and usually with persistent leafy bracts.

36 GERBERA Flower-heads solitary, with both ray- and disk-florets. Florets two-lipped: ray-florets with strap-shaped outer lip and a bi-lobed inner lip, disk-florets with nearly equal lips. Fruit rough, flattened, ribbed; pappus-hairs many. 4 spp.

742 **G. gossypina** (Royle) Beauv. (*G. lanuginosa* (Wallich ex DC.) Sch. Bip.)
Pakistan to W. Nepal. 1200–2400 m. Rocks, open slopes; common in pine forests. Mar.–May.

Flower-heads white tinged with pink, solitary, borne on long cottony leafless stems, and with entire or lobed elliptic basal leaves which are white-woolly beneath. Flower-head 2.5–4 cm across, with many small spreading ray-florets; involucre white-woolly, with narrow lanceolate acute bracts. Leaves mostly 5–15 cm, entire, toothed to pinnately lobed near the base, with a short or long winged leaf-stalk; stem often solitary, 10–30 cm, with few tiny bracts. **Pl. 65**.

743 **G. nivea** (DC.) Sch. Bip.
W. Nepal to S.W. China. 3300–4500 m. Shrubberies, open slopes. Jul.–Oct.

Flower-heads pale yellow, solitary, drooping and borne on a long cottony leafless stem, longer than the lobed leaves which are densely white-cottony-haired beneath. Flower-heads *c.* 3 cm across; ray-florets oblong, about one and a half times as long as the involucral bracts which are hairless and have long points, the inner larger and broader. Leaves obovate, all regularly pinnately-lobed with rounded obscurely-toothed lobes, hairless above, narrowed to a short leaf-stalk. Flowering stem usually solitary, 4–30 cm.

37 CICHORIUM Flower-heads blue, with ray-florets only present; invo-lucral bracts in two ranks, the outer shorter. 1 sp.

744 **C. intybus** L. CHICORY
Pakistan and Kashmir. W. Asia. Europe. 1500–2400 m. Cultivated areas; quite common in the Kashmir valley. Jul.–Sep.

Flower-heads bright blue, solitary and terminal, or clustered and axillary and ranged along the stout stiff green nearly leafless spreading stems. Flower-heads 2.5–4 cm across, with spreading ray-florets; involucral bracts green, the outer lanceolate spreading, the inner longer erect. Leaves oblong-lanceolate, the basal leaves pinnately lobed, the upper entire bract-like, clasping; stems grooved, with wide-spreading branches, 30–120 cm.

Used medicinally.

38 PICRIS Ray-florets only present, yellow; involucre of many rows of bracts; receptacle without scales. Fruit beaked; pappus in two rows, the inner white, feathery. 1 sp.

745 **P. hieracioides** L. HAWKWEED OX-TONGUE
Afghanistan to Bhutan. Temperate Eurasia. 1800–3600 m. Field weed; common. Apr.–Oct.

Flower-heads yellow, many, usually in a lax branched flat-topped cluster; leaves lanceolate toothed; stem erect robust 30–120 cm. Flower-heads 1.3–2 cm across, with many spreading ray-florets; involucral bracts 1–1.5 cm, with bristly black hairs, the outer bracts much shorter, spreading. Leaves 7.5–20 cm, often with wavy margins, the upper smaller, clasping, usually entire; stem branched above, rough with stiff bristly hairs. Fruit ribbed and transversely wrinkled. **Page 481.**

39 SCORZONERA Ray-florets only present, yellow; involucral bracts in many rows, thin green, the outer smaller. Fruit linear, not beaked; pappus-hairs feathery, in several rows, often with the outer hairs simple. Plants with milky latex. 2 spp.

746 **S. virgata** DC. (*S. divaricata* Hook. f.)
Pakistan to Himachal Pradesh. 2700–4200 m. Stony slopes; common in Ladakh. Jul.–Aug.

A slender perennial, much-branched from the woody rootstock and forming quite a large clump, with very narrow leaves, and with solitary terminal long-stalked bright yellow flower-heads with narrow cylindrical involucres. Florets 5–8, 2.5–3 cm long; involucral bracts slender, blunt, with papery margins, the inner to 2 cm and much longer than the broader outer bracts. Leaves grey-green, entire, 1–4 mm broad, with margins inrolled, 5–15 cm long; stems usually to 35 cm. **Pl. 71.**

40 TRAGOPOGON Like *Scorzonera*, but fruit with a beak; pappus-hairs in one or two rows, the inner feathery. Involucral bracts in one row. Plants with milky latex. 3 spp.

747 **T. gracilis** D. Don
Afghanistan to E. Nepal. 1500–3000 m. Open slopes, grassland. Apr.–Jul.

A nearly hairless perennial, with tufted stems often branched from the base, with narrow leaves with broader sheathing bases, and with solitary terminal yellow flower-heads 2.5–4 cm across. Involucral bracts green, 5–8, 1.3–2 cm long, shorter than the florets, enlarging in fruit. Leaves linear entire, acute, 7.5–20 cm by 5 mm broad, the basal leaves sometimes as long as the stems, which are usually 6–20 cm long.

Young fruit heads are eaten. **Page 481**.

748 **T. dubius** Scop. GOATSBEARD
Pakistan to Kashmir. 1500–3600 m. Meadows, cultivated areas. Jun.–Aug.

Differs from 747 in having involucral bracts as long or longer than the yellow florets, and fruit with a long beak as long as the nut (beak short in 747). Flower-heads yellow; involucral bracts lengthening to 4 cm in fruit. Leaves mostly broader 5–10 mm wide, and sheathing bases larger; plant 20–30 cm or more.

41 LACTUCA LETTUCE Flower-heads yellow, blue or whitish, with many ray-florets; involucre cylindrical, bracts in 3–4 rows; receptacle naked. Fruit flattened and usually ribbed, with a pale slender beak; pappus hairs simple, in two rows, white or straw-coloured. *c.* 11 spp.

Tall erect perennials with much-branched inflorescence
a *Leaves mostly pinnately lobed*

749 **L. decipiens** (Hook.f. & Thoms.) C. B. Clarke
Afghanistan to Kashmir. 2100–4000 m. Open slopes, forest clearings; common in Kashmir. Jul.–Aug.

A tall hairless perennial, with pinnately lobed leaves, and with a lax much-branched terminal cluster of many drooping blue or violet flower-heads each *c.* 2.5 cm long. Florets 6–8; involucre cylindrical, the inner bracts several times longer than the outer, hairless or with stiff bristly hairs. Leaves lanceolate in outline, the lower with narrow backward-pointing lobes and a terminal broadly arrow-shaped toothed lobe, and with leaf-stalks dilated and eared at the base; upper leaves sometimes entire and toothed, stalkless; stem branched above, 30–100 cm. Fruit with a short beak and dirty white pappus.

b *Leaves mostly entire or toothed*

750 **L. dolichophylla** Kitam. (*L. longifolia* DC.)
Afghanistan to S.W. China. 1300–3300 m. Cultivated areas. Jul.–Sep.

Differs from 749 in having long narrow, usually entire leaves, and erect blue to violet flower-heads in a broad or narrow branched terminal cluster. Flower-heads 1–1.5 cm long, with 12–20 florets; involucre with fewer much shorter ovate outer bracts and with narrow-oblong inner bracts. Leaves 7–15 cm long by 3–10 mm broad, lanceolate to linear-lanceolate, long-pointed, entire or sparingly lobed, stalkless, with enlarged or eared basal lobes; stem branched above, 1–2 m.

751 **L. bracteata** Hook.f. & Thoms. ex C. B. Clarke
W. Nepal to Arunachal Pradesh. 2400–3600 m. Shrubberies, open slopes. Aug.–Sep.

A bristly-haired perennial, branched above, and with rather numerous erect or inclined bell-shaped blue or mauve flower-heads, which are often subtended by rather conspicuous oblong-ovate bracts. Flower-heads 1.5–3 cm across, with many florets; involucral bracts papery, oblong-ovate acute, speckled with red. Leaves oblong-ovate 6–15cm, toothed, with a heart-shaped clasping base; stem hairy, 30–90 cm. Fruit with a slender white beak and white pappus.

Short erect perennials to 30 cm, with a dense domed inflorescence

752 **L. lessertiana** (DC.) C. B. Clarke
Pakistan to S.E. Tibet. 3000–4800 m. Open slopes; common. Jul.–Sep.

Flower-heads blue or purple, erect or drooping, stalked and borne on a relatively stout rather leafless stem, or stems branched from the base and bearing terminal flower-heads. Heads *c.* 1.5 cm long; involucral bracts linear-oblong blunt, bristly-haired; stalks of flower-heads often blackish, sometimes densely hairy. Leaves mostly 5–15 cm, usually oblanceolate, entire or variably toothed or shallow-lobed; stem usually 3–20 cm; rootstock stout. Fruit black, strongly ribbed; pappus white or yellowish. **Pl. 67.**

42 PRENANTHES Differs from *Lactuca* in having fruit without a beak. 2 spp.

753 **P. brunoniana** Wallich ex DC. (*Lactuca b.* (Wallich ex DC.) C. B. Clarke)
Afghanistan to C. Nepal. 1800–3600 m. Forests, shrubberies, meadows; common. Jul.–Oct.

An erect perennial with variously dissected leaves, and with a lax terminal branched cluster of slender blue, purple or less commonly white flower-heads each with 3–5 florets. Flower-heads 1.3–2 cm long by 2–3 mm broad; involucral bracts hairless, the outer broader and less than one third as long as the inner oblong blunt bracts. Leaves very variable; leaf-stalks long or short, winged or not, with enlarged or winged base; blade 10–20 cm, triangular heart-shaped, or with cut-off base, toothed or pinnately lobed, the lobes further lobed or toothed; stem 30–200 cm, simple or branched above, often glandular-hairy above. **Page 480.**

43 DUBYAEA Like *Lactuca* but differing in having fruit rounded in section (not flattened) with 5–10 principal ribs. Corolla-tube shorter than corolla-lobes. 2 spp.

754　**D. hispida** DC. (*Lactuca dubyaea* C. B. Clarke)
Uttar Pradesh to S.W. China. 3000–4300 m. Shrubberies, open slopes, rock-ledges; quite common. Jul.–Sep.

A bristly-hairy erect perennial, with upper leaves mostly simple, clasping, and with a widely branched inflorescence of few large nodding yellow flower-heads *c*. 2.5 cm across. Flower-stalks stout curved, with black glandular hairs; florets many, 2–2.5 cm long; involucre bell-shaped 1.5–2 cm long, densely covered with black bristly hairs; bracts linear-oblong. Leaves 7–15 cm, the lower with a winged leaf-stalk, sometimes pinnately lobed at base, the upper lanceolate shallowly lobed, stalkless and with clasping eared base; stem 15–45 cm; root spindle-shaped. Fruit black, with a slender pale beak; pappus persistent. **Page 481**.

755　**D. oligocephala** (Sch. Bip.) Stebbins
Pakistan to Himachal Pradesh. 2400–3300 m. Open slopes. Aug.–Sep.

Differs from 754 in having erect, not nodding, flower-heads, and numerous evenly overlapping involucral bracts; and fruit straw-coloured (involucral bracts relatively few; fruit dark in 754). Flower-heads yellow, with 70–80 florets. Leaves toothed or pinnately lobed, the upper lanceolate; stems 15–45 cm.

44 SOROSERIS　Dwarf alpine perennials with a thick often hollow stem, and lower leaves often bract-like. Florets all strap-shaped, yellow or white; involucral bracts in 2–3 ranks. Fruit with bristly-haired pappus in many rows. 4 spp.

756　**S. hookerana** (C. B. Clarke) Stebbins
Himachal Pradesh to S.E. Tibet. 4300–5500 m. Stony slopes, screes; drier areas. Jul.–Aug.

A high-alpine plant with a compact, usually almost stalkless inflorescence of many yellow flower-heads, and a rosette of narrow leaves. Ray-florets 4, oblong 1–1.7 cm, conspicuously 5–toothed at apex; involucral bracts linear blunt, with woolly hairs at base, usually hairless towards apex. Leaves oblong and shallowly lobed or toothed, to lanceolate entire, stalked; flowering stem 3–10 cm, sometimes more. Fruit cylindrical, with long grey pappus. **Pl. 67**.

757　**S. pumila** Stebbins
C. Nepal to S.E. Tibet. 4000–5000 m. Screes. Aug.–Oct.

Like 756, but with numerous ovate to linear bract-like leaves on the lower part of the short stem, and upper leaves spathulate to oblanceolate, purplish-green, coarsely toothed, hairy. Flower-heads numerous, in a flat-topped cluster, each with 4 yellow florets with broad, toothed limbs. Leaves surrounding flower-cluster long-stalked, with blade 0.5–2 cm; whole plant small, with stem 2.5–10 cm.

45 CICERBITA　Flower-heads blue to mauve; involucre of several rows of unequal bracts. Fruit flattened, usually not beaked; pappus of two rows of simple hairs, the outer shorter. 4 spp.

758 **C. macrantha** (C. B. Clarke) Beauv.
C. Nepal to Bhutan. 3300–4300 m. Shrubberies, open slopes. Jul.–Sep.
Flower-heads blue, inclined or nodding, few borne at the ends of the branches on a stout smooth stem, and with large pinnately lobed leaves. Flower-heads 3–5 cm across; ray-florets to 2.5 cm; involucre of many large overlapping ovate acute fringed bracts, the inner to 2.3 cm, twice as long as the outer. Leaves 7–23 cm, lanceolate in outline, glaucous beneath, with usually a triangular-ovate terminal lobe, and deep triangular lateral lobes, and a broad-eared base, lobes all toothed; stem leafy, 30–90 cm.
Page 481.

759 **C. macrorhiza** (Royle) Beauv. (*Lactuca m.* C. B. Clarke)
Pakistan to Bhutan. 1800–4300 m. Rocks, steep banks. Jul.–Sep.
Flower-heads mauve to blue, often drooping, in a terminal domed branched cluster, borne on slender prostrate or pendulous stems, and with usually pinnately lobed leaves. Flower-heads 1.3–2 cm; ray-florets strap-shaped, finely toothed at apex; involucral bracts bristly-haired, the outer lanceolate much shorter than the narrower inner. Leaves very variable, usually with distinctive rounded lobes, the terminal lobe largest; lower leaves with a winged or smooth leaf-stalk, sheathing or lobed at the base; stems tufted, much branched, usually to 60 cm; rootstock thick woody.
Pl. 61.

760 **C. cyanea** (D. Don) Beauv. (*Lactuca hastata* DC.)
Kashmir to S.W. China. 2100–3900 m. Forests, shrubberies; quite common. Sep.
Very like 753 *Prenanthes brunoniana* with purple florets, but flower-heads with 10–30 florets (3–5 florets in 753), and fruit distinctly beaked. A robust perennial 120–225 cm, with a long terminal branched cluster of many flower-heads each 1–2 cm long, with very unequal sparsely bristly-hairy involucral bracts. Leaves variable, blade 30–60 cm, pinnately lobed, the terminal lobe broad often triangular, leaf-stalks usually winged, inflated or eared at the base; upper leaves lanceolate entire. Plant sometimes dwarf and slender. Fruit black, beak half as long as nut.

46 CREPIS HAWKSBEARD Flower-heads with yellow to reddish ray-florets; involucre with two rows of bracts, the outer shorter than the inner. Fruit ribbed, with or usually without a beak; pappus-hairs simple, white. 7 spp.

761 **C. flexuosa** (DC.) Benth.
Pakistan to C. Nepal. Tibet. C. Asia. 3000–4200 m. Stony hillsides; prominent in Ladakh. Jul.
A much-branched perennial, forming rounded tufts to 30 cm across, with slender rigid nearly leafless forking branches terminating in small yellow flower-heads. Flower-heads 8–13 mm long by 3–4 mm broad;

outer involucral bracts minute, the inner 6–8, linear, hairless. Upper leaves few, linear entire; basal leaves broader, long-stalked, toothed or pinnately lobed; stems 5–20 cm. **Page 481**.

762 **C. sancta** (L.) Babc.
Afghanistan to Uttar Pradesh. 1500–3800 m. Cultivated ground; a common weed in Kashmir. Apr.–Jul.

Leaves all basal in a cluster or a lax rosette, and with slender erect sparingly-branched stems bearing rather few small yellow flower-heads *c.* 1 cm long. Outer involucral bracts very short, the inner linear acute 8 mm, with papery white margins. Leaves oblanceolate to spathulate 3–10 cm, with few teeth or shallow lobes, narrowed to the winged leaf-stalk; stems several, widely branched, with bristly hairs, to 30 cm.

Also cultivated in gardens and sometimes escaped and naturalized round villages are: **Tagetes patula** L. FRENCH MARIGOLD Flowers golden-yellow; much used to decorate shrines. **Zinnia elegans** Jacq. YOUTH AND AGE Flowers scarlet, violet, pink or white. **Dahlia imperialis** Roezl A shrubby plant flowering in the autumn, 2–6 m, with nodding white flowers tinged with pink; prominent in parts of C. & E. Nepal.

CAMPANULACEAE   Bellflower Family

A medium-sized family mostly of herbaceous perennials and annuals, rarely shrubs, largely of the North Temperate Zone. Leaves usually alternate, simple, without stipules. Flowers often relatively large and showy, often blue. Floral parts in fives; calyx-tube fused with the ovary, with 5 free lobes or teeth; corolla usually bell-shaped with 5 short lobes, but occasionally petals not fused; stamens 5. Ovary mostly inferior, but sometimes half-inferior; in *Cyananthus* it is superior; ovary of 5, 3, or 2 fused carpels with as many chambers, each with numerous ovules; stigmas as many as carpels. Fruit a capsule, opening by pores or valves.

Flowers regular; stamens free
  a Corolla divided nearly to base into narrow lobes   1 ASYNEUMA
  b Corolla bell- or funnel-shaped, usually with 5 short lobes
    i Capsule usually splitting by lateral pores   2 CAMPANULA
    ii Capsule splitting by terminal valves   3 CODONOPSIS
                                                 4 CYANANTHUS

Flowers irregular, with a 3–lobed lip; stamens fused   5 LOBELIA

1 ASYNEUMA Corolla with narrow lobes fused only at the base. Capsule 3–chambered, opening by 3 pores. 2 spp.

763 **A. fulgens** (Wallich) Briq. (*Campanula f.* Wallich)
W. Nepal to S.W. China. Burma. Sri Lanka. 1500–3000 m. Grassy slopes. Jul.–Aug.

Flowers blue, usually in a slender interrupted spike, with corolla shorter than the linear calyx-lobes. Corolla erect, *c.* 6 mm long with narrow lobes fused only at the base; calyx-lobes linear 8–10 mm. Basal leaves short-stalked, lanceolate-elliptic, to 5 cm, the upper stalkless, hoary or not; stem usually unbranched, 20–75 cm. Capsule ellipsoid, ribbed, with persistent calyx.

764 **A. thomsonii** (Hook.f. & Thoms.) Bornm. (*Phyteuma t.* (Hook. f. & Thoms.) C. B. Clarke)
Afghanistan to Kashmir. 1500–2400 m. Shrubberies, grassy banks. May–Jul.

Like 763, but differing in its broader triangular-ovate, coarsely saw-toothed leaves, which have a rounded base and a distinct leaf-stalk 1–2 cm (leaves lanceolate-elliptic, narrowed below, stalkless or with a short winged stalk in 763). Flowers short-stalked, in a lax often branched cluster; corolla blue, to 1 cm; calyx-lobes linear, shorter than the corolla. An erect nearly hairless perennial to 1 m. **Page 482.**

2 **CAMPANULA** BELLFLOWER Corolla bell-shaped or funnel-shaped, fused for at least one third its length and with short lobes. Capsule 3–5–celled, opening by apical or basal pores. 11 spp.

Flower 4 cm or more

765 **C. latifolia** L. LARGE BELLFLOWER
Pakistan to C. Nepal. W. & N. Asia. Europe. 2100–3600 m. Forests, shrubberies. Jul.–Aug.

A handsome robust leafy perennial, with large nodding dark blue-purple bell-shaped flowers 4–5 cm long, borne in a long spike-like terminal cluster on stems 60–200 cm. Flowers axillary, short-stalked, the stalk down-curved in fruit; corolla-lobes triangular-lanceolate; calyx-lobes narrow lanceolate, much shorter than corolla. Stem leaves mostly stalkless, 5–12 cm, ovate or lanceolate with rounded base, margin toothed, hairy, leaves becoming progressively smaller above. Capsule globose. **Pl. 72.**

Flowers less than 3 cm
a *Flowers solitary; stems unbranched* (see also 769)

766 **C. aristata** Wallich
Afghanistan to Bhutan. 3300–4500 m. Alpine meadows; common. Jun.–Aug.

A slender hairless perennial with small solitary terminal, often nodding, blue funnel-shaped flowers, and linear stem leaves. Corolla 8–10 mm long lobed to nearly halfway; calyx with linear lobes about as long as the corolla, or longer. Basal leaves long-stalked, elliptic, almost entire; stems unbranched, 8–30 cm. Capsule narrow-cylindrical to 3 cm, with persistent calyx.

CAMPANULACEAE

767 **C. modesta** Hook.f. & Thoms.
Kashmir to Bhutan. 4000–4800 m. Open slopes. Jun.–Aug.
Like 766 with solitary terminal somewhat inclined small conical blue flowers, but differing in its capsule which is globular, broadly winged and narrowed towards the base. Corolla 6–8 mm long, lobed to halfway. Stem leaves smaller mostly *c.* 1.5 cm; stem unbranched, 8–12 cm. **Page 482**.

b *Flowers several or many; stems branched above*
i *Leaves linear, usually less than 3 mm broad*

768 **C. sylvatica** Wallich
Pakistan to Bhutan. 1200–2400 m. Banks, pine forests. Apr.–May.
Flowers blue, narrow-funnel-shaped, usually few in a lax branched cluster and borne on slender erect hairy stems 10–30 cm. Corolla 0.8–1.3 cm long, 5–lobed; calyx-lobes hairy, linear-lanceolate, 6–8 mm long. Leaves linear to 5 cm, shaggy-haired. Capsule obconical, crowned with the somewhat elongated calyx-lobes. **Page 482**.

ii *Leaves lanceolate to ovate, more than 3 mm broad*

769 **C. argyrotricha** Wallich ex A. DC.
Pakistan to Bhutan. 3000–4700 m. Rock ledges, stony slopes; drier areas. Aug.–Oct.
A delicate often tufted perennial, with many spreading slender stems arising from the rootstock and bearing small blue funnel-shaped flowers 1–1.5 cm long. Flowers long-stalked, solitary or few; corolla hairy outside; calyx-lobes elliptic *c.* 5 mm. Leaves 8–12 mm, ovate, nearly entire or toothed, softly and densely silvery-haired; stems hairy, 8–15 cm.

770 **C. pallida** Wallich (*C. colorata* Wallich)
Afghanistan to S.W. China. 1200–4500 m. Cultivated areas, stone walls, banks. Jun.–Sep.
A much branched hairy-stemmed plant, with lax terminal clusters of small pale lilac or purple flowers, and narrow-elliptic softly hairy leaves. Corolla hairy, 8–15 mm long, lobed to less than half its length; calyx-lobes lanceolate to triangular 4–6 mm, hairy. Leaves stalkless, toothed, *c.* 2.5 cm; stems 15–60 cm. Capsule obconical, nodding.

771 **C. cashmeriana** Royle
Afghanistan to Uttar Pradesh. 2100–3600 m. Rocks. Aug.–Sep.
An attractive perennial usually found in rock crevices and cliffs, with rather large bright blue broadly bell-shaped flowers, and small softly hairy greyish leaves. Flowers few, in widely branched terminal clusters; corolla 2–2.5 cm long, shallowly 5–lobed; calyx-lobes broadly lanceolate, 6–10 mm, hoary or finely hairy. Leaves elliptic blunt 1.5–3.5 cm, entire or sparingly toothed, rather thick, hoary or softly hairy; stems often tufted, usually 15–20 cm, zigzag or flexuose, hairy. Capsule 5 mm, broader than long. **Page 482**.

3 CODONOPSIS Like *Campanula*, but usually strong-smelling plants. Capsule beaked, splitting by 3 valves (not pores). 14 spp.

Climbing or twining herbaceous plants
a *Flowers blue*

772 **C. convolvulacea** Kurz
W. Nepal to S.W. China. Burma. 2400–3600 m. Shrubberies. Aug.–Sep.

A perennial with twining stems, and with large broadly bell-shaped blue flowers, often with a reddish ring within, with spreading narrow ovate acute corolla-lobes 2–3 cm long. Flowers 4–5 cm across; calyx with 5 triangular acute lobes to 2 cm; base of corolla densely hairy within; style long, stigma 3–lobed. Leaves ovate to broadly lanceolate 1.3–5 cm, stalked, entire or shallow-toothed, smooth, shining; stems hairless. Capsule top-shaped, with persistent calyx. **Pl. 72**.

b *Flowers greenish, with purple veins or blotches, or purplish lobes*
  i *Calyx-lobes oblong to linear*

773 **C. affinis** Hook. f. & Thoms.
C. Nepal to Sikkim. 2100–3300 m. Shrubberies. Jul.–Sep.

A twining plant with tubular purplish or green flowers with purple tips to the shallow corolla-lobes, and with ovate acute leaves with deeply heart-shaped bases. Flowers *c*. 2 cm across, axillary or terminal, often in branched clusters; corolla-lobes broadly triangular, recurved; calyx-lobes widely spaced, oblong to narrow triangular *c*. 8 mm, finely hairy. Leaves stalked, blade 5–10 cm, sparsely hairy at least when young. Capsule hemispherical, beaked, with persistent calyx. **Pl. 72**.

774 **C. viridis** Wallich
Uttar Pradesh to Bhutan. 1200–2700 m. Shrubberies. Aug.–Sep.

A twining perennial with large tubular greenish-yellow flowers blotched or streaked with purple within, and with elliptic-oblong leaves with cut-off or wedge-shaped bases. Flowers 2–3 cm long, with a long stalk 2–7 cm long; flower-buds with dense shaggy hairs; calyx-lobes very widely spaced, linear, hairy. Leaves 5–7.5 cm, paler and finely shaggy-haired beneath, stalked. Capsule hemispherical, to 2.5 cm, beaked. **Pl. 72**.

  ii *Calyx-lobes large and elliptic*

775 **C. rotundifolia** Benth.
Pakistan to C. Nepal. 1800–3600 m. Shrubberies. Jul.–Aug.

A twining perennial with broadly bell-shaped greenish-white flowers veined with purple, with large conspicuous green calyx, and with alternate, triangular-ovate, toothed leaves. Flowers 2–3 cm long, with broadly triangular corolla-lobes, stalked, axillary; calyx-lobes elliptic *c*. 2 cm, often toothed. Leaves stalked, very variable in size; blade often *c*. 5 cm, with rounded base, hairless or sparsely hairy. Capsule hemispherical, beaked, with persistent calyx.

CAMPANULACEAE

Erect herbaceous perennials
a *Flowers red-purple*

776 **C. purpurea** Wallich
Uttar Pradesh to S.W. China. 1800–3000 m. Rocks, banks. Aug.–Sep.

A spreading weak-stemmed perennial, often hanging from cliffs and banks, with large terminal and axillary, dark red-purple, bell-shaped flowers with large triangular corolla-lobes. Flowers *c.* 3 cm across, stalked; corolla-lobes *c.* 2 cm; calyx-lobes triangular-ovate *c.* 1.3 cm, glaucous. Leaves opposite, *c.* 6 cm, oblong-elliptic with wedge-shaped or cut-off base (never heart-shaped), stalked, glaucous beneath; stems branched, to 60 cm. Capsule obconical to 3 cm, beaked, with persistent calyx.

b *Flowers blue or white*
  i *Calyx-lobes elliptic*

777 **C. clematidea** (Schrenk) C. B. Clarke
Afghanistan to Himachal Pradesh. 2400–4200 m. Cultivated areas, alpine slopes; common on irrigated land in Ladakh and Lahul. Jul.–Sep.

A strong-smelling perennial with long-stemmed solitary, often nodding bell-shaped pale blue flowers with a brown ring within, and with small obovate to heart-shaped leaves. Corolla 1.5–2.5 cm long, with shallow recurved lobes; calyx-lobes elliptic, recurved, usually more than half as long as corolla. Leaves short-stalked, 1.5–2.5 cm, grey-hairy; stems branched below, 30–60 cm. **Pl. 72**.

778 **C. ovata** Benth.
Pakistan to Kashmir. 3000–4200 m. Rocks, alpine slopes. Jul.–Aug.

Like 777 with small ovate hairy leaves, and erect stems bearing solitary nodding long-stemmed milky-blue broadly funnel-shaped flowers, but with calyx-lobes spreading, less than half as long as corolla. Corolla 1.5–3.5 cm long, gradually widening to the mouth, lobes broad-triangular, blunt, out-curved; calyx-lobes oblong-elliptic, 8–10 mm, minutely hairy. Leaves 1–3 cm, short-stalked, with a very pungent odour when crushed; flowering-stems 15–20 cm. Capsule obconical, beaked. **Pl. 72**.

  ii *Calyx-lobes linear or oblong*

779 **C. thalictrifolia** Wallich
C. Nepal to S.E. Tibet. 3600–4500 m. Shrubberies, open slopes. Jul.–Sep.

Distinguished by its solitary funnel-shaped pale blue flowers 2.5–3.5 cm long, with the corolla-tube narrow-cylindrical at base and widening abruptly to twice as broad at the mouth, and by its tiny rounded to heart-shaped leaves. Calyx-lobes oblong acute, *c.* 6 mm, minutely hairy. Leaves small, rarely more than 8 mm long and wide, shaggy-haired; stems flexuose, sparsely leafy above, to 40 cm; rootstock woody. Capsule hemispherical, beaked, *c.* 1 cm.

780 **C. dicentrifolia** (C. B. Clarke) W. W. Smith
C. Nepal to Sikkim. 3300–4000 m. Cliffs, rocky slopes. Jul.–Sep.

An almost hairless perennial with slender branched stems bearing large solitary broadly funnel-shaped milky-blue flowers, and with distinctive linear, widely spaced calyx-lobes. Corolla to 3 cm long, widening gradually to the mouth, the lobes with fine points; calyx-lobes linear, *c*. 1 mm broad. Leaves broadly elliptic blunt, short-stalked, usually 1.5–3 cm; stems widely branched, to 40 cm. Capsule elliptic, nodding.

4 CYANANTHUS Distinguished from *Codonopsis* by its superior ovary, and by its capsule which is nearly or quite surrounded by the persistent calyx, and which splits usually into 4–5 valves. 9 spp.

Leaves less than 1 cm long, entire or obscurely toothed or lobed

781 **C. incanus** Hook.f. & Thoms.
C. Nepal to S.W. China. 2100–4500 m. Rocks, open slopes. Jul.–Sep.

A low spreading perennial, with short leafy stems each bearing a solitary terminal azure-blue funnel-shaped flower with oblong acute spreading lobes, and with long white hairs in the throat. Flowers 2–2.5 cm across; corolla-tube *c*. 1.5 cm long, the lobes 5–10 mm; calyx lobed to halfway or less, the lobes 8–10 mm, usually ciliate. Leaves elliptic, entire or very shallowly lobed, 6–10 mm, softly white-haired; flowering stems 6–10 cm; rootstock stout. Capsule *c*. 8 mm, surrounded by the enlarged and inflated calyx. **Pl. 72**.

782 **C. microphyllus** Edgew.
Uttar Pradesh to S.E. Tibet. 3000–4800 m. Rocks, banks, open slopes. Aug.–Oct.

Distinguished from 781 by the conspicuous long dense dark hairs on the calyx. Flowers solitary, blue-violet, funnel-shaped and with narrow obovate acute spreading lobes 1–1.5 cm long, and with a tuft of white hairs within the corolla-tube; calyx half as long as the corolla-tube. Leaves numerous, alternate, oblong to elliptic 5–8 mm long; stems many tufted, trailing or ascending, 5–15 cm; rootstock stout above, branched.

Leaves usually 1.5 cm or more, conspicuously lobed

783 **C. lobatus** Wallich ex Benth.
Himachal Pradesh to S.W. China. 3300–4500 m. Shrubberies. Jul.–Sep.

Distinguished from 782 by its larger obovate to wedge-shaped, deeply lobed and rather fleshy leaves, and by its larger bright blue-purple flowers with conspicuous calyx covered with short blackish hairs. Flowers to 3 cm long; corolla-lobes obovate, spreading, corolla-tube hairy in the throat; calyx 1.5–2 cm, lobed to one third. Leaves 1–2.5 cm, deeply lobed, becoming hairless; stems several, 10–35 cm. Capsule elliptic, *c*. 1.5 cm. **Pl. 71**.

5 LOBELIA Flowers conspicuously irregular, with a 3–lobed lower lip and oblique and curved corolla-tube; stamens fused into a single column, with hairy apex; stigma 2–lobed. Capsule 2–celled. 8 spp.

784 **L. pyramidalis** Wallich
Uttar Pradesh to Arunachal Pradesh. Burma. S.E. Asia. 1200–2100 m.
Forests, shrubberies. Apr.–May.

An erect often tall perennial, widely branched above, with many white
or pale pink two-lipped flowers in rather dense spike-like clusters at the
ends of the branches. Flowers 1.5–2.5 cm long, with two narrow up-
curved lobes, with three broader down-curved lobes; stamens in a conspicuous curved column; calyx-lobes entire, linear. Leaves *c.* 15 cm, linear-
lanceolate, finely toothed, hairless above; stem 60–220 cm. Capsule
elliptic *c.* 1 cm, with persistent calyx.

785 **L. seguinii** Léveillé & Vaniot var. **doniana** (Skottsb.) F. Wimmer
Uttar Pradesh to S.W. China. Burma. 1800–3200 m. Shrubberies. Aug.–
Oct.

Like 784 with distinctive 3–lobed lower lip and a 2–lobed upper lip, but
with larger purple to mauve flowers which appear in the autumn, and
calyx-lobes which have scattered gland-tipped teeth. Flowers *c.* 2.5 cm
long, axillary in long lax spikes. Leaves elliptic, finely toothed, to 15 cm;
stem 1–2.5 m, branched above. **Page 482.**

ERICACEAE Heath Family

A large nearly cosmopolitan family, mainly of shrubs, or shrublets. Leaves
simple, usually alternate, commonly evergreen, without stipules. Inflorescence very variable, ranging from umbels, racemes, to solitary flowers.
Flowers regular; calyx of 4–5 sepals fused at their bases (sometimes reduced
to a rim); petals 4–5, fused to form usually a funnel-shaped, flask-shaped, or
bell-shaped corolla with short lobes; stamens usually 8 or 10 borne on the
receptacle; anthers opening by pores, and often with outgrowths or *awns*.
Ovary superior, or inferior as in *Vaccinium*, of 4–5 fused carpels, usually with
1–5 chambers each with many ovules. Fruit a capsule; or a fleshy berry as in
Vaccinium.

Fruit a dry capsule
  a Flowers axillary, usually solitary   1 CASSIOPE
  b Flowers in umbels or domed clusters   2 ENKIANTHUS
                                     3 RHODODENDRON
  c Flowers in elongate racemes or spike-like clusters   4 LYONIA
                                     5 PIERIS

Fruit fleshy   6 GAULTHERIA   7 AGAPETES   8 VACCINIUM

1 **CASSIOPE** Dwarf evergreen shrublets with tiny overlapping leaves.
Flowers solitary, nodding; fruit a capsule. 2 spp.

786 **C. fastigiata** (Wallich) D. Don
Pakistan to S.E. Tibet. 2800–4500 m. Shrubberies, open slopes. Jun.–Aug.

A small much-branched tufted shrublet to 30 cm, with numerous tiny 4–ranked overlapping closely adpressed thick leaves, and with small axillary pendent white bell-shaped flowers. Corolla 6–8 mm long, with short out-curved lobes; sepals brown, elliptic, to 4 mm. Leaves 3–5 mm, lanceolate, grooved, stalkless. Capsule globular erect, calyx persistent. **Pl. 73.**

2 ENKIANTHUS Flowers in pendulous umbels; corolla bell-shaped. Fruit a woody capsule. 1 sp.

787 **E. deflexus** (Griffith) Schneider (*E. himalaicus* Hook. f. & Thoms.)
E. Nepal to S.W. China. Burma. 2500–3300 m. Forests, shrubberies. May–Jun.

A shrub or small tree to 6 m, with pendulous umbels of dull orange-red, broadly bell-shaped flowers, and with terminal clusters of elliptic pointed fine-toothed leaves. Corolla 1–2 cm across, with 5 short blunt lobes; calyx-lobes lanceolate, 2–3 mm. Leaves *c.* 5cm, narrowed at both ends, stalked, finely hairy beneath at least when young. Capsule erect with curved stalk, broadly elliptic *c.* 6 mm, splitting into 5 valves. **Pl. 73.**

3 RHODODENDRON Flowers usually in umbel-like clusters. Corolla bell- or funnel-shaped with 5 spreading lobes; stamens usually 5–16. Ovary superior; fruit a 4–20-chambered capsule. The main concentration of the genus is in the E. Himalaya and S.W. China. In our area 30 spp.

COLOUR OF FLOWERS

Red, scarlet: 797 **arboreum**, 796 **barbatum**, 801 **cinnabarinum**, 791 **fulgens**, 788 **thomsonii**.

Magenta or mauve-purple: 797 **arboreum**, 790 **campanulatum**, 816 **cowania-num**, 795 **hodgsonii**, 807 **nivale**, 809 **virgatum**.

Pink: 797 **arboreum**, 790 **campanulatum**, 802 **ciliatum**, 813 **glaucophyllum**, 794 **grande**, 795 **hodgsonii**, 805 **lepidotum**, 807 **nivale**, 803 **pumilum**, 806 **setosum**, 808 **vaccinioides**.

Yellow, greenish-yellow: 804 **anthopogon**, 789 **campylocarpum**, 799 **dalhousiae**, 793 **falconeri**, 805 **lepidotum**, 814 **lowndesii**, 815 **trichocladum**, 810 **tri-florum**, 792 **wightii**.

White: 804 **anthopogon**, 797 **arboreum**, 812 **camelliiflorum**, 790 **campanu-latum**, 802 **ciliatum**, 799 **dalhousiae**, 793 **falconeri**, 813 **glaucophyllum**, 794 **grande**, 798 **griffithianum**, 800 **lindleyi**, 811 **pendulum**.

LEAF CHARACTERS

With brown or silvery indumentum on undersides of leaves: 797 **arboreum**, 790 **campanulatum**, 793 **falconeri**, 791 **fulgens**, 794 **grande**, 795 **hodg-sonii**, 811 **pendulum**, 792 **wightii**.

ERICACEAE

Leaves glaucous beneath: 789 **campylocarpum**, 801 **cinnabarinum**, 799 **dal-housiae**, 813 **glaucophyllum**, 798 **griffithianum**, 800 **lindleyi**, 788 **thomsonii**.

Leaves scaly beneath: 804 **anthopogon**, 812 **camelliiflorum**, 802 **ciliatum**, 801 **cinnabarinum**, 799 **dalhousiae**, 805 **lepidotum**, 800 **lindleyi**, 807 **nivale**, 803 **pumilum**, 806 **setosum**, 815 **trichocladum**, 810 **triflorum**, 808 **vaccinioides**, 809 **virgatum**.

DISTRIBUTION

Widespread Himalayan species spreading west of Nepal to the W. Himalaya: 804 **anthopogon**, 797 **arboreum**, 796 **barbatum**, 790 **campanulatum**, 805 **lepidotum**.

Species endemic to Nepal: 816 **cowanianum**, 814 **lowndesii**

E. Himalayan species: all other species named above have an East Himalayan distribution, and many of them occur within our area only in extreme East Nepal.

Evergreen trees and shrubs
a *Flowers many (more than 6) in each cluster*
  i *Corolla bell-shaped, about as long as broad*
    x *Leaves glaucous and almost hairless beneath*

788 **R. thomsonii** Hook. f.
E. Nepal to S.E. Tibet. 3000–3800 m. Forests, shrubberies. May–Jun.
Readily distinguished by its deep blood-red fleshy waxy flowers with large conspicuous calyx, and by its neat rounded-oval leaves which are dark green above and glaucous beneath. Flowers bell-shaped 5–6 cm long, darker spotted within above, corolla-lobes rounded, notched and overlapping; calyx 1–2 cm, cup-shaped, red-tinged. Leaves 5–10 cm, with rounded or shallow heart-shaped base. An erect shrub 3–5 m. Capsule broadly oblong 1.5–2 cm, half-encircled by persistent calyx. **Pl. 76**.

789 **R. campylocarpum** Hook. f.
E. Nepal to S.E. Tibet. Burma. 3300–4000 m. Forests, shrubberies. May–Jun.
Like 788 in leaf, but flowers pale or bright yellow, often with crimson blotching at base, widely bell-shaped, *c.* 4 cm long and broad; corolla-lobes rounded, notched. Calyx small *c.* 5 mm, glandular. Capsule cylindric 2–2.5 cm, curved, covered with glands.

xx *Leaves woolly-haired beneath*

790 **R. campanulatum** D. Don
Kashmir to S.E. Tibet. 3000–4400 m. Forests, alpine shrubberies; common and gregarious. Apr.–Jun.
Flowers pale mauve to rosy-purple, or rarely white, purple-spotted within above, broadly bell-shaped, in a lax cluster. Corolla *c.* 4 cm long

and broad, lobes rounded; calyx usually 1–2 mm. Leaves broadly elliptic to oval, 8–14 cm, with a rounded or shallowly heart-shaped base, dark glossy green above and with brown felted woolly hairs beneath. A widely branched, spreading shrub 2–6 m.

Leaves and wood used medicinally. **Pl. 74**.

791 **R. fulgens** Hook. f.
E. Nepal to S.E. Tibet. 3300–4200 m. Forests, alpine shrubberies. May–Jun.

Distinguished by its blood-red tubular-bell-shaped flowers with chocolate-brown anthers, borne in compact rounded clusters. Corolla *c.* 4 cm across, unspotted, with conspicuous rounded dark glands at the base within; corolla-lobes slightly notched; calyx very small, shallowly lobed. Leaves oblong-oval to broadly ovate, 6–10 cm, with rounded or shallow heart-shaped base, deep glossy green above, with thick reddish-brown felt beneath. A shrub or small tree with peeling bark, to 4 m. Capsule oblong-cylindrical, purplish, glaucous, 3 cm.

792 **R. wightii** Hook. f.
E. Nepal to S.E. Tibet. 3600–4300 m. Forests, shrubberies. May–Jun.

Flowers pale yellow, usually blotched or spotted with crimson at the base within, bell-shaped 4–5 cm across, in a rather loose umbel of 12–20 flowers with persistent, sticky bud-scales at its base. Calyx very small; ovary densely woolly; style much longer than stamens. Leaves oblong to elliptic or oblanceolate, 15–20 cm, blunt or acute, bright green above, with reddish-brown or sometimes greyish felty hairs beneath. A shrub or small tree to 4 m. Capsule cylindrical, slightly curved, 10–chambered. **Pl. 76**.

ii *Corolla obliquely bell-shaped and swollen on lower side; mostly trees*
x *Flowers white or cream; leaves very large, usually more than 15 cm*

793 **R. falconeri** Hook. f.
E. Nepal to Arunachal Pradesh. 2400–3300 m. Forests. Apr.–May.

Flowers obliquely bell-shaped, creamy-white to pale yellow, 20 or more in a large rounded cluster 15–23 cm across. Corolla 4–6 cm long, purple blotched at base within, 8–10 lobed; stamens 12–16; style stout, longer than stamens; stigma disk-like. Leaves very large, 20–30 cm, oblong-oval to obovate, wrinkled and with deep-set veins above, with rusty woolly hairs and raised veins beneath; young shoots stout, woolly. A tree to 16 m, of fine stature, with distinctive large leaves and large flowers clusters. Capsule large, woody, 4–6 cm. **Pl. 76**.

794 **R. grande** Wight
E. Nepal to S.E. Tibet. 1700–3000 m. Forests. Mar.–Apr.

Flowers pale pink in bud, later white with purple blotches at the base, 20–25 in large rounded clusters to 18 cm across. Corolla obliquely bell-shaped, 5–8 cm long and wide; stamens 16, unequal; ovary glandular and hairy; stigma large disk-like. Leaves very large, 15–38 cm, oblong to oblanceolate, shining deep green above, and with a thin covering of silvery-white hairs beneath. A handsome tree to 12 m, with distinctive large leaves and large flower clusters. Capsule *c.* 4 cm, woody, with persistent style.

xx *Flowers pink, purple, magenta or red; leaves usually less than 20 cm*

795 **R. hodgsonii** Hook. f.

E. Nepal to S.E. Tibet. 3000–3800 m. Forests. Apr.–May.

Flowers pale-pink to magenta-pink, somewhat obliquely bell-shaped, 15–20 in a compact cluster 15–20 cm across. Corolla rather fleshy, usually with a few darker blotches within, 4–5 cm across, usually 7–8–lobed; stamens usually 16, style longer; ovary and flower-stalk downy. Leaves large to 28 cm, oblong-elliptic to oblanceolate, very leathery, dark dull green above, grey- or brown-woolly beneath, leaf-stalk very stout. A rounded shrub or small tree to 7 m, with smooth pinkish bark. Capsule narrow-cylindrical, woolly, 4 cm. **Pl. 75.**

796 **R. barbatum** Wallich ex G. Don

Uttar Pradesh to Bhutan. 2400–3600 m. Forests; often gregarious. Apr.– Jun.

Flowers blood-red, with 5 black-crimson blotches at the base within, closely packed in a compact umbel 10–13 cm across. Corolla tubular-bell-shaped, 3 cm long, 5–lobed; stamens 10, filaments white, anthers purple-black. Leaves elliptic-lanceolate 10–23 cm; upper surface shining, mid-vein deeply grooved above and lateral veins deeply impressed; lower surface at first woolly, becoming hairless, leaf-stalk with distinctive long glandular bristles. A shrub or tree to 6 m, with smooth greyish peeling bark and bristly young shoots. Capsule oblong-cylindrical 2–2.5 cm, bristly, 6–8–chambered. **Pl. 74.**

797 **R. arboreum** Smith

Pakistan to S.E. Tibet. Sri Lanka. 1500–3600 m. Forests, shrubberies; common. Feb.–May.

The most widely distributed tree-rhododendron in the Himalaya, found over a wide range of altitudes, and in many different forms, some of which have been described as separate species. Flowers blood-red, pink to white, the latter usually found at higher altitudes, in large compact clusters of about 20 flowers 10–13 cm across. Corolla tubular-bell-shaped, 4–5 cm long and wide, 5–lobed; stamens 10, filaments white. Leaves 10–20 cm, oblong to lanceolate, with grooved mid-vein and lateral veins deeply impressed above, glossy-green, undersides with thin or thick felted hairs either white, fawn, cinnamon or rusty-brown. A robust tree to 15 m, with reddish-brown bark. Capsule *c.* 3 cm, ribbed, 10–chambered.

The national flower of Nepal. The young leaves are poisonous. Wood used for turning, but mainly for fuel and charcoal. Flowers presented as offerings in hill temples. **Pl. 74.**

b *Flowers few in each cluster, usually 5 or less*
  i *Stamens not longer than corolla*
    x *Shrubs more than 1 m; flowers white or pale yellow*

798 **R. griffithianum** Wight
E. Nepal to S.E. Tibet. 2100–3000 m. Forests, shrubberies; rare in Nepal but prominent when flowering in the valley below Topke Gola. May.
A beautiful rhododendron with very large white, very fragrant, widely bell-shaped flowers in a lax cluster of 3–5. Corolla 3.5–7 cm long and as much as 15 cm across; calyx large, hairless, saucer-shaped, greenish or pinkish, to 5 cm across; stamens 12–18. Leaves oblong, 15–30 cm, hairless, rather glaucous beneath. A shrub or small tree to 6 m. Capsule purplish-brown, 3.5 cm, 10–chambered.

799 **R. dalhousiae** Hook. f.
C. Nepal to Arunachal Pradesh. 2000–2500 m. An epiphyte, or growing on rocks. May.
Flowers large funnel-shaped, yellow then turning white, fragrant, in clusters of 3–5. Corolla 7–10 cm long and wide, 5–lobed; calyx deeply 5–lobed, hairless; stamens 10, filaments downy, anthers large, brown. Leaves obovate to oblanceolate 8–15 cm, veins deeply impressed above, glaucous and scaly beneath. An epiphytic shrub to 2.5 m, with papery bark and bristly-scaly young shoots. Capsule 4.5 cm, with persistent calyx. **Pl. 75**.

800 **R. lindleyi** T. Moore
E. Nepal to S.E. Tibet. Burma. 2100–3300 m. Epiphyte, or growing on rocks. May.
An epiphytic shrub, very like 799 with large white fragrant funnel-shaped flowers, but calyx fringed with soft white hairs; anthers short *c.* 8 mm; young shoots and leaf-stalks not bristly. Capsule 5 cm, scaly, with persistent calyx. **Pl. 75**.

    xx *Shrubs more than 1 m; flowers pink-flushed or red* (see also 802)

801 **R. cinnabarinum** Hook. f.
E. Nepal to S. E. Tibet. 3000–3600 m. Forests, shrubberies. May–Jun.
Easily distinguished by its long deep red tubular, drooping, waxy flowers, in clusters of about 5. Corolla 3–5 cm long, narrow below and widening gradually to the mouth, with 5 short rounded, erect not spreading lobes; calyx unequally lobed; stamens 10. Leaves obovate, elliptic to oblanceolate, often blunt, 5–10 cm, greyish-green above, scaly and rather glaucous beneath. A distinctive and beautiful shrub to 2.5 m. Capsule *c.* 1.3 cm, scaly. **Pl. 74**.

    xxx *Shrublets to 1 m, or less; flowers pink, white or yellow*

802 **R. ciliatum** Hook. f.
E. Nepal to S.E. Tibet. 2700–3900 m. Shrubberies, steep banks. May–Jun.

229

A small shrub, often procumbent on rocks, with short clusters of 2–4 widely funnel-shaped pink-flushed flowers, gradually turning white, 3.5–5 cm long, with bristly flower-stalks. Corolla-tube wide, with 5 notched overlapping lobes; calyx deeply 5–lobed; stamens 10. Leaves elliptic 5–9 cm, with conspicuous bristly margins and scaly beneath. A widely branched shrub to 2 m, with young shoots, leaves, leaf-stalks and calyx all more or less bristly-haired. Capsule 1.6 cm, with persistent calyx. **Pl. 75.**

803 **R. pumilum** Hook. f.
E. Nepal to S.W. China. Burma. 3600–4300 m. Rocks, banks; rare in Nepal; prominent in flower in Barun valley. Jun.–Jul.

A slender spreading shrublet with stems 8–10 cm long, and wide tubular nodding pink flowers, borne on long scaly stalks in clusters of 2–3, or solitary. Corolla 1.3–2 cm long, 5–lobed, downy and sometimes slightly scaly outside; stamens 10. Leaves tiny 8–20 mm, elliptic with a rounded or blunt apex with a short acute tip, hairless above, with loose scales below. Capsule *c*. 1 cm, scaly. **Page 483.**

804 **R. anthopogon** D. Don
Pakistan to S.E. Tibet. 3000–4800 m. Alpine shrubberies, open slopes; common and gregarious. May–Jul.

A small strongly aromatic shrublet to 60 cm, with compact clusters of 4–6 white or yellow flowers tinged with pink. Flowers *c*. 2 cm across; corolla with a narrow tube and 5 rounded spreading lobes; stamens 5–8, included in corolla-tube. Leaves oval to obovate 2.5–4 cm, densely scaly beneath; young shoots brown-scaly. Capsule *c*. 3 mm, encircled by the persistent calyx. Var. **hypenanthum** (Balf. f.) Hara has yellow flowers, and winter bud-scales persisting for several years; it occurs in the W. Himalaya.

Used mixed with Juniper for incense in Buddhist monasteries. **Pl. 76.**

ii *Stamens exserted, longer than corolla*
x *Flowers pink to magenta, or red* (see also 813)

805 **R. lepidotum** Wallich ex G. Don
Pakistan to S.W. China. 2400–4500 m. Forests, shrubberies, alpine slopes; common. Jun.–Jul.

A low shrublet with clusters of usually 2–4, slender-stalked pink, dull purple, or pale yellow flowers each 2–2.5 cm across. Corolla shortly and broadly tubular, with 5 spreading rounded lobes, scaly and glandular outside; stamens exserted, 8, filaments hairy below. Leaves small, narrow-oblanceolate 1.2–4 cm, scaly above and beneath. A resinous shrub to 1 m, very variable in size, habit, and flower-colour. Capsule *c*. 2 cm, densely scaly. **Pl. 73.**

806 **R. setosum** D. Don
C. Nepal to S.E. Tibet. 3600–4800 m. Alpine slopes; gregarious. May–Jul.

A small shrub, with pink-purple, widely funnel-shaped and deeply

5–lobed flowers, with longer spreading stamens. Corolla *c*. 2.5 cm across; flowers usually in clusters of 3, or sometimes more; calyx *c*. 6 mm, deeply lobed, scaly; stamens 10. Leaves very small, oblong-elliptic blunt, 8–12 mm, often with bristly inrolled margins, scaly on both sides. A low dense shrub to 50 cm, with densely bristly-hairy branchlets. Capsule *c*. 6 mm, scaly, encircled by the persistent calyx. **Pl. 73**.

807 **R. nivale** Hook.f.
W. Nepal to S.E. Tibet. 4500–5500 m. High-altitude alpine slopes, in drier areas only; gregarious. Jun.–Jul.

A small spreading shrub found growing further into the dry zone than any other *Rhododendron* species. Flowers lavender, magenta, or pink, 2 cm across, solitary or paired; corolla with a very short tube and 5 elliptic spreading lobes; stamens 10, exserted; calyx *c*. 3 mm, narrow-lobed, scaly. Leaves very small to 13 mm, elliptic to rounded, scaly on both sides, leaf-stalk very short. Stems to 50 cm; branchlets densely scaly. Capsule ovoid *c*. 4 mm, scaly. **Pl. 73**.

808 **R. vaccinioides** Hook.f.
E. Nepal to S.W. China. Burma. 2400–3000 m. Forest epiphyte, and on rocks. Jun.

A small compact epiphytic shrub, with 1–2 small pink, or white and pink-tinged bell-shaped flowers with 5 spreading lobes and with exserted stamens. Corolla 6–12 mm long; calyx sparsely scaly. Leaves 1.2–3 cm, spathulate to oblanceolate, with a notched apex and a fine point, veinless and with a few scattered scales below, tapering to a stout winged leaf-stalk. Stems to 50 cm; branchlets with many leaves, warted. Capsule cylindrical *c*. 2 cm. **Page 483**.

809 **R. virgatum** Hook.f.
E. Nepal to S.E. Tibet. 2100–2700 m. Forests, shrubberies; rare in Nepal. Apr.–May.

The only species in our area which has axillary flowers; they are borne singly or in pairs, from the uppermost leaves of the previous year's growth. Flowers pale purple, 2.5–4 cm across, funnel-shaped, with 5 rounded spreading lobes; stamens 10, exserted; calyx deeply 5–lobed, scaly outside. Leaves broadly lanceolate 2.5–8 cm, with dense shining brown scales below. An erect shrub to 1.5 m or more, with young branches covered with brown scales. Capsule *c*. 12 mm, densely scaly.

xx *Flowers yellow* (see also 805)

810 **R. triflorum** Hook.f.
E. Nepal to S.E. Tibet. 2400–3300 m. Shrubberies. Apr.–May.

Flowers pale yellow spotted with green, fragrant, in clusters of 3, set amongst the new year's leaves. Corolla 4–5 cm across, with 5 spreading elliptic lobes, scaly; stamens 10, exserted; calyx shortly 5–lobed, fringed with stout hairs. Leaves lanceolate and very acute, 2–8 cm, shiny hairless above, densely glandular and scaly beneath. A shrub to 3 m, with older bark peeling off in cinnamon-coloured flakes. Capsule *c*. 12 mm, covered with scales. **Pl. 75.**

xxx *Flowers white, often tinged with pink* (see also 808)

811 **R. pendulum** Hook.f.
E. Nepal to S.E. Tibet. 2700–3600 m. Epiphytic, and on rocks. May.
A small shrub to 1 m, with clusters of 2–3 broadly tubular white flowers with yellow throats and with spreading corolla-lobes. Corolla *c*. 4 cm across, more or less scaly outside; stamens slightly exserted; calyx deeply lobed, fringed with hairs and scaly. Flower-stalks and leaf-stalks woolly-haired. Leaves small 3–5 cm, elliptic to oblong, leathery, hairless above with impressed veins, densely brown-woolly beneath and with glandular spots. Branchlets trailing, densely covered with brown or fawn woolly hairs. Capsule *c*. 12 mm, hairy and scaly. **Page 483.**

812 **R. camelliiflorum** Hook.f.
E. Nepal to S.E. Tibet. 2700–3600 m. Epiphytic, or a weak shrub in forests. Jun.–Jul.
Flowers white tinged with pink, usually paired, with a short broad tubular fleshy corolla with 5 rounded spreading overlapping lobes. Corolla *c*. 4 cm across; stamens 12–16, exserted; calyx deeply 5–lobed, with few scales at base. Leaves narrowly oblanceolate pointed, 6–10 cm, dark green above, with glistening brown scales beneath. A straggling shrub to 1.5 m. Capsule *c*. 12 mm, densely scaly. **Page 483.**

813 **R. glaucophyllum** Rehder (*R. glaucum* Hook.f.)
E. Nepal to S.E. Tibet. 2700–3600 m Rocks and open slopes; rare in Nepal. May.
Flowers white and marked with pink, or rosy-red, usually in clusters of 5, but sometimes more. Flowers 2–3 cm across; corolla-tube short, with 5 spreading rounded, overlapping lobes; stamens exserted; calyx-lobes elliptic, scaly. Leaves lanceolate pointed, 3–8 cm, dull green above and glaucous and scaly beneath, margins recurved. A shrub mostly 30–60 cm,but sometimes up to 2 m; young branches, underside of leaves, flower-stalks and calyx covered with reddish-brown scales; foliage very aromatic. Capsule *c*. 8 mm, encircled by persistent calyx. **Page 483.**

Deciduous shrubs
a *Flowers yellow*

814 **R. lowndesii** Davidian
Endemic to W. & C. Nepal. 3300–4300 m. Banks, rock ledges. Jun.–Jul.
A slender deciduous alpine shrublet *c*. 10 cm, similar to 805 *R. lepidotum*, with solitary or paired long-stalked pale yellow flowers spotted with carmine within, but readily distinguished by the bristly-haired leaf-margins. Flowers 1.5–2 cm across; corolla-tube short, lobes ovate, spreading;

calyx deeply lobed, scaly. Leaves elliptic *c*. 1.5 cm; leaf-stalks and calyx with bristly spreading hairs. **Pl. 73**.

815 **R. trichocladum** Franchet
E. Nepal to S.W. China. Burma. 3000 m. Shrubberies; very rare in Nepal. Jun.

Flowers greenish-yellow spotted with dark green, 3–5 in a close cluster, with widely funnel-shaped corolla *c*. 2.5 cm long with spreading overlapping lobes, woolly in throat; stamens downy, shortly exserted. Leaves oblong 2.5–4 cm, hairy above, paler and minutely scaly beneath, bristly on margins and on leaf-stalks. A stiff deciduous shrub to 1 m, with bristly branchlets. Capsule *c*. 6 mm, scaly.

b *Flowers magenta*

816 **R. cowanianum** Davidian
Endemic to C. Nepal, where it is not uncommon. 3000–4000 m. Light forests, shrubberies. Jun.

Flowers magenta, appearing before or with the new leaves, with a short corolla-tube and 5 rounded spreading lobes and with stamens exserted. Flowers *c*. 3 cm across, 2–5 in each cluster; flower-stalks and calyx scaly; filaments of stamens woolly-haired. Leaves ovate-elliptic 5–6 cm, margins of young leaves bristly-haired and with scattered scales beneath. A spreading shrub to 2 m. Capsule *c*. 1 cm, scaly. **Pl. 75**.

4 **LYONIA** Flowers in axillary spike-like clusters. Capsule globular with prominent thickened ribs. Leaves opposite. 2 spp.

817 **L. ovalifolia** (Wallich) Drude (*Pieris o.* (Wallich) D. Don)
Pakistan to S.W. China. Burma. S.E. Asia. 1500–3000 m. Forests, shrubberies; very common in oak forests and in association with *Rhododendron arboreum*. Apr.–May.

A small deciduous tree to 10 m, with brown bark peeling in narrow strips, rather leathery ovate leaves, and with white flowers in long nearly horizontal axillary clusters, 5–15 cm. Flowers numerous, 6–12 mm long, flask-shaped, constricted at the mouth and with 5 short recurved lobes, finely hairy; calyx-lobes triangular-lanceolate 2–3 mm long. Leaves 8–15 cm, acute or long-pointed, hairless above, often shaggy-haired beneath when young, short-stalked. Capsule globular *c*. 4 mm, 5–valved.
Leaves and buds poisonous to animals; used to kill insects. **Pl. 76**.

818 **L. villosa** (Hook.f.) Hand.–Mazz.
Uttar Pradesh to S.W. China. Burma. 2100–3800 m. Forests, shrubberies. Jun.–Jul.

A shrub or small tree similar to 817, but flower-clusters much shorter 2.5–5 cm, fewer-flowered, and filaments of stamens without apical horns (filaments with 2 horns in 817). Corolla white, broadly bell-shaped *c*. 6 mm; calyx-lobes narrow lanceolate. Leaves smaller, mostly 4–5 cm, more shaggy-haired beneath, and with blunt apex and a short narrow tip. **Page 483**.

5 **PIERIS** Like *Lyonia*, but differing in having terminal flower-clusters, and capsules without prominent thickened ribs. 1 sp.

819 **P. formosa** (Wallich) D. Don
C. Nepal to S.W. China. 2100–3300 m. Forests, shrubberies. Mar.–May.
   A small evergreen shrub, with dark green leathery lanceolate leaves, and with terminal erect branched dense spike-like clusters of small white, globular, pendent flowers. Corolla 5–8 mm, constricted at the mouth and shallowly lobed; calyx-lobes triangular, glandular-hairy. Leaves brilliant red when young in spring; mature leaves 6–15 cm long, stalked, finely toothed, smooth, becoming quite hairless; young twigs downy; shrub to 3 m. Capsule globular, 5–valved. **Pl. 76**.

6 **GAULTHERIA** Like *Vaccinium*, but ovary superior, and fruit readily distinguished in having the appearance of a berry, comprising a dry 5–valved capsule enclosed in a large succulent calyx. 7 spp.

Procumbent or low shrublets
a *Leaves less than 2 cm long*

820 **G. nummularioides** D. Don
Himachal Pradesh to S.W. China. Burma. 2100–4000 m. Rocks, banks. Jul.–Sep.
   A prostrate shrublet with opposite, tiny, ovate leaves, with brown bristly stems, and with black berry-like fruits. Flowers *c.* 4 mm, reddish to nearly white, globular, axillary, solitary and pendent; calyx-lobes broadly ovate. Leaves 1–1.8 cm, with rounded bases, hairless above, with bristly brown hairs beneath and often on margins. Stems densely leafy, much-branched and spreading over banks. Fruit *c.* 6 mm, succulent.

821 **G. trichophylla** Royle
Pakistan to S.W. China. 2700–4500 m. Rocks, banks; very common in Nepal. May–Jul.
   A spreading shrublet like 820, but leaves smaller, elliptic with a wedge-shaped base, and succulent fruit sky-blue. Flowers widely bell-shaped, 5–6 mm long and broad, red, pink or nearly white; calyx-lobes triangular, 2 mm. Leaves *c.* 6 mm, hairless above and below, with conspicuous bristles on margin. Fruit edible. **Pl. 77**.

b *Leaves more than 2 cm long*

822 **G. pyroloides** Hook. f. & Thoms. ex Miq. (*G. pyrolaefolia* Hook. f. ex C. B. Clarke)
E. Nepal to S.E. Tibet. Burma. 3300–4000 m. Rocks, peaty banks. Jun.
   A low shrublet covering the ground with spreading rooting stems 10–20 cm, with leathery elliptic-obovate leaves, and with small clusters of white

or pinkish flask-shaped flowers. Corolla *c.* 6 mm, constricted at the mouth and with minute lobes; calyx-lobes triangular, becoming enlarged and fleshy in fruit. Leaves 2–4 cm, clustered towards ends of shoots, with bristly-toothed margin and rounded or blunt apex. Fruit blue-black, with ovate pointed fleshy lobes. **Pl. 77.**

Erect shrubs to 1 m or more

823 **G. fragrantissima** Wallich
Uttar Pradesh to S.E. Tibet. Burma. 1500–2700 m. Forests, shrubberies; very common in Nepal. Apr.–May.

A robust shrub with ovate to lanceolate evergreen leaves, and with numerous axillary spike-like clusters of small fragrant white or pink globular flowers. Flower clusters 2.5–8 cm long, shorter than the leaves; flowers short-stalked; corolla 4 mm by 3 mm broad. Leaves acute, 5–10 cm long, smooth above, dotted with the bases of bristles beneath, margins bristly-toothed; a branched shrub to 1–1.75 m. Fruit *c.* 6 mm, with dark violet-blue fleshy pointed calyx-lobes.

Oil extracted from leaves used medicinally. **Pl. 77.**

824 **G. griffithiana** Wight
E. Nepal to S.E. Tibet. Burma. 2100–3000 m. Forests, shrubberies. May–Jun.

A stout shrub to 4 m, like 823 but differing in its larger elliptic leaves 10–15 cm with long-pointed tips and closely and finely toothed margins, and in its black fleshy fruit with blunt calyx-lobes. Flower-clusters 5–8 cm; flowers bell-shaped, *c.* 6 mm across, pale green, white or pink.

825 **G. hookeri** C. B. Clarke
E. Nepal to S.W. China. 2700–3000 m. Forests. May–Jun.

A shrub like 823, but distinguished by its conspicuously bristly-haired branches, broader elliptic leaves, and its axillary clusters of pink or white globular flowers. Inflorescence 2.5–5 cm long, hairy; corolla *c.* 4 mm; bracts 2–6 mm, elliptic. Leaves *c.* 5 cm by 2–4 cm broad; a branched shrub to 1 m. Fruit fleshy, blue.

7 **AGAPETES** Shrubs with leathery leaves, usually epiphytic. Flowers tubular. Ovary inferior; fruit a berry. 5 spp.

826 **A. serpens** (Wight) Sleumer (*Pentapterygium s.* (Wight) Klotzsch)
E. Nepal to Arunachal Pradesh. 1500–2700 m. Forest epiphyte, or sometimes on damp banks. Feb.–Jun.

A shrub with a large swollen woody stem-base, and with pendulous bristly-haired branches with two regular ranks of numerous small evergreen leaves. Flowers conspicuous 2–2.5 cm long, axillary, pendulous, scarlet or sometimes yellow, tubular 5–angled, hairy, with transverse v-shaped veins and with 5 short triangular lobes; calyx-lobes lanceolate blunt, glandular-hairy, enlarging in fruit. Leaves *c.* 2 cm, elliptic, with in-rolled margins and with a fine point, leathery, hairless. Berry top-shaped, conspicuously 5–winged. **Pl. 77**.

827 **A. incurvata** (Griffith) Sleumer (*A. hookeri* (C. B. Clarke) Sleumer)
E. Nepal to Arunachal Pradesh. 2100–2900 m. Forest epiphyte. Jun.–Jul.

Differs from 826 in its larger broadly lanceolate toothed leaves 5–10 cm, and its yellow angled tubular flowers *c*. 2 cm, with conspicuous ovate calyx-lobes. Not so common as the previous species. **Pl. 77**.

8 VACCINIUM Evergreen shrubs with small alternate leaves. Distinguished from *Gaultheria* by its inferior fruit which is a fleshy berry crowned by 5 persistent calyx-lobes. 6 spp.

Leaves small, 3 cm or less, entire

828 **V. nummularia** Hook. f. & Thoms. ex C. B. Clarke
C. Nepal to S.E. Tibet. Burma. 2400–4000 m. Forest epiphyte, and on rocks. Apr.–Jun.

A small rigid usually epiphytic shrub with pendulous conspicuously densely hairy branches, with small broad-elliptic leathery leaves, and with terminal clusters of pale pink flowers. Flower-clusters 2.5–5 cm long; corolla flask-shaped, shortly lobed, 5–6 mm; calyx-lobes short, broadly triangular. Leaves 1.3–2 cm, blunt, margins bristly-haired and incurved; stems branched, to 1 m. Berries red then black, globular 6 mm, crowned with the lobed calyx. **Pl. 77**.

829 **V. retusum** (Griffith) Hook. f. ex C. B. Clarke
C. Nepal to S.E. Tibet. Burma. 1400–3600 m. Forest epiphyte. Apr.–May.

Like 828, but young branches finely hairy and older branches hairless, and leaves larger, to 3 cm, oblong-obovate, narrowed to a short stalk. Flower-clusters borne at ends of branches, with conspicuous pale bracts as long as the flowers; corolla pink, with 5 reddish stripes, *c*. 4 mm. Berries deep purple.

Leaves large, more than 5 cm, toothed

830 **V. vacciniaceum** (Roxb.) Sleumer
E. Nepal to Arunachal Pradesh. Burma. 1500–2800 m. Epiphyte, and growing on rocks; common in Arun and Tamur valleys. Apr.–Jun.

Flowers greenish, pale yellow, white, or pinkish, borne in one or several stalked one-sided clusters at the ends of the leafy branches. A straggling, usually branched epiphytic shrub, with leathery elliptic, toothed leaves often in a cluster at the ends of the branches. Flowers numerous, long-stalked, with cylindrical corollas 6–8 mm long with short recurved lobes. Leaves mostly 6–8 cm, stalkless; stems to 2 m. **Pl. 77**.

PYROLACEAE   Wintergreen Family

A small family of evergreen perennials with creeping rhizomes, largely of the cool Temperate and Arctic Zones. Leaves entire or toothed, usually alternate. Flowers in terminal clusters, or solitary. Sepals and petals 4 or 5, usually free; stamens 8 or 10; anthers opening by pores. Ovary superior, of 4–5 fused carpels and similar number of chambers each with many ovules; style unbranched. Fruit a splitting capsule.

PYROLA   3 spp.

831   **P. karakoramica** Krīśa (*P. rotundifolia* L. subsp. *indica* Andersson)
Pakistan to Kashmir. 2600–3600 m. Alpine slopes, shrubberies; very local. Jul.–Aug.

A small perennial with stalked leaves with rounded blades in a basal rosette, and with a slender leafless stem bearing a spike of drooping white or pink flowers. Petals spreading, rounded or obovate, *c.* 6 mm; sepals elliptic; style long down-curved; bracts narrow-elliptic, often longer than flower-stalks. Leaves with entire leathery blades 2.5–5 cm, leaf-stalks about as long; flowering stem 15–20 cm. Capsule globular, 5–valved, capped with the persistent style. **Page 484**.

MONOTROPACEAE   Easily distinguished by its lack of green chlorophyll, having only brown scale-like leaves; often placed in the Ericaceae. Distribution is in the North Temperate Zone and in tropical mountains.

MONOTROPA   BIRD'S NEST   2 spp.

832   **M. hypopitys** L. YELLOW BIRD'S NEST
Afghanistan to S.E. Tibet. N. Temperate Zone. 2400–3600 m. Saprophyte in coniferous forests. Jul.–Aug.

A yellowish or whitish plant with erect fleshy stems with many scale-like leaves, and with a dense terminal one-sided drooping cluster of yellowish-white flowers. Inflorescence densely hairy; petals 4–5, narrow-obovate *c.* 1.3 cm; sepals 4–5, papery, soon falling; bracts ovate. Leaves papery, yellow to brownish, ovate-oblong 1–2 cm; stem 5–20 cm. Capsule 4–5–celled, erect in fruit. **Page 484**.

DIAPENSIACEAE   Closely related to the Ericaceae, but stamens 5, alternating with petals; anthers opening by slits (not pores). Ovary superior, with 3 chambers; style 1. Fruit a capsule. Distribution: Arctic, N. America and E. Asia.

DIAPENSIA   1 sp.

833   **D. himalaica** Hook.f. & Thoms.

E. Nepal to S.W. China. 3600–4400 m. Steep banks; very local in Nepal, quite prominent on route to Barun Valley. May–Jun.

A small densely tufted hairless perennial with tiny crowded hairless leathery leaves, and with solitary white or pink funnel-shaped flowers with large spreading rounded lobes, borne almost stalkless at the ends of the leafy stems. Corolla-tube c. 5 mm long, lobes obovate, spreading to 1 cm; sepals ovate, half as long as corolla-tube, but enlarging in fruit. Leaves c. 4 mm, oblong-obovate; stems 2–3 cm or more. Capsule ovoid, stalked, 3–4 mm. **Page 484**.

PLUMBAGINACEAE   Sea Lavender Family

A medium-sized family of herbaceous plants, shrubs and climbers, particularly characteristic of dry and saline areas of the world. Leaves simple, alternate, or in rosettes. Flowers in branched clusters, spikes, or heads; bracts papery. Sepals 5, fused into a tube and often papery, ribbed, and coloured; petals 5, fused into a short or long tube; stamens 5. Ovary superior, of 5 fused carpels, with a single chamber with 1 ovule; styles 5. Fruit dry papery, 1–seeded, enclosed by calyx, usually not splitting.

ACANTHOLIMON   Spiny cushion-forming shrublets. Flowers in a spike-like cluster; calyx with a papery limb. 1 sp.

834   **A. lycopodioides** (Girard) Boiss.
Afghanistan to Kashmir. C. Asia. 3000–4500 m. Dry slopes; common in Ladakh. Jul.–Aug.

A low tufted, mat-forming, extremely spiny-leaved shrublet, with short-stemmed dense clusters of pale pink flowers, with conspicuous papery calyx-limb. Petals 5, c. 6 mm long, notched, spreading, fused below; calyx cylindrical below, 10–ribbed, with a large white-papery funnel-shaped limb; bracts with papery margins. Leaves linear-lanceolate 1–1.5 cm, with needle-sharp tips; flowering stems leafless, 2.5–5 cm. **Pl. 78**.

CERATOSTIGMA   Herbaceous perennials or shrubs. Flowers in compact terminal and axillary clusters. 1 sp.

835   **C. ulicinum** Prain
W. Nepal (N. of Dhaulagiri only). Tibet. S.W. China. 3500–4000 m. Alpine shrubberies, open slopes; gregarious but local. Jul.–Sep.

A semi-prostrate much-branched shrub forming mats 1 m or more across and c. 15 cm high, with dense terminal clusters of deep blue flowers with red calyces, and with spathulate leaves with conspicuous brown bristly-hairy margins. Flowers c. 1 cm across, with a narrow corolla-tube and spreading rounded lobes with pointed tips; calyx tubular, papery, with bristly tips; bracts bristly. Leaves 2–3 cm, with spine-tipped apex; stems with bristly hairs; lateral buds bristly. **Page 484**.

PRIMULACEAE   Primula Family

A medium-sized cosmopolitan family of herbaceous plants, with its main concentration in the North Temperate Zone. Leaves commonly in a rosette, entire, often bearing simple or compound glandular hairs. Flowers often borne on leafless stems, or *scapes*, solitary, in umbels, or in branched clusters. Sepals 5, fused into a calyx-tube with 5 pointed lobes or teeth; petals 5, fused into a short or long corolla-tube with usually 5 spreading, or sometimes reflexed lobes; stamens 5, fused to the corolla-tube. Ovary of 5 carpels, fused into a single chamber, with many ovules arising from a central column; style single. Fruit a 5–valved capsule.

(The lobes of the corolla are described as petals in the following descriptions.)

ANDROSACE   Distinguished from *Primula* by its smaller flowers, usually less than 1.3 cm across, and its very short corolla-tube shorter than the calyx which is narrowed and thickened at the mouth; style short. Many species are mat- or cushion-forming. *c.* 22 spp.

Leaves mostly more than 1 cm long; plant not mat- or cushion-forming
a *Leaf-blade rounded, toothed or lobed*

836   **A. rotundifolia** Hardw.
Afghanistan to Bhutan. 1500–3600 m. Open slopes; very common in W. Himalaya. Jun.–Jul.

Distinguished by its rounded, deeply-lobed, long-stalked basal leaves, and its lax umbel of long-stalked pink flowers, with leafy calyx usually longer than the corolla. Flowers to 10 mm across; petals notched; calyx-lobes elliptic, toothed, the bracts similar. Leaves with blade usually 1.5–2 cm across, with rounded, toothed lobes; flowering stems 6–15 cm. **Pl. 78.**

837   **A. geraniifolia** Watt
Uttar Pradesh to S.W. China. 2100–3300 m. Forests, shrubberies. Apr.–Jun.

Like 836, but plants with stolons, and calyx-lobes not leafy, and shorter than the corolla, and bracts small, entire. Flowers pink or white, with a yellow eye, to 10 mm across, with entire or notched petals; individual flower-stalks *c.* 1.5 cm (usually longer in 836). Leaves like 836, but more deeply lobed to one third the width of the blade, or more.

b *Leaf-blade much longer than broad* (see also 849)
  i *Plants with stolons* (see also 841)

838   **A. sarmentosa** Wallich
Kashmir to Sikkim. 2700–3900 m. Shrubberies, open slopes, light forests. May–Jul.

Usually a densely white-woolly plant with leaves in rosettes, with stolons with reddish hairs, and with an umbel of small bright pinkish flowers borne on long flowering stems. Flowers *c.* 8 mm across; calyx-lobes elliptic 2–3 mm, densely silky-haired; bracts unequal, the outer linear, hairy. Leaves elliptic-lanceolate to oblanceolate variable in size, outer rosette-leaves spreading, the inner much larger, erect, to 3 cm; flowering stems usually 10–15 cm; stolons 5–10 cm. **Pl. 79**.

839 **A. primuloides** Duby
Endemic to Kashmir. 3000–4000 m. Alpine slopes; very common. Jun.–Aug.

Very similar to 838, but distinguished by the outer bracts of the umbels which are lanceolate (not linear), and plant usually with more numerous and longer stolons to 12 cm, with reddish-brown hairs. Flowers with entire petals, *c*. 8 mm across, many in a dense hairy umbel 2–3 cm across. Rosette-leaves spathulate, numerous, overlapping in a dense white-woolly rosette; inner leaves erect, 2–4–times as long as the outer, narrow elliptic, stalked. **Pl. 79**.

ii *Plants usually without stolons*

840 **A. strigillosa** Franchet
W. Nepal to S.E. Tibet. 2400–4300 m. Shrubberies, open slopes, light forests. May–Jul.

An erect rather tufted plant with branched rootstock, with often relatively large elliptic stalked leaves, and with a lax umbel of numerous pink flowers borne on slender flower-stalks which elongate in fruit. Flowers 6–15 mm across, white or pink; petals rounded, usually paler on the upper surface and deeper red on the underside; calyx *c*. 3 mm, with ovate blunt 3–ribbed lobes; bracts elliptic; flower-stalks to 5 cm. Leaves variable, the inner much longer, to 6 cm or more, the outer smaller in a rosette, the outermost bract-like; flowering stem hairy, 12–30 cm. **Pl. 78**.

841 **A. lanuginosa** Wallich
Himachal Pradesh to W. Nepal. 1800–3000 m. Grassy slopes. May–Aug.

Plants with densely silky-haired grey rosettes and with distinctive spreading leafy stems, and with compact umbels of pink or white flowers borne on woolly-haired flowering stems. Umbels *c*. 2 cm across, with white-woolly calyx, bracts and short flower-stalks; flowers 5–10 mm across; petals usually not notched. Leaves narrow elliptic with a fine point, mostly 1–2 cm; flowering stems usually 4–10 cm; non-flowering stems leafy, often creeping; short stolons sometimes present.

Leaves 1 cm long or less; plants usually mat- or cushion-forming
a *Flowers solitary*
  i *Rosettes lax, not silvery-haired; leaves linear pointed*

842 **A. lehmannii** Wallich ex Duby (*A. nepalensis* Derg.)
W. Nepal to Sikkim. 3500–5500 m. Open slopes; a common cushion-forming species of the humid alpine zone of Nepal. Jun.–Aug.

A rather lax or dense cushion- or mat-forming plant to *c*. 30 cm across, with solitary pink or less commonly white flowers, and with distinctive linear leaves with bristly hairs and a bristle-tipped apex, in relatively lax rosettes. Flowers 6–8 mm across, with rounded petals, stalkless or short-stalked; calyx *c.* 2 mm. Summer leaves to 10 mm, winter leaves elliptic *c.* 2 mm, with silky hairs; the densely tufted stems are covered below with old leaves. **Pl. 78**.

ii *Rosettes dense, globular, silvery-haired; all leaves blunt*
    x *Rosettes 5 mm or more across*

843 **A. delavayi** Franchet
Uttar Pradesh to S.W. China. 4300–5200 m. Stony slopes, dry areas; quite common. Jun.–Aug.

A dense cushion-forming plant with tiny silvery globular rosettes *c.* 5 mm across, and solitary stalkless white flowers 7–8 mm across, with a greenish to yellow eye. Leaves obovate 4–5 mm, incurved, with long silvery-white marginal bristles; cushions mostly 4–8 cm across. **Pl. 78**.

844 **A. globifera** Duby
Pakistan to Bhutan. 3600–4500 m. Stony slopes in dry areas; common. Jun.–Jul.

Like 843 with dense cushions of tiny silvery-leaved rosettes, but flowers usually pink or mauve and with an orange eye, and leaves elliptic, rather blunt, silvery-haired. Flowers 5–10 mm across, stalkless or short-stalked; petals rounded or notched, overlapping, rarely white. Rosettes globular; leaves to 5 mm; old stems with persistent old globular rosettes below. **Pl. 78**.

xx *Rosettes mostly 2–4 mm across*

845 **A. tapete** Maxim. (*A. sessiliflora* Turrill)
W. & C. Nepal. Tibet. 3800–5500 m. Stony slopes, dry areas; Tibetan borderlands only. May–Jun.

One of the smallest-flowered species in this genus, with tiny silvery-haired rosettes mostly 2–4 mm across, very densely clumped together in a tight domed cushion usually 6–10 cm across, or sometimes to 20 cm across. Flowers solitary, white, 3–4 mm across. Leaves *c.* 4 mm; rosettes appearing stuffed with woolly hairs and surrounded with oval leaves which become hairless outside; stems densely covered below with old hardened rosettes of the previous years. **Pl. 79**.

b *Flowers in stalked umbels.*
    x *Leaves covered in silvery hairs*

846 **A. muscoidea** Duby
Pakistan to E. Nepal. Tibet. 3600–5200 m. Stony slopes, dry areas; common. Jun.–Aug.

A lax cushion-forming plant, with globular silvery-haired rosettes 5–10 mm across, often borne one above the other on reddish stems, and with small compact umbels of few mauve to lilac flowers with orange or yellow eyes. Flowers 2–8, borne on stems 1–6 cm; corolla *c.* 10 mm across; petals usually rounded. Leaves oblanceolate to elliptic to 8 mm, silvery-haired above and below.

847 **A. zambalensis** (Petitm.) Hand.-Mazz.
C. Nepal to S.W. China. Tibet. 3700–5500 m. Stony slopes, dry areas; rare in Nepal. Jun.–Jul.

Like 846, but cushions more compact, and rosettes slightly larger to 15 mm across. Flowers several, at first white then turning pink, almost stalkless, in compact umbels borne on stems *c.* 1 cm. Corolla *c.* 8 mm across, petals entire; calyx and bracts with dense woolly hairs. Leaves elliptic *c.* 1 cm, bristly-haired all over; stems with compact globular hardened old rosettes below. **Pl. 79**.

    xx *Leaves with bristly marginal teeth only*

848 **A. mucronifolia** Watt
Afghanistan to Kashmir. 3300–4300 m. Rocks, screes, alpine slopes; common in Kashmir. Jun.–Aug.

A lax mat-forming plant, with usually pink, or less commonly white, scented flowers, and distinctive oval leaves with marginal bristles and apex but otherwise hairless. Flowers several, in a short-stemmed (*c.* 1 cm) compact umbel, rarely solitary; corolla 6–8 mm across, petals rounded or notched; calyx and bracts bristly-hairy. Leaves 2–4 mm; rosettes 5–8 mm across. **Pl. 79**.

849 **A. sempervivoides** Jacquem. ex Duby
Pakistan to Himachal Pradesh. 3000–4000 m. Alpine slopes; common in Kashmir. Jun.–Aug.

Distinguished by its relatively large compact rosettes usually 1.5–2.5 cm across, borne at the ends of spreading branches, with spathulate to obovate leaves with bristly marginal hairs. Flowers pink with a yellow or orange eye, 8–10 mm across, in compact umbels borne on flowering stems *c.* 5 cm; calyx *c.* 4 mm; bracts narrow elliptic, covered with glandular hairs like the calyx. Leaves mostly *c.* 1 cm or less, but some to 1.5 cm; short stolons bearing terminal rosettes present. **Pl. 79**.

PRIMULA A very distinctive genus, with its greatest concentration of species in the Himalaya. Usually with an umbel of colourful flowers borne on a leafless flowering stem, or *scape*, and with a basal cluster or rosette of usually entire leaves. Corolla-tube long, widening at the throat, with often spreading lobes (usually described in the following descriptions as petals). The white or yellow powder, or *farina*, is characteristic of many species. Many species, though restricted in distribution, are locally prominent. *c.* 66 spp.

Key to Primula sections:
  1 Flowers in 2 or more superposed whorls; bracts leafy
                                        5 FLORIBUNDAE
  1 Flowers in compact clusters or simple umbels; bracts not leafy

2 Flowers bell-shaped, without a distinct eye
  3 Flowers stalkless   11 SOLDANELLOIDES
  3 Flowers individually stalked   10 SIKKIMENSIS
2 Flowers funnel-shaped, with a distinct eye
  4 Plants without farina   2 CORTUSOIDES
      (see also 4 FARINOSAE   8 PETIOLARES)
  4 Plants with farina (sometimes scanty)
   5 Flowers stalkless, or almost so; bracts without a pouch at base
    6 Corolla-throat closed with a dense cluster of hairs
         6 MINUTISSIMAE
    6 Corolla-throat open
     7 Most flowers erect   3 DENTICULATA
     7 Most flowers pendent   1 CAPITATAE
   5 Flowers with individual stalks; bracts sometimes pouched at base
    8 Capsule round, crumbling, not opening by teeth
         8 PETIOLARES
    8 Capsule round or cylindrical, opening by teeth
     9 Leaf-blade rounded-heart-shaped, long-stalked
          9 ROTUNDIFOLIA
     9 Leaf-blade not rounded, tapering to the leaf-stalk
      10 Bracts more or less pouched   4 FARINOSAE
      10 Bracts never pouched   7 NIVALES
          (see also 6 MINUTISSIMAE)

1 CAPITATAE

850 **P. glomerata** Pax

W. Nepal to S.E. Tibet. 3000–5000 m. Shrubberies, open slopes. Aug.–Nov.

Flowers blue, pendent, many in a compact globular head, borne on flowering stems 10–30 cm, and with a rosette of oblong to oblanceolate, raggedly toothed leaves. Flowers funnel-shaped, with the corolla-tube 8–15 mm, and spreading deeply notched obcordate petals 6–10 mm; individual flower-stalks short; calyx with farina. Leaves 3–10 cm, including winged leaf-stalk which is usually reddish at the base. Leaves usually covered with white farina when young. **Pl. 82.**

2 CORTUSOIDES

851 **P. geraniifolia** Hook. f.

C. Nepal to S.W. China. 2700–4500 m. Forests, shrubberies. May–Jul.

Distinguished by its leaves which have rounded blades with heart-shaped bases, and with 7–9 lobes which are irregularly and coarsely sharp-toothed. Flowers rose to purple, semi-pendent, 2–12 in a lax umbel, borne on a slender hairy stem. Flowers 1–2 cm across; corolla-tube slender, petals spreading, notched; calyx salver-shaped with broad spreading lobes; individual flower-stalks hairy, 5–12 mm. Leaves hairy, leaf-stalk slender, much longer than the blade which is 3–8 cm long and wide; flowering stem 10–30 cm. **Pl. 80**.

3 DENTICULATA

852 **P. denticulata** Smith
Afghanistan to S.E. Tibet. Burma. 1500–4500 m. Shrubberies, open slopes; very common. Apr.–Jun.

Flowers purple to mauvish-blue, or occasionally white, in compact globular heads, borne on stout flowering stems 3–15 cm at flowering, lengthening in fruit. Flowers 1–2 cm across, with deeply bi-lobed petals; corolla-tube 0.5–2 cm; flower-stalks to 5 mm. Leaves in a compact rosette at flowering, oblong to oblanceolate, toothed, blunt, tapering to a broadly winged leaf-stalk; leaves much enlarging after flowering, to 30 cm. Wintering bud with broad rather fleshy scales which persist at base of rosette during flowering. **Pl. 82**.

853 **P. atrodentata** W. W. Smith
Uttar Pradesh to S.E. Tibet. 3600–4900 m. Alpine slopes. May–Jun.

Very like 852, but a smaller plant usually less than 10 cm, with fragrant lavender-coloured flowers *c.* 2 cm across, and with dark nearly black calyx-lobes. Leaves spathulate 1–4 cm, with dense white or yellow farina beneath. Basal bud scales not persisting. **Pl. 83**.

4 FARINOSAE

Leaves with farina, at least on undersides

854 **P. concinna** Watt
W. Nepal to S.E. Tibet. 4000–5000 m. Rocks, screes, banks. Jun.–Aug.

A tiny plant with yellow farina, and with pink, mauve or white, yellow-eyed flowers 4–8 mm across, 1–4 borne on a very short flowering stem. Corolla-tube about as long as calyx, petals narrow-obovate notched; flower-stalks *c.* 8 mm, bracts shorter, linear. Leaves spathulate 1–2.5 cm, blunt or acute, entire or toothed; flowering stem *c.* 1 cm.

855 **P. sharmae** Fletcher
Endemic to W. and C. Nepal. 2500–4800 m. Open slopes, rock ledges; drier areas; very local. Jun.–Aug.

Flowers mauve-purple or bluish-purple, 1.5–2 cm across, 3–8 borne on a short flowering stem 5–10 cm. Corolla-tube slender 1–3 cm, 2–3 times as long as calyx; petals broadly obcordate, deeply notched. Leaves 3–6 cm including the winged leaf-stalk, spathulate to oblong-elliptic, acute or blunt, usually finely toothed, undersides with white farina.

Leaves without farina
   a *Leaves more or less entire; capsule longer than calyx*

856 **P. involucrata** Wallich ex Duby

Pakistan to S.W. China. 3000–4500 m. Streamsides, marshes; common. Jun.–Aug.

Flowers white with a yellow eye, less commonly tinged purple, nodding, long-stalked, in a lax umbel of 2–6 flowers, borne on a slender flowering stem 10–30 cm. Flowers 1.2–2 cm across; corolla-tube purple, 1–1.5 cm, petals shallowly notched; calyx 5–ribbed, with lanceolate lobes often with recurved tips, hairless; bracts oblong to lanceolate prolonged beneath insertion into a broad blunt pouch 4–5 mm. Leaves with leaf-stalk longer than the ovate or oblong blunt blade which has a wedge-shaped or rounded base, and is entire or obscurely toothed, 3–8 cm. **Pl. 83**.

857 **P. sibirica** Jacquem.

Pakistan to Himachal Pradesh. 3600–4500 m. Bogs, damp flushes; common in Ladakh. Jun.–Aug.

Like 856, but a smaller plant with pink, mauve, or purple flowers, with a yellow eye, and petals more deeply notched. Flowers 1–10, 8–12 mm across, borne on short or long stalks 0.5–2 cm; calyx tubular with blunt lobes, 5–ribbed; bracts with a short basal pouch. Leaves with an ovate to oblong blade and with long narrowly winged leaf-stalk; flowering stem usually 3–20 cm.

858 **P. tibetica** Watt

Uttar Pradesh to S.E. Tibet. 3000–4500 m. Streamsides, marshes; drier Tibetan borderlands. Jun.–Jul.

Like 857, but plant smaller and with smaller pink flowers with a yellow eye, usually less than 1 cm across, and corolla-tube less than 1 cm long. Calyx often with darker ridges, lobes more or less acute; flower-stalks long, often to 3 cm or more; flowering stems variable, 2–15 cm. Leaves mostly with narrow-elliptic blade, long-stalked.

b *Leaves distinctly toothed; capsule equalling the calyx*

859 **P. elliptica** Royle

Pakistan to Uttar Pradesh. 3600–4300 m. Rocks, alpine slopes; common in Kashmir. Jun.–Aug.

Flowers pinkish-purple to mauve, 3–10 in a lax umbel, with individual flower-stalks 5–10 mm and linear bracts as long or longer. Corolla-tube as long or longer than calyx, petals obovate, deeply notched, spreading to 1–1.5 cm across; calyx-lobes narrow triangular. Leaves nearly erect, with rounded to elliptic, sharply toothed blade usually 1–3 cm, narrowed to a broad hairless leaf-stalk as long as blade; flowering stems 5–15 cm. **Pl. 83**.

860 **P. rosea** Royle

Pakistan to Himachal Pradesh. 2700–4000 m. Damp places, snow melts, stream-sides; common. Jun.–Aug.

Flowers rose-pink to red with a yellow eye, 1–2 cm across, borne in a lax umbel of 4–12 flowers, on flowering stems 3–10 cm at flowering. Petals obcordate, deeply notched; calyx 8 mm long, with broadly lanceolate

245

lobes; bracts long-pointed, pouched at base; individual flower-stalks 1 cm. Leaves partially developed at flowering, enlarging to 20 cm in fruit; blade obovate, finely toothed, narrowed to a winged leaf-stalk. **Pl. 83**.

5 FLORIBUNDAE

861 **P. floribunda** Wallich
Afghanistan to W. Nepal. 800–2000 m. Shady damp cliffs. Apr.–Jul.

A softly hairy plant with golden-yellow flowers, in usually 2–6 superposed umbels each with 3–6 flowers, borne on a downy flowering stem 5–15 cm. Flowers *c*. 1 cm across; corolla-tube long slender, hairy; calyx hairy, 6 mm, with ovate acute lobes; individual flower-stalks unequal, spreading, to 2 cm; bracts leafy, toothed. Leaves ovate to elliptic, coarsely and unequally toothed; leaf-stalk winged. Calyx enlarging in fruit, longer than capsule.

6 MINUTISSIMAE

Flowers solitary, or few, stemless.

862 **P. minutissima** Jacquem. ex Duby
Kashmir to C. Nepal. 3600–5200 m. Stony slopes, drier areas. Jun.–Jul.

A tiny densely tufted plant with rosettes of leaves and with stemless bright purple flowers much larger than the tiny leaves. Flowers 8–13 mm across; corolla-tube much longer than the hairless calyx, petals deeply notched. Leaves densely crowded, spathulate to oblanceolate 6–13 mm, toothed, with white farina beneath. Fruiting stems to 5 cm.

863 **P. reptans** Hook. f. ex Watt
Pakistan to C. Nepal. 3600–5500 m. Rocks, open slopes. Jun.–Aug.

Like 862 forming mats of rosettes, but with even smaller leaves without farina, and with larger solitary stalkless pale purple to pink flowers to 1.5 cm across. Corolla-tube narrow funnel-shaped, 4 times as long as the calyx, petals obcordate, deeply notched. Leaves only 4–6 mm, blade ovate to rounded, deeply toothed and with recurved margin, leaf-stalk short. **Pl. 82**.

863a **P. walshii** Craib
E. Nepal to S.W. China. 4200–5000 m. Open slopes, peaty turf; not common in Nepal. May–Jun.

A tiny glandular-hairy plant with a rosette of oblanceolate leaves, and 1–4 stalkless pink flowers with a yellow eye. Flowering stem very short, hidden amongst the leaves; flowers *c*. 8 mm across; petals bi-lobed; calyx 4–5 mm, as long or little shorter than the corolla-tube, rough, glandular. Leaves 8–15 mm, without farina; fruiting stem to 2 cm. **Pl.79**.

Flowers several, borne on slender flowering stems

864 **P. primulina** (Sprengel) Hara (*P. pusilla* Wallich)
Uttar Pradesh to S.E. Tibet. 3600–4500 m. Rocks, open slopes. Jun.–Aug.

A tiny plant with a rosette of numerous toothed leaves, and with a dense umbel of 2–4 purple, violet, or rarely white flowers, readily distinguished by the tuft of white hairs in the throat of the corolla. Flowers nearly stalkless, 8–12 mm across, with deeply bi-lobed petals; calyx densely softly hairy outside, about as long as corolla-tube; flowering stem slender, to 8 cm. Leaves spathulate to oblanceolate, 1–3 cm, including winged leaf-stalk, coarsely and deeply rounded-toothed, softly hairy. **Pl. 82.**

7 NIVALES

Flowers purple to pinkish

865 **P. macrophylla** D. Don (*P. nivalis* var. *macrophylla* (D.Don) Pax, *P. moorcroftiana* Wallich ex Klatt, *P. stuartü* var. *purpurea* (Royle) Watt)
Afghanistan to S.E. Tibet. 3300–4800 m. Open slopes, damp places; common, and often gregarious. Jun.–Aug.

A rather robust plant with narrow lanceolate or strap-shaped leaves, usually with white farina, and with a moderately dense head of 5–25 purple, violet, or lilac flowers with usually a darker eye. Petals elliptic to ovate, entire, corolla-tube 1–3 cm; calyx cylindrical, with linear-lanceolate lobes; flower-stalks to 2.5 cm, with farina; bracts awl-shaped; flowering stem 12–25 cm. Leaves erect, 10–30 cm, entire or toothed, with white farina beneath, gradually tapering to a long sheathing leaf-stalk. Capsule cylindrical to 2.5 cm. Plants very variable, and different forms have the appearance of distinct species. **Pl. 81.**

866 **P. megalocarpa** Hara (*P. nivalis* var. *macrocarpa* (Watt) Pax)
C. Nepal to S.E. Tibet. 3600–4800 m. Open slopes, damp places. Jun.–Aug.

Very like 865, but distinguished by its paler pinkish, mauve, lilac or white flowers with obcordate petals; by its blunt calyx-lobes; and its large thick capsule to 3 cm long.

Flowers yellow or white (see also 866)
a *Bracts awl-shaped*

867 **P. obliqua** W. W. Smith
C. Nepal to S.E. Tibet. 3600–4400 m. Alpine shrubberies, open slopes; gregarious on heavily dunged grazing sites. Jun.–Jul.

Flowers pale yellow to white, sometimes pink-flushed, in an umbel of 5–10 drooping flowers, each flower 2.5–3 cm across. Corolla-tube 1.5–3 cm long, petals spreading, shallowly notched; calyx with blunt lobes and much farina; bracts awl-shaped, shorter than calyx. Leaf-blade lanceolate to obovate 10–30 cm, finely rounded-toothed, with yellow farina beneath, narrowed to a winged leaf-stalk; flowering stem stout, 30–45 cm; base of plant with ovate reddish bracts. **Pl. 82.**

868 **P. stuartii** Wallich
Himachal Pradesh to E. Nepal. 3600–4500 m. Open slopes; gregarious on grazing grounds. Jul.

Flowers golden-yellow, 2–3 cm across, usually with rounded, conspicuously toothed petals, many in a lax umbel, or sometimes with 2 superposed umbels, borne on a flowering stem 20–40 cm. Corolla-tube usually longer than calyx which has lanceolate lobes to 1 cm; bracts awl-shaped; individual flower stalks to 2.5 cm; calyx, bracts and flowerstalks densely covered with yellow farina. Leaves lanceolate 10–30 cm, sharply toothed, with yellow farina beneath, narrowed to a sheathing leaf-stalk. **Pl. 81.**

b *Bracts leafy*

869 **P. poluninii** Fletcher
Endemic to Sisne Himal in W. Nepal. 4500–5000 m. Open slopes, snow-melt areas; gregarious. Jun.–Jul.

Differing from 868 in having narrow leafy bracts to 3 cm, without farina beneath, and oblong blunt calyx lobes. Flowers yellow, 2–2.5 cm across, up to 12 in each umbel; petals overlapping obcordate notched, *c.* 1 cm; calyx 1–1.5 cm, with farina. Leaves lanceolate to oblong, to 18 cm, blunt, margin with rounded teeth, and with farina below; flowering stem as long as leaves. **Pl. 84.**

8 PETIOLARES

Stem 5 cm or more at flowering

870 **P. calderana** Balf.f. & Cooper (*P. obtusifolia* auct. non Royle)
W. Nepal to S.E. Tibet. 3600–4800 m. Shrubberies, open slopes. Jun.–Jul.

Flowers usually dark purple with a golden eye, erect or drooping, 2–25 in an umbel borne on flowering stems 5–30 cm; a strong-smelling plant. Flowers 1.5–2 cm across with broad obcordate overlapping petals, corolla-tube 1–2 cm, twice as long as calyx; calyx, flower-stalks, and underside of petals with pale yellow farina. Leaves spathulate to oblanceolate 10–15 cm, toothed, well developed at flowering. This form occurs only from E. Nepal eastwards. **Pl. 82.**

Subsp. **strumosa** (Balf.f. & Cooper) A.Richards has golden-yellow flowers with orange-yellow eye, and with a longer corolla-tube 2–3 times as long as calyx. Leaf-blade mostly elliptic, some narrowed abruptly at base, margin with acute or rounded teeth, with or without yellow farina. Common on grazing grounds and in rhododendron shrubberies thoughout Nepal; gregarious. **Pl. 80.**

Flowering stem absent, or less than 5 cm at flowering
a *Flowers yellow*

871 **P. aureata** Fletcher
Endemic to C. Nepal. 3600–4300 m. Rock crevices. Apr.–May.

Very distinctive with its rosettes of broad toothed leaves covered with silvery-white farina, lining overhanging rock-crevices, and its cluster of almost stalkless creamy-yellow flowers with a central flush of golden-yellow spreading halfway along the petals. Flowers 2.5–3 cm across, in an umbel of 3–10, borne on a short flowering stem hidden amongst the leaves. Petals rounded, toothed or finely cut, spreading, corolla-tube 1.5–2 cm; calyx 8–10 mm, with ovate long-pointed, recurved lobes; bracts linear-lanceolate. Leaves to 10 cm, broadly spathulate, irregularly toothed, leaf-stalk winged. **Pl. 84**.

b *Flowers pink, rose or magenta*
  i *Whole plant lacking farina (except on young calyx)*
   x *Petals with a narrow pointed tip, often 4*

872 **P. sessilis** Royle ex Craib
Kashmir to W. Nepal. 2100–3600 m. Forests, shady banks; common in the Kulu valley, Himachal Pradesh. Apr.–May.
Distinguished by its pale lilac-pink flowers with an orange eye, with usually 4 entire narrow petals which are abruptly narrowed to a fine point. Flowers *c*. 2.5 cm across, very short-stalked; calyx cylindrical, with narrow keeled lobes (sometimes with farina). Leaves dark green, to 6 cm, with rounded blade and short leaf-stalk.

   xx *Petals toothed, always 5*

873 **P. deuteronana** Craib
C. Nepal to Sikkim. 3500–4500 m. Open rocks, damp ground, ledges; a common early flowering alpine species. Apr.–Jun.
Flowers pale lilac with purple streaks outside, 1.5–2.5 cm across and with a deep yellow eye; distinguished by the shaggy hairs in the corolla-tube, and the widely spaced narrow acute petals which are irregularly toothed at the apex. Flower-stalks to 3 cm; corolla-tube *c*. 2 cm; calyx cylindrical, with narrow pointed lobes (with farina in bud). Leaves in a dense rosette, oblong-spathulate, irregularly and finely toothed, tapering to a very short broad leaf-stalk.

874 **P. petiolaris** Wallich
Uttar Pradesh to Sikkim. 2400–3600 m. Forests; common. Apr.–May.
Flowers pink with a yellow eye with a thin white border, and with rounded irregularly toothed petals. Calyx-lobes with distinctive ribs. Flowers *c*. 2 cm across, nearly stalkless. Leaves at flowering in a tight crisped rosette, later elongating, spathulate and distinctly stalked; plant without farina.

  ii *Farina present, at least on buds and calyx*
   x *Leaves at flowering 6–12 cm; calyx-lobes toothed*

875 **P. irregularis** Craib
W. Nepal to Sikkim. 2700–3500 m. Forests; common in E. Nepal. Apr.–May.
Flowers pink with an orange-yellow eye with a broad white border, and with stiff waxy, deeply and regularly toothed petals. Distinguished from

other species by its blunt, irregularly-toothed calyx-lobes with farina; flower-stalks and calyx with stalked glands. Leaves at flowering of two kinds, some almost stalkless, others long-stalked; fruiting stalk elongating. **Pl. 83**.

xx *Leaves at flowering rarely more than 6 cm; calyx-lobes not toothed*

876 **P. edgeworthii** (Hook.f.) Pax
Himachal Pradesh to C. Nepal. 2100–3600 m. Forests; common. Apr.–May.

Flowers blue, lilac, pink or white, with an orange-yellow eye with a white border, to 3 cm across, and petals varying from almost entire to deeply cut into long teeth. Calyx broadly cylindrical, lobes broad, blunt or acute, with farina; flower-stalks to 2 cm, with farina. Leaves spathulate entire, greyish-green, to *c*. 5 cm at flowering, becoming larger, green and without farina; later leaves toothed; fruiting stem elongating. **Pl. 83**.

877 **P. gracilipes** Craib
C. Nepal to S.E. Tibet. 2700–4100 m. Forests, open slopes; common in the lower alpine zone. Apr.–May.

Flowers bright pink-purple, with small orange-yellow eye surrounded by a narrow white border, with obovate petals which are irregularly toothed and spreading to 2–4 cm. Flowering stem absent; individual flower-stalks 1–6 cm; calyx about one third as long as corolla-tube, with triangular-acute farinose lobes. Leaves oblong-spathulate to elliptic, irregularly toothed, and with distinct leaf-stalks.

9 ROTUNDIFOLIA

878 **P. rotundifolia** Wallich
W. Nepal to Sikkim. 3600–4500 m. Always beneath overhanging rocks; quite common. Jun.–Jul.

Distinguished by its rounded heart-shaped, or kidney-shaped, toothed leaves with copious yellow farina on the underside, and its lax umbel of pinkish-purple flowers with a golden-yellow eye. Corolla 1.5–2 cm across, with rounded usually entire petals; calyx with lanceolate teeth about one third as long as the slender corolla-tube; bracts awl-shaped; inflorescence with dense farina. Leaves long-stalked, blades mostly to 5 cm across; flowering stem 10–30 cm. **Pl. 81**.

879 **P. gambeliana** Watt
E. Nepal to S.E. Tibet. 3600–4300 m. Rock ledges. Jun.–Jul.

A smaller and more delicate plant than 878, with an umbel of few flowers with notched petals, and with the inflorescence without farina. Flowers pink-purple or violet-purple, with a yellow throat, 1.5–2.5 cm across. Leaf-blade ovate to rounded, with a heart-shaped base, 2–4 cm across, toothed, and with a long slender leaf-stalk; flowering stem mostly 6–12 cm.

880 **P. caveana** W. W. Smith
E. Nepal to S.E. Tibet. 4500–5600 m. Beneath overhanging rocks; quite common in Rolwaling and Khumbu, Nepal. Jun.–Jul.

A small plant with pale purple flowers with a yellow eye, each *c*. 1 cm across, with broadly ovate usually entire petals, and borne in an umbel of few flowers on slender stems 3–5 cm. Distinguished from 879 by its leaves which taper abruptly to a broad winged leaf-stalk; blade oblong to rounded, sharply toothed, 1–3 cm across, with dense white farina beneath. Inflorescence with farina. **Pl. 84**.

10 SIKKIMENSIS

881 **P. reticulata** Wallich
C. Nepal to S.E. Tibet. 3300–4800 m. Shrubberies, open slopes. Jun.–Jul.

Flowers yellow or white, few or many in a lax umbel, the outer flowers long-stalked generally nodding, the inner erect, borne on flowering stems 18–45 cm. Flowers 1–2 cm across; petals obovate, shallowly notched or wavy at apex; corolla-tube up to twice as long as the farinose calyx which has short lobes with recurved apices. Leaves with a long stalk to 30 cm, and broadly oblong blunt toothed blade with a heart-shaped base, mostly 4–15 cm.

882 **P. sikkimensis** Hook.f.
W. Nepal to S.W. China. 3300–4400 m. Streamsides, open slopes; quite common in wet places in drier inner valleys; usually gregarious. May–Jul.

A robust yellow-flowered perennial like 881, but differing in having leaves with blades which taper to the shortly winged leaf-stalk. Flowers scented, numerous, in one or sometimes two superposed umbels, with long arching or pendulous individual flower-stalks which become erect in fruit. Corolla 1.5–3 cm across, with entire or shallowly notched petals. Flowering stem 15–90 cm. **Pl. 81**.

Var. **hopeana** (Balf.f. & Cooper) W. W. Smith & Fletcher has white drooping flowers, and is a more slender plant with smaller leaves. It is found from C. Nepal eastwards. **Pl. 84**.

11 SOLDANELLOIDES

Flowers predominantly white

883 **P. buryana** Balf.f.
W. to E. Nepal. Tibet. 3300–5000 m. Open slopes in drier areas. Jun.–Jul.

Flowers white, sweet-scented, stalkless and semi-pendent in a dense head of 4–7, borne on a slender flowering stem 8–18 cm. Corolla 8–20 mm across, with tube gradually widening to the mouth and with broad petals with deeply notched apices, softly hairy. Calyx to 6 mm, lobes elliptic, softly hairy; bracts elliptic, hairy. Leaves 2–5 cm, in a basal tuft, ovate to oblong-ovate, toothed, tapering to winged leaf-stalk, densely hairy.

A form with blue flowers is found locally in C. Nepal, on rock ledges. It is possibly a distinct species.

884   **P. reidii** Duthie
Kashmir to C. Nepal. 3300–4500 m. Rock ledges, peaty banks. Jun.–Jul.
Flowers ivory-white, fragrant, widely bell-shaped, in a compact umbel of 3–10 pendent flowers, borne on a slender flowering stem 6–15 cm. Corolla 1.8–2.5 cm long and broad, the tube narrow below and abruptly widened above and with broad notched lobes; calyx cup-shaped, lobed, 6–8 mm, green, with farina; bracts 1–3 mm, often lobed. Leaves 3–20 cm, in a lax rosette, with distinct leaf-stalk as long or longer than the oblong to oblong-lanceolate, shallowly lobed or toothed blade, which has long hairs and no farina. **Pl. 84**.
Var. **williamsii** Ludlow has blue flowers, and is found in W. & C. Nepal.

885   **P. wigramiana** W. W. Smith
Endemic to C. Nepal. 3600–4400 m. Open slopes. Jun.–Jul.
Like 884 with fragrant pendent white flowers, but more robust and leaves in a tight rosette, and leaf-stalks not distinct or not more than half as long as the blade; calyx frequently purple. Flowers 6–7, 2.5 cm across and *c*. 3 cm long; corolla-lobes toothed; flowering stem 10–25 cm. **Pl. 80**.

Flowers blue or violet (see also 883, 884)

886   **P. wollastonii** Balf. f.
C. to E. Nepal. Tibet. 3600–4800 m. Open slopes in dry heads of inner valleys. Jun.–Jul.
Flowers light or dark purple, or blue, bell-shaped, with much farina, in a compact head of 2–5 drooping flowers, borne on flowering stems 12–20 cm. Flowers 1.5–2 cm long, with shallow entire lobes; calyx cup-shaped, with short green or purple triangular-pointed lobes. Leaves in a compact rosette, oblanceolate to obovate 2.5–5 cm, toothed, densely hairy, with copious farina on the lower surface, short-stalked. **Pl. 83**.

CORTUSA   Like *Primula*, but corolla bell-shaped and deeply 5–lobed. Leaves palmately-lobed. 1 sp.

887   **C. brotheri** Pax ex Lipsky (*C. matthioli* Hook. f. non L.)
Afghanistan to Uttar Pradesh. C. Asia. 3000–4000 m. Rocks, forests, shady places; common in Kashmir. Jun.–Aug.
Flowers rosy-purple, drooping, with long flower-stalks in a lax umbel of 5–12 , borne on a slender flowering stem longer than the leaves. Corolla *c*. 1.5 cm long, bell-shaped, divided at least to the middle into 5 oblong blunt lobes; calyx *c*. 5 mm, with 5 triangular lobes; bracts elliptic, toothed. Leaves long-stalked, hairy, with a rounded-heart-shaped blade 2.5–10 cm across and with 5–7 shallow lobes which are toothed; flowering stem 15–25 cm. Fruit cluster erect. **Pl. 80**.

OMPHALOGRAMMA   Flowers large solitary, tubular. 1 sp.

888   **O. elwesiana** (King ex Watt) Franchet
E. Nepal to Arunachal Pradesh. 3600–4000 m. Wet open slopes; prominent on route to Barun valley; otherwise rare. Jun.–Jul.
A beautiful and unusual plant with large inclined solitary violet tubular flowers, borne on stout brownish-hairy stems with large rounded overlapping leafy bracts at their bases. Corolla-tube stout, hairy, *c.* 2.5 cm long and with 5–6 squarish, toothed lobes spreading to 3 cm; calyx 5–lobed, *c.* 8 mm, densely hairy; bracts absent. Leaves all basal, appearing after the flowers; blade ovate-lanceolate 5–12 cm, leathery, hairless, narrowed to a winged leaf-stalk; flowering stems lengthening in fruit; rootstock stout. Capsule *c.* 2 cm, splitting into 6 or more elliptic teeth. **Pl. 84.**

MYRSINACEAE

A medium-sized family of trees and shrubs of the warm Temperate, Tropical and Sub-tropical Zones. Leaves simple, leathery, alternate, dotted with glands, or with conspicuous resin-ducts. Flowers small, usually borne in clusters, either on scaly short shoots in the leaf-axils, or in terminal branched clusters. Sepals and petals 4–6, each whorl fused by their bases; stamens equal in number to petals. Ovary superior, or half-inferior, with usually 1 chamber. Fruit a fleshy drupe or berry.

MAESA   Fruit a berry with many seeds. 5 spp.

889   **M. chisia** Buch.–Ham. ex D. Don
W. Nepal to S.E. Tibet. Burma. 1200–2400 m. Forests, shrubberies. Mar.–Apr.
A gregarious shrub or small tree to 10 m, with lanceolate long-pointed evergreen leaves, and lax spikes of small scented white flowers less than half as long as the leaves. Flowers *c.* 4 mm across, cup-shaped, with spreading rounded, veined petals; calyx-lobes triangular, veined like the petals. Leaves 7–13 cm, weakly toothed, hairless. Fruit a many-seeded berry, pale brown, *c.* 4 mm, crowned by the calyx-lobes. **Pl. 85.**

890   **M. macrophylla** (Wallich) A. DC.
Uttar Pradesh to Bhutan. 300–1800 m. Shrubberies, open slopes; common and often gregarious in hot lower valleys. Mar.–Apr.
Distinguished from 889 by its much larger broader conspicuously toothed leaves which are softly hairy, and by its velvety-haired branches. Flowers cream to yellowish, in lax much-branched clusters often as long or longer than the leaves. A shrub 2–5 m, with large elliptic or rounded leaves 15–20 cm, with more or less heart-shaped bases, regularly toothed. Berry pinkish-white.

MYRSINE   Differs from *Maesa* in its 1-seeded berry. 3 spp.

891   **M. africana** L.

Afghanistan to C. Nepal. China. Arabia. Africa. 900–1800 m. Forests, shrubberies, open slopes; often gregarious. Feb.–Apr.

A small densely branched evergreen shrub to 1 m, with branchlets and leaf-stalks with rusty hairs, and with small sharply toothed, almost stalk-less, lanceolate to elliptic leaves mostly *c.* 2 cm. Flowers 2 mm across, white, corolla glandular; female flowers almost stalkless, in clusters of 4–6 in axils of the leaves; male flowers with 4 stamens with conspicuous purple anthers. Fruit a globular red berry *c.* 4 mm, becoming dark purple or black when fully ripe.

The fruit is used medicinally.

892   **M. semiserrata** Wallich

Pakistan to S.W. China. Burma. 1000–2700 m. Forests, shrubberies; very common in Nepal. Nov.–Apr.

A small tree or large shrub to 5 m, with hairless branches, and elliptic long-pointed, toothed, distinctly stalked leaves 5–12 cm. Flowers reddish to white, 3 mm across, stalked, in many-flowered stalkless axillary clusters. Berry bright blue to reddish-purple, *c.* 6 mm across, edible. **Page 485.**

SYMPLOCACEAE

A small family of the Tropical and Sub-tropical Zones, often placed in the Theaceae. Distinguished by its corolla with 5–10 petals; stamens 5–10, or more. Ovary inferior or half-inferior, of 2–5 carpels with usually 2 ovules in each chamber. Fruit a berry or drupe.

SYMPLOCOS   10 spp.

Inflorescence rounded; flowers hairy at base

893   **S. theifolia** D.Don

W. Nepal to S.W. China. Burma. S.E. Asia. 1500–3000 m. Forests. Oct.–Apr.

A medium-sized tree, flowering unlike other members of the genus in the autumn and winter, with very short compound clusters of white flowers in the axils of elliptic to lanceolate long-pointed leaves. Inflores-cence 1.3–2.5 cm across, often 3–branched; flowers to 10 mm, with ovate petals and with long white hairs at the base of the corolla within; stamens numerous; calyx with rounded lobes with marginal hairs. Leaves to 8 cm, obscurely and finely toothed, and with rounded to wedge-shaped base, hairless. Fruit cylindrical 8 mm long, with conspicuous incurved calyx-lobes. **Page 485.**

Inflorescence elongate, branched or not

894 **S. paniculata** (Thunb.) Miq. (*S. crataegoides* Buch.–Ham. ex D. Don)
Pakistan to S.W. China. Burma. Japan. S.E. Asia. 1000–2700 m. Forests,
shrubberies. Apr.–Jun.

A large deciduous shrub or medium-sized tree, with hairy branchlets
and leaves, and with fragrant snow-white flowers borne in cylindrical
branched clusters on lateral stems. Flowers 8–10 mm across, with 5 oblong-
elliptic spreading petals; calyx with rounded, ciliate lobes; stamens many,
exserted. Leaves membraneous, with a broad-elliptic to ovate blade 4–7 cm,
sharply toothed, impressed with veins above, short-stalked. Bark light grey,
corky. Fruit globular *c.* 5 mm.

Leaves lopped for fodder; bark gives a yellow dye. **Pl. 85**.

895 **S. ramosissima** Wallich ex G. Don
Uttar Pradesh to S.W. China. Burma. S.E. Asia. 1500–2700 m. Forests; in
E. Nepal often gregarious and forming a dense understory. May–Jul.

A small tree with lanceolate long-pointed finely toothed papery leaves,
and numerous lax spike-like clusters of usually 4–8 white flowers borne in the
axils of the lower leaves. Spikes 1.3–3 cm; flowers 3–6 mm across; stamens
20, anthers exserted. Leaves with blade 6–10 cm, and leaf-stalk to 10 mm.
Fruit ellipsoid 6–8 mm long, with a narrow calyx-rim.

896 **S. sumuntia** Buch.–Ham. ex D. Don
C. Nepal to S.W. China. Burma. S.E. Asia. Japan. 1400–2700 m. Forests.
Mar.–Apr.

A small tree with elliptic-lanceolate long-pointed very stiff leathery
leaves, and with dense unbranched clusters of white flowers 3–8 cm long,
with conspicuously hairy bracts which envelope the flower-buds. Flowers 4–6
mm across; petals elliptic, veined; stamens exserted, 35–40. Leaves toothed,
blade usually 8–10 cm. Fruit ovoid 8 mm long, with a wide calyx-rim.

OLEACEAE Olive Family

A medium-sized cosmopolitan family of trees, shrubs and woody climbers,
with characteristic silvery-grey scale-like hairs, in addition to normal hairs, on
leaves and twigs. Leaves simple, trifoliate, or pinnate, usually opposite.
Flowers in branched or spike-like clusters. Flowers bisexual or unisexual;
sepals usually 4; petals 4, usually united into a long tube below; stamens 2–4,
attached to the corolla-tube. Ovary superior, with 2 fused carpels, usually
with 2 ovules per chamber. Fruit various: berry, drupe, capsule or winged
nut, either dry or fleshy.

Leaves pinnate
  a Fruit a winged nut   1 FRAXINUS
  b Fruit a berry   2 JASMINUM

OLEACEAE

Leaves simple, undivided
  a Fruit a dry splitting capsule   3 SYRINGA
  b Fruit a berry or drupe
    i Flowers in terminal clusters   2 JASMINUM   4 LIGUSTRUM
    ii Flowers in axillary clusters   5 OLEA   6 OSMANTHUS

1 FRAXINUS Ash Fruit a winged nut. Leaves pinnate; usually deciduous trees. 4 spp.

Inflorescence terminal, on young shoots

897 **F. floribunda** Wallich
Himachal Pradesh to S.W. China. 1200–2700 m. Forests; planted around villages. Apr.–May.
    A large deciduous tree with grey bark, and with numerous white flowers in large branched clusters to 25 cm across, distinctive when in flower in having linear-oblong petals 3–4 mm long. Leaves pinnate, with 7–9 elliptic-lanceolate long-pointed, saw-toothed, stalked leaflets 6–15 cm long. Fruit with a long narrow wing 2.5–4 cm long by 3–4 mm broad with blunt or notched apex.
    Wood used for oars, poles, and ploughs.

898 **F. micrantha** Lingelsh.
Himachal Pradesh to C. Nepal. 1200–2700 m. Forests, planted around villages. Apr.–May.
    Distinguished from 897 by the absence of petals, by its smaller erect branched flower clusters, and by its young rusty-haired shoots. Fruits *c.* 4.5 cm by 8 mm broad. A large tree, with leaves with 5–9 elliptic-lanceolate leaflets.

Inflorescence lateral, on shoots of previous year

899 **F. xanthoxyloides** (Wallich ex G. Don) DC.
Afghanistan to Himachal Pradesh. 1000–2700 m. Inner dry valleys; common. Apr.
    A small tree, with pinnate leaves with a finely winged mid-rib and 5–11 small ovate-lanceolate, weakly toothed leaflets, 3–4 cm long. Flowers in dense stalkless clusters appearing with or before the young leaves; bracts woolly-brown. Calyx absent in male flowers, small in bisexual flowers; petals absent. Fruit in dense stalkless clusters, each 3–4 cm long.
    Wood hard and close-grained, used for walking sticks and tool handles.
Page 485.

900 **F. excelsior** L. Common Ash
Pakistan to Kashmir. W. Asia. Europe. 1000–3000 m. Native, and cultivated. Flowering in spring.
    Distinguished from 899 by the absence of calyx both in flower and fruit, and by the flower clusters which are solitary or paired (clusters many, dense in 899). A large tree with pinnate leaves, usually with 5–7 pairs of

leaflets 5–12 cm long. Fruit 3.5–5 cm long, wing with rounded or notched apex.

Yields valuable timber to make carts.

2 JASMINUM JASMINE Fruit a lobed or paired berry. Shrubs or woody climbers; leaves usually pinnate. 13 spp.

Flowers yellow; erect shrubs

901 **J. humile** L.
Afghanistan to S.W. China. India. Burma. C. Asia. 1500–3000 m. Shrubberies, forests; common. Apr.–Jun.

A small erect much-branched shrub to 1 m or more, with green angular twigs, dark green leathery pinnate leaves, and terminal lax clusters of yellow tubular flowers. Corolla-tube slender 1–2 cm long, with 5 rounded lobes spreading to 6 mm; calyx cup-shaped, very short *c*. 3 mm, with triangular teeth. Leaflets 3–4, ovate to lanceolate entire, the terminal leaflet sometimes larger. Fruit a single or paired black berry 8 mm, with crimson juice.

Flower, roots, and juice used medicinally. **Pl. 85.**

Flowers white to pinkish; climbing shrubs
a *Calyx-teeth short, triangular*

902 **J. dispermum** Wallich
Kashmir to S.W. China. S.E. Asia. 900–2500 m. Shrubberies, ravines. Apr.–May.

A woody climber, with many fragrant pinkish-white flowers in terminal and axillary pendulous clusters, and with pinnate leaves with leathery leaflets. Flowers with a pink corolla-tube 1–1.5 cm long, and 5 shorter oblong-rounded lobes spreading to 8 mm; calyx *c*. 3 mm, cup-shaped with triangular teeth. Leaflets 3–5, the terminal leaflet much longer 8–10 cm, but leaves at ends of branches often with only one leaflet. Fruit elliptic *c*. 1.3 cm long, dark purple.

b *Calyx-teeth long, linear*

903 **J. officinale** L. WHITE JASMINE
Afghanistan to S.W. China. 1200–3000 m. Shrubberies, forests. May–Jul.

A large climbing shrub with dark green twigs and pinnate leaves, differing from 902 in its linear calyx-teeth 5–10 mm long (teeth triangular; 1 mm long in 902). Flowers white, fragrant, in axillary umbel-like clusters; corolla-tube pink, 15–17 mm long, the lobes ovate, 9–12 mm long, overlapping; flower-buds pink. Leaves with 3–7 ovate to lanceolate leaflets, the terminal larger. Fruit black, 8–10 mm long, with crimson juice.

The national flower of Pakistan, and frequently cultivated there. Valued as a medicinal plant. The perfume from the flowers is highly prized. **Pl. 85.**

904 **J. multiflorum** (Burm. f.) Andrews (*J. pubescens* (Retz.) Willd.)
CHINESE JASMINE

Kashmir to Bhutan. India. Burma. China. 300–1500 m. Hedges, shrubberies, forests; common in outer foothills and cultivated in the plains. Mar.–Oct.

Distinguished by its entire elliptic long-pointed leaves, by its dense umbels of relatively large white tubular flowers, and by its densely rusty-hairy calyx with linear teeth to 1.5 cm or more. Corolla-tube to 2.5 cm, with 6–9 lanceolate pointed spreading lobes which are shorter than the tube. Leaves 3–8 cm, opposite, stalked, hairy; a shrub or climber to 2 m or more. Carpels ellipsoid *c.* 1.3 cm long, surrounded by the long-hairy calyx.

3 SYRINGA Lilac Fruit a splitting capsule. Leaves entire. Small trees or shrubs. 1 sp.

905 **S. emodi** Wallich ex Royle
Afghanistan to C. Nepal. 2100–3600 m. Forests, shrubberies, drier areas. Jun.–Jul.

A large deciduous shrub to 5 m, with oblong-elliptic entire often rather large leaves, and with dense terminal branched clusters of white scented flowers. Corolla-tube *c.* 8 mm long, the lobes oblong, hooded at the tip, shorter, spreading. Leaf-blade 6–10 cm, finely hairy at first beneath, becoming hairless, leaf-stalk to 2.5 cm. Capsule cylindrical to 1.5 cm, opening by valves. **Page 486**.

4 LIGUSTRUM Privet Fruit a black berry. Leaves entire. 4 spp.

906 **L. compactum** (Wallich ex DC.) Hook. f. & Thoms. ex Brandis
Uttar Pradesh to S.W. China. 2000–2400 m. Forests, damp gulleys; common in W. Nepal. May–Jul.

A shrub to 3 m, or a small tree to 10 m or more, with large dark green lanceolate pointed leaves, and large branched terminal pyramidal clusters of small creamy-white flowers. Flowers *c.* 5 mm across, with a short tube and 4 blunt-triangular lobes; calyx cup-shaped, half as long as corolla-tube, not or scarcely lobed; stamens 2. Leaves entire, paler beneath, variable, mostly 8–15 cm, short-stalked. Fruit glaucous blue-black, ovoid, in dense clusters.

Much lopped for fodder. **Page 485**.

907 **L. indicum**(Lour.) Merr. (*L. nepalense* Wallich)
Uttar Pradesh to Bhutan. S.E. Asia. 1200–2700 m. Damp gulleys, wet oak forests; common. Jun.–Jul.

Distinguished from 906 by its twigs with many close lenticles, by its finely hairy inflorescences, and by its globular fruits. Flower clusters smaller, mostly 3–7 cm long. Leaves smaller, elliptic pointed, entire, 2–6 cm. A shrub or small tree.

5 OLEA Olive Fruit a one-seeded drupe, with somewhat fleshy outer coat. Leaves entire. Small evergreen trees. 3 spp.

908　**O. ferruginea** Royle (*O. cuspidata* Wallich ex G. Don)
Afghanistan to W. Nepal. 500–2600 m. Inner dry valleys; very common in lower hills, gregarious. Mar.–Sep.

A greyish-green tree or shrub, with smooth grey bark which strips off when old, and with dark green leathery oblong-lanceolate leaves with a dense film of scales beneath which turn reddish-brown on older leaves. Flowers whitish, in axillary clusters 2–4 cm long; corolla-tube very short, lobes 4, spreading. Leaves variable 3–10 cm, short-stalked. Fruit a drupe *c*. 8 mm long, black when ripe.

Wood hard and heavy, much prized for turning, and used for making implements. Fruits eaten, and oil extracted from them. **Page 486**.

6 OSMANTHUS　Evergreen shrubs or trees. Fruit a drupe. 2 spp.

909　**O. suavis** King ex C. B. Clarke
Uttar Pradesh to S.W. China. 2100–3600 m. Forests. Apr.–May.

An evergreen shrub or small tree to 7 m, with leathery narrow-elliptic acute, finely toothed leaves, and with short axillary clusters of small scented white tubular 4–lobed flowers. Flowers with corolla-tube *c*. 8 mm long, and with spreading recurved blunt lobes; calyx lobed to halfway, with oval blunt papery lobes. Leaves mostly 5–8 cm, stalked, hairless. Fruit a drupe, elliptic, 8–13 mm long. **Pl. 86**.

APOCYNACEAE　Oleander Family

A large tropical family, mainly comprising large rain-forest trees, but also some shrubs, climbers, and a few herbaceous plants in the temperate regions. Leaves simple, usually opposite, or whorled or spirally arranged, without stipules; sap a milky latex. Flowers in flat-topped branched clusters, often large showy and fragrant. Sepals 5, fused into a tube; petals 5, fused into a tube and with spreading lobes; stamens 5, with the anthers fused to each other and to the head of the style in many species. Ovary superior, or half-inferior, of 2 free or fused carpels. Fruit paired, either fleshy and not splitting, or dry and splitting; seeds with or without hairs.

BEAUMONTIA　Flowers funnel-shaped, white. Evergreen woody climbers with opposite leaves. 1 sp.

910　**B. grandiflora** (Roxb.) Wallich
W. Nepal to Bhutan. 150–1400 m. Climbing on trees, scrambling on banks and rocks. Mar.–Apr.

A large woody climbing or trailing shrub, with clusters of very large white, broadly funnel-shaped flowers with short triangular-ovate acute-tipped lobes. Corolla 8–12.5 cm long and 10 cm across at mouth; calyx-lobes leafy, variable, from oblong-lanceolate to obovate 2.5–5 cm long, velvet-haired. Leaves opposite, papery, with elliptic to obovate blade 18–30 cm, short-stalked; shoots with rusty hairs and with milky juice; bark thick corky. Fruit usually paired, narrow cylindrical 13–30 cm by 5 mm, at length dividing into valves; seeds *c.* 2 cm, with a tuft of hairs twice as long. **Pl. 86**.

NERIUM OLEANDER Evergreen shrubs with leaves usually in threes. Corolla funnel-shaped, with 5 toothed or lobed scales in the throat; anthers included in the corolla-tube. 1 sp.

911 **N. indicum** Miller (*N. odorum* Solander)
Afghanistan to C. Nepal. India. 600–1500 m. Open hillsides; common in foothills of W. Himalaya. Apr.–May.

An evergreen shrub with silvery-grey bark, numerous long narrow pointed leathery leaves, and with large terminal branched clusters of fragrant red, pink, or white funnel-shaped flowers. Flowers to 4 cm across, with 5 obovate lobes with irregularly cut scales at their bases; calyx-lobes lanceolate, hairy; stamens with anthers fused round stigma. Leaves in whorls of 3, linear-lanceolate to oblong, fine-pointed, entire, mostly 10–20 cm, shining dark green above, rough beneath. Fruit paired, narrow cylindrical, 13–18 cm by 6–8 mm broad; seeds linear, ribbed, with a tuft of greyish-brown hairs.
The plant is poisonous. The leaves and root are used medicinally. **Pl. 86**.

TRACHELOSPERMUM Evergreen climbing shrubs with opposite leaves. Flowers with a slender corolla-tube which is enlarged above; anthers spurred at base. 2 spp.

912 **T. lucidum** (D. Don) Schumann (*T. fragrans* (Wallich ex G. Don) Hook. f.)
Pakistan to Bhutan. 600–2100 m. Climber in forests and shrubberies. Apr.–Jul.

Usually a tall woody evergreen climber, sometimes creeping over rocks, with shining elliptic pointed leaves, and with small fragrant white tubular flowers in loosely 3–branched clusters. Corolla-tube slender *c.* 1 cm, enlarged above, with 5 wedge-shaped lobes spreading diagonally from the centre to *c.* 2 cm across; calyx-lobes elliptic, about one quarter as long as corolla-tube. Leaves opposite, mostly 8–12 cm, short-stalked. Fruit of paired, slender cylindrical in-curved carpels 15–30 cm long; seeds with a crown of long white hairs. **Pl. 86**.

PLUMERIA Small deciduous trees with thick but weak branches; leaves spirally arranged; sap milky. Corolla flat, with spreading lobes. 1 sp.

913 **P. rubra** L. (*P. acutifolia* Poiret) FRANGI-PANI, TEMPLE-TREE
Native of tropical America. Cultivated in the Himalaya to 1200 m, and semi-naturalized. Flowering throughout the year.
A small deciduous tree with thick fleshy branches, with large lanceolate

leaves with conspicuous parallel lateral veins, and with terminal clusters of many very fragrant white, pink, or yellow flowers, often flowering when tree is leafless. Flowers 5–10 cm across, with a slender corolla-tube and spreading elliptic lobes overlapping to the left; calyx with short blunt lobes. Leaves entire 20–35 cm, spirally arranged at ends of branches, short-stalked; stem with milky juice. Fruit paired, cylindrical *c*. 12 cm long.

The root, bark and latex are used medicinally.

ASCLEPIADACEAE   Milkweed Family

A fairly large family of herbaceous plants, shrubs, trees, and woody climbers, of the Tropical and Sub-tropical Zones, usually with milky latex. Leaves usually opposite or whorled, commonly simple and entire. Inflorescence flat-topped or spike-like, arising from the side of the leaf-axil. Sepals 5, fused below; petals 5, fused below into a short tube; stamens 5, often united into a fleshy column, anthers fused to style. The petals and stamens may bear additional appendages forming a single or double *corona*. Ovary superior, of 2 almost separate carpels, each with a style but with a common 5–lobed stigma. Fruit usually of paired many-seeded carpels, though only one may develop; seeds flattened, with a long crown of silky hairs.

Flowers with corolla-lobes spreading
  a Climbing plants, or epiphytes
    i Flowers white or yellow   1 CYNANCHUM   2 HOYA
    ii Flowers orange   3 MARSDENIA   2 HOYA
  b Shrubs or herbaceous perennials   4 CALOTROPIS
                                                      5 VINCETOXICUM

Flowers with corolla-lobes attached by their tips   6 CEROPEGIA

1 CYNANCHUM Stranglewort Climbing shrubs. Corona variable, either a ring, cup-shaped, or 5–lobed. 2 spp.

914 **C. auriculatum** Wight
Pakistan to S.W. China. 1800–3600 m. Climber on shrubs. Jun.–Jul.
   A slender climbing shrub, with numerous small greenish-white flowers sometimes tinged with pink, in long-stalked branched rounded clusters, and with ovate pointed very deeply heart-shaped leaves. Flowers 8–12 mm across, with 5 narrow lanceolate spreading lobes, and a much shorter deeply 5–lobed cup-shaped corona; calyx-lobes lanceolate, hairy. Leaves with blade 5–12 cm and leaf-stalk nearly as long, hairless or nearly so. Carpels cylindrical straight, *c*. 10 cm.

2 HOYA Epiphytic or climbing shrubs. Corona of 5 fleshy horn-like processes spreading in a star. 7 spp.

915 **H. fusca** Wallich
C. Nepal to S.E. Tibet. Burma. S. China. 900–2400 m. Forest epiphyte. Aug.–Sep.

An epiphyte with axillary umbels of tubular cream, yellow to pale orange flowers with spreading lobes, and with large narrow-oblong long-pointed leaves. Flowers with slender corolla-tube to 2.5 cm, and triangular lobes spreading to 1.3 cm, and with 5 short thick blunt coronal lobes with out-curved conical often pink horns. Leaves 10–23 cm by 3–5 cm broad, hairless, with thick fleshy mid-vein. Carpels slender 10–13 cm, straight.

916 **H. lanceolata** Wallich ex D. Don
Uttar Pradesh to Bhutan. Burma. 1000–1800 m. Forest epiphyte. May–Jun.

An epiphyte with long pendulous branches with small fleshy nearly stalkless closely-set leaves, and with terminal short-stalked umbels of 6–10 fleshy tubular white flowers with pink centres. Flowers stalkless, 12 mm across, with triangular spreading lobes; corolla-tube slender to 1.5 cm; lobes of corona short, thick and blunt; sepals hairy. Leaves lanceolate to ovate-lanceolate, 2.5–5 cm. Carpels usually hairy, slender 13–15 cm. **Pl. 85**.

3 **MARSDENIA** Climbing shrubs. Corona of 5 slender horn-like processes longer than the stamens. 5 spp.

917 **M. roylei** Wight
Pakistan to Arunachal Pradesh. Burma. 900–2400 m. Climber on shrubs. May–Jun.

A softly hairy climbing shrub, with heart-shaped leaves, and with small orange flowers crowded into branched flat-topped stalked axillary clusters 2.5–4 cm across. Flowers *c.* 6 mm across, corolla-lobes 5 fleshy, spreading diagonally from the centre, densely hairy within except on margin; lobes of corona linear erect, much longer than the anthers, with converging tips; calyx and flower-stalks with a dense felt of hairs. Leaves stalked, blade 6–15 cm, acute. Carpels hairy, deeply wrinkled, *c.* 8 cm by 2.5–4 cm, beaked. **Pl. 85**.

4 **CALOTROPIS** Shrubs. Corona of 5 fleshy laterally flattened horns radiating from the stamen column. 3 spp.

918 **C. gigantea** (L.) Dryander
India. China. S.E. Asia. Throughout Himalaya to 1200 m. Wasteland, river gravel; common in lower Nepal valleys. Mar.–Oct.

A small shrub with large opposite leathery leaves which are white-cottony beneath, and with large simple or compound axillary clusters of dull mauve, purple, or white flowers. Inflorescence white-cottony; flowers bell-shaped 1.2–2.5 cm across, with reflexed ovate-lanceolate corolla-lobes 1.5 cm, hairy on the outside; horns of corona hairy, up-curved.

Leaves ovate to oblong with heart-shaped base, 10–20 cm; young stems covered with adpressed white-cottony hairs, and exuding white latex when damaged. Carpels paired, curved, 8–10 cm by 2 cm broad; seeds flattened, with long bright silky-white hairs.

Hairs of seeds used for stuffing; the fibre of stems used for fishing-lines and nets. The juice and roots are used medicinally. **Pl. 85**.

5 VINCETOXICUM Herbaceous perennials. Corona deeply 5–lobed. 1 sp.

919 **V. hirundinaria** Medicus (*Cynar.chum vincetoxicum* (L.) Pers.)
Pakistan to S.E. Tibet. W. Asia. Europe. 2300–3600 m. Open slopes, wasteland. May–Jul.

An erect perennial with stems 15–25 cm, with broadly elliptic to ovate acute leaves, and with terminal and axillary umbels of small yellowish-green flowers. Corolla *c.* 5 mm across, with nearly hairless triangular lobes; corona deeply 5–lobed; calyx-lobes triangular, nearly as long as corolla. Leaves opposite, *c.* 5 cm, short-stalked, often with a rounded base; lower leaves often blunt. Carpels 4–6 cm, cylindrical, broader at base and tapering gradually to apex.

A poisonous plant not grazed by animals. **Page 486**.

6 CEROPEGIA Climbing or terrestrial herbaceous perennials. Flowers very distinctive, having a long tubular corolla-tube with a swollen base, and with up-curved lobes which remain attached at their tips. 6 spp.

Climbing plants

920 **C. longifolia** Wallich
Himachal Pradesh to Arunachal Pradesh. Burma. 1000–2400 m. Climber on shrubs. Aug.

Flowers 3–4 cm long, few borne in stalked axillary clusters, green spotted with purple. Corolla-tube long and curved, with broad ovate bristly-hairy lobes attached by their tips; lobes of corona 10, hairy, lanceolate, with linear horns. Calyx-lobes linear pointed. Leaves linear-lanceolate long-pointed, entire, 12–15 cm, short-stalked; branches climbing. Carpels narrow-cylindrical, to 15 cm.

921 **C. pubescens** Wallich
C. Nepal to S.W. China. Burma. 1000–2700 m. Climber on shrubs. Jul.–Aug.

Flowers green spotted with purple, in stalked axillary clusters. Corolla-tube curved 3–4 cm long, and lobes elliptic at the base with long linear tips cohering from the middle, and about one third as long as the tube. Leaves long-stalked, ovate long-pointed 8–18 cm, very membraneous, sparsely hairy, or nearly hairless; branches slender climbing, nearly hairless. **Page 486**.

Plants not climbing

922 **C. hookeri** C. B. Clarke ex Hook.f.

W. Nepal to S.E. Tibet. 2300–3600 m. Grassy slopes. Jun.–Jul.

Flowers dark purple to greenish-purple, 2–3 cm, stalked, usually 1–2 borne from the leaf-axils. Corolla-tube straight, with swollen base, and narrowed above to the linear to lanceolate curved lobes which are attached by their tips and which are one third as long as the tube. Leaves ovate long-pointed, 1–5 cm, stalked; stem slender, minutely hairy; a straggling herb to c. 30 cm, with long white suckers.

LOGANIACEAE Buddleia Family

A rather small family of trees, shrubs, and climbers, mainly of the Tropical and Sub-tropical Zones, but also of the Temperate Zone. Leaves opposite, entire. Flowers in terminal clusters, or rarely solitary. Sepals 4–5, fused into a tube below; corolla tubular with 4–5 lobes, but lobes sometimes more; stamens 4–5, attached in a ring to the corolla-tube. Ovary superior, of 2 fused carpels, with many ovules; style 2–lobed. Fruit a capsule or a fleshy berry.

BUDDLEJA 5 spp.

Flowers large, 2 cm or more long, funnel-shaped

923 **B. colvilei** Hook.f. & Thoms.

E. Nepal to Arunachal Pradesh. 2100–3600 m. Forests, shrubberies; not common in Nepal. Jun.–Jul.

A large shrub or small tree, with very showy long drooping terminal spike-like clusters of large deep crimson to pink flowers with white centres. Corolla funnel-shaped c. 2.5 cm, with short spreading rounded lobes; calyx broadly bell-shaped, c. 5 mm, with short hairy lobes. Leaves elliptic-lanceolate, toothed, with rusty hairs beneath, or nearly hairless; blade 15–18 cm, narrowed to a short stalk. Capsule woolly-haired, 1–2 cm, splitting into 2 valves.

Flowers small, less than 1 cm long, not funnel-shaped

924 **B. paniculata** Wallich

W. Nepal to Bhutan. 900–2400 m. Forests, shrubberies; common. Feb.–May.

A shrub or small tree with peeling bark, and with fragrant pale mauve, pink or white flowers in short dense spikes arranged in large terminal leafy branched clusters. Flowers to 8 mm long; corolla-tube very densely woolly-haired, with orange throat, with 4 spreading rounded lobes; calyx urn-shaped, woolly-haired, with rounded lobed. Leaves very variable, mostly 10–18 cm, those at ends of branches oblong-lanceolate entire, those lower down arrow-shaped and deeply toothed; young branches and leaves very densely covered with red-brown hairs. Capsule woolly-haired, *c*. 4 mm.

925 **B. asiatica** Lour.

Pakistan to Bhutan. C. & S. China. India. Burma. S.E. Asia. 300–1800 m. Shrubberies, open slopes; common. Nov.–Apr.

An evergreen shrub or small tree, with fragrant white flowers in long dense cylindrical terminal and axillary clusters to 15 cm. Flowers tubular, 4–lobed, the lobes *c*. 6 mm long and wide, corolla-tube at least twice as long as the densely hairy calyx. Branches, undersides of leaves, and inflorescence covered with a dense grey felt of star-shaped hairs. Leaves lanceolate, mostly 6–13 cm, entire or minutely toothed, usually hairless above. Capsule hairless, *c*. 4 mm.

926 **B. crispa** Benth. (*B. tibetica* W. W. Smith, incl. var. *grandiflora* Marquand)

Pakistan to S.W. China. 1200–4000 m. Shrubberies, open slopes; common in W. Himalaya. Apr.–Jun.

A very variable densely woolly-haired shrub, with interrupted spikes of dense clusters of small fragrant purple flowers (or at higher altitudes, frequently with tight heads of flowers), often flowering before the leaves develop. Flowers with slender corolla-tube *c*. 8 mm long and with short rounded lobes with undulate margins, throat white; calyx very densely covered with star-shaped hairs. Leaves variable 4–12 cm, lanceolate to ovate, sometimes narrowed abruptly and with an enlarged base, conspicuously toothed or not, densely hairy on both sides with star-shaped hairs, silvery-grey when young; stems 3–4 m. Fruit woolly, *c*. 5 mm. **Page 486**.

GENTIANACEAE   Gentian Family

A cosmopolitan family of herbaceous plants, particularly characteristic of mountains and northern regions, marshes, and salt-rich areas. Leaves usually opposite, entire, without stipules. Flowers usually in branched clusters, or solitary. Calyx and corolla 4–5–lobed, each fused at their bases into a tube; corolla often bell-shaped or saucer-shaped; stamens attached to corolla-tube, and equal in number to corolla-lobes. Ovary superior, of 2 fused carpels, usually with one chamber with many ovules which are borne on the walls of the ovary; style simple, stigma 2–lobed. Fruit usually a capsule.

(Petals are referred to as corolla-lobes, sepals as calyx-lobes in the following descriptions.)

Corolla-tube shorter than lobes

  a Ovary 2–chambered; stigma entire   1 EXACUM
  b Ovary 1–chambered; stigma 2–lobed   2 HALENIA
                       3 LOMATOGONIUM   4 SWERTIA

GENTIANACEAE

Corolla-tube longer than lobes

a Twining plants   5 CRAWFURDIA

b Herbaceous annuals and perennials   6 GENTIANA

      7 GENTIANELLA   8 MEGACODON   9 JAESCHKEA

1 EXACUM   Corolla-tube inflated, much shorter than the lobes. Ovary 2–chambered; stigma entire. 2 spp.

927   **E. tetragonum** Roxb.

Uttar Pradesh to Bhutan. India. S. China. S.E. Asia. 1000–1800 m. Open slopes, damp places. Aug.–Oct.

Flowers blue, in terminal spike-like clusters borne on branching stems; leaves opposite, clasping at base and broadly lanceolate. Flowers 3 cm across, with 4 spreading ovate acute lobes much longer than the corolla-tube, and with long-exserted anthers and style; calyx with 4 ovate long-pointed keeled lobes; bracts lanceolate, the lower larger. Leaves 3.5–12.5 cm, usually conspicuously 3–veined; stem 4–angled, erect 30–120 cm. Capsule globular, with persistent calyx and corolla. **Pl. 88**.

2 HALENIA   Corolla-lobes spurred at the base, corolla-tube very short. Ovary 1–celled; stigma 2–lobed. 1 sp.

928   **H. elliptica** D. Don

Kashmir to S.W. China. 1800–4500 m. Forests, open slopes, damp places. Jul.–Oct.

Flowers small, pale blue, readily distinguished by the 4 short blue spurs projecting backwards and outwards beyond the calyx. Flowers in axillary and terminal branched clusters; corolla *c.* 8 mm across, bell-shaped, deeply 4–lobed, spurs to 7 mm; calyx about one third as long as corolla and with 4 ovate-lanceolate lobes; stamens 4. An erect hairless plant, with 4–angled narrowly winged stems, and opposite narrow-elliptic stalkless leaves 2.5–5 cm; stem 15–60 cm, branched above. Capsule ovoid. **Page 487**.

3 LOMATOGONIUM   Annuals. Corolla-tube very short, with 5 spreading lobes and with distinctive fringed nectaries; stamens 5. Capsule cylindrical. 8 spp.

929   **L. carinthiacum** (Wulfen) Reichb.

Afghanistan to S.W. China. W. & N. Asia. Europe. 3000–4800 m. Open slopes. Aug.–Oct.

A small annual with long-stalked blue flowers in a lax cluster, each flower with 5 elliptic spreading corolla-lobes with small pale fringed nectaries at their bases. Corolla-lobes *c.* 1.5 cm, green-veined; calyx-lobes narrow-elliptic, unequal, half as long as corolla, calyx-tube very short; stamens and ovary blue, projecting. Basal leaves elliptic *c.* 1.5 cm, persisting or not at flowering, stem-leaves elliptic, smaller, stalkless; stems solitary or several branching from the base, to 15 cm. **Pl. 87**.

930 **L. caeruleum** (Royle) Harry Smith ex B. L. Burtt
Pakistan to Kashmir. 3300–4000 m. Open slopes. Aug.–Sep.
Like 929, but a perennial with oblanceolate to spathulate lower stem leaves, and with smaller flowers with corolla-lobes to 1.3 cm, or often less. Flowers mauve to violet, 1–3 on most stems. Basal leaves to 5 cm, narrowed to a stalk, upper leaves lanceolate, stalkless or nearly so; stems several 10–20 cm. **Page 487**.

4 SWERTIA Calyx and corolla both with very short tubes and with 4–5 lobes free almost to the base; corolla-lobes each with 1 or 2 usually fringed nectaries at their base; stigma 2–lobed. Capsule oblong. c. 28 spp.

Perennials; corolla-lobes usually more than 1 cm long
a *Stems numerous, usually 12 cm long or less*

931 **S. multicaulis** D. Don
C. Nepal to S.E. Tibet. 4000–4800 m. Open slopes. Jun.–Sep.
Distinguished from other species by its many short spreading stems which arise directly from the stout rootstock. Flowers slaty-blue, long-stalked, in a much-branched inflorescence; corolla-lobes 8–13 mm, blunt, with a single nectary at the base fringed with hairs; calyx-lobes narrow-oblong, 6–8 mm. Leaves narrowly spathulate c. 5 cm, narrowed to a long winged leaf-stalk; stems numerous, branched, 5–12 cm.

b *Stem erect, solitary, branched above, usually 15 cm or more*
i *Corolla-lobes usually less than 1.3 cm*

932 **S. cuneata** D. Don
Uttar Pradesh to S.E. Tibet. 3600–5000 m. Alpine slopes. Aug.–Oct.
An erect perennial with stem mostly 10–25 cm, with lurid blue flowers 2.5–3.5 cm across, and with spathulate stalked leaves. Corolla-lobes 5, narrow-elliptic, with 2 linear basal nectaries surrounded by long hairs, or hairless; calyx-lobes narrow-oblong to spathulate c. 1 cm; flowers long-stalked. Leaves 3.5–7.5 cm., the lower long-stalked, the upper nearly stalkless. Capsule c. 2 cm, narrowly oblong.

933 **S. petiolata** D. Don
Afghanistan to W. Nepal. S.E. Tibet. 3300–4500 m. Open slopes; common in Kashmir. Jul.–Aug.
Like 935 with lurid grey, or nearly white flowers with blue-green veins, but flowers smaller, and seeds not winged. Flowers in lax spike-like terminal cluster; corolla-lobes narrow-elliptic c. 1 cm, with 1 or 2 basal lanceolate acute fringed nectaries. Basal leaves lanceolate, long-stalked and clasping the stem at the base, uppermost leaves narrow-lanceolate acute, stalkless; stem erect, usually 20–60 cm. **Pl. 88**.

ii *Corolla-lobes 1.5 cm or more*

934 **S. hookeri** C. B. Clarke
E. Nepal to S.E. Tibet. 3600–4300 m. Open slopes. Jun.–Sep.

A very robust erect plant, with maroon and darker-veined, bell-shaped flowers with ascending ovate corolla-lobes, borne in dense short-stalked whorls and forming a terminal interrupted spike-like cluster. Corolla-lobes *c.* 1.8 cm, blunt, with 1 large basal nectary without a fringe; calyx-lobes ovate acute, green, more than half as long as corolla. Basal leaves spathulate-elliptic 15–20 cm, stem leaves ovate, the uppermost purple-flushed and little longer than the whorls of flowers. Stem stout, hollow, to 1 m or more; rootstock stout. Capsule shining brown, to 2 cm. **Pl. 88**.

935 **S. speciosa** D. Don (*S. perfoliata* Royle ex G. Don)
Pakistan to Bhutan. 2700–4000 m. Open slopes. Jul.–Sep.

Flowers lurid grey, with 5 narrow elliptic spreading corolla-lobes each with paired long-fringed nectaries at its base. Flowers 2–2.5 cm across, in a long narrow terminal cluster; corolla-lobes 1.5–2 cm long; calyx-lobes narrow elliptic acute, toothed, much overlapping at base, to 1.5 cm. Stem leaves opposite, with elliptic stalkless blades 12–15 cm, basal leaves stalked and clasping stem; stem rather thick, hollow, mostly 60–140 cm.

936 **S. alternifolia** Royle
Uttar Pradesh to C. Nepal. 3000–4000 m. Shrubberies, alpine slopes. Aug.–Oct.

Distinctive in having a few large pale yellow flowers with blue stamens, borne in a lax branched cluster. Flowers spreading to 2.5 cm; corolla-lobes narrow-elliptic pointed, *c.* 1.8 cm, with paired, fringed nectaries; calyx-lobes lanceolate *c.* 1 cm, enlarging to 2 cm in fruit. Leaves elliptic pointed, the upper only 2–3, clasping, much shorter than the lower; basal leaves with blade to 10 cm; stem slender, branched only above and with terminal flowers, 15–25 cm. **Pl. 86**.

Annuals; corolla-lobes less than 1 cm long

937 **S. angustifolia** Buch.–Ham. ex D. Don
Pakistan to S.E. Tibet. Burma. 600–2700 m. Cultivated areas, grazing grounds, shrubberies; very common. Sep.–Oct.

A slender erect branched annual, with small white or pale blue flowers with darker dots; corolla-lobes oblong pointed 6–9 mm, with one green rounded basal nectary. Calyx with 4 lobes as long or a little longer than the corolla. Leaves linear-lanceolate acute, 5.5 cm; stem branched above with angled narrowly winged branches, usually 15–60 cm. Capsule ovoid.

937a **S. racemosa** (Griseb.) C. B. Clarke
W. Nepal to S.E. Tibet. 3300–5000 m. Open slopes, shrubberies; common. Aug.–Oct.

A rather robust annual, branched above and with a lax cluster of many small lilac or pale blue flowers. Corolla-lobes oblong acute, 6–8 mm, with a single basal nectary covered with a scale with linear-lanceolate marginal teeth, not hairy; calyx with a distinct tube at the base and widely spaced linear very acute lobes, 3–4 mm. Leaves elliptic-lanceolate, 3–veined, to 4 cm; stem 4–lined, to 45 cm.

5 CRAWFURDIA Herbaceous climbers, with large tubular corollas; calyx tubular 5-lobed. Seeds disk-like, broadly winged. 1 sp.

938 **C. speciosa** Wallich
C. Nepal to S.E. Tibet. Burma. 2400–3300 m. Climber on shrubs; not common. Sep.–Oct.

Flowers blue-purple, large drooping; corolla tubular-bell-shaped, narrow cylindrical below and enlarging to twice its diameter or more above, and with 5 triangular widely spaced spreading lobes. Flowers borne in axillary stalked clusters of 1–3; corolla to 5 cm; calyx *c.* 1.5 cm, with 5 tiny triangular out-curved lobes. Leaves opposite, elliptic long-pointed, short-stalked, mostly 5–8 cm, strongly 3–veined; stems slender, twining, to 2–3 m. Capsule to 4 cm, with conspicuous bi-lobed style.

(Some botanists place this species in *Gentiana*) **Pl. 87**.

6 GENTIANA Corolla with a funnel-shaped to cylindrical tube much longer than the lobes. Lobes often with folds or smaller *lobules* in their angles; calyx tubular, usually 5–lobed; nectaries at base of ovary; stigmas 2, persistent. *c.* 60 spp., mostly autumn-flowering.

Flowers 2 cm or more long
a *Flowers usually 2–4 in branched clusters*
  i *Leaves spathulate, lanceolate to linear*

939 **G. algida** Pallas (*G. nubigena* Edgew.)
Kashmir to S.W. China. N. Asia. 4000–5600 m. Open slopes, drier areas. Aug.–Oct.

Flowers yellowish-white spotted with blue, and blue-ribbed outside, 1–3 in terminal and axillary stalked clusters, borne on erect stems 7.5–15 cm. Corolla 3.5–4.5 cm, funnel-shaped, with erect triangular widely spaced lobes with or without distinctive lobules; calyx tubular *c.* 2 cm long, with purple-tinged unequal oblong widely separated lobes. Basal leaves spathulate to linear-oblong blunt, thick and fleshy, 2.5–5 cm; stem leaves of several pairs, lanceolate, shorter; rootstock often covered with old leaf-bases. **Page 488**.

940 **G. kurroo** Royle
Pakistan to Uttar Pradesh. 1800–2700 m. Grassy slopes. Aug.–Oct.

Flowers deep blue, paler in the throat and spotted with green and white, usually 2 or more on each stem, but sometimes solitary. Corolla narrow funnel-shaped to 5 cm, with ovate acute spreading lobes and small triangular lobules; calyx tubular, with narrow linear lobes 8–12 mm one-half to two-thirds as long as the calyx-tube. Basal leaves lanceolate, usually 10–12 cm, stem leaves 2–3 pairs, narrow linear, to 2 cm; stems 5–30 cm, several, unbranched, arising from the stout rootstock. **Page 488**.

GENTIANACEAE

ii *Leaves ovate*

941 **G. cachemirica** Decne.
Pakistan to Kashmir. 2400–4000 m. Always on rocks; common in Kashmir. Aug.–Oct.

A tufted perennial growing on rock ledges, with many leafy stems each with few or solitary pale blue funnel-shaped terminal flowers with spreading lobes and with distinctive short triangular toothed lobules. Corolla-tube long slender 2.5 cm or more, dark purple outside, and with roundish-ovate to elliptic pointed lobes; calyx tubular *c.* 1.3 cm, with elliptic acute lobes. Stem leaves glaucous, ovate 8–18 mm; stems spreading 10–15 cm. **Pl. 86**.

b *Flowers several to many in a dense head*

942 **G. tianschanica** Rupr. ex Kusn.
Pakistan to Uttar Pradesh. C. Asia. 3000–4500 cm. Alpine slopes. Jul.–Sep.

Flowers with a purple corolla-tube and spreading blue triangular lobes, distinctive in being clustered together in a dense terminal rounded head, with long spreading bracts. Corolla to 2.5 cm, tubular-funnel-shaped, with triangular entire or 2–lobed lobules alternating with the lobes; calyx divided to halfway or more, with unequal linear lobes. Basal leaves linear-lanceolate to 7.5 cm or more, stem-leaves 2–4 pairs, oblong 2.5–5 cm; bracts elliptic, broader than the stem-leaves; stem erect or spreading, 10–25 cm. **Pl. 88**.

943 **G. tibetica** King ex Hook.f.
W. Nepal to S.E. Tibet. 3700–4500 m. Open slopes, drier areas. Jul.–Sep.

A robust plant with many large greenish-white flowers in a dense terminal cluster encircled by an involucre of longer uppermost leaves. Corolla tubular-funnel-shaped 2.5–3 cm long, with triangular-ovate lobes alternating with short triangular lobules; calyx papery, *c.* 12 mm, lobes minute. Basal leaves broadly lanceolate to 30 cm, narrowed below to a winged leaf-stalk; stem-leaves narrower, fused below and encircling the stem; stem robust, 20–90 cm.

c *Flowers solitary* (see also 940)
  i *Flowering stems erect*

944 **G. phyllocalyx** C. B. Clarke
C. Nepal to S.W. China. 3600–5500 m. Open slopes, rhododendron shrubberies. Jun.–Aug.

Flower pale to deep blue with corolla-tube 2.5–3 cm, inflated in the middle and with erect broad-triangular lobes with short points alternating with irregular triangular lobules. Calyx tubular *c.* 12 mm, with oblong acute widely separated lobes half as long. Basal leaves in a rosette, broadly ovate to rounded *c.* 2 cm; stem with several pairs of shorter broader rounded leaves, the upper encircling the base of the corolla-tube, all leaves

270

thick and fleshy; stem unbranched, 2–10 cm. Capsule long-stalked, carried above the persistent perianth. **Pl. 87**.

945 **G. tubiflora** (G. Don) Griseb.
Himachal Pradesh to S.E. Tibet. 4000–5000 m. Screes, open slopes. Jul.–Sep.

Flower solitary, deep blue, narrow tubular to 2.5 cm, with erect triangular to ovate lobes and short entire lobules. Calyx about half as long as corolla, with erect, short, narrow-ovate lobes. A tufted cushion-forming leafy perennial, with tiny spathulate to oblong leaves in rosettes 1–2 cm across; leaves to 7 mm long; stems erect, leafy, 2.5–5 cm. Capsule cylindrical *c.* 1 cm, with long style, borne on a long stalk projecting from the persistent perianth. **Pl. 87**.

946 **G. depressa** D. Don
C. Nepal to S.E. Tibet. 3300–4300 m. Open slopes. Sep.–Nov.

A low tufted plant with spreading stems, and with solitary pale blue to greenish-blue large broadly bell-shaped flowers with triangular blue lobes alternating with nearly as long broadly triangular, toothed paler lobules. Corolla-tube 2–3 cm, greenish-white streaked with purple outside, and dotted with purple within; calyx tubular, with ovate acute, erect, white-margined lobes *c.* 6 mm, as long as tube. Leaves on barren stems small, on flowering stems broadly ovate, overlapping, encircling the calyx; rosettes lax, 1–2 cm across, somewhat glaucous. **Pl. 87**.

947 **G. urnula** Harry Smith
E. Nepal to S.E. Tibet. 4500–6130 m. High altitude screes, rocks. Sep.–Oct.

A low matted perennial forming clumps, distinguished by its broadly ovate tightly overlapping leaves, and its solitary broad tubular flowers with dark blue tubes, and blue rounded lobes alternating with pale rounded fringed lobules. Corolla to 2.5 cm long; calyx with large rather rectangular lobes, winged on the mid-vein and much shorter than the corolla-tube. Upper leaves very broadly obovate and surrounding base of flowers, lower leaves smaller, to 5 mm; stems very short. Fruiting capsule 2.5 cm, borne on long stalks to twice as long as the persistent perianth. **Page 488**.

ii *Tufted plants with short spreading flowering stems* (see also 947, 946)

948 **G. ornata** (G. Don) Griseb.
C. Nepal to S.W. China. 3600–5500 m. Open slopes. Sep.–Nov.

Flower solitary, widely bell-shaped 2.5–3.5 cm long, with broadly triangular or ovate blue lobes and ragged lobules nearly as large. Corolla-tube with broad bands of purplish-blue with creamy-white between; calyx bell-shaped *c.* 12 mm, with narrow linear acute lobes as long as the tube. A low matted plant with leaves in rosettes, and with spreading leafy stems to 10 cm, ascending at the tips. Leaves linear acute, the basal leaves 2.5 cm, the stem leaves 1.5 cm. **Pl. 87**.

949 **G. prolata** Balf.f.
C. Nepal to S.E. Tibet. 3300–5500 m. Open slopes. Aug.–Oct.

Like 948, but differing in its smaller broader elliptic stem-leaves 6–10 mm. Flowers blue streaked with yellowish-white; lobes of corolla erect, broadly triangular acute, lobules short, toothed; calyx reddish, with oblong lobes half as long as the tube; stems many, spreading to 10 cm. Stalk of capsule about twice as long as persistent perianth.

950 **G. venusta** (G. Don) Griseb.
Pakistan to C. Nepal. 3000–5800 m. Open slopes, peaty ground. Aug.–Oct.

A tufted perennial with blue tubular, usually solitary flowers, with roundish-ovate lobes and smaller irregularly toothed lobules. Flowers 2–2.5 cm long; corolla-tube yellow; calyx 6 mm, with oblong acute lobes with wide pale folds between, nearly as long as the calyx-tube. Leaves of sterile stems broadly spathulate c. 6 mm, on flowering stems broader, to 10 mm long; stems several, spreading to 2–8 cm, up-turned and leafy at apex and bearing solitary flowers. **Page 488**.

951 **G. stipitata** Edgew.
Uttar Pradesh to C. Nepal. 3600–4500 m. Stony slopes. Aug.–Oct.

A tufted plant with a few short spreading stems, and terminal pale blue to pale mauve tubular flowers, often with darker lines in the tube within. Flowers c. 2.5 cm long, with broad fine-pointed lobes and with blunt lobules a little shorter; calyx with spathulate lobes c. 6 mm. Stem-leaves elliptic c. 8 mm, basal leaves in rosettes, elliptic pointed, the outer to 1 cm; stems spreading, mostly 3–5 cm.

Flowers small, less than 1.5 cm long
a *Leaves broadly ovate*

952 **G. capitata** Buch.-Ham. ex D.Don
Pakistan to S.E. Tibet. 1500–4500 m. Forest clearings, shrubberies; common. Dec.–Apr.

Distinctive in having green, broadly ovate leaves closely clustered under the flower-head, and stem usually leafless below. Flowers blue or white, c. 6 mm long, crowded in a rounded terminal leafy head 1.3–2 cm or more across; calyx nearly as long as corolla and with triangular acute lobes. Leaves c. 6 mm, basal leaves soon disappearing; stem erect, usually simple 1.5–10 cm. **Page 487**.

b *Leaves elliptic to lanceolate*
i *Plants with densely clustered branches, or stems erect simple*

953 **G. argentea** (D. Don) C.B. Clarke
Pakistan to C. Nepal; common. 2100–4300 m. Open slopes, shrubberies. Apr.–Aug.

A tiny annual, distinguished by its shining silvery, narrow curved, finely pointed leaves, and its small blue flowers in dense terminal leafy heads.

Flowers stalkless, *c.* 8 mm long, with 5 triangular spreading lobes and shorter blunt lobules; calyx nearly as long as corolla, lobes finely pointed. Basal leaves in a rosette, to 13 mm long, recurved at the tips, stem-leaves elliptic, shorter; stem erect, usually simple, 2.5–10 cm.

954 **G. carinata** Griseb.
Pakistan to Uttar Pradesh. 3000–4300 m. Open slopes; common in Kashmir. May–Aug.

A small compact annual, with dark blue flowers clustered at the ends of the short branches, and with spreading ovate blunt corolla-lobes with lobules as long, thus appearing 10–lobed. Corolla-tube *c.* 1.3 cm long, narrowly funnel-shaped, the mouth of the tube with scales, lobules bi-lobed; calyx with lanceolate acute erect lobes, shorter than the calyx-tube. Basal leaves oblong to lanceolate, stalkless, 2.5–3.5 cm long, stem-leaves smaller, the upper curved outwards; stem erect, branched, mostly 2–5 cm. **Pl. 89**.

955 **G. marginata** (G. Don) Griseb.
Afghanistan to Uttar Pradesh. 2700–4300 m. Open slopes. May–Aug.

Like 954, but differing in its more crowded out-curved sickle-shaped leaves, its conspicuous out-curved calyx-lobes, and its corolla-tube without scales in the throat. Flowers bright to pale blue, to 12 mm long. Plant usually densely branched and forming a nearly stemless domed cluster usually 2–5 cm across, but sometimes more.

ii *Plants with short spreading branches*

956 **G. pedicellata** (D. Don) Griseb.
Pakistan to Bhutan. India. Burma. 750–3800 m. Cultivated areas, shrubberies; common in Nepal on heavily grazed ground. Jan.–Jun.

A small much-branched annual with a lax form, with tiny pale blue flowers at the ends of terminal and lateral branches. Flowers 6–8 mm long, with shallow triangular lobes with fine points and with lobules a little shorter and broader; calyx about half as long as corolla, with out-curved lobes. Leaves lanceolate 5–10 mm, the basal leaves when present much larger and broader; stems branched above, to 8 cm but often much shorter. **Page 487**.

7 **GENTIANELLA** Like *Gentiana*, but without lobules or scales between the corolla-lobes, and usually with hairs or lobes in the throat. Annuals or biennials. 12 spp.

Flowers small, usually less than 1.5 cm long at flowering; corolla with tufts of narrow lobes in throat

957 **G. falcata** (Turcz. ex Karelin & Kir.) Harry Smith
Pakistan to S.W. China. 3600–5200 m. Open slopes, dry areas. Aug.–Sep.

A small annual, branched from the base, and with deep blue-violet flowers with a conspicuous tuft of many white linear lobes projecting from the throat of the corolla. Corolla variable 1–2 cm long, corolla-lobes ovate, usually 5, spreading; calyx with nearly equal linear-lanceolate spreading lobes as long as the corolla-tube. Leaves almost all basal, elliptic *c.* 1 cm; stems many, 6–10 cm, ascending and ending in a solitary flower.

958 **G. pedunculata** (D. Don) Harry Smith
Pakistan to S.E. Tibet. 3000–5000 m. Open slopes. Aug.–Oct.

Very like 957 and not easily distinguished from it, but calyx-lobes unequal, some ovate, some lanceolate, and calyx usually less than a third as long as the corolla-tube. Flowers often pale blue or nearly white, mostly 1.3–1.5 cm long.

Flowers large, 2 cm or more at flowering; corolla without lobes in throat

959 **G. moorcroftiana** (Wallich ex G. Don) Airy Shaw
Pakistan to C. Nepal. 2700–4800 m. Open slopes, irrigated areas; common in Ladakh and Lahul. Aug.–Oct.

Usually a much-branched annual, with many pale blue to dark mauve tubular flowers in a lax cluster, and with lanceolate leaves. Corolla *c.* 1.5 cm at opening but enlarging to 3 cm as the flower matures; corolla-tube funnel-shaped with a wide mouth and with conspicuous yellow anthers slightly exserted from it, corolla-lobes 5, spreading, elliptic *c.* 8 mm; calyx-tube short, the lobes linear blunt, enlarging to 1.5 cm. Leaves 1.5–3 cm, the basal leaves absent at flowering; stem branched below, mostly 5–20 cm. **Pl. 89**.

960 **G. paludosa** (Hook.) Harry Smith
Pakistan to S.W. China. 3000–4500 m. Open slopes, alpine shrubberies. Jul.–Aug.

A slender erect annual, with usually solitary long-stemmed blue to white erect or spreading flowers, with a long pale corolla-tube to 3.5 cm, and 4 rounded to broadly elliptic blue spreading lobes which have conspicuous finely toothed margins at their bases. Throat of corolla without hairs or scales and stamens included; calyx green, enlarging to 3 cm, with narrow-elliptic erect lobes shorter than the calyx-tube, and with a winged mid-vein; corolla lengthening to 6 cm in fruit. Leaves narrow-elliptic blunt 2–4 cm; stems to 30 cm. **Pl. 88**.

8 **MEGACODON** Like *Gentiana*, but with a long style. Flowers and fruit large. 1 sp.

961 **M. stylophorus** (C. B. Clarke) Harry Smith (*Gentiana s.* C. B. Clarke)
E. Nepal to S.W. China. 3300–4300 m. Shrubberies, open slopes. Jun.–Jul.

A robust plant with large elliptic conspicuously 5–veined entire leaves, and with large pale yellow bell-shaped flowers, netted with green within.

Flowers few, short-stalked and often paired from the axils of the upper leaves. Corolla 6–8 cm long and broad, with a short wide tube and much longer broadly ovate, overlapping, ascending lobes with out-curved apices; calyx-lobes to 2 cm. Lower leaves long-stalked, blade elliptic to 30 cm, the upper smaller opposite and shortly fused at the base; stems stout, hollow, 1–2 m. Capsule cylindrical to 5 cm, with persistent style. **Pl. 89**.

9 JAESCHKEA Like *Gentiana* but differing in its capsule which is separated from the base into 2 carpels each with half of the style; stamens arising from the mouth of the corolla-tube and alternating with the corolla-lobes. 2 spp.

962 **J. canaliculata** (Royle) Knobloch
Pakistan to Kashmir. 2700–4000 m. Shrubberies, open slopes; common in Kashmir. Jul.–Sep.

An erect annual, with many small pale mauve flowers, in axillary widely spaced clusters borne in a terminal lax inflorescence, and with narrow clasping leaves. Flowers opening when about 6 mm, enlarging gradually to 13 mm; corolla-tube globular, the lobes triangular, pressed together, not spreading, style later exserted; calyx-lobes unequal, ovate or rounded with a fine point, *c.* 8 mm. Leaves oblong and narrrowed to a long apex, 2–5 cm; stem 20–50 cm.

963 **J. oligosperma** (Griseb.) Knobloch
Afghanistan to Kashmir. 2700–4300 m. Meadows, open slopes. Aug.–Sep.

Very like 962, but differing in its lanceolate calyx-lobes (obovate or rounded in 962), and its smaller narrower lanceolate blunt leaves. Flowers bluish- to reddish-purple, in a spike-like cluster, the lower axillary flowers widely spaced. Leaves mostly 1–3 cm, stalkless. A smaller plant, usually 10–20 cm. **Pl. 88**.

MENYANTHACEAE   Bogbean family

Often included in the Gentianaceae, but differing in being largely aquatic plants with underwater creeping and rooting stems, and having alternate simple, or compound leaves. A small more or less cosmopolitan family.

NYMPHOIDES Fringed water-lily 3 spp.

964 **N. peltata** (S. Gmelin) Kuntze (*Limnanthemum p.* S. Gmelin)
Temperate Eurasia. To 1500 m. Shallow waters, ditches; common in Kashmir valley. May–Jun.

A submerged aquatic plant, with rounded deeply heart-shaped floating leaves, and with bright yellow flowers borne above water, with rounded petals with fringed margins. Flowers 3–4 cm across, stalked, in axillary clusters of 2–5; calyx-lobes oblong-lanceolate. Blade of leaf glossy, 3–10 cm across, with sheathing leaf-stalks; stems slender, spreading underwater to 160 cm. **Pl. 89**.

POLEMONIACEAE   Phlox Family

A small family of wide distribution, predominant in the Americas, mostly of herbaceous plants. Sepals and petals 5, each fused below into a tube; stamens 5, fused to the corolla-tube. Ovary superior, inserted on the basal disk, with 3 fused carpels; style simple, stigma 3–lobed. Fruit a capsule, opening by 3 valves.

POLEMONIUM   Jacob's ladder   1 sp.

965   **P. caeruleum** L. subsp. **himalayanum** (Baker) Hara
Pakistan to W. Nepal. 2400–3700 m. Forests, open slopes, damp places; common in Kashmir. Jun.–Sep.
A leafy perennial, with a terminal branched cluster of pale blue flowers, and with pinnate leaves with many pairs of oblong-lanceolate leaflets. Flowers with very short corolla-tube, and spreading ovate lobes to 2 cm; calyx bell-shaped with narrow-triangular hairy lobes, enlarging in fruit to 1 cm. Leaves 5–15 cm; with many pairs of leaflets 1–3 cm; stem erect, branched only in inflorescence, 30–100 cm. **Page 489**.

BORAGINACEAE   Borage Family

A large family mostly of herbaceous plants but with some trees and shrubs, occurring throughout the Temperate and Sub-tropical Zones, with a high concentration in the Mediterranean region. Plants usually rough-hairy, with simple, usually entire, alternate leaves, without stipules. Inflorescence characteristic, consisting of often paired outward-coiled branches with flowers on the upper side which open progressively from the base to the apex of each branch. Sepals 5, fused into a tube below, or sometimes free; corolla-tube saucer-shaped or bell-shaped, 5–lobed, often with scales in the mouth; stamens 5, attached to corolla-tube. Ovary superior, of 2 fused carpels with 2 or 4 chambers, with 1 ovule in each chamber, style usually simple. Fruit usually of 4 nutlets, or a drupe.

Stamens longer than corolla   1 SOLENANTHUS
Stamens shorter than corolla
  a Flowers yellow or orange   2 ARNEBIA   3 ONOSMA
  b Flowers purple or red   2 ARNEBIA   3 ONOSMA
                                             4 MAHARANGA   7 LINDELOFIA
  c Flowers blue or white
    i Fruit with hooked bristles   5 CYNOGLOSSUM   6 HACKELIA
                                        7 LINDELOFIA   8 ERITRICHIUM

276

ii Fruit without hooked bristles   4 MAHARANGA
8 ERITRICHIUM 9 CHIONOCHARIS
10 PSEUDOMERTENSIA 11 MICROULA
12 MYOSOTIS 13 TRIGONOTIS

1 SOLENANTHUS Corolla-tube with 5 sac-like folds in the throat; stamens longer than corolla; nutlets with hooked spines. 1 sp.

966 **S. circinnatus** Ledeb.
Afghanistan to Himachal Pradesh. W. Asia. 2100–3000 m. Shrubberies, open slopes. May–Jun.

An erect hairy perennial to 1 m, with large oval-heart-shaped to elliptic leaves, and with a terminal cylindrical inflorescence of numerous incurved branches with many small ashy-blue to dark purplish flowers with long yellow protruding stamens. Corolla *c.* 8 mm, with triangular erect lobes; calyx densely woolly-haired, lobes elongating in fruit to 8 mm. Lower leaves long-stalked, to 30 cm; stem-leaves elliptic, clasping, becoming progressively smaller above. Branches elongating in fruit. Nutlets densely covered with hooked spines. **Pl. 90**.

2 ARNEBIA Corolla-tube without scales in the throat, but with a ring of hairs at the base; style divided into 2 or 4. Nutlets covered with small swellings. 4 spp.

Flowers purple

967 **A. benthamii** (Wallich ex G. Don) I. M. Johnston (*Macrotomia b.* (Wallich) A. DC.)
Pakistan to W. Nepal. 3000–4300 m. Open slopes, shrubberies. May–Jul.

A very striking plant, with a large very dense shaggy-haired cylindrical spike of red-purple flowers showing between much longer linear grey hairy drooping bracts, and borne on a stout leafy hairy stem. Flowers with corolla-tube *c.* 2.5 cm, and with 5 triangular lobes spreading to *c.* 10 mm; calyx-lobes linear 2.5–4 cm, tipped with purple. Leaves linear to narrow-lanceolate, to 30 cm, bristly-haired, the upper numerous; bracts smaller to 5 cm. Stem 15–60 cm; rootstock very stout, covered with bases of old leaves. Fruiting spike elongating to 30 cm. **Pl. 91**.

968 **A. euchroma** (Royle ex Benth.) I. M. Johnston
Afghanistan to C. Nepal. W. & C. Asia. 3300–4500 m. Open slopes, rocks, drier areas only. Jun.–Aug.

Differing from 967 in having rounded clusters of pale pink-purple flowers which turn blackish-purple, and with slender corolla-tubes usually longer than the subtending bracts and calyx. Corolla-tube to 1.5 cm, but often shorter, lobes spreading to *c.* 12 mm. Leaves linear, with conspicuous long bristly hairs; stem leaves many, stalkless, mostly 5–8 cm; stem one or several, to *c.* 30 cm; rootstock stout. **Pl. 91**.

Flowers yellow

969 **A. guttata** Bunge
Afghanistan to Kashmir. C. Asia. 2100–4500 m. Stony slopes in Ladakh.
Jul.–Sep.

A low tufted very bristly-haired perennial, with leafy stems arising from
a stout rootstock, and with terminal coiled inflorescences with bright
yellow tubular flowers. Corolla-tube slender to 1.5 cm, much longer than
the calyx, corolla-lobes blunt, spreading to 10 mm; calyx-lobes linear,
bristly-hairy. Leaves strap-shaped blunt, covered with bristly white hairs
with swollen bases; basal leaves mostly similar to stem-leaves; stems often
many and spreading to 15 cm. Fruiting clusters elongating; nutlets conical,
hairless.

3 ONOSMA Corolla-tube cylindrical, with 5 very short blunt lobes and
without scales in throat. Nutlets smooth, or with small swellings. 7 spp.

970 **O. bracteatum** Wallich
Uttar Pradesh to C. Nepal. 3300–5000 m. Rocky slopes, dry areas.
Jun.–Jul.

A very hairy perennial, with narrow leaves, and with dense globular
silky heads of dark red to mauve tubular flowers surrounded and almost
hidden by longer linear woolly-haired bracts. Corolla-tube *c.* 1.5 cm long,
hairy outside, with broad triangular blunt spreading lobes; calyx-lobes
linear, enlarging in fruit to 2.5 cm. Lower leaves stalked, narrow lanceolate to 15 cm, rough bristly-hairy above, paler silky-white beneath; upper
leaves smaller; stem stout erect, leafy, to 40 cm. Nutlets ovoid, rough.

971 **O. hispidum** Wallich ex G. Don
Afghanistan to Himachal Pradesh. 1000–4000 m. Shrubberies, open
slopes, rocks. Jun.–Aug.

Flowers pale yellow, long-tubular, often in forked elongated clusters,
with leaf-like bracts. Corolla to 3 cm long, with very shallow lobes; calyx
half to one third as long as corolla-tube, lobes oblong acute, bristly-haired.
A very bristly perennial with several stems 20–50 cm, with linear basal
leaves to 13 cm, and with stem-leaves often broader and shorter. Nutlets
white, smooth, shining. **Pl. 90**.

4 MAHARANGA Like *Onosma*, but corolla enlarged below and distinctly narrowed at the mouth; stamens much enlarged below. 5 spp.

972 **M. emodi** (Wallich) A. DC. (*Onosma e.* Wallich)
Uttar Pradesh to S.E. Tibet. 2200–4500 m. Rocks, open slopes. Jun.–
Sep.

Flowers ranging from red, purple, to blue on the same plant, and corolla
ovoid and distinctly narrower at the mouth, borne in rounded nodding
heads which become flat-topped in fruit. Corolla *c.* 9 mm long, lobes

triangular, recurved, style longer; calyx fused below into a short tube which widens in fruit, lobes long-bristly, little shorter than the finely hairy corolla-tube. Lower leaves oblanceolate to 15 cm, rough hairy, the upper elliptic, shorter; stems 15–45 cm. Nutlets ovoid, rough. **Page 489**.

5 **CYNOGLOSSUM** HOUND'S-TONGUE Corolla-tube short and throat closed with scales; corolla-lobes spreading. Nutlets with bristles or hooks. 8 spp.

973 **C. glochidiatum** Wallich ex Benth.
Afghanistan to Bhutan. 1200–4000 m. Cultivated areas, grazing grounds; a common weed. Jun.–Aug.

Flowers bright blue, *c.* 8 mm across, with spreading rounded lobes, many in dense one-sided elongated branches which are terminal or axillary. Corolla-tube 2–3 mm, furnished at the throat with 5 almost square scales; calyx-lobes elliptic, densely hairy. Lower leaves lanceolate to 5 cm or more, stalked, the upper stem-leaves smaller, stalkless, all covered with long bristles arising from chalky basal swellings. An erect plant with simple or branched stems, 20–40 cm. Nutlets covered with hooked bristles, which become attached to one's socks and are very difficult to remove. **Page 489**.

974 **C. zeylanicum** (Vahl) Thunb. ex Lehm.
Afghanistan to Bhutan. India. Sri Lanka, China. Japan. S.E. Asia. 1200–4000 m. Cultivated areas, wasteland; common. May–Aug.

Like 973, but a larger more branched plant with larger leaves covered with short soft adpressed hairs. Inflorescence often widely branched with many lateral branches; flowers usually bright blue, *c.* 8 mm across, with blue scales in the throat. Leaves oblong-elliptic, long-stalked, the uppermost oblong-lanceolate, stalkless; stem stout, erect, 30–90 cm. Nutlets with hooked bristles, but without distinct marginal wings as in 973.

6 **HACKELIA** Distinguished from *Cynoglossum* by its inflorescence in which the flowers are subtended by bracts, and by its nutlets which are attached to the receptacle for part of their length (nutlets attached to the receptacle for the whole of their length in *Cynoglossum*). Throat of corolla with 5 swollen scales; nutlets with hooked bristles. 6 spp.

975 **H. uncinata** (Royle ex Benth.) C. Fischer
Pakistan to S.W. China. 2700–4200 m. Forests, shrubberies; common. Jun.–Aug.

Flowers blue, to 1.3 cm across, with rounded overlapping corolla-lobes and with a raised usually yellow ring of blunt scales, borne in lax branched clusters. Leaves distinctive, broadly elliptic narrow-pointed with rounded or shallow heart-shaped base, and with 2–3 pairs of incurved lateral veins arising from near the base of the blade, all leaves stalked; stems leafy, 30–60 cm. Nutlets with long hooked bristles. **Pl. 91**.

976 **H. macrophylla** (Brand) I. M. Johnston
Pakistan to Kashmir. 2100–3000 m. Forests. May–Jun.

Distinguished from 975 by its larger white flowers often with a blue base, to 1.5 cm across, and its much larger leaves, with the lower leaves with heart-shaped blades, 10–15 cm long and broad, or sometimes larger. A leafy perennial to 1 m or more, with a lax inflorescence. Nutlets with long hooked spines.

7 LINDELOFIA Corolla-tube with scales in throat forming a cone, and with spreading lobes. Nutlets with hooked bristles. 3 spp.

Upper leaves linear to lanceolate, narrowed to the base but not clasping

977 **L. anchusoides** (Lindley) Lehm.
Afghanistan to Himachal Pradesh. 2100–3600 m. Stony slopes; common in Lahul and Ladakh. Jun.–Aug.

Flowers bright blue, funnel-shaped, with triangular blunt lobes, clustered at the ends of slender branched stems, and with narrow lanceolate, silvery-grey leaves. Corolla *c.* 12 mm long, wider at the throat, style protruding, stamens included; calyx-lobes oblong blunt, very woolly-haired, usually about half as long as corolla. Leaves with adpressed greyish hairs, the lower long-stalked *c.* 20 cm, the upper narrower, much smaller, stalkless; stems with adpressed hairs, to *c.* 50 cm; rootstock stout. Nutlets with large hooked bristles.

978 **L. stylosa** (Karelin & Kir.) Brand (*L. angustifolia* Brand)
Pakistan to Himachal Pradesh. 3300–4700 m. Stony slopes, steppes; common in Ladakh, forming large clumps. Jun.–Aug.

Flowers claret-coloured, tubular, with protruding styles, borne in lax nodding clusters, and with narrow lanceolate leaves. Corolla 12 mm, lobes *c.* 3 mm, erect or spreading; calyx-lobes narrow-lanceolate, *c.* 8 mm at flowering, densely hairy. Leaves rough-hairy, basal leaves long-stalked, the upper stalklesss; stems 60–90 cm. Nutlets forming a pyramid, with dense hooked hairs particularly on the margins. **Pl. 91**.

Upper leaves lanceolate, with clasping heart-shaped bases

979 **L. longiflora** (Benth.) Baillon
Pakistan to W. Nepal. 3000–3600 m. Open slopes; common in Kashmir. Jun.–Aug.

Flowers deep blue to purple, very variable in size, to 1.5 cm across, with spreading rounded lobes, borne in dense or lax elongated clusters. Calyx-lobes oblong-elliptic, shorter than the corolla-tube, hairy particularly on margins. Basal leaves lanceolate long-stalked, upper leaves stalkless clasping stem, 4–8 cm; stems solitary, or several, 15–60 cm. Nutlets with stout hooked bristles. **Pl. 92**.

8 ERITRICHIUM Corolla-tube with 5 scales in throat. Nutlets ovoid, 3-angled with winged margins, with or without hooked bristles. 5 spp.

980 **E. canum** (Benth.) Kitam. (*E. strictum* Decne., *E. rupestre* Bunge)
Afghanistan to C. Nepal. Tibet. C. Asia. 2400–4500 m. Rock ledges,
screes, drier areas. Jun.–Aug.

A small silky-white perennial, with several or many erect unbranched
stems with many linear leaves, and with branched clusters of tiny blue
flowers. Flowers *c.* 6 mm across; corolla-lobes rounded, spreading; calyx
silky-haired, lobes *c.* 2 mm long in fruit. Leaves to 4 cm by 2–4 mm, with
adpressed silky hairs; stems mostly to 15 cm. Nutlets in a pyramid, with
margins with stout hooked bristles.

981 **E. nanum** (Vill.) Schrader
Pakistan to Uttar Pradesh. 3600–5000 m. Open slopes, drier areas.
Jul.–Aug.

A small loosely tufted plant with rosettes of silvery-haired narrow-
elliptic leaves, and erect leafy stems with rather compact clusters of bright
blue flowers with yellow eyes. Flowers *c.* 8 mm across, with rounded lobes;
calyx 3 mm, lobes narrow-elliptic, densely bristly-haired. Lower leaves
narrow-elliptic acute 1–3 cm, with long white shaggy hairs, stem-leaves
narrower; stems to *c.* 10 cm. Nutlets with bristles at apex. **Page 490**.

9 CHIONOCHARIS Cushion plants. Close to *Myosotis* but differing in
its spathulate calyx-lobes, its overlapping corolla-lobes, and its laterally
attached, lanceolate, very hairy nutlets. 1 sp.

982 **C. hookeri** (C. B. Clarke) I. M. Johnston (*Eritrichium h.* (C. B. Clarke)
Brand)
E. Nepal to S.W. China. 4300–5200 m. Rocks, screes. Jun.–Jul.

A rare but beautiful high-alpine plant, with very dense hard rounded
cushions formed of tiny rosettes of leaves which are covered with long
white-woolly hairs, and with stalkless sky-blue flowers with yellow eyes.
Flowers with rounded lobes, spreading to *c.* 6 mm. Leaves spathulate, in
rosettes about 6 mm across; cushions 3–10 cm across; stems short, covered
with old leaves. **Page 490**.

10 PSEUDOMERTENSIA Corolla-tube usually much longer than calyx,
with or without scales in throat. Nutlets smooth, slightly angular, with a
triangular scar at base of inner angle. 6 spp.

Flowers blue
a *Style longer than corolla*

983 **P. echioides** (Benth.) Riedl
Pakistan to Himachal Pradesh. 2700–3600 m. Rocks, banks; common in
Lahul. Jun.–Aug.

A low softly hairy perennial, with a dense many-flowered cluster of
blue-purple tubular flowers; distinguished by their ascending, not spreading corolla-lobes and their long protruding styles. Corolla-tube slender *c.* 7
mm, without scales in throat, one third as long again as the linear

281

bristly-haired calyx-lobes. Inflorescence remaining dense in fruit. Leaves oblong blunt to oblanceolate, to 18 cm, the lower narrowed to a long stalk, the upper few, smaller, stalkless; stem slender, unbranched, usually 15 cm, or more. Nutlets *c.* 2 mm, white, shining. **Pl. 90**.

b *Style included in corolla*

984 **P. nemerosa** (DC) R. R. Stewart & Kazmi
Afghanistan to Kashmir. 1500–2400 m. Forests, shrubberies; common in Kashmir. May.

Flowers blue, in a lax cluster, with corolla-tube at least twice as long as the linear calyx-lobes, and with ascending, not spreading, obovate corolla-lobes. Leaves mostly basal, blade elliptic 3–4 cm, leaf-stalk as long; flowering stems 10–12 cm, with several narrow leaves and bracts. **Pl. 90**.

985 **P. moltkioides** (Royle) Kazmi
Pakistan to Kashmir. 3600–4500 m. Rocks, open slopes, coniferous forests. Jul.–Sep.

A rather tufted plant, with lax rosettes of elliptic leaves, and with short leafless stems bearing a terminal cluster of long tubular flowers with spreading blue rounded lobes and with a yellow or white eye. Corolla-tube reddish-purple, to 2 cm, many times longer than the narrow-oblong bristly white-haired calyx-lobes. Outer leaves spathulate, narrowed to a stalk, usually 1–3 cm, inner leaves stalkless, all with adpressed bristly hairs. Flowering stems 2–10 cm. Nutlets brown-black. **Pl. 90**.

Flowers white

986 **P. racemosa** (Benth.) Kazmi
Himachal Pradesh to W. Nepal. 1800–3000 m. Forests, damp rocks; common in cedar forests around Simla. Mar.–May.

Flowers white, *c.* 1 cm across, in a terminal cluster, with blunt spreading lobes and with 5 short scales in the throat. A softly hairy, weak, spreading or ascending perennial, with all leaves stalked and with a heart-shaped to elliptic blade 2–3 cm; stems leafy, 8–15 cm. Nutlets smooth, brown-black.

11 **MICROULA** Corolla-tube short, with 5 scales in the throat. Nutlets flattened dorsally, attached to the receptacle by a small scar above the middle. 3 spp.

987 **M. sikkimensis** (C. B. Clarke) Hemsl.
W. Nepal to S.W. China. 3500–4600 m. Open slopes, shepherds' camping places. Jul.–Aug.

A bristly-haired erect plant, with a terminal often widely branched inflorescence of bright blue flowers with yellow eyes. Flowers *c.* 6 mm across; calyx with bristly-haired narrow triangular lobes, enlarging in fruit. Leaves elliptic pointed, the upper stalkless, the lower long-stalked 4–8 cm; stem 10–40 cm. **Page 490**.

12 MYOSOTIS Corolla-tube short straight, throat closed by 5 short notched scales, corolla-lobes spreading; stamens included. Nutlets brown to black, small, smooth, shining. 7 spp.

988 **M. alpestris** F. W. Schmidt ALPINE FORGET-ME-NOT
Pakistan to Bhutan. Temperate Eurasia. 3000–4300 m. Open slopes. Jun.–Aug.

A low softly-hairy perennial, with rather short dense terminal clusters *c.* 1 cm across of blue flowers with yellow eyes, the clusters elongating after flowering, and with narrow stem leaves. Corolla with rounded spreading lobes, and tube longer than the silvery-haired, narrow-triangular calyx-lobes. Basal leaves rather tufted, oblanceolate 4–8 cm, distinctly or indistinctly stalked, stem leaves narrower, linear-oblong, stalkless; stems to 20 cm, branched only in the inflorescence. **Page 490**.

989 **M. silvatica** Ehrh. ex Hoffm. WOOD FORGET-ME-NOT
Pakistan to Bhutan. Temperate Eurasia. N. Africa. 1800–4000 m. Shrub-beries, meadows. Jun.–Aug.

Like 988 with bright blue flowers to 8 mm across or often smaller, but nutlets more or less acute, with a very small attachment-area and without lateral folds; calyx rounded at base and deciduous (in 988 nutlets obtuse, attachment-area large and with lateral folds; calyx narrowed to the base in fruit, never deciduous). A branched leafy perennial or biennial to 50 cm, with narrowly ovate to elliptic leaves, the basal leaves very shortly stalked.

13 TRIGONOTIS Very like *Myosotis* but nutlets 3–angled. 5 spp.

990 **T. rotundifolia** (Wall. ex Benth.) Benth. ex C. B. Clarke
Himachal Pradesh to S.E. Tibet. 3600–5600 m. Rock ledges, screes. Jun.–Aug.

A small tufted or mat-forming forget-me-not-like plant, with bright blue flowers with yellow eyes, 6–8 mm across, in terminal clusters, borne on short spreading stems. Calyx *c.* 3 mm in fruit. Basal leaves elliptic to rounded and with a heart-shaped base, 1–2 cm, stalked, with bristly adpressed hairs; stem-leaves stalkless, often with heart-shaped blade; stems 8 cm or more. Nutlets shining black, 3–sided, 1–1.5 mm long. **Pl. 90**.

CONVOLVULACEAE   Convolvulus Family

A family of often climbing herbaceous or woody plants, with an almost world-wide distribution, but absent in the cool temperate and arctic regions. Leaves alternate, usually simple, rarely with stipules. Flowers often axillary, regular and with an involucre of bracts. Sepals 5, usually free; petals 5, usually fused into a funnel-shaped corolla with 5 shallow lobes; stamens 5, fused to the base of corolla. Ovary superior, of usually 2 fused carpels with 2 chambers each with 1–2 ovules. Fruit a capsule, usually splitting into 2 valves, but sometimes a fleshy berry.

CUSCUTA   Herbaceous parasitic climbing plants, with leaves reduced to scales. *c.* 6 spp.

991   **C. reflexa** Roxb.
Afghanistan to S.W. China. S.E. Asia. 600–3300 m. Climber on shrubs; covers many of the bushes at village level in Nepal in late monsoon. Aug.–Oct.
Forming dense interlacing masses of stout yellow to purplish stems with small fleshy bracts, and with lax clusters of very fragrant white to pink flowers. Clusters 2.5–10 cm long; flowers bell-shaped, 6–8 mm long, with short triangular reflexed lobes; calyx-lobes rounded, pale green; stamens yellow. Fruit fleshy, 4–seeded.

992   **C. europaea** L. var. **indica** Engelm. LARGE DODDER
Pakistan to C. Nepal. 2700–4000 m. Shrubberies, grassy slopes. Jul.–Aug.
A pinkish parasite, smaller and more delicate than 991, with tiny pinkish-white flowers in stalkless globular clusters *c.* 1 cm across. Flowers *c.* 2 mm long, with blunt spreading triangular lobes. Stems to 1 mm thick, red or pink. **Pl. 92**.

ARGYREIA   Large woody climbers with pink funnel-shaped flowers; fruit a fleshy globular berry. 5 spp.

993   **A. hookeri** C. B. Clarke
C. Nepal to Bhutan. 300–1500 m. Rambling on banks. Jun.–Sep.
A large woody climber or trailing plant, with ovate-heart-shaped leaves, and very long-stalked, branched, dense axillary clusters of few large pink funnel-shaped flowers 5–7 cm long to 8 cm across. Sepals *c.* 8 mm, bristly-hairy, enlarging in fruit. Leaves with blade 10–20 cm, with a deep heart-shaped base and with an acute apex, long-stalked; young leaves silvery-hairy. Fruit globular, to *c.* 2 cm. **Pl. 92**.

IPOMOEA   Usually herbaceous climbers, distinguished from *Convolvulus* by their stigmas which have 1–3 globular lobes; style long slender; stamens unequal. *c.* 16 spp.

Leaves pinnate

994   **I. quamoclit** L.
Native to Tropical America; naturalized in the Himalaya to 1200 m. Cultivated areas; climber on shrubs and herbaceous plants. Aug.–Oct.
A slender hairless climber, with pinnate leaves with numerous narrow leaflets, and with long-stemmed axillary clusters of few or solitary, crimson or rarely white, showy narrow tubular flowers. Flowers 2.5–3 cm long,

with short triangular lobes spreading to 1.5 cm; sepals elliptic, with papery margins. Leaves 7–12 cm; leaflets linear *c.* 1 mm broad, numerous, distant. Capsule ovoid, 8 mm, with long persistent styles. **Page 490**.

Leaves 3–lobed

995 **I. nil** (L.) Roth   MORNING GLORY
Tropical and sub-tropical regions of both hemispheres; in the Himalaya to 1800 m. Cultivated areas. Jun.–Sep.

A hairy climbing perennial, with ovate-heart-shaped 3–lobed leaves, and with large purple funnel-shaped flowers with pale bases and often tinged with pink. Flowers solitary or several, stalked, axillary; corolla 3.5–6 cm long, tube rather slender; sepals lanceolate long-pointed, hairy particularly at base, to 3 cm. Leaves stalked, deeply triangular-lobed, 5–12 cm broad; stems hairy, climbing to 3 m or more. Ovary 3–celled; capsule 6–seeded, surrounded by much longer calyx-lobes.

Leaves entire

996 **I. purpurea** (L.) Roth
Native of C. America; naturalized from Pakistan to Sikkim. 100–2400 m. Cultivated areas, wasteland. Apr.–Oct.

Distinguished by its ovate-heart-shape pointed entire leaves, and its large dark purple funnel-shaped flowers with paler pinkish centre, 3.5–5 cm long. Sepals unequal, lanceolate, rough hairy, 8–13 mm. Leaves variable, 3–10 cm, stalked. An annual climber to 3 m. **Pl. 92**.

997 **I. carnea** Jacq.
Native of S. America; naturalized in the Himalaya to 1400 m. Wasteland, hedges and ditches; often gregarious; common in Jammu and Kangra. May–Jun.

Differs from 996 in being a non-climbing shrub, and having white to pinkish flowers with a narrow tube and a wide-spreading limb with shallow rounded lobes, and with the centre of the corolla darker pink. Calyx with rounded, hairless lobes *c.* 6 mm. Leaves long-stalked, blade elliptic long-pointed with a heart-shaped base, to 15 cm, hairless; stems stout, to 2 m. **Pl. 92**.

PORANA   Flowers white or mauve. Climbing shrubs or herbs, distinguished from *Ipomoea* by their sepals which become much enlarged and papery in fruit. 3 spp.

998 **P. grandiflora** Wallich
C. Nepal to Bhutan. 1500–2400 m. Climber on shrubs; common. Aug.–Oct.

A slender extensive climber, with stalked axillary clusters of several mauve funnel-shaped flowers, and with ovate deeply heart-shaped long-pointed leaves. Corolla to 3 cm long, with a narrow corolla-tube and wide-spreading limb to 4.5 cm across; calyx with linear-oblong lobes to 5 mm, becoming papery and much enlarged to 4 cm in fruit; stamens and style included in the enlarged base of the corolla-tube. Leaves long-stalked, silvery-hairy beneath when young, blade to *c.* 12 cm. Capsule globular *c.* 1.3 cm, surrounded by unequal enlarged calyx. **Pl. 93.**

999 **P. racemosa** Roxb. SNOW CREEPER
Kashmir to Bhutan. China. S.E. Asia. 100–2400 m. Climber on trees and shrubs. Aug.–Oct.
Flowers white, in long branched clusters 'resembling dazzling patches of snow in the jungle'. Flowers relatively small, in long slender axillary clusters; corolla 8–13 mm long, lobed to nearly halfway, lobes rounded, spreading to 12 mm; sepals oblong, hairy, 2 mm, enlarging and papery in fruit to 15 mm. Leaves with blade 3–8 cm, ovate-pointed, with a deeply heart-shaped base; bracts heart-shaped, clasping. A lofty climber.

CONVOLVULUS Distinguished by its branched style with 2 slender stigmas. 1 sp.

1000 **C. arvensis** L. FIELD BINDWEED
Pakistan to C. Nepal. N. & S. Temperate Zones. 1000–4100 m. Cultivated areas, shrubberies, open slopes. May–Sep.
A slender spreading or climbing perennial, with distinctive arrow-shaped leaves, and with broadly funnel-shaped, pink or purple flowers with white or pale yellow centres. Flowers usually solitary, axillary, borne on stalks longer than the leaves; corolla 2–2.5 cm across; sepals oval, blunt or notched, unequal, hairless. Leaves 2–7 cm, ovate or lanceolate with two spreading triangular basal lobes, stalked; stems to 2 m. Capsule opening by 4 valves, hairless; seeds 4.

SOLANACEAE Potato Family

A rather large cosmopolitan family, mainly of herbaceous plants, which is of considerable importance to man. Leaves very variable, entire or compound, usually alternate and without stipules. Inflorescence usually a flat-topped cluster (*cyme*), or flowers rarely solitary. Flowers regular; sepals 5, fused at the base, usually persistent and often considerably enlarged in fruit; petals 5, fused to a greater or lesser extent, thus corolla varying from flat with spreading lobes to bell-shaped, or tubular; stamens 5, attached to corolla-tube. Ovary superior, of 2 fused carpels with 2 chambers each with many ovules; style 1. Fruit either a berry, or a capsule.

Fruit a fleshy or dry berry
a Herbaceous perennials or annuals   1 ATROPA
          2 MANDRAGORA   4 NICANDRA
b Shrubs   3 CESTRUM

Fruit a dry splitting capsule
  a Flowers in branched or spike-like clusters   5 HYOSCYAMUS
                                                          6 PHYSOCHLAINA
  b Flowers solitary, paired, or in small clusters   7 DATURA
                                                          8 SCOPOLIA

1 ATROPA BELLADONNA, DEADLY NIGHTSHADE Calyx bell-shaped, somewhat enlarging in fruit; corolla bell-shaped. Fruit a berry. 1 sp.

1001 **A. acuminata** Royle (*A. belladonna* (non L.) C. B. Clarke)
Pakistan to Himachal Pradesh. 1500–3000 m. Forests, shrubberies; cultivated in Kashmir herb gardens. Jun.–Aug.

An erect herbaceous perennial, with large ovate-lanceolate pointed leaves, and pale purple drooping bell-shaped flowers tinged with yellow or green. Flowers axillary, one or several on short stalks; corolla to 2.5 cm, with 5 short broad blunt spreading lobes; calyx with broadly spreading rather leafy lobes 1.5 cm; filaments hairy and swollen at base. Leaves entire, stalked, mostly 10–20 cm, the upper smaller; stem to 1 m. Berry globular *c.* 1.3 cm, black-purple, surrounded by the persistent calyx.

Berries poisonous. Roots and leaves used medicinally.

2 MANDRAGORA MANDRAKE Rootstock fleshy; stem absent at flowering; leaves in a rosette. Calyx bell-shaped, enlarging in fruit. Fruit a berry. 1 sp.

1002 **M. caulescens** C. B. Clarke
W. Nepal to S.W. China. 3000–4500 m. Open slopes, shrubberies. May–Jul.

An unusual-looking plant with many nodding bell-shaped yellowish to greenish-purple flowers, usually borne on short stalks arising directly from the rootstock among the young developing leaves. Flowers *c.* 2.5 cm across, with triangular lobes, the corolla often veined with purple within; calyx green, often deeply lobed with 5 ovate conspicuously veined lobes. Mature leaves large, obovate entire, with wavy ciliate margins, narrowed to winged leaf-stalks, to 28 cm by 5 cm; leafy stems at length to 30 cm; rootstock long and stout. **Pl. 93.**

3 CESTRUM Foetid shrubs. Flowers tubular-funnel-shaped, with spreading lobes. Fruit a berry. 4 spp.

1003 **C. nocturnum** L. NIGHT BLOOMING JESSAMINE
Native of tropical America; commonly cultivated in India and the Himalaya to 1500 m. Semi-naturalized in cultivated areas, hedges; common. Aug.–Oct.

An evergreen shrub, with ovate-lanceolate long-pointed shining leaves, and with long tubular greenish-yellow flowers which are very sweet-scented at night, borne in lax mostly axillary branched clusters. Corolla-tube slender to 2.5 cm, with 5 narrow-triangular lobes spreading to *c.* 1.3 cm across; calyx 3 mm, with short triangular lobes. Leaves entire, 8–12 cm, stalked; stem to 4 m, branches green, slender. **Pl. 92**.

4 NICANDRA Flowers bell-shaped; calyx divided almost to base, enlarging in fruit. Fruit a dry berry. 1 sp.

1004 **N. physalodes** (L.) Gaertn. APPLE OF PERU
Native of Peru; naturalized from Himachal Pradesh to Sikkim. 800–2200 m. Cultivated areas. May–Oct.

A foetid annual, with solitary axillary drooping blue or violet bell-shaped flowers with white centres, opening only for a few hours. Corolla 2.5–4 cm across, with 5 spreading blunt lobes; calyx with 5 oval lobes each with 2 reflexed blunt lobules, enlarging in fruit. Leaves ovate-lanceolate, irregularly shallow-lobed and toothed, stalked, 5–20 cm; stem branched above, 30–150 cm. Berry globular, encircled by the papery net-veined, 5–angled persistent calyx which is up to 3 cm long.

5 HYOSCYAMUS Corolla broadly funnel-shaped with 5 blunt lobes. Fruit a capsule opening by a lid, included in the enlarged calyx. 2 spp.

1005 **H. niger** L. HENBANE
Pakistan to S.W. China. Temperate Eurasia. N. Africa. N. America. 2100–3300 m. Cultivated areas, wasteland; common in drier areas. May–Sep.

Readily distinguished by its dull yellow cup-shaped flowers which are conspicuously netted with purple veins and with darker purple centres. A coarsely hairy, sticky, unpleasant-smelling annual or biennial, with irregularly-lobed elliptic leaves. Flowers 2–3 cm across, with rounded spreading lobes; calyx *c.* 1.5 cm, funnel-shaped, with triangular-pointed lobes, enlarging conspicuously to 2.5 cm or more and becoming papery rigid and spine-tipped in fruit. Basal leaves stalked, 15–30 cm, often absent at flowering; stem-leaves stalkless 5–8 cm; stem robust to 1 m. Capsule encircled by the globular base of the enlarged calyx.

A very poisonous plant. The drug hyoscamine is obtained from the leaves; it is used medicinally. **Pl. 93**.

6 PHYSOCHLAINA Inflorescence leafless. Corolla funnel-shaped; calyx tubular-bell-shaped, enlarging in fruit. Fruit a capsule opening by a lid. 1 sp.

1006 **P. praealta** (Decne.) Miers
Pakistan to C. Nepal. Tibet. C. Asia. 2400–4600 m. Stony slopes; common in Ladakh and Lahul. Jun.–Aug.

A glandular downy leafy herbaceous perennial, with a terminal branched cluster of greenish-yellow funnel-shaped flowers with purple veining. Flower cluster 5–10 cm across; corolla 2.5–3 cm long, with 5 shallow spreading lobes, stamens and style exserted; calyx bell-shaped, 5–lobed, *c.* 6 mm, enlarging in fruit to 2.5 cm. Leaves stalked, ovate or wedge-shaped, entire or with wavy margins, 5–15 cm; stem 60–120 cm, branched above. Fruit conspicuous, with tubular calyx encircling and much longer than the capsule.

A poisonous plant. **Pl. 92**.

7 DATURA Calyx often 5–angled, the upper part deciduous after flowering, the lower part persistent. Corolla tubular to funnel-shaped. Fruit a capsule, 4–valved. 4 spp.

1007 **D. stramonium** L. Thorn-Apple
Native of Tropical America. Naturalized in the Himalaya, 200–2200 m. Cultivated areas, wastelands. Mar.–Sep.

A strong-smelling hairless annual, with few large erect white or rarely violet narrow funnel-shaped flowers, and with ovate, coarsely lobed or toothed leaves. Flowers 7.5–12 cm long by 2.5–7.5 cm across, with 5 lobes with long-pointed apices; calyx tubular, 5–ribbed and lobed, about half as long as corolla. Leaves *c.* 18 cm; stem 60–120 cm. Capsule ovoid, 4–valved, 3.5–7 cm long, covered with sharp slender spines, and with the enlarged base of the calyx below.

A very poisonous plant containing drugs which produce hallucinations and dilation of the pupils. The leaves, seeds and fruits are used medicinally. **Pl. 92**.

1008 **D. suaveolens** Humb. & Bonpl. ex Willd. Angel's Trumpet
Native of Tropical America; naturalized in the Himalaya to 1700 m. Cultivated areas; sometimes a hedge plant. May–Nov.

A large shrub, with large leaves, and very large, white, pendulous funnel-shaped flowers to 30 cm long, which are very sweet-scented at night. Corolla with slender tube enlarging above, with 5 short lobes with out-curved apices; calyx to 12 cm, inflated, angled, and with 5 elliptic lobes. Leaves ovate-oblong entire, to 30 cm, sometimes downy beneath; stem stout, 2–5 m. Fruit *c.* 1 cm, oblong, smooth. **Pl. 93**.

8 SCOPOLIA Calyx bell-shaped enlarging in fruit; corolla cylindrical to bell-shaped, shallowly 5–lobed. Fruit a capsule opening by a lid, surrounded by the calyx. 2 spp.

1009 **S. stramonifolia** (Wallich) Shrestha (*S. lurida* (Link & Otto) Dunal)
Kashmir to S.W. China. 2700–4300 m. Wasteland, cultivated areas. Jun.–Jul.

Distinguished by its green, yellow, or greenish-purple widely bell-shaped drooping solitary axillary flowers. Corolla 4 cm long and broad, with shallow blunt lobes little longer than calyx; calyx bell-shaped, shortly lobed, enlarging to 5 cm and much longer than the capsule. Leaves elliptic entire, undersides often white-woolly, stalked, to 25 cm, much longer than the short-stalked flowers; stem branched above, 1–2 m.

The fresh leaves are poisonous; often grown in Tibetan borderlands, and dried for winter fodder for yaks. **Page 491**.

Commonly cultivated in the hills are the following:

Capsicum frutescens L. CHILLI   It is grown up to 2400 m in the Himalaya. The bright scarlet fruit, drying in the sun outside the houses, is a characteristic feature of villages in autumn.

Nicotiana tabacum L. TOBACCO   Plants with either pink or yellow flowers are commonly grown in gardens.

Solanum tuberosum L. POTATO   An important crop in the Himalaya; it is grown in winter in the plains and outer foothills, and in the higher valleys to 4400 m as a summer crop.

SCROPHULARIACEAE   Figwort Family

A large family, mainly of herbaceous plants, typical of the North Temperate Zone but occurring throughout the world. Leaves opposite or alternate, simple or variously lobed or dissected. Inflorescence spike-like or a flat-topped cluster, with both bracts and bracteoles variously developed and sometimes conspicuous. Flowers usually irregular; calyx 5–lobed; corolla 5–lobed and often 2–lipped; stamens commonly 4, two long and two short, attached to the corolla-tube. Ovary superior, of 2 carpels with 2 many-seeded chambers. Fruit usually a dry capsule, splitting in a number of ways, and with persistant calyx. There are many variations of this general structure, often resulting in the reduction of the numbers of parts of the flower.

Stamens 5 or more   1 VERBASCUM   9 ELLISIOPHYLLUM
Stamens 2
  a Corolla regular, not 2–lipped   13 WULFENIA   15 VERONICA
  b Corolla 2–lipped   12 LAGOTIS   18 CALCEOLARIA
Stamens 4
  a Corolla regular, not 2–lipped   8 DIGITALIS
                                    9 ELLISIOPHYLLUM
           10 HEMIPHRAGMA   11 PICRORHIZA
  b Corolla 2–lipped
    i Flowers yellow, or white or green   2 SCROPHULARIA
           4 MIMULUS   7 LINDENBERGIA
           14 OREOSOLON   16 EUPHRASIA
                     17 PEDICULARIS
    ii Flowers pink, purple or blue   2 SCROPHULARIA
         3 WIGHTIA   5 MAZUS   6 LANCEA
         11 PICRORHIZA   17 PEDICULARIS

1 VERBASCUM Mullein Corolla-tube short; corolla-lobes 5, spreading, equal. Stamens usually 5, some or all with hairy filaments. 3 spp.

1010 V. thapsus L.
Afghanistan to S.W. China. Temperate Eurasia. 1800–4000 m. Stony slopes, cultivated areas. May–Sep.

A very distinctive plant, with an erect leafy stem with a slender woolly spike of many yellow flowers, and with oblanceolate pale yellowish-grey woolly leaves. Flower-spikes 10–30 cm; flowers 2–2.5 cm across, with a short corolla-tube and 5 rounded spreading lobes; stamens with woolly-hairy filaments; bracts woolly, longer than flowers. Upper leaves with base of blades continuing in a wing down stem, 5–15 cm, basal leaves stalked, to 30 cm; stem unbranched, 1–2 m.

The plant is used medicinally.

2 SCROPHULARIA Figwort Flowers with an almost globular corolla-tube, 2–lipped, with 5 small blunt spreading unequal lobes. Fertile stamens 4, and one sterile scale-like staminode. Stems quadrangular; leaves opposite. *c.* 12 spp.

Leaves broad, ovate-heart-shaped, toothed
a *Flower clusters dense*

1011 S. pauciflora Benth.
W. Nepal to S.E. Tibet. 2500–4000 m. Forests, shrubberies, herdsmens' camping sites. Jun.–Jul.

Flowers green, in a terminal spike composed of few widely spaced, compact rounded clusters, and with ovate-heart-shaped coarsely toothed long-stalked leaves. Flowers short-stalked; corolla ovoid, 7–9 mm, with short lobes, the upper longer and slightly hooded; calyx-lobes narrow-triangular acute, *c.* 6 mm. Leaves with blades 3–10 cm, stalked; stem usually not branched, 60–90 cm. Capsule to 7 mm.

b *Flower clusters lax*

1012 S. urticifolia Wallich ex Benth.
W. Nepal to Bhutan. 1800–2800 m. Forests, shrubberies. Jul.–Sep.

Distinguished from 1011 by its smaller green flowers borne in lax leafy clusters consisting of many slender branched lateral branches, and by its rounded calyx-lobes. Corolla 5–6 mm, the upper 2 lobes distinctly longer than the lower; staminode obovate. Leaves elliptic to ovate, coarsely toothed, blade 3.5–10 cm, much smaller in inflorescence; stem slender, 30–100 cm. Capsule 5 mm.

1013 S. elatior Benth.
Uttar Pradesh to Bhutan. 1600–3800 m. Forests, shrubberies, grazing grounds. Jun.–Sep.

Like 1012 with spreading long-stalked branched clusters of green flowers, but sepals elliptic, and stamens and style twice as long as the corolla. Corolla 5 mm; calyx-lobes nearly equal, not margined. Leaf-blades large, variable, ovate or lanceolate 13–25 cm, coarsely toothed, stalked; stem stout, 4–winged. Capsule 6 mm. **Page 491**.

Leaves oblong in outline, pinnately cut towards base

1014 **S. koelzii** Pennell
Afghanistan to W. Nepal. 2600–5000 m. Stony slopes; common in Ladakh and Lahul. Jun.–Aug.
   Distinguished by its dark purple flowers in slender spike-like clusters, and by its narrow leaves. Flowers *c*. 5 mm across, the upper lobes longer and broader than the lower; styles exserted; calyx with rounded lobes. Leaves 3–4 cm, oblong to spathulate, with few widely spaced, oblong-pointed lobes; stems many, arising from the rootstock, 20–30 cm. Capsule conical, *c*. 6 mm, with style as long.

3 WIGHTIA   Trees, often epiphytic. Flowers funnel-shaped, 2–lipped; stamens and style exserted. 1 sp.

1015 **W. speciosissima** (D. Don.) Merr. (*W. gigantea* Wallich)
C. Nepal to S.W. China. Burma. S.E. Asia. 1200–2600 m. Forests, rocky slopes; rare in Nepal. Nov.
   An evergreen tree to 10 m, often epiphytic, with horizontal aerial roots clasping the forest tree on which it grows, with spike-like clusters of numerous pink tubular 2–lipped flowers, and with large elliptic entire leaves. Flowers 3 cm long, finely hairy, in clusters 10–20 cm; corolla-tube funnel-shaped, with an oblique 2–lipped mouth; upper lip 3–fid with rounded spreading lobes, finely hairy; stamens and style much exserted; calyx cup-shaped, with short blunt lobes. Leaves leathery entire, with 4–6 pairs of conspicuous lateral veins, short-stalked, 12–25 cm. Capsule elliptic, to 3 cm, splitting by two valves. **Page 491**.

4 MIMULUS   Corolla-tube long, 2–lipped, hairy within. Calyx 5–angled and 5–lobed. Stamens 4. Leaves opposite. 2 spp.

1016 **M. nepalensis** Benth.
C. Nepal to S.W. China. Burma. 1200–3000 m. Forests, damp slopes. Apr.–Sep.
   Flowers yellow, tubular, weakly 2–lipped, the upper lip erect, 2–lobed, the lower spreading, 3–lobed. Flowers solitary, axillary, borne on stalks usually longer than the leaves. Corolla 1.5–2.5 cm long; calyx *c*. 1 cm, tubular, 5–angled, with 5 very short lobes, but variable. A straggling perennial, with usually many spreading stems rooting at the nodes, or erect leafy stems 15–25 cm, and opposite elliptic, coarsely toothed, stalked leaves, mostly 2–2.5 cm. Calyx inflated in fruit. **Page 491**.

5 MAZUS   Corolla 2–lipped, upper lip 2–lobed, the lower larger 3–lobed. Capsule globular, splitting into 2 valves. 4 spp.

Runners absent; flowers large, 1.5 cm or more

1017 **M. dentatus** Wallich ex Benth.
Uttar Pradesh to Arunachal Pradesh. 1800–2400 m. Forests, rocks. Apr.–Jun.

Flowers horizontal or drooping, purple or white, funnel-shaped and 2–lipped, the upper lip erect, the lower much larger, spreading, 3–lobed, with a paler throat with a 2–lobed swelling within. Flowers to 2 cm, borne on few-flowered stems; calyx tubular, with triangular-elliptic lobes. Basal leaves long-stalked, the blade oblong-elliptic, toothed or wavy-margined, 2.5–10 cm; flowering stems leafy only at base, erect or spreading, 8–15 cm.

Runners present, rooting; flowers small, 12 mm or less

1018 **M. surculosus** D. Don
Pakistan to S.W. China. 1000–3300 m. Weed of cultivated areas; common. Apr.–Jul.

Flowers pale blue or white, tubular 6–12 mm long, in short terminal stalked clusters arising direct from the rootstock, and leaves in a lax rosette, and with long slender runners. Corolla 2–lipped, the upper lip darker, with 2 short triangular lobes, the lower lip much larger, with pale rounded lobes and the throat with 2 orange swellings within; calyx-lobes triangular blunt. Runners several, rooting at the nodes. Basal leaves broadly obovate, rounded-toothed, 2.5–8 cm, sometimes pinnately lobed near the base, narrowed to a short leaf-stalk; flowering stems mostly 6–8 cm. **Pl. 93.**

6 **LANCEA**   Corolla 2–lipped. Fruit globular, not splitting. 1 sp.

1019 **L. tibetica** Hook.f. & Thoms.
Pakistan to S.W. China. Tibet. 3000–4800 m. Damp places and grazing grounds in Tibetan borderlands; common. May–Aug.

A small stemless plant with a lax rosette of small obovate leaves, and a central cluster of 2–lipped bright mauve flowers. Corolla *c.* 2 cm, the tube narrow but enlarged above, the upper lip 2–lobed, the lower lip broader, 3–lobed, hairy within; calyx-tube short, lobes elliptic acute. Leaves oblong-ovate to spathulate, entire, 2–5 cm, narrowed to a short half-clasping leaf-stalk; flowering stems 0–2 cm; underground runners long slender. Fruit to 1 cm, encircled by persistent enlarged calyx. **Pl. 94.**

7 **LINDENBERGIA**   Corolla 2–lipped, yellow. Capsule ovoid, 2–valved. 4 spp.

1020 **L. grandiflora** (Buch.–Ham. ex D. Don) Benth.
Uttar Pradesh to S.W. China. 700–2400 m. Shrubberies, banks of paddy-fields. Dec.–Apr.

Flowers bright yellow, tubular, 2–lipped, 2.5–3 cm long, in axillary

293

clusters forming a dense or lax leafy spike, or sometimes in a branched inflorescence. A softly hairy herbaceous plant. Corolla-tube slender, with a small upper lip with 2 rounded lobes, and a much larger 3–lobed lower lip with rounded overlapping lobes, and swellings in the throat; calyx-lobes blunt; bracts ovate, to 1.2 cm. Leaves ovate, coarsely toothed, with conspicuous lateral veins, 5–12 cm, stalked; stems mostly 20–30 cm. **Pl. 93**.

8 DIGITALIS FOXGLOVE No native Himalayan species. **D. purpurea** L. COMMON FOXGLOVE, with a spike of rosy-purple spotted tubular flowers, is locally naturalised from gardens in Kashmir. **D. lanata** Ehrh. is another escape in Kashmir; it has brownish flowers with a white lip.

9 ELLISIOPHYLLUM Creeping herbaceous plants; leaves pinnately lobed. Corolla with 4–8 equal spreading lobes. 1 sp.

1021 **E. pinnatum** (Wallich ex Benth.) Makino (*Sibthorpia pinnata* (Wallich ex Benth.) Benth. ex Hook.f.)
C. Nepal to S.W. China. S.E. Asia. 1800–2700 m. Forests, shrubberies. May–Jun.
  A slender creeping hairy herbaceous plant, with small white axillary flowers often with yellow centres, with the corolla with 4–8 spreading equal spathulate lobes, and with pinnately lobed leaves. Corolla-tube short, lobes spreading to 1.5 cm, stamens exserted; calyx hairy, to 6 mm, with ovate acute lobes; flower-stalks very slender, often longer than the leaves. Leaf-blade broadly oblong or ovate in outline, 2.5–4 cm, with 5–7 broadly wedge-shaped lobes which are further toothed or lobed, long-stalked; stems very slender, rooting at the nodes, 10–20 cm. **Page 492**.

10 HEMIPHRAGMA Creeping herbaceous plants with leaves of two forms. Corolla with 5 nearly equal lobes; calyx with narrow lobes. Fruit fleshy. 1 sp.

1022 **H. heterophyllum** Wallich
Uttar Pradesh to S.W. China. S.E. Asia. 1800–3600 m. Forests, shrubberies; common. Mar.–May.
  A creeping hairy plant often carpeting the ground, with small solitary axillary, usually stalkless pink flowers *c*. 8 mm across, with a corolla with 5 broad spreading lobes. Distinctive in fruit with its shining red globular berry-like fruits *c*. 8 mm across. Leaves of two forms, those on main stems opposite, short-stalked, rounded-heart-shaped, toothed, 8–16 mm across, those on branches linear pointed numerous, tufted, mostly *c*. 5 mm long; stems spreading 30–60 cm. **Page 492**.

11 PICRORHIZA Flowers in dense spikes; corolla either with 5 equal lobes, or 2–lipped; stamens 4, exserted. Capsule ovoid, swollen. 2 spp.

1023 **P. scrophulariiflora** Pennell
Uttar Pradesh to S.W. China. 3600–4800 m. Screes, rocky slopes. Jun.–Aug.

Flowers dark blue-purple, in a dense cylindrical head, borne on a stout stem arising from a rosette of conspicuously toothed leaves. Corolla *c.* 1.5 cm, with a long 3–lobed upper lip and a short lower lip; stamens and style exserted; calyx hairy, nearly as long as the corolla-tube, calyx-lobes 5, lanceolate blunt. Leaves 2–6 cm, oblanceolate and narrowed below to a winged leaf-stalk; stems creeping; rootstock stout, covered with old leaf-bases above. Fruiting stem 5–10 cm; capsule 6–10 mm.
Used as a febrifuge.

1024 **P. kurrooa** Royle ex Benth.
Pakistan to Uttar Pradesh. 3300–4300 m. Rocky slopes. Jun.–Aug.

Like 1023, but flowers pale or purplish blue, and corolla much smaller, *c.* 8 mm, 5–lobed to the middle and with very much longer stamens. Leaves 5–15 cm, almost all basal, spathulate to narrow elliptic, coarsely saw-toothed, narrowed to a winged stalk; flowering stems usually longer than leaves. **Pl. 98**.

12 LAGOTIS Flowers in spikes, blue, purple or white, with broad overlapping bracts; corolla 2–lipped; stamens 2; anthers kidney-shaped. Fruit a drupe. 4 spp.

1025 **L. kunawurensis** (Royle ex Benth.) Rupr.
Pakistan to S.E. Tibet. 3900–5600 m. Open slopes, damp places; common. Jun.–Aug.

Flowers white, pale mauve or blue, numerous, often little longer than the elliptic, densely overlapping, subtending bracts, in a very dense spike, borne on short stems arising from the rootstock, and with a tuft of pale, hairless rounded leaves. A very variable species both in form, shape of leaves, and colour and size of flowers. Spikes 4–10 cm; flowers with slender corolla-tube to *c.* 8 mm, 2–lipped, with strap-shaped, out-curved lobes, or lobes short rounded; calyx spathe-like. Leaves rather fleshy, 5–15 cm, mostly basal, narrow-elliptic to obovate, entire or with blunt teeth, stalked; stem-leaves bract-like, stalkless, progressively smaller above. Fruit *c.* 5 mm, included in the calyx. **Pl. 94**.

1026 **L. cashmeriana** (Royle) Rupr.
Pakistan to Himachal Pradesh. 3300–4500 m. Alpine slopes; common. Jun.–Aug.

Like 1025, but a much smaller plant with dense cylindrical spikes of dark blue tubular flowers 3–4 cm long. Corolla 4–5 mm, with rounded lobes, and much longer than the small bracts. Leaves with blade 3–6 cm, oblong to elliptic, rounded-toothed, stalked; stem 3–15 cm, leafless below, with clasping elliptic leaves which become progressively smaller above.
Plant used medicinally.

13 WULFENIA Flowers in spikes; corolla with 4 equal lobes; stamens 2. Fruit a capsule. 1 sp.

1027 **W. amherstiana** Benth.
Pakistan to C. Nepal. 1500–3000 m. Shady rocks. Jul.–Aug.
Flowers tubular, blue-purple, many in long one-sided spikes; individual flowers drooping, *c*. 8 mm long, corolla with 4 nearly straight acute lobes, and exserted style. Flowering spike 8–15 cm; calyx half as long as corolla-tube, calyx-lobes lanceolate acute. Leaves all basal, crowded, blade oblong-obovate, narrowed below, 5–10 cm, coarsely and irregularly rounded-lobed or toothed, stalked; flowering stems leafless, unbranched, 10–25 cm. Capsule 2–chambered, oblong. **Page 492**.

14 OREOSOLEN Rosette plants. Flowers yellow; corolla 2–lipped, 5–lobed; calyx 4–lobed. Capsule with a long slender style. 3 spp.

1028 **O. wattii** Hook.f.
W. Nepal to Bhutan. Tibet. 4500–5500 m. Open slopes; drier Tibetan borderlands. Jun.–Aug.
Flowers tubular, bright yellow, densely clustered at the centre of a compact rosette of broadly ovate, toothed leaves which spread over the ground. Corolla-tube *c*. 1.5 cm, with 5 rounded spreading nearly equal lobes; calyx-lobes 4, linear. Leaves 2–4 cm, narrowed to a wide stalk, coarsely toothed, paler beneath. Whole plant almost stemless, arising from a stout rootstock covered with scales and the bases of old leaves. **Page 492**.

15 VERONICA SPEEDWELL Corolla-tube very short, with 4 usually spreading often unequal lobes, the upper larger and the lowest smaller. Stamens 2. Fruit a capsule with 2 valves. *c*. 28 spp.; a difficult genus, with some species not easily distinguished.

Plants of ditches, marshes and damp ground

1029 **V. anagallis-aquatica** L. WATER SPEEDWELL
Pakistan to Bhutan. North Temperate Zone. 1500–4000 m. Marshes; common. May–Aug.
Flowers pale blue to pink, in slender lax axillary spikes 8–15 cm long, or more. Distinguished by its stalkless oblong-lanceolate leaves with usually heart-shaped bases, and with entire or toothed margins. Flowers 4–8 mm across. Leaves 5–15 cm; stem rather fleshy, erect, 15–45 cm.

1030 **V. beccabunga** L. BROOKLIME
Pakistan to W. Nepal. Temperate Eurasia. N. Africa. 2100–4400 m. Damp places; common in Kashmir. May–Sep.
Distinguished from 1029 by its broad glossy elliptic to oblong, blunt leaves 3–5 cm, and its shorter lax, often few-flowered, axillary clusters of usually darker blue flowers. Flower clusters mostly 2–4 cm long; flowers *c*.

6 mm across. A rather fleshy plant with creeping hollow stems rooting at the nodes, 15–45 cm.

Plants of open slopes, meadows, shrubberies
a *Flower-clusters axillary*
  i *Leaves stalked*

1031 V. cana Wallich ex Benth.
Kashmir to S.W. China. 2100–3000 m. Forests, shrubberies; common. May–Aug.

A grey-hairy perennial, with erect stems, and ovate toothed leaves, and with usually 1–3 long spikes of pale blue flowers borne from the axils of the upper leaves. Flowers to 10 mm across, in spikes 3–6 cm long; sepals linear-oblong. Leaves stalked, 2–5 cm, with rounded teeth; stem unbranched, usually 10–20 cm. Capsule laterally flattened into 2 triangular wings, 6–8 mm across, notched.

  ii *Leaves stalkless*

1032 V. laxa Benth. (*V. melissaefolia* auct. non Poiret)
Pakistan to C. Nepal. Tibet. China. Japan. 2100–3400 m. Shrubberies, meadows. May–Aug.

Differs from 1031 in having stalkless sparsely hairy leaves, and longer slender axillary stalked clusters of blue flowers 8–12 cm long. Flowers 8–10 mm across; bracts longer than flower-stalks (shorter in 1031). Leaves 2–3.5 cm, ovate or heart-shaped, coarsely toothed; stems 15–40 cm. Capsule broadly obcordate, *c.* 4 mm across.

1033 V. himalensis D. Don
Uttar Pradesh to S.E. Tibet. Burma. 3000–4000 m. Shrubberies, open slopes; common in Nepal. Jun.–Aug.

A rather stout-stemmed perennial, with somewhat distant pairs of stalkless ovate leaves, and with long-stalked terminal and axillary clusters of relatively large blue flowers. Petals *c.* 1 cm long, often not spreading widely; sepals 4, unequal; stamens often blue; bracts linear-oblong. Leaves 2.5–5 cm, ovate acute with rounded base, coarsely sharp-toothed; stems 15–50 cm. Capsule ovoid *c.* 8 mm. **Page 493.**

b *Flower clusters terminal*

1034 V. lanuginosa Benth. ex Hook.f.
C. Nepal to S.E. Tibet. 4500–5200 m. Screes, drier areas; rare. Jun.–Aug.

A low-growing densely woolly-haired perennial, with rounded overlapping leaves, and with a terminal woolly head of blue flowers. Flowers 10 mm across; corolla with the upper lobe rounded and much larger and broader than the spathulate lower lobes; sepals 4, oblong. Leaves *c.* 12 mm, woolly on both sides; stems several, slender, usually 3–8 cm. Capsule elliptic, notched. An attractive and unusual plant of high altitudes. **Page 493.**

1035 **V. lanosa** Royle ex Benth.
Pakistan to Uttar Pradesh. 2400–3600 m. Rocky slopes, crevices; quite common. Jun.–Aug.

An erect perennial with usually many flowering stems, with paired narrow-elliptic stalkless leaves, and with terminal spike-like clusters of blue flowers. Flowers *c.* 12 mm across; calyx-lobes narrow-elliptic, blunt, *c.* 4 mm, enlarging in fruit; bracts oblong, the upper shorter than the flower-stalks. Leaves 1–4 cm, narrowed to a blunt apex, toothed, almost hairless; stems 15–40 cm. Capsule ellipsoid *c.* 6 mm.

c *Flowers solitary, axillary*

1036 **V. persica** Poiret
Pakistan to C. Nepal. N. & W. Asia. Europe. 1500–2800 m. Cornfield weed; common. Apr.

A slender spreading plant, with solitary blue axillary flowers, borne on long stalks longer than the small subtending ovate leaves. Flowers 8–12 mm across; calyx-lobes ovate-lanceolate 6–7 mm; filaments of stamens white, anthers blue. Leaves alternate, 0.5–2 cm, broadly ovate, toothed, short-stalked, the lowest leaves opposite; stems 10–60 cm. Capsule with two widely divergent lobes, strongly keeled and with glandular hairs.

16 EUPHRASIA EYEBRIGHT Corolla-tube straight, 2–lipped, with an open throat; calyx bell-shaped, 4–lobed; stamens 4. A very difficult genus with *c.* 13 spp, all very similar in general appearance. The Himalayan species need further critical studies.

1037 **E. himalayica** Wettst.
Afghanistan to Bhutan. Wider distribution not known. 2700–4200 m. Open slopes, damp places. May–Jul.

A small erect often branched annual, with terminal spike-like clusters of small white and purple tinged 2–lipped flowers with yellow throats. Corolla *c.* 8 mm, the lower lip with 3 oblong bi-lobed lobes; calyx with narrow triangular, bristly-hairy lobes; bracts slightly larger than the leaves, ovate with coarse triangular teeth. Leaves ovate, saw-toothed, 4–10 mm; stems slender with erect branches, but sometimes unbranched, 8–20 cm. **Page 493**.

17 PEDICULARIS LOUSEWORT Flowers 2–lipped, the upper lip hooded and often prolonged into a straight or curved beak, the lower lip broad, 3–lobed; calyx tubular or bell-shaped, usually 4–lobed. An important and difficult genus of semi-parasitic plants, often conspicuous in the high summer flora of the alpine zone, and in clearings in the upper forests; they are not grazed by animals. *c.* 77 spp.

Flowers yellow to white
a *Upper lip of flower without a beak* (see also 1045)

1038　**P. oederi** Vahl

Pakistan to S.W. China. Europe. 3600–4800 m. Open slopes; a common alpine. Jun.–Aug.

Flowers uniformly lemon-yellow, or with a purplish or reddish tip to the upper lip, borne in a dense oblong cluster, and with broadly linear pinnately cut leaves which are mostly basal. Corolla-tube up to twice as long as calyx, the upper lip oblong with a rounded apex and longer than the broad 3–lobed lower lip; calyx densely hairy, bell-shaped, with entire or toothed lanceolate lobes. Leaves stalked, the blade 5–10 cm, with numerous pairs of short pointed toothed lobes; stem solitary, erect, 2.5–10 cm; tubers several, spindle-shaped. **Pl. 94**.

b　*Upper lip of flower with a coiled bi-lobed beak*

1039　**P. bicornuta** Klotzsch

Pakistan to Uttar Pradesh. 2700–4400 m. Alpine slopes, irrigated meadows; common in Kashmir. Jul.–Aug.

A robust handsome plant, with a terminal cylindrical cluster of usually many large pale yellow flowers, each up to 2 cm across. Corolla with a distinctive upper lip which is s–shaped, or spirally curved, with a slender deeply bifid beak; lower lip with two lateral rounded lobes twice as large as the oblong mid-lobe; corolla-tube hairy, 2–3 cm; calyx inflated, hairy, with small irregular lobes. Leaves to 15 cm alternate, stalked, linear-oblong, and pinnately cut into rounded, toothed or lobed segments; stem stout, usually 15–60 cm. **Pl. 95**.

1040　**P. scullyana** Prain ex Maxim.

W. Nepal to Sikkim. 3300–4800 m. Open slopes, shrubberies. Jul.–Aug.

Very like 1039, with a dense terminal cluster of numerous pale yellow or white flowers, but corolla-tube about as long as the calyx (corolla-tube a little longer or up to twice as long as calyx in 1039), and flowers larger 3.5–4 cm long and to 3 cm across, with mid-lobe of lower lip much smaller than the lateral lobes; beak curved, often purple-tinged, bi-lobed at apex; calyx *c*. 2 cm, with unequal ovate lobes. Leaves linear-lanceolate, with numerous paired oblong coarsely toothed lobes. Plant variable in height, 6–45 cm, or more. **Pl. 95**.

c　*Upper lip of flowers with a long curved entire beak*
　i　*Corolla-tube little longer than calyx*
　　x　*Upper leaves not whorled*

1041　**P. klotzschii** Hurusawa

Uttar Pradesh to C. Nepal. 2100–4500 m. Open slopes, damp ground; common in the Jumla region of Nepal. Jun.–Jul.

Flowers yellow (occasionally purple) with a stout violet or mauve curved beak, and with a large 3–lobed lower lip, many in a dense terminal cluster. Corolla 2–2.5 cm long; lobes of lower lip nearly equal; calyx 10–12 mm, densely woolly-haired, lobes elliptic, coarsely toothed. Leaves oblong-lanceolate in outline, with numerous narrow-elliptic deeply toothed lobes, long-stalked; lower bracts similar to the leaves but smaller; stem mostly 10–30 cm.

xx *Upper leaves whorled*

1042 **P. tenuirostris** Benth.
Pakistan to Uttar Pradesh. 2400–3700 m. Meadows. Jul.–Aug.

A robust plant with distinctive pale yellow flowers, with the upper lip inflated below and terminating in a long flexuous curved beak, and with the lower lip very broadly heart-shaped with a small mid-lobe. Calyx *c.* 10 mm, very hairy, with triangular lobes; bracts hairy, ovate long-pointed, longer than the calyx. Upper leaves in whorls of 3–5, stalkless, lanceolate and pinnately lobed, the lobes finely toothed, blade 5–10 cm; basal leaves stalked; stem stout erect, leafy, 60–90 cm.

ii *Corolla-tube very much longer (2–6 times) than the calyx*

1043 **P. longiflora** Rudolph var **tubiformis** (Klotzsch) Tsoong
Pakistan to S.W. China. 2700–4800 m. Damp places, bogs; common in Tibetan borderlands. Jun.–Aug.

Distinguished by its terminal cluster of golden-yellow flowers, often with dark brown or red markings, with very long slender corolla-tubes mostly 2–4 cm, and sometimes more, and with their upper lips curved into long beaks, the lower lips much broader, shallowly 3–lobed. Corolla-tube hairy below; calyx *c.* 1 cm, with 2–3 leafy toothed lobes. Leaves oblong blunt in outline, 3–5 cm, with many oblong, toothed lobes 2–4 cm; bracts similar; stems often several, erect or spreading 5–10 cm. **Pl. 96**.

1044 **P. hoffmeisteri** Klotzsch
Himachal Pradesh to E. Nepal. 2500–4500 m. Shrubberies, open slopes; common in wetter parts of Nepal. Jun.–Aug.

Flowers very pale yellow or almost cream, in a terminal cluster borne on a very short stem, each flower with a very distinctive long slender corolla-tube many times longer than the calyx, and with what appears to be a hooded nodding bell-shaped corolla. Upper and lower lip of corolla rounded, equal, encircling the long slender beak which is curved in a circle, lateral lobes small, triangular; corolla-tube 4–5 cm; calyx 1.5–2 cm, inflated, papery, with leafy toothed lobes. Leaves narrow-elliptic, deeply lobed, coarsely toothed, lower leaves stalked, 10–15 cm; stems variable, mostly 15–60 cm. **Pl. 95**.

Flowers predominantly pink or purple
a *Upper lip of flowers with a short beak, or beak absent*
i *Flowers more than 1 cm long*

1045 **P. cheilanthifolia** Schrenk
Pakistan to C. Nepal. 3300–4500 m. Alpine slopes, screes, meadows; common in Ladakh. Jul.–Aug.

Flowers pink (sometimes white) in a lax or dense very hairy cluster, with the corolla-tube longer than the calyx, and with the upper lip arched with a blunt reflexed tip, often much longer than the short 3–lobed lower lip. Calyx with ribs densely hairy, lobes blunt, toothed. Stem-leaves in whorls of 3–6, stalked, pinnately lobed, 1.2–2.5 cm; basal leaves long-stalked; stems usually many, hairy, spreading, 10–20 cm.

1046 **P. roylei** Maxim.
Pakistan to S.W. China. 4000–4800 m. Open slopes; common. Jun.–Aug.

A low plant with many stems, and with reddish-purple flowers in many whorls forming a broad short cluster, the lower whorls often distant. Flowers 1.5–2.3 cm; corolla-tube much longer than calyx, the upper lip nearly straight, much smaller than the 3–lobed lower lip; calyx hairy, bell-shaped, 5–lobed, toothed. Stem-leaves few, whorled, c. 2.5 cm, pinnately cut with pointed deeply toothed lobes; basal leaves long-stalked, linear in outline, 5–8 cm; stems 5–15 cm.

ii *Flowers less than 1 cm long*

1047 **P. mollis** Wallich ex Benth.
Ladakh. W. Nepal to S.E. Tibet. 2700–4300 m. Shrubberies, open slopes. Jun.–Aug.

Distinguished by its small dark red or purple flowers borne in long spikes 15–30 cm, with the upper whorls of flowers densely crowded and the lower whorls distant. Flowers 7–8 mm; the upper lip of corolla straight, much longer than the small 3–lobed lower lip; calyx ovoid, with oblong, toothed lobes; bracts pinnately cut, nearly as long as the flowers. Leaves 2.5–5 cm, in whorls of 3–5, ovate or oblong and pinnately lobed with lobes which are further toothed or pinnately cut; stem stout hollow, simple or with whorled branches, 30–70 cm. Often a densely woolly-haired plant.

b *Upper lip of flower with a long tail-like beak*
  i *Beak slender curved*
    x *Corolla-tube shorter or a little longer than calyx*

1048 **P. pectinata** Wallich ex Benth.
Pakistan to W. Nepal. 2400–4000 m. Forests, shrubberies, open slopes; common in Kashmir. Jul.–Aug.

A tall erect unbranched plant, with a lax or dense spike of pink flowers with long sickle-shaped beaks with twisted tail-like tips, and with whorled and often doubly pinnately cut leaves. Flowers c. 2 cm; corolla-tube short, lower lip 3–lobed, the middle lobe much smaller; calyx-lobes acute, entire, calyx inflated in fruit. Stem leaves c. 8 cm, lanceolate, stalked, the lobes further toothed or lobed; stem mostly 15–70 cm, usually hairless, or with 4 rows of hairs; flowering spike hairy. **Pl. 96**.

1049 **P. pyramidata** Royle
Pakistan to Himachal Pradesh. 2100–5000 m. Forests, meadows; common in Kashmir. Jul.–Sep.

Very like 1048, but a more showy plant with dense spikes of pink flowers, and with whorled and once- or twice-cut stem leaves. Distinguished by its stamens which are inserted about the middle or at the top of the corolla-tube (stamens inserted at the bottom in 1048), and its conspicuous broadly lanceolate to elliptic hairy bracts 1.5 cm which overlap and conceal the calyx (bracts, smaller often toothed, not concealing calyx in 1048). Corolla-tube little longer than the hairy calyx, upper lip inflated at base and curved inwards above and with a long horizontal flexuous tail-like beak to 2 cm. Leaves whorled above; stem one or several, mostly 30–40 cm. **Pl. 94.**

1050 **P. trichoglossa** Hook.f.
Uttar Pradesh to S.W. China. 3000–4800 m. Open slopes. Jul.–Aug.
Distinguished by its reddish-purple flowers with the upper lip conspicuously and very densely covered with woolly red hairs, and with a hairless incurved beak, and by its alternate stem leaves. Flowers in a slender lax spike, with lanceolate bracts; calyx shaggy-haired, the lobes toothed. Leaves stalkless, lanceolate 4–8 cm, deeply cut into numerous close oblong rounded-toothed lobes; stem erect, leafy, 20–40 cm. **Pl. 96.**

1051 **P. elwesii** Hook.f.
C. Nepal to S.W. China. 3600–4800 m. Open slopes. Jun.–Jul.
Flowers large dark purple, in dense terminal heads, distinguished by the upper lip which is inflated sickle-shaped, and suddenly contracted to a short incurved beak. Flowers to 4 cm long, in axils of leafy pinnately lobed bracts; corolla-tube not longer than calyx, the lower lip very large and encircling the upper lip with two very large lateral lobes and a smaller mid-lobe; calyx cleft to a quarter its length into two toothed lobes. Leaves alternate, oblong in outline, deeply lobed and toothed, mostly 5–10 cm; stems several, spreading, 10–18 cm.

xx *Corolla-tube more than twice as long as calyx*

1052 **P. megalantha** D. Don.
Uttar Pradesh to S.E. Tibet. 3000–4300 m. Shrubberies, open slopes; common in wetter parts of Nepal. Jun.–Sep.
Flowers large, pink to reddish-purple, with a very slender corolla-tube as long as 7 cm, with a slender curved tail-like beak, and with a large 3–lobed lower lip 2.5 cm broad. Inflorescence usually in a dense terminal head, sometimes a spike, with corolla-tubes much longer than the pinnately lobed bracts; calyx inflated, mostly hairy, strongly veined, with crested lobes. Leaves pinnately lobed, 5–25 cm, long-stalked; stem leaves alternate. Stems one or several, stout hollow, erect, mostly 30–60 cm.

1053 **P. rhinanthoides** Schrenk
Pakistan to S.W. China. 3300–4800 m. Damp meadows; in Nepal only in dry Tibetan borderlands. Jul.–Sep.
Flowers pale or bright pink, in a very short cluster, with pinnately lobed

leaf-like bracts. Corolla-tube slender 2–3 times as long as calyx, the upper lip sickle-shaped inflated in the middle, and with a slender incurved or S-shaped beak, the lower lip 3–lobed, 1.6–2.5 cm broad; calyx with 4 short, toothed lobes, hairy. Basal leaves mostly *c.* 5 cm, oblong in outline, stalked, pinnately cut into short blunt toothed lobes; stems often several, 10–25 cm, each with 1–2 leaves.

1054 **P. siphonantha** D. Don
Uttar Pradesh to S.E. Tibet. 3300–4500 m. Open slopes; common. Jun.–Aug.

Stems usually many, tufted, with rather dense terminal clusters of bright red or purplish-pink flowers, with prominent white throats and with very slender corolla-tubes 3–6 times as long as the calyx. Upper lip of corolla with a slender horn gradually narrowed from the base upwards into a long curved bifid beak; lower lip with broad rounded lateral lobes and a notched mid-lobe; corolla-tube 3–5 cm. Leaves mostly 3–5 cm, oblong in outline, pinnately cut into many short oblong blunt and toothed lobes; stems 5–15 cm, sometimes solitary.

1055 **P. punctata** Decne.
Pakistan to Kashmir. 2700–4500 m. Alpine slopes, damp places; common in Kashmir. Jul.–Sep.

Flowers pinkish-red with a white spot in the throat, with a long slender corolla-tube, and with a slender curved bifid beak which is abruptly enlarged into a knee-like base. Corolla-tube to 2 cm; lower lip *c.* 1.5 cm across, 3–lobed, the middle lobe smaller. Flowers in rather dense short spikes, with pinnately lobed bracts. Leaves oblong to elliptic in outline, with rather broad ovate, finely toothed lobes; stems usually several, 10–25 cm.

ii *Beak slender, straight*

1056 **P. bifida** (D. Don) Pennell
Himachal Pradesh to Bhutan. 1000–2700 m. Shrubberies, open slopes, cultivated areas. Jun.–Aug.

Distinguished by its entire, not lobed, leaves which are narrow-elliptic with coarse rounded teeth, and by its pink flowers with corolla-tube about as long as calyx and with the upper lip with a slender straight beak, enlarged below into a curved base 2–3 times as long. Flowers in a lax leafy spike; lower lip of corolla shallow-lobed; calyx narrowly tubular, deeply cleft, finely hairy; bracts like the leaves but smaller. Leaves alternate, mostly 2–3 cm; stem 10–30 cm.

1057 **P. gracilis** Wallich ex Benth.
Afghanistan to S.W. China. 2100–4300 m. Shrubberies, open slopes. Jun.–Sep.

Stem slender, much branched, with small opposite and whorled stem-leaves, and with lax clusters of pink-purple flowers, or flowers axillary. Upper lip of corolla sickle-shaped, swollen in the middle and with a long almost straight beak. Flowers small 1–1.5 cm; corolla-tube slender, longer than calyx, lower lip very variable, rounded or 3–lobed, and smaller than the upper lip; calyx-lobes rounded, wider than long, margins ciliate. Stem-leaves mostly 1–3 cm, deeply pinnately-lobed, the lobes toothed or deeply cut; basal leaves stalked, larger; stem 15–80 cm.

18 CALCEOLARIA Flowers 2–lipped, the lower lip inflated bag-like, the upper smaller; stamens 2. 1 sp.

1058 **C. gracilis** Kunth (*C. mexicana* Benth.)
Native of Mexico; naturalized in E. Nepal. India. S.E. Asia. 1800–2400 m. Shrubberies, shady places. Apr.–Sep.

An annual, unmistakable with its bright yellow bag-like flowers, formed by the much enlarged spherical lower lip of the corolla, and its very much smaller upper lip. Flowers *c.* 1 cm or more, long-stalked, in branched leafy clusters. Leaves often variable, pinnate, pinnately lobed, or entire and coarsely toothed, stalked, glandular-hairy, blade 2.5–5 cm. Stems hairy, commonly branched, to 30 cm or more.

OROBANCHACEAE Broomrape Family

A family of parasitic plants lacking green leaves, mostly of Temperate Eurasia; closely related to the Scrophulariaceae and often included in it. Underground stems parasitic on the roots of host plants; aerial stems usually solitary and usually bearing a spike-like cluster of flowers, and with scale-like leaves which are often fleshy at first and colourless or brownish. Calyx cup-shaped usually 5–lobed, or 2–lipped; corolla 5–lobed, either tubular or 2–lipped; stamens 4. Ovary superior, 1–chambered, seeds many; stigma usually 2–lobed. Fruit a capsule.

AEGINETIA Calyx like a spathe, split in front nearly to the base. Corolla with broad nearly equal spreading lobes. 1 sp.

1059 **A. indica** L.
Uttar Pradesh to Arunachal Pradesh. India. China. Japan. S.E. Asia. 600–1700 m. Forests, shrubberies. Aug.–Oct.

A leafless plant with one or several slender stems, and large solitary inclined broadly tubular, red, purple or sometimes white flowers with broad shallow rounded lobes, and with distinctive spathe-like calyx encircling much of the corolla. Corolla-tube 2.5–5 cm long and 2 cm across; calyx 2–4 cm long, enlarging somewhat in fruit. Stems without bracts, 10–30 cm; roots of fleshy interlaced fibres. Capsule enclosed in calyx, 2 cm; seeds yellowish-white. **Page 493**.

BOSCHNIAKIA Corolla 2–lipped, the upper lip erect, hooded, almost entire, the lower very short; calyx cup-shaped. 1 sp.

1060 **B. himalaica** Hook.f. & Thoms. ex Hook.f.
Uttar Pradesh to S.W. China. 2700–4000 m. Rhododendron shrubberies.
Jun.–Aug.

A robust yellowish to brown leafless plant, with a stout erect stem covered with overlapping scale-leaves, and with a dense stout spike of yellowish-brown flowers. Flowers 2 cm long, in a spike 8–20 cm; corolla-tube straight, the upper lip boat-shape entire, the lower lip with small lateral lobes; stamens exserted; calyx cup-shaped; bracts mottled purple, broadly ovate, overlapping, half as long as the corolla, or more. Stem 15–45 cm, with numerous rigid oblong to broadly lanceolate scale-leaves, and with a swollen globular base. Capsule ovoid, to 2 cm, 4–valved. **Pl. 97**.

OROBANCHE BROOMRAPE Corolla, 2–lipped, the upper lip 2–lobed, the lower 3–lobed. *c.* 10 spp.

1061 **O. cernua** Loefl.
Pakistan to C. Nepal. W. Asia. S. Europe. 1500–3300 m. Rocky slopes; prominent in Lahul. Jun.–Aug.

Flowers mauve, in a dense spike, borne on a stout brown or purplish stem, with ovate pointed scale-leaves. Corolla 1.3 cm long, tube curved, with rounded-toothed lobes, the upper 2 broad, the lower 3 narrower; calyx deeply lobed, with lanceolate pointed lobes half as long as the corolla, hairy; bracts broadly lanceolate. Flowering spikes usually 8–10 cm, borne on a stem 15–30 cm. **Pl. 96**.

1062 **O. alba** Stephen ex Willd. (*O. epithymum* DC.)
Afghanistan to C. Nepal. Temperate Eurasia. 2400–3300 m. Stony slopes, edges of cultivation, shrubberies. Jun.–Aug.

Flowers yellowish- to reddish-brown, or pinkish, in a short spike to 7.5 cm. Corolla-tube curved, glandular-hairy, 1.5–2 cm, with rounded, toothed lobes; calyx-lobes ovate-lanceolate; bracts lanceolate. Stem 15–30 cm, with a few scattered triangular scale-leaves, finely hairy, often swollen below. **Pl. 97**.

LATHRAEA Calyx 4–lobed; corolla 2–lipped, the upper lip entire, the lower lip entire or 3–lobed. Often placed in the Scrophulariaceae. 1 sp.

1063 **L. squamaria** L. TOOTHWORT
Pakistan to Himachal Pradesh. Temperate Eurasia. 1800–3000 m. Coniferous forests. Jun.–Aug.

Whole plant white or cream-coloured, with numerous whitish to pinkish spreading or drooping flowers, in a dense one-sided spike, borne on a stout stem with scale-leaves. Corolla 1.4–1.7 cm, creamy-white and tipped with pink, with a cylindrical tube and 2 lips which only slightly diverge; style and anthers exserted, the latter hairy; calyx bell-shaped; bracts rounded. Stem stout, erect 15–30 cm, with sparse papery scale-leaves, arising from a branched underground rhizome covered with rounded fleshy scale-leaves. Capsule globular *c*. 1 cm. **Pl. 98**.

LENTIBULARIACEAE   Butterwort Family

A small family of carniverous plants, which entrap their prey either by means of glandular hairs as in *Pinguicula*, or by bladders as in *Utricularia*. Leaves simple and in a rosette in *Pinguicula*, or divided into narrow segments and with bladders in *Utricularia*. Flowers irregular, 2–lipped, arranged in spikes or racemes. Calyx fused below, 2–5–lobed, 2–lipped; corolla 2–lipped, the lower lip often 3–lobed and with a spur projecting backwards; stamens 2, borne on the corolla-tube. Ovary superior, of 2 fused carpels, 1–chambered, with many ovules. Fruit a capsule splitting into 2 or 4 valves, or not splitting.

PINGUICULA BUTTERWORT Terrestrial plants with a basal rosette of entire leaves. Flowers solitary, 2–lipped, borne on leafless stems. 1 sp.

1064   **P. alpina** L.
Kashmir to S.W. China. Temperate Eurasia. 3000–4400 m. Streamsides, boggy ground; in drier areas only. May–Jun.
Leaves in a flat rosette flush with the ground, entire, pale yellowish-green, sticky-glandular. Flowers 2–lipped, usually whitish with yellow spots, occasionally pale mauve, and with a short conical down-curved spur, borne singly on short slender stems. Corolla 8–10 mm, the upper lip 2–lobed, the lower lip 3–lobed; calyx 2–lipped, lobes blunt. Leaves 2–4 cm, elliptic; stems several, 5–10 cm. **Page 493**.

UTRICULARIA BLADDERWORT Usually submerged aquatics with usually much-divided leaves and with small animal-catching bladders. Flowers borne above water, 2–lipped. 11 spp., mostly very similar and not easily distinguished.

1065   **U. brachiata** Oliver
C. Nepal to S.W. China. 3000–3800 m. Damp rocks, mossy banks. Jul.–Sep.
A very delicate plant growing in wet places, with small white flowers usually with yellow spots, either solitary or paired and borne on very slender leafless stems. Flowers *c*. 8 mm, with a 3–lobed lower lip and a cylindrical-blunt pale purple spur, *c*. 4 mm; calyx violet. Leaves entire with kidney-shaped blade 3–6 mm broad, short-stalked; bladders *c*. 1 mm; stems 4–5 cm, with paired lanceolate bracts, often 2–flowered.
Several other species are quite common growing in water, in irrigation channels and paddy-fields at low altitudes in the Himalaya.

GESNERIACEAE   Gloxinia Family

A large family of mostly tropical herbs and shrubs – the tropical counterpart of the Scrophulariaceae. Leaves opposite or alternate, usually simple, entire or toothed. Flowers usually in spike-like or flat-topped clusters; calyx of 5 sepals fused at the base; corolla tubular, with 5 often obliquely placed lobes, often 2–lipped; stamens 2 or 4, often cohering in pairs. Ovary superior, or inferior, with a single chamber with numerous ovules. Fruit usually a capsule with many seeds.

Epiphytic shrubs	1 AESCHYNANTHUS
Terrestrial herbaceous plants	
a Leaf solitary	2 PLATYSTEMMA
b Leaves 2 or more	
i Flowers 2 cm or more	3 CHIRITA
ii Flowers less than 1.5 cm	
x Leaves in rosettes; flowering stems leafless	
	4 CORALLODISCUS
xx Plants with leafy stems	5 DIDYMOCARPUS
	6 RHYNCHOGLOSSUM

1 AESCHYNANTHUS   Epiphytes; leaves leathery. Stamens 4, exserted. Seeds tipped with long hairs. 5 spp.

1066   **A. sikkimensis** (C. B. Clarke) Stapf
   E. Nepal to Bhutan. 1200–2100 m. Forest epiphyte. May–Jun.
   Flowers crimson, striped black at apex, few in dense terminal clusters, with a long curved corolla-tube with short rounded lobes, and with a long-projecting style and slightly exserted stamens. Corolla *c.* 3 cm; calyx 5–lobed, *c.* 5 mm. Leaves thick-leathery, opposite, lanceolate long-pointed entire, to 12 cm. A pendulous epiphytic shrub to *c.* 1 m, or more, with green stems. Capsule linear 30–50 cm; seeds with very long hairs. **Pl. 98.**

2 PLATYSTEMMA   Leaf solitary. Stamens 4, included in corolla. 1 sp.

1067   **P. violoides** Wallich
   Himachal Pradesh to Bhutan. 1500–3000 m. Rocks in forests. Jul.–Sep.
   A slender hairy erect perennial of damp shaded rocks, with a single large broadly ovate leaf, and one or usually few blue-violet funnel-shapeo flowers with paler centres, borne on slender stalks. Corolla-tube very short, with 4 blunt spreading lobes *c.* 8 mm. the upper lobe 2–lobed, the lower 3 lobes oblong blunt spreading and forming a lower lip; calyx widely bell-shaped, deeply 5–lobed; anthers golden; style long curved. Leaf 4–6.5 cm, coarsely toothed, hairy, rarely with a second much smaller leaf opposite; stem 5–15 cm, the leaf placed two-thirds of the way up stem. Capsule elliptic *c.* 5 mm, with persistent style. **Pl. 98.**

3 CHIRITA Leaves opposite, unequal. Stigma shortly bi-lobed; fertile stamens 2, staminodes 2. 5 spp.

1068 **C. pumila** D. Don.
Himachal Pradesh to S.W. China. Burma. S.E. Asia. 1000–2400 m. Rocks, shady banks. Jul.–Sep.
An erect hairy herbaceous plant, with rather large solitary or several drooping funnel-shaped pale purple flowers tinged with yellow, or flowers white, and with 2–6 unequal lanceolate stalked leaves. Corolla hairy, 3.5 cm long, with 5 short nearly equal rounded spreading lobes; calyx tubular, with dense white hairs and with narrow recurved lobes. Leaves broadly lanceolate acute, toothed, the largest 6–10 cm; stems mostly 5–15 cm, but in larger plants branched and to 50 cm. Capsule 10 cm or more.

1069 **C. urticifolia** Buch.–Ham. ex D. Don
W. Nepal to S.W. China. Burma. 1000–2400 m. Shrubberies. Aug.–Sep.
Flowers red-purple streaked with yellow, funnel-shaped, 2–lipped, drooping, to 5 cm long, one or several borne on sparingly hairy stalks, with ovate bracts. Corolla-tube inflated above, sparingly hairy, lobes rounded; calyx 2–3 cm, lobes lanceolate pointed, bristly-haired. Leaves to 13 cm, elliptic long-pointed, saw-toothed, stalked. A herbaceous plant with un-branched stem, 20–45 cm. **Pl. 97**.

4 CORALLODISCUS Leaves several, all basal. Fertile stamens 4. 1 sp.

1070 **C. lanuginosus** (Wallich ex DC.) B. L. Burtt (*Didissandra l.* (DC.) C. B. Clarke)
Himachal Pradesh to Bhutan. 1000–2400 m. Rocks; common in Nepal. Jul.–Sep.
Flowers pale purple or white, tubular 2–lipped, long-stalked, few in a lax terminal cluster, borne on slender stems arising direct from the rootstock, and with a lax basal rosette of leaves. Flowers *c.* 12 mm long, cylindrical, the upper lip erect the lower lip longer, 3–lobed; calyx *c.* 3 mm, the lobes elliptic acute; flowering stems several, 5–10 cm. Leaves crowded, with ovate blade, variably rounded-toothed to almost entire 1.5–4 cm, and with leaf-stalk nearly as long, densely rusty-haired when young. Capsule cylindrical 2 cm, with persistent style. Unlike other members of the family, growing on rocks which are dry much of the year when the plant becomes shrivelled and appears lifeless. **Page 494**.

5 DIDYMOCARPUS Like *Chirita*, but stigma oblong, notched; fertile stamens 2, and 2–3 stamens rudimentary. 12 spp.

1071 **D. oblongus** Wallich ex D. Don
Himachal Pradesh to Arunachal Pradesh. 1000–3000 m. Rocks in forests. Jun.–Jul.
Flowers small, dark purple, with a cylindrical corolla-tube and 5 more or less equal spreading lobes, several or many borne in a terminal branched

cluster. Corolla-tube to 1 cm, twice as long as the red bell-shaped, hairless calyx which has rounded lobes; bracts broadly ovate, fused below, 4–6 mm long. Leaves often in an apparent whorl of 4 borne at the top of the stem, or leaves opposite and arranged up stem; leaves oblong or elliptic, to 10 cm, coarsely lobed or toothed, with spreading hairs above and hairless beneath except on the veins, leaf-stalk hairy; stem hairy, 5–15 cm. Capsule 1–2 cm, smaller than others of the genus.

1072 **D. primulifolius** D. Don (*D. aromatica* Wallich)
Uttar Pradesh to Sikkim. 2000–3000 m. Rocks in forests; sometimes epiphytic. Jun.–Jul.
Like 1071, but purple flowers larger *c.* 1.6 cm long and nearly hairless, and pink calyx larger, cup-shaped, *c.* 6 mm, with shallow blunt lobes. Stems with 3–4 unequal leaves borne at the top, ovate to elliptic, coarsely toothed, densely softly hairy above and often so beneath; stem 8–15 cm. Capsule 2–4 cm. **Pl. 98**.

1073 **D. aromaticus** Wallich ex D. Don
Uttar Pradesh to Sikkim. 1800–3000 m. Shady wet rocks. Jun.–Aug.
Like 1072, but differing in its smaller calyx *c.* 4 mm, with triangular blunt lobes, and glandular-hairy like the inflorescence. Flowers deep purple, *c.* 1.4 cm long; bracts elliptic *c.* 4 mm. Leaves several on stem as well as in a terminal cluster of 4, broadly elliptic, stalked, finely toothed, densely woolly-haired; stem 4–30 cm.

6 RHYNCHOGLOSSUM Capsule ellipsoid or ovoid, hardly longer than calyx. Fertile stamens 2. 1 sp.

1074 **R. obliquum** Blume
Uttar Pradesh to S.W. China. India. Burma. S.E. Asia. 600–2100 m. Forests, shady banks. Sep.–Nov.
Flowers small, blue, with a cylindrical corolla-tube constricted at the mouth, 2–lipped, with upper lip shortly bi-fid and lower lip much longer and 3–lobed with the outer lobes curved. Inflorescence usually a spike-like cluster to 12 cm, but often much smaller. Corolla 8–10 mm; calyx 3–4 mm, enlarging to 8 mm in fruit and encircling capsule. Leaves alternate, blade 4–10 cm, elliptic long-pointed, very unequal-sided with a heart-shaped base on one side, long-stalked; stem variable, 5–50 cm. **Page 494**.

BIGNONIACEAE   Bignonia Family

A mainly tropical family centred in S. America, largely woody, with some climbers and with a few herbaceous species in *Incarvillea*. Leaves opposite, in 4 ranks, usually compound. Flowers showy, usually in flat-topped clusters. Calyx 5–lobed; corolla larger, funnel- or bell-shaped, and 5–lobed; stamens usually 4, attached to corolla-tube. Ovary superior, of 2 fused carpels and a single style. Fruit a 2–celled capsule, with many flat, winged seeds.

BIGNONIACEAE

INCARVILLEA Herbaceous perennials with pinnate leaves. Capsule linear, splitting down one side. 4 spp.

Stems leafy

1075 **I. arguta** (Royle) Royle
Himachal Pradesh to C. Nepal. S.E. Tibet. S.W. China. 1800–3500 m. Rock ledges, stony slopes. Jun.–Aug.

A shrubby often pendulous perennial growing on rock-faces, with many stems with pink funnel-shaped flowers in long terminal clusters, and with pinnate leaves. Corolla 2.5–3.8 cm, with a narrow cylindrical base, abruptly widening to a broad tube and with 5 rounded nearly equal lobes; calyx *c*. 7 mm, with short awl-shaped lobes. Leaves with 5–9 elliptic, saw-toothed leaflets 2–3 cm; stems to 150 cm. Capsule linear 8–20 cm. **Pl. 99**.

1076 **I. emodi** (Wallich ex Royle) Chatterjee
Afghanistan to W. Nepal. 600–2500 m. Rocks. Mar.–Apr.

A handsome plant, with terminal one-sided clusters of 6–8 large pinkish-purple tubular flowers which are orange-yellow within. Differs from 1075 in having unbranched stems to 50 cm, with leaves aggregated at the base, and calyx with rounded fine-pointed lobes. Corolla-tube 3.5–5.8 cm, lobes rounded; calyx 4–7 mm. Leaves pinnate with 9–11 ovate, toothed leaflets 1–4 cm long, or leaflets rarely lobed. Capsule linear, to 18 cm.

Leaves all basal, flowering stems leafless

1077 **I. mairei** (Léveillé) Grierson
W. Nepal to S.W. China. Tibet. 3600–4500 m. Stony slopes; dry Tibetan border country. Jun.

Flowers 1 or several, borne on leafless stems 3–25 cm which lengthen to 50 cm in fruit. Flowers large mostly 4–6 cm, tubular-funnel-shaped, crimson outside, yellow, grey or white within, lobes 5, rounded, spreading; calyx bell-shaped, 1–2 cm, with triangular teeth. Leaves all basal, pinnate, the terminal leaflet much larger 3–6.5 cm, the lateral leaflets progressively smaller towards the base; leaflets ovate, toothed, hairless above. Capsule covered with black dots, 5–9 cm, nearly straight. **Pl. 98**.

1078 **I. younghusbandii** Sprague
W. Nepal. Tibet. 4400–5500 m. Stony slopes in dry Tibetan borderlands. Jun.

Like 1077, but differing in its corolla-tube which is narrow-cylindrical at base and widening only well above the calyx and becoming bell-shaped above. Flowers pinkish-purple with white lines in the throat, appearing with the young wrinkled leaves; corolla-tube 3–5 cm long. Leaves pinnate, 1.5–2.5 cm at flowering, enlarging to 4–5.5 cm at maturity. Capsule curved.

Jacaranda mimosifolia D. Don (*J. ovalifolia* R. Br.)
A native of Brazil; commonly planted by roadsides in the outer foothills of the W. Himalaya, and in the valley of Kathmandu where it gives a fine

display of blue flowers in April and May. A small tree, with twice-pinnate leaves, and with erect pyramidal clusters of drooping tubular blue flowers, *c.* 5 cm.

PEDALIACEAE   Sesame Family

A small family mainly of herbaceous plants of Africa, S.E. Asia and Australia. Calyx of 5 fused sepals; corolla tubular, 5–lobed; fertile stamens 2 long 2 short, and 1 sterile staminode. Ovary superior, with 2–4 chambers, with one or many ovules. Fruit a capsule or a nut.

MARTYNIA   Flowers in small axillary clusters. Leaves heart-shaped. 1 sp.

1079   **M. annua** L. Unicorn Plant, Devil's Claw
Native of America; a common tropical weed. In Himalaya to 1200 m. Shrubberies and waste land. Jun.–Oct.
   A robust plant, with large heart-shaped leaves, and with small axillary clusters of several large pink and white tubular 2–lipped flowers, each 4–5 cm long. Corolla-tube broad, pinkish, variegated with saffron marks and violet dots, or with bright magenta blotches on the rounded lobes; calyx pink, with ovate lobes much shorter than the corolla-tube. Leaves rounded-heart-shaped, obscurely lobed, blade 10–20 cm broad, long-stalked; stems stout erect, branched above, 60–120 cm; plant glandular-hairy. Fruit ellipsoid, hairy, *c.* 3 cm. **Pl. 97**.

SESAMUM   Flowers axillary, stalkless. Leaves not heart-shaped, the lower usually lobed. 1 sp.

1080   **S. orientale** L. (*S. indicum* L.) Sesame
Probably native of Africa; cultivated throughout the Tropics, and in the lower valleys of the Himalaya to 1200 m. Sep.–Oct.
   An erect hairy annual, with usually narrow alternate upper leaves and opposite lobed lower leaves, and with tubular 2–lipped white or pinkish solitary axillary flowers with purplish or yellow markings. Flowers *c.* 3 cm; calyx with linear lobes *c.* 9 mm. Lower leaves *c.* 10 cm, usually lobed and stalked, the upper usually entire; stem simple or branched 30–60 cm. Capsule 2 cm, cylindrical, 4–chambered, hairy.
   The seeds are pressed to obtain a valuable edible oil, and are used in sweetmeats. Seeds and leaves used medicinally. **Page 494**.

ACANTHACEAE   Acanthus Family

A medium-sized family mainly of the Tropics, largely of shrubs and herbaceous perennials. Closely related to the Scrophulariaceae, having flowers with fused sepals and petals, and 5 or fewer stamens, and a superior ovary of 2 carpels. Flowers usually solitary, or in few-flowered clusters; sepals and petals 4–5, fused at the bases; corolla often 2–lipped; stamens usually 2 or 4. Fruit a 2–celled capsule. The genera are often difficult to distinguish from each other.

Fertile stamens 4

a Shrubs, undershrubs, or herbs   1 AECHMANTHERA

2 ASYSTASIA   3 GOLDFUSSIA   4 PTERACANTHUS

b Woody climbers   5 THUNBERGIA

Fertile stamens 2

a Corolla with 5 nearly equal lobes   6 BARLERIA

7 ERANTHEMUM

b Corolla 2–lipped   8 JUSTICIA   9 PHLOGACANTHUS

1 AECHMANTHERA Corolla with 5 equal spreading lobes; fertile stamens 4. Capsule linear, 6–8–seeded. 2 spp.

1081 **A. gossypina** (Wallich) Nees

Pakistan to Bhutan. 300–2400 m. Grass slopes, cultivated areas. Aug.–Oct.

A small shrub to 120 cm, with densely white-felted lower branches, and with elongated clusters of pale purplish to blue tubular flowers, borne in a lax leafy spike-like inflorescence. Corolla-tube to 2.5 cm, enlarged upwards, lobes 5, rounded and spreading to *c*. 1 cm; calyx densely glandular-hairy, with linear lobes, *c*. 1 cm. Leaves opposite, elliptic acute, toothed, the undersides often densely white-woolly, stalked, 5–10 cm. **Page 494**.

2 ASYSTASIA Corolla inflated above, with 5 equal lobes; sepals 5, nearly equal; fertile stamens 4. 1 sp.

1082 **A. macrocarpa** Nees

Uttar Pradesh to Bhutan. 300–2100 m. Forests, shrubberies. Mar.–Apr.

A rambling often lax shrub, with usually short clusters of large dull pink-purple cylindrical flowers netted with purple within, somewhat 2–lipped, usually appearing on leafless branches. Flowers *c*. 3 cm long; corolla-tube narrow below and abruptly enlarged to a longer upper part; sepals linear acute, finely hairy. Leaves entire, oblong to elliptic 10–15 cm, with 8 pairs of lateral veins; stems to 150 cm. Capsule cylindrical 3.5 cm, broader above; seeds 2, disk-like. **Pl. 99**.

3 GOLDFUSSIA Corolla straight or curved, with tube enlarged above, with 5 ovate or rounded nearly equal lobes; calyx 5–lobed. Fertile stamens 4. Capsule 2- or 4–seeded, seeds much flattened. 5 spp.

1083 **G. pentastemonoides** Nees

Himachal Pradesh to Bhutan. S.W. China. S.E. Asia. 1000–2700 m. Shrubberies, dense herbage; common in Nepal. Aug.–Oct.

A much-branched shrub 1–3 m across, with mauvish-blue to white tubular flowers, borne in axillary long-stalked rounded heads, which when young are encircled by the outer rounded pale bracts which soon fall. Corolla curved *c.* 4 cm, nearly hairless; calyx with linear lobes, hairy, to 1 cm. Leaves elliptic long-pointed, to 18 cm, closely toothed, gradually narrowed to a winged leaf-stalk. Capsule cylindrical *c.* 1.8 cm. **Page 495**.

4 PTERACANTHUS Like *Goldfussia*, but differing in its curved and inflated corolla. Bracts deciduous. Leaves saw-toothed and often narrowed into a long winged leaf-stalk. 5 spp.

1084 **P. alatus** (Wallich ex Nees) Bremek. (*Strobilanthes wallichii* Nees)
Pakistan to Bhutan. 2700–3600 m. Forests, shrubberies; common. Jun.–Jul.

Flowers blue, in lax branched leafy clusters with leaf-like persistent bracts. Corolla-tube 2.5–4.5 cm, pale blue or nearly white, curved and broadly dilated from a short cylindrical base, and with dark blue or mauve rounded lobes spreading to 1.3–1.6 cm. Leaves 3–8 cm, elliptic long-pointed, toothed, the lower stalked, the upper stalkless; stems weak, not woody, quadrangular, 15–60 cm. Capsule cylindrical to 2 cm; calyx-lobes lengthening to 2 cm in fruit.

1085 **P. urticifolius** (Kuntze) Bremek. (*Strobilanthes alatus* Nees)
Kashmir to C. Nepal. 1200–2700 m. Forests, shrubberies; very common. Jul.–Oct.

Like 1084, but differing in its bracts which fall before the flowers expand, and its larger hairy, ovate-heart-shaped toothed leaves, with long stalks winged at least above. Flowers bright blue or purple with a pale throat, in glandular-hairy usually branched inflorescences, with small narrowly oblong bracts; corolla-tube *c.* 3.5 cm, curved and gradually dilated from the base, lobes rounded and spreading to 2 cm. Leaves with long-pointed blades 5–18 cm, the upper smaller, stalkless. A sticky-hairy, rather shrubby perennial 60–120 cm. **Pl. 99**.

5 THUNBERGIA Climbers. Corolla with curved swollen tube and with 5 equal lobes twisted to the left in bud; calyx tiny. Capsule globular, beaked. 4 spp.

1086 **T. fragrans** Roxb.
C. Nepal to Sikkim. India. China. S.E. Asia. 500–2600 m. Climber on shrubs and lush monsoon herbage. Aug.–Sep.

A slender climbing perennial with fragrant white, curved funnel-shaped flowers, borne on long stalks from the axils of ovate to arrow-shaped leaves. Corolla with a narrow tube to 3 cm, expanding abruptly above into a wide funnel-shaped 5–lobed mouth to 5 cm across; calyx *c.* 2.5 cm. Leaves with blade 5–8 cm, often with two triangular basal lobes, or blade ovate acute, becoming hairless; stem slender, with backward-pointing hairs, or nearly hairless. Capsule distinctive, globular at base, with a long sabre-like beak which splits into two when ripe.

1087 **T. coccinea** Wallich ex D. Don
Uttar Pradesh to S.W. China. 300–2100 m. Climber on shrubs. Oct.–Mar.
A large climber with pendent branches, bearing long lax terminal clusters of scarlet flowers with large conspicuous scarlet calyx. Corolla-tube *c*. 2.5 cm, lobes rounded; calyx 2–lobed, lobes ovate nearly as long as corolla. Leaves with ovate-heart-shaped blade with wavy margins, 8–15 cm, slender-stalked; stems 3–8 m long. Fruit *c*. 4 cm.

6 BARLERIA Corolla with 5 nearly equal lobes; calyx with 2 opposite outer lobes much longer than the 2 inner. Fertile stamens 2. 2 spp.

1088 **B. cristata** L.
Pakistan to Bhutan. India. Burma. S.E. Asia. 600–2000 m. Light forests, shrubberies, open slopes. Jul.–Oct.
Flowers violet with a white throat, or pink, in dense rounded terminal and axillary clusters, with slender funnel-shaped corolla *c*. 4 cm long and with rounded lobes spreading to 2.5 cm. Calyx-lobes unequal, the outer ovate slender-pointed, toothed, the inner awl-shaped entire. An erect, stiffly hairy, bushy perennial 60–120 cm, with branched stems and narrow-elliptic entire, shortly stalked leaves 5–10 cm, with fine bristly hairs. Capsule oblong *c*. 1.3 cm. **Pl. 99**.

7 ERANTHEMUM Corolla with a slender tube and 5 equal lobes; calyx with 5 equal lobes. Fertile stamens 2. Capsule rounded; seeds flattened. 3 spp.

1089 **E. pulchellum** Andrews (*E. nervosum* (Vahl) R. Br. ex Roemer and Schultes)
Himachal Pradesh to Bhutan. S.W. China. Burma. S.E. Asia. To 1200 m. Forests, shrubberies. Feb.–Mar.
Flowers bright blue, in short spikes 2.5–8 cm, with conspicuous elliptic concave overlapping bracts, the spikes often in threes and forming close branched terminal clusters. Corolla *c*. 3.5 cm, with a slender tube and broad elliptic spreading lobes 8–12 mm long; calyx *c*. 6 mm, whitish with minute hairs, lobed to half-way. Leaves elliptic long-pointed to 20 cm, entire or obscurely toothed, short-stalked; stems 60–200 cm, hairless or nearly so. Capsule *c*. 12 mm; bracts persistent in fruit. **Page 494**.

8 JUSTICIA Corolla 2–lipped, tube straight. Capsule usually 4–seeded. 4 spp.

1090 **J. adhatoda** L. (*Adhatoda vasica* Nees)
Pakistan to Bhutan. India. S.E. Asia. 500–1600 m. Wasteland, shrubberies; common; often used as a hedge plant. Feb.–Apr.

A large strong-smelling gregarious shrub, not browsed by cattle, with white 2–lipped flowers with red spots and streaks within. Flowers in short compact terminal and axillary spikes and with conspicuous ovate overlapping bracts. Corolla *c.* 3 cm long, the tube short, the upper lip narrow, notched, incurved, the lower much broader, deeply 3–lobed; calyx with 5 lanceolate lobes; stamens 2, exserted. A hairless shrub 1.25–2.75 m, with short-stalked elliptic-lanceolate entire leaves 13–25 cm. Capsule club-shaped 3 cm, hairy, 4–seeded.

Plant used medicinally. **Pl. 98**.

9 PHLOGACANTHUS Corolla curved, 2–lipped; stamens 2. Capsule many-seeded. 2 spp.

1091 **P. pubinervius** T. Anderson
Uttar Pradesh to Bhutan. N. India. Burma. 200–1700 m. Forests. Feb.–Mar.

Showy in early spring with its dense cylindrical spikes of brick-red velvety, 2–lipped tubular flowers 8–25 cm long, and with conspicuous bracts in bud. An evergreen hairless shrub to 3 m, with 4–angled grey branches, and with drooping leaves with oblanceolate blades 15–25 cm, narrowed to a stalk. Corolla broad-tubular, curved, 2–2.5 cm, 2–lipped with 5 nearly equal lobes, stamens exserted; calyx-lobes 6–8 mm, bristly-haired; bracts 6–12 mm long. Capsule cylindrical 4–angled, hairless, to 4 cm; seeds disk-like, finely hairy. **Pl. 99**.

VERBENACEAE Verbena Family

A large family of trees, shrubs, herbaceous plants and climbers, with a mostly tropical and sub-tropical distribution. Leaves usually opposite, or whorled, and often entire. Flowers usually in racemes or cymes. Calyx 4–5–lobed; corolla tubular, 2–lipped or with 4–5 nearly equal lobes; stamens usually 4, alternating with the corolla-lobes. Ovary superior, of 2 fused carpels, soon divided into 4 chambers, each with 1 ovule. Fruit usually a drupe, less commonly a capsule.

Leaves compound, digitate   1 VITEX
Leaves simple
  a Flowers in axillary ovoid heads   2 LANTANA
  b Flowers in branched clusters
    i Fruit with 4 nutlets   3 CLERODENDRUM
                                                       4 HOLMSKIOLDIA
    ii Fruit breaking into 4 valves   5 CARYOPTERIS
    iii Fruit a fleshy drupe; 8–seeded   6 DURANTA

1 VITEX Leaves with 3–5 leaflets. Flowers small, 2–lipped. Fruit a drupe. 2 sp.

1092 **V. negundo** L.

Afghanistan to Bhutan. India. China. S.E. Asia. To 2000 m. Cultivated areas, wasteland; very common. Apr.–Oct.

A large deciduous shrub or small tree to 6 m, with whitish hairy branches, digitate leaves with 3–5 leaflets, and with small pale mauve flowers in branched clusters forming a long terminal branched pyramidal inflorescence. Flowers *c*. 8 mm long; corolla hairy outside, obscurely 2–lipped with 5 short unequal lobes, the lowest much larger stamens and style exserted, the former in unequal pairs, calyx bell-shaped, white-woolly, 5–toothed. Leaflets lanceolate long-pointed, margins entire, shiny above, paler with dense grey-matted hairs beneath, stalked, the longest leaflet 5–10 cm; leaves opposite, long-stalked; bark thin grey. Drupe black when ripe.

Readily grown from cuttings; often planted as a hedge-plant. Branches used for wattle-work, and baskets; leaves used in storage of grain to keep off insects. Leaves, roots and fruit used medicinally. **Page 495**.

2 **LANTANA** Flowers in dense heads; corolla cylindrical with spreading equal lobes. Fruit a fleshy drupe with 2 nutlets. 2 spp.

1093 **L. camara** L.

Native of tropical America; semi-naturalized and common in cultivated areas and hedges. To 1500 m. Aug.–Oct.

A rambling rough-hairy evergreen shrub with 4–sided branches, with ovate toothed leaves, and with long-stalked rounded heads 2–3 cm across of numerous white, pale purple, or commonly orange or yellow flowers. Flowers with a curved corolla-tube to 1 cm and with 4 spreading rounded lobes *c*. 3 mm; stamens not exserted, calyx *c*. 4 mm, with lanceolate lobes. Leaves opposite, shortly stalked, 2.5–7.5 cm; twigs often more or less prickly; stems 1.25–2.25 m. Fruit in a stalked cylindrical cluster; drupes black shining, *c*. 5 mm.

The plant is used medicinally.

3 **CLERODENDRUM** Usually distinguished by its long slender corolla-tube, with much longer stamens projecting conspicuously from the tube. Fruit a fleshy drupe, separating into 4 nutlets. 7 spp.

Flowers in long terminal clusters, with conspicuous bracts

1094 **C. serratum** (L.) Moon

Uttar Pradesh to Bhutan. India. Burma. To 1600 m. Shrubberies; quite common. Jun.–Oct.

A shrubby plant or small tree, with large narrow-oblong acute to obovate nearly stalkless leaves, and with long slender or narrow pyramidal clusters of bluish-purple or white tubular 2–lipped flowers. Flower-clusters 10–25 cm long; corolla-tube narrow 8–13 mm, with an oblique mouth with narrow lobes to 1.5 cm, stamens exserted, at least twice as long as corolla; calyx cup-shaped, bluntly 5–lobed; bracts narrow-elliptic. Leaves coarsely toothed, to 15 cm; stems usually 3–8 m. Fruit succulent, black-purple, *c.* 8 mm.

Roots and leaves used in local medicine.

Flowers in terminal branched often rounded heads
a *Flowers white*

1095 **C. viscosum** Vent. (*C. infortunatum* auct. non L.)
Uttar Pradesh to Sikkim. India. S.E. Asia. To 1500 m. Forests, shrubberies; very common; a coloniser of deforested ground. Mar.–Apr.

A densely hairy shrub with hairy twigs, with large ovate pointed leaves, and with large lax terminal branched clusters of white sweet-scented flowers with pink centres and long exserted stamens. Corolla-tube slender to 16 mm, densely hairy, with broadly elliptic to oblong lobes to 1.5 cm; calyx silky-haired, with elliptic lobes, much enlarged in fruit; stamens about three times as long as corolla. Leaves opposite, long-stalked, with a rounded or slightly heart-shaped base, densely rough-hairy on both sides; branches densely hairy, stems with very large corky lenticels *c.* 1–2 m. Fruit black when ripe, seated in the enlarged red calyx.

Leaves and juice used in local medicine. **Page 495.**

1096 **C. indicum** (L.) Kuntze
W. Nepal to Sikkim. India. Burma. S.E. Asia. To 1400 m. Cultivated areas, edges of rice fields.

Flowers white, in a dense terminal pyramidal cluster, with very long slender down-curved corolla-tubes to 10 cm long, with elliptic lobes 1 cm or more. Calyx 1–1.5 cm, with 5 elliptic lobes, enlarging to 2 cm in fruit. Leaves narrow-elliptic, entire, to 20 cm, the upper in whorls of 4. A hairless shrub to 2 m. Fruit fleshy, blue-green, 2–4–lobed, encircled by the much enlarged leathery red calyx.

Root and juice of plant used medicinally. **Pl. 100.**

Flowers red or scarlet

1097 **C. japonicum** (Thunb.) Sweet
Native of tropical Asia; naturalized in cultivated areas and also in gardens in Nepal. To 1800 m. Jul.–Sep.

A shrub with large heart-shaped leaves, and with conspicuous terminal branched clusters of scarlet flowers. Corolla-tube *c.* 2 cm, with oblong spreading lobes, stamens long-exserted; calyx red, ovoid, nearly half as long as corolla-tube. Leaves with toothed blade 8–30 cm or more, leaf-stalk nearly as long; a shrub to 2 m. **Pl. 100.**

4 HOLMSKIOLDIA Corolla cylindrical, curved, with 5 short lobes, anthers exserted. Fruit a drupe included in the enlarged calyx, with 1–4 nutlets. 1 sp.

1098 **H. sanguinea** Retz.
Uttar Pradesh to Bhutan. India. To 1500 m. Shrubberies. Aug.–Dec.

A straggling climbing shrub with broadly ovate leaves, and with scarlet tubular flowers with large conspicuous red cup-shaped calyx, borne in crowded axillary clusters. Corolla-tube 2–2.5 cm, slender curved, obscurely 2–lipped, with 5 unequal blunt lobes, the lowest longer, stamens exserted; calyx brightly coloured, with a very short tube with a wide circular papery limb to 2.5 cm across. Leaves 7.5–10 cm, with a rounded or shallow heart-shaped base, stalked, obscurely toothed, dotted with minute glands on both sides; twigs quadrangular, hairy when young but becoming hairless; stems 3–10 m. Fruit 4–lobed, dry, *c.* 6 mm across. **Page 495**.

5 **CARYOPTERIS** Corolla 2–lipped, stamens exserted. Fruit a globular capsule splitting in 4 valves with incurved margins retaining the seeds. 3 spp.

1099 **C. odorata** (D. Don) Robinson (*C. wallichiana* Schauer)
Pakistan to Bhutan. 300–2100 m. Shrubberies, hedges, open slopes; common in outer foothills. Feb.–May.

A shrub to 3 m, with numerous pale mauve flowers in axillary and terminal dense spike-like clusters, and with elliptic pointed leaves. Corolla-tube hairy *c.* 8 mm, little longer than the calyx, with narrow-ovate lobes to 8 mm, stamens and style much exserted; calyx densely hairy like the flower-stalks, with elliptic lobes. Flower spikes 5–15 cm. Leaves short-stalked, nearly entire or toothed, mostly 5–12 cm. Fruit dark blue when ripe, densely hairy. **Page 495**.

6 **DURANTA** Ovary 8–celled. Fruit a drupe with 4 2–seeded nutlets. 1 sp.

1100 **D. repens** L.
Native of C. America. 200–1600 m. Very common as a hedge-plant in the Nepal valley, and many other places in W. Himalaya. May–Aug.

A spiny shrub to 3 m, with bright lavender-blue flowers with pale centres, borne in terminal pyramidal branched clusters. Flowers *c.* 1.5 cm, with corolla-tube about twice as long as calyx and with rounded spreading lobes, stamens not exserted; calyx 4 mm, lobes tiny, linear. Leaves opposite, broadly elliptic entire, stalked, mostly 3–5 cm; spines stout, to 1.5 cm. Ripe fruit orange, globular *c.* 5 mm.

LABIATEAE   Mint Family

A cosmopolitan family, well represented in temperate climates, and particularly so in the Mediterranean region. Herbaceous plants, or small shrubs, rarely trees, often with quadrangular stems, and often glandular and aromatic, with opposite leaves placed successively at right angles to each other. Flowers usually clustered into whorls, borne from the upper leaves or bracts. Flowers commonly irregular, 2–lipped; calyx of 5 fused sepals, sometimes 2–lipped; corolla funnel- or bell-shaped usually conspicuously 2–lipped; stamens 4 or 2, attached to the corolla-tube. Ovary superior, of 2 fused carpels with 4 distinct one-seeded units, and with a style arising from the base. Fruit of 4 nutlets.

Corolla with 4–5 nearly equal lobes, or with upper lip absent or very short
   1 AJUGA 2 TEUCRIUM 3 LEUCOSCEPTRUM
      25 MENTHA 28 ELSHOLTZIA

Corolla 2–lipped
 a Stamens 4
  i Lower lip of corolla boat-shaped, entire 4 ORTHOSIPHON
             5 COLEUS 6 RABDOSIA
  ii Lower lip of corolla 3–lobed
   x Calyx with 5 or 10 more or less equal lobes 8 MARRUBIUM
     9 PHLOMIS 10 LAMIUM 11 LEONURUS
    12 STACHYS 13 ERIOPHYTON 14 LEUCAS
   15 ROYLEA 16 COLQUHOUNIA 17 NEPETA
      22 MICROMERIA 23 ORIGANUM
  xx Calyx 2–lipped   7 SCUTELLARIA 18 GLECHOMA
     19 DRACOCEPHALUM 20 PRUNELLA
     21 CLINOPODIUM 24 THYMUS
 b Fertile stamens 2   26 SALVIA 27 PEROVSKIA

1 AJUGA Bugle Corolla with a very short upper lip and a large 3–lobed lower lip, and corolla-tube with a ring of hairs within; calyx with 5 nearly equal lobes. 8 spp.

Plants with ascending stems only

1101 **A. bracteosa** Wallich ex Benth.
Afghanistan to Bhutan. India. China. Japan. S.E. Asia. 1000–4000 m. Open slopes; common. Jun.–Sep.
 Flowers pale blue to whitish, with corolla-tube nearly twice as long as calyx, 2–lipped with upper lip very short 2–lobed, the lower lip 3–lobed, borne in crowded whorls forming leafy spike-like clusters. Corolla *c.* 1 cm; calyx densely hairy; bracts often blue, ovate to elliptic, coarsely toothed, overlapping, longer than the flowers. Leaves 2.5–3 cm, oblanceolate blunt, wavy-toothed, hairy, the lower short-stalked, the upper stalkless; stems many spreading, leafy, hairy, 10–30 cm.

1102 **A. lupulina** Maxim.
W. & C. Nepal. Tibet. 2400–4300 m. Open slopes. Jun.–Jul.
 A densely hairy perennial with white flowers with mauve markings and with a pale blue tube, borne in whorls aggregated into a dense ovoid spike, with long overlapping pale bracts spreading much beyond the flowers.

Corolla *c*. 1 cm, lower lip with bi-lobed central lobe and 2 rather smaller lateral lobes; calyx hairy; bracts ovate-lanceolate to 3 cm, often conspicuously net-veined. Leaves obovate, stalked, coarsely toothed, to 10 cm; stem densely woolly-haired, erect 10–25 cm.

Plants with creeping stems rooting at the nodes

1103 **A. lobata** D. Don
C. Nepal to S.W. China. Burma. 1500–3300 m. Forests, shrubberies. Apr.–May.

Flowers lilac, *c*. 1.5 cm, with corolla-tube nearly three times as long as calyx, and with a very large 3–lobed lower lip, borne in axillary whorls, sometimes forming a short terminal cluster. Stems long slender 30–60 cm, prostrate creeping and rooting at the nodes; leaves long-stalked, with rounded or oblong, shallowly lobed blade with a heart-shaped base, 2.5–4 cm. **Page 496.**

2 **TEUCRIUM** G<small>ERMANDER</small> Upper lip of corolla absent or very short, the lower lip 5–lobed; calyx somewhat 2–lipped, or with 5 equal lobes. 4 spp.

1104 **T. quadrifarium** Buch.-Ham. ex D.Don
Kashmir to Sikkim. India. China. S.E. Asia. 1000–2400 m. Shrubberies. Jul.–Sep.

A shortly hairy perennial, with pink or purple flowers almost concealed by the conspicuous often red-purple bracts, borne in long spike-like often branched clusters up to 15 cm. Flowers *c*. 1.3 cm, with corolla-tube hardly longer than the calyx, and lower lip 5–lobed; calyx with a rounded upper lobe and 2 acute or obtuse lower lobes; bracts broadly ovate long-pointed, *c*. 1 cm, overlapping and in 4 ranks when young. A stout erect herbaceous perennial 60–120 cm, with oval-heart-shaped, toothed, short-stalked leaves 4–6 cm. **Page 496**.

3 **LEUCOSCEPTRUM** Flowers tiny, 5–lobed, in cylindrical spikes; stamens exserted. 1 sp.

1105 **L. canum** Smith
Uttar Pradesh to S.W. China. 1500–2400 m. Forests. Oct.–Apr.

A shrub or small tree to 10 m, with dense erect cylindrical spikes of tiny whitish flowers with long projecting stamens and styles. Spikes 10–15 cm long and 2.5 cm in diameter; flowers densely packed, with tiny 5–lobed corolla, *c*. 8 mm; calyx *c*. 7 mm, densely downy, lobes short triangular; bracts overlapping and covering buds but soon falling. Leaves ellipticlanceolate 15–30 cm, entire or finely toothed, hairless above, with dense white downy or sometimes brown hairs beneath, like the young branches and leaf-stalks. Not at all typical of this family. **Pl. 103.**

4 **ORTHOSIPHON** Corolla 2–lipped, upper lip with 3–4 lobes, the lower entire concave; calyx 2–lipped. 3 spp.

1106 **O. incurvus** Benth.
C. Nepal to Bhutan. 400–2000 m. Forests, shrubberies. Feb.–Nov.

A hairy usually sparingly branched undershrub with 4–angled stems, and with whorls of pale pink to mauve flowers in terminal interrupted spike–like clusters. Corolla *c.* 1.5 cm, with a straight or curved tube and with a 3–4–lobed upper lip and an entire lower lip. Calyx *c.* 8 mm, bell-shaped, 2–lipped, the upper lip broad and papery, the lower with 2 awned teeth and longer curved basal lobes; bracts tiny, ovate. Leaves 8–15 cm, ovate to ovate-lanceolate, coarsely toothed, narrowed to a short or long winged leaf-stalk; stems finely hairy, 15–90 cm. **Pl. 104**.

5 COLEUS Corolla 2–lipped, upper lip 3–4–lobed, the lower entire boat-shaped; calyx 2–lipped. Distinguished by its stamens with the filaments fused in a sheath round the style. 1 sp.

1107 **C. barbatus** (Andrews) Benth. (*C. forskahlii* Briq.)
Uttar Pradesh to Bhutan. India. Sri Lanka. E. Africa. 1200–2400 m. Pine forests, open slopes. Sep.–Oct.

A densely hairy herbaceous perennial, with pale blue flowers in whorls, forming long leafless interrupted spikes. Flowers to 2 cm; corolla-tube deflexed, longer than calyx, 2–lipped, the upper lip short reflexed 3–lobed, the lower much longer, boat-shaped entire acute; calyx hairy, bell-shaped, with lanceolate bristle-tipped lobes deflexed in fruit; bracts broadly ovate, pointed, overlapping in bud, soon falling. Leaves ovate to oblong blunt, rounded-toothed, short-stalked, 5–8 cm; stem erect 30–90 cm. **Page 496**.

6 RABDOSIA Stamens free (in *Coleus* stamens fused into a tube below); calyx usually equally 5–lobed; corolla-tube curved. Shrubs, with branched inflorescence with leafy bracts. 11 spp.

Leaves distinctly paler and densely hairy beneath

1108 **R. rugosa** (Wallich ex Benth.) Hara (*Plectranthus rugosus* Wallich ex Benth.)
Afghanistan to S.W. China. 1000–2700 m. Shrubberies, open slopes; common except in wet areas. May–Sep.

An erect branched leafy shrub or undershrub 60–150 cm, with ovate wrinkled hairy leaves, and with numerous tiny white 2–lipped flowers spotted or streaked with purple, in narrow leafy clusters. Flowers *c.* 4 mm; corolla-tube little longer than calyx, the upper lip with 3–4 lobes, the lower longer, boat-shaped entire; calyx grey-woolly, bell-shaped at flowering, elongate and tubular in fruit, with 5 nearly equal lobes. Leaves coarsely toothed, 2.5–4 cm, conspicuously white or grey-woolly beneath in contrast to the green wrinkled upper surface. **Pl. 105**.

1109 **R. pharica** (Prain) Hara (*Plectranthus p.* Prain)
W. to C. Nepal. Bhutan. Tibet. 2700–4000 m. Dry scrub, cultivated areas

of the Tibetan borderlands. Jul.–Sep.

An attractive blue-flowered shrub to 1 m or more, with stalked clusters of flowers forming a lax interrupted spike-like inflorescence to 20 cm. Flowers c. 1 cm; calyx 5 mm, with equal triangular pointed lobes, hairy; bracts ovate, c. 5 mm. Leaves distinctly paler and densely felted beneath, broadly ovate, coarsely toothed, short-stalked, mostly 1.5–3 cm; stems often much-branched. **Page 496**.

Leaves green beneath, hairless or sparsely hairy

1110 **R. coetsa** (Buch.–Ham. ex D. Don) Hara
Pakistan to S.W. China. S.E. Asia. 1000–3300 m. Forests, shrubberies, damp places. Aug.–Dec.

Like 1109, but differing in having lavender-blue flowers with a down-curved corolla-tube much longer than the bristly-haired calyx, and leaves with undersides green, hairless or nearly so, the larger 5–10 cm. Corolla c. 7 mm. A strong-smelling shrub, to 2 m or more.

7 **SCUTELLARIA** Corolla-tube long, 2–lipped, the upper lip broad, hooded, the lower 3–lobed with mid-lobe notched. Calyx 2–lipped, the upper lip with a distinctive small transverse scale on back, the lips closing after flowering. c. 10 spp.

Flowers blue or purple

1111 **S. discolor** Colebr. (*S. indica* Blume)
Uttar Pradesh to S.W. China. India. S.E. Asia. 1000–2400 m. Shady banks, edge of cultivation. Jul.–Nov.

Flowers blue, tubular 2–lipped, in long slender lax leafless spikes 8–25 cm, with tiny bracts mostly shorter than the short flower-stalks. Flowers 1.3–1.8 cm long, with a long slender curved corolla-tube, 2–lipped with the upper lip entire, hooded, the lower lip broad and 3–lobed, often paler; calyx c. 2 mm, enlarging in fruit and encircling the nutlets. Leaves mostly basal, with an elliptic blunt-toothed blade 2–8 cm and with a rounded or heart-shaped base and blunt apex, often purple beneath, long-stalked; flowering stems ascending, 30–60 cm, often with small paired leaves. **Pl. 104**.

1112 **S. linearis** Benth.
Afghanistan to Uttar Pradesh. 600–1800 m. Open slopes. May–Jun.

Flowers 2–2.5 cm, pale purple with the tip of the lower lip yellow, borne in short lax clusters. Upper lip of corolla curved, hooded, lower lip 2–lobed, shorter; calyx c. 2 mm, enlarging in fruit; bracts elliptic, hairy, c. 6 mm. Leaves linear c. 2–3 mm broad, stalkless. A hairy tufted slender spreading perennial, with stems 10–15 cm.

Flowers yellow or white, often with purple tips
a *Prostrate alpine plants*

1113 **S. prostrata** Jacquem. ex Benth.
Pakistan to C. Nepal. 2400–4500 m. Screes, open slopes; often forming loose mats. Jun.–Sep.

A dwarf spreading much-branched perennial, with relatively large yellow flowers tipped with violet, in short dense leafy terminal clusters which are 4–angled in bud. Flowers c. 2.5 cm long; corolla-tube many times longer than calyx, upper lip curved forward, blunt, lower lip broad triangular; bracts ovate acute entire, sometimes purplish beneath. Leaves with blade 6–10 mm, ovate, deeply coarsely toothed, softly hairy on both sides, stalked; stems many, spreading to 20 cm or more. **Pl. 101**.

b *Rambling or ascending plants of medium altitudes*

1114 **S. repens** Buch.–Ham. ex D. Don
Kashmir to Sikkim. Burma. 600–2000 m. Rambling on rocks and banks; common. Sep.–Oct.

A hairy perennial, with axillary whorls of few creamy yellow flowers tinged with purple, in a leafy terminal often unbranched cluster. Flowers c. 2 cm, often paired, with slender corolla-tube, 2–lipped with entire upper and lower lips; calyx hairy, c. 2 mm. Leaf-blade ovate-heart-shaped 1.2–4 cm, entire or coarsely toothed, with a stalk nearly as long; stems usually slender, spreading to 60 cm.

1115 **S. scandens** Buch.–Ham. ex D. Don
Kashmir to E. Nepal. 1200–2400 m. Shrubberies. Mar.–May.

A tall sparsely hairy perennial, with long rambling acutely 4–angled branches, and with lax spikes of pale yellow or nearly white flowers with purple upper lips. Flower spikes 8–10 cm long, glandular; corolla 2–2.5 cm, the upper lip hooded with margins in-curved, the lower lip broader than long, notched; bracts leafy, the upper smaller. Leaves 2.5–8 cm, ovate or lanceolate, coarsely toothed, stalked, often purple beneath; stems spreading to 1 m. **Pl. 104**.

8 MARRUBIUM Corolla-tube short, 2–lipped, the upper lip flat and often notched, the lower lip usually 3–lobed; calyx with 5 or 10 strong veins and 5 or 10 often recurved lobes. 2 spp.

1116 **M. vulgare** L. WHITE HOREHOUND
Pakistan to Kashmir. W. Asia. Europe. 1500–1800 m. Wasteland; common in Kashmir valley. May–Jun.

A robust, densely shortly woolly-haired, strong-smelling perennial, with rounded wrinkled and toothed leaves, and with many globular stalkless axillary whorls of small whitish 2–lipped flowers, forming a long inter-rupted leafy spike. Flowers c. 1.5 cm, the upper lip flat, 2–lobed, the lower broad entire; calyx densely hairy, with 10 small hooked spines. Leaves stalked, blade mostly 2–3 cm; stems several, forming large clumps, white-woolly, 60–120 cm.

Plant used medicinally.

9 PHLOMIS Corolla-tube shorter than calyx and with a ring of hairs within, strongly 2–lipped, the upper lip large and hooded, the lower usually 3–lobed. Calyx with 5 often spine-tipped lobes. 8 spp.

Almost stemless plants (stem if present not more than 15 cm)

1117 **P. rotata** Benth. ex Hook.f.
W. Nepal to S.W. China. Tibet. 4500–5200 m. Open slopes; Tibetan border areas. Jun.–Aug.
An unusual-looking plant, with a rosette of rounded leathery woolly leaves pressed flat to the ground, and with a dense usually stemless cluster of mauve flowers at the centre. Flowers *c.* 1.5 cm, with a densely hairy hooded upper lip and longer nearly entire lower lip; calyx *c.* 8 mm, hairy, with tiny bristle-pointed teeth; bracts with spiny tips. Whorls of flowers usually compact and nearly stemless but sometimes interrupted, and borne on a stout stem to 15 cm. Leaves kidney-shaped 4–13 cm across, wrinkled with deeply impressed veins above, toothed, suddenly contracted into a very broad woolly leaf-stalk. **Pl. 101.**

Plants with erect stems at least 30 cm
a *Stems and branches more or less rounded (not 4–angled)*

1118 **P. cashmeriana** Royle ex Benth.
Afghanistan to Kashmir. 1800–3300 m. Wasteland, open slopes; common in Kashmir valley. Jun.–Aug.
An erect perennial with several stems 60–90 cm, with many whorls of relatively large pale purple or pink flowers, and with numerous long awl-shaped and very hairy bracts. Flowers *c.* 2.5 cm, with a very large woolly, hooded upper lip, and a darker broad entire lower lip; calyx *c.* 1.5 cm, furrowed, with awl-shaped spine-tipped lobes. Leaves lanceolate to oblong with irregular heart-shaped base, blunt, 10–30 cm, obscurely toothed, wrinkled above, densely white-woolly beneath; stems, calyx and leaf-stalks white-woolly with star-shaped hairs. **Pl. 102.**

1119 **P. spectabilis** Falc. ex Benth.
Afghanistan to W. Nepal. 1600–4000 m. Stony slopes; not common. Jun.–Jul.
A stout hoary perennial with a stem 30–200 cm, with whorls of rosy-purple flowers in a terminal interrupted spike, and with the hooded upper lip beautifully fringed with long silvery hairs. Whorls of flowers 4–5 cm across; corolla *c.* 1.5 cm, the tube shorter than the calyx; calyx with spreading spiny awl-shaped lobes about one third as long as the calyx-tube; bracts awl-shaped, curved. Leaves broadly ovate-heart-shaped, to 30 cm, long-stalked, coarsely toothed or shallowly lobed, wrinkled, paler with star-shaped hairs beneath; stem stout, hoary or downy. **Pl. 104.**

b *Stems and branches conspicuously 4–angled*

1120 **P. bracteosa** Royle ex Benth.
Afghanistan to S.W. China. 1200–4000 m. Open slopes; common in alpine Kashmir. Jul.–Aug.

An erect hairy plant 20–80 cm, with heart-shaped toothed leaves, and with pink-purple flowers crowded into a few large whorls and forming an interrupted spike. Whorls 2.5–4 cm across; corolla 1.5–2 cm, the tube shorter than the calyx, upper lip larger hooded, very hairy and with a fringe of white hairs, the lower lip smaller, 3–lobed; calyx hairy, with 5 narrow awl-shaped teeth, much shorter than the calyx-tube; bracts linear-lanceolate, bristly-haired, without a spiny tip. Leaves 5–10 cm, stalked, hairy. **Pl. 102.**

1121 **P. macrophylla** Wallich ex Benth.
Uttar Pradesh to Bhutan. 3300–4000 m. Open slopes. Jul.–Aug.

A large leafy perennial to 2 m, with long-stalked heart-shaped toothed leaves, and with whorls of small purple flowers c. 2.5 cm across. Flowers c. 2 cm; corolla-tube short, the upper lip shaggy-haired, the lower 3–lobed, smaller; calyx with spine-tipped lobes about one third as long as the calyx-tube, which has few scattered hairs and is often purple; bracts awl-shaped rigid, ciliate, with spiny tips. Leaves large, hairy, blade 10–20 cm, with a stout leaf-stalk nearly as long; stem stout, 4–angled.

1122 **P. setigera** Falc. ex Benth.
Pakistan to S.E. Tibet. 2400–3600 m. Forest clearings, shrubberies. Jun.–Jul.

Like 1121, but differing in its shorter leaf-stalks, its upper leaves which are 1.2–2.5 cm, and its much shorter spine-tipped bracts which are shorter than the calyx. Flowers pink to mauve, c. 2 cm, many in each whorl. Nutlets only 4 mm (6 mm in 1121).

10 LAMIUM Corolla-tube widening above, 2–lipped, the upper lip hooded over the stamens, the lower lip with 2 teeth-like lateral lobes and a broad notched mid-lobe. Calyx 5–veined; anthers diverging, usually hairy. 6 spp.

Flowers white

1123 **L. album** L. WHITE DEAD NETTLE
Pakistan to C. Nepal. Temperate Eurasia. 1500–3700 m. Shrubberies, damp places. Apr.–Jul.

A hairy rather slender perennial, with white, sometimes lilac-tinged flowers, in few distant whorls forming a leafy inflorescence. Flowers to 2.5 cm, with a curved corolla-tube with a swollen base, and with an arched, hairy upper lip, and a small bi-lobed lower lip; calyx c. 1 cm, lobes with long linear apices and longer than the calyx-tube, usually hairy. Leaves ovate-heart-shaped, 2.5–8 cm, coarsely toothed, the lower stalked; stem ascending, 15–45 cm.

Flowers purple or pink

LABIATEAE

1124 **L. rhomboideum** (Benth.) Benth.
Afghanistan to Himachal Pradesh. 3600–4500 m. Screes; rather rare. Jul.–Aug.

A remarkable plant, with large crowded densely velvety-haired wrinkled, rhomboid, toothed leaves with short thick woolly leaf-stalks, in a lax rosette, and with large purplish flowers at the centre. Flowers to 3 cm, almost stalkless; corolla-tube straight, with a broad woolly curved upper lip and with a 3–lobed lower lip; calyx densely woolly-haired, *c*. 1.3 cm, with awl-shaped lobes with bristly tips. Leaves to 5 cm across; stem very stout, to 5 cm. **Page 497**.

1125 **L. tuberosum** Hedge
Endemic to C. Nepal. 3600–4500 m. Stony slopes; drier areas. Jul.–Aug.

A low loosely-matted plant, with lax clusters of basal leaves, and several relatively large pink flowers arising from each leaf-cluster. Flowers *c*. 2.5 cm long; corolla-tube hairy, much longer than the calyx, with a rounded concave upper lip, and a longer bi-lobed lower lip; calyx *c*. 8 mm, with unequal triangular bristly-tipped lobes. Leaves with ovate blade 2–3 cm, coarsely toothed, stalked; stems slender spreading, to 20 cm; rootstock stout woody. **Pl. 101**.

11 LEONURUS Corolla-tube shorter than calyx, not widening at throat, 2–lipped, the upper lip hooded, the lower 3–lobed. Calyx 5–veined, with 5 spiny-pointed lobes; anthers parallel or nearly so. 2 spp.

1126 **L. cardiaca** L. MOTHERWORT
Pakistan to W. Nepal. Temperate Eurasia. 2400–3600 m. Shrubberies, forest clearings. Jun.–Aug.

An erect leafy perennial, with small pinkish, mauve or white flowers in numerous dense rounded whorls, forming a long interrupted leafy spike with distinctive lobed bracts and leaves. Flowers *c*. 12 mm, hairy; corolla 2–lipped, the upper concave, densely hairy, the lower 3–lobed; calyx with 5 spreading awl-shaped spine-tipped teeth; anthers purple; bracts spine-tipped. Leaves ovate to lanceolate, green above and paler beneath, variously cut, but usually with 3–7 deep or shallow triangular, toothed lobes, the upper leaves entire or 3–lobed, stalked; stem 60–150 cm.
The plant has been used medicinally. **Pl. 100**.

12 STACHYS WOUNDWORT Corolla 2–lipped, the upper lip flat or arched, the lower 3–lobed, with the mid-lobe largest. Calyx 5- or 10–veined. 7 spp.

Herbaceous perennials

1127 **S. sericea** Wallich ex Benth.
Afghanistan to W. Nepal. Bhutan. 2100–4000 m. Open slopes, meadows; a common Kashmir alpine. Jun.–Aug.

An erect silky-haired perennial, with pink purple-spotted flowers in dense whorls, forming a long more or less interrupted terminal leafy spike. Corolla c. 1.5 cm, hairy, corolla-tube little longer than the calyx, the upper lip concave, the lower lip larger, 3–lobed with the mid-lobe longer and notched; calyx longer than broad, c. 1 cm, bristly-hairy, the lobes triangular spine-tipped, sometimes red. Leaves stalked, oblong-heart-shaped, toothed, blade 3–7 cm; stems several, unbranched, 60–120 cm. **Pl. 103**.

1128 S. melissaefolia Benth.
Kashmir to S.E. Tibet. 2400–4000 m. Open slopes, forest clearings. Jun.–Aug.
Like 1127, but flowers smaller c. 8 mm; calyx with narrow triangular lobes thickened at the tip (not spine-tipped), and calyx-tube as broad as long. An erect shaggy or woolly-haired perennial, with pink flowers spotted with purple, in distant many-flowered whorls.

Small shrubby plants

1129 S. tibetica Vatke
Pakistan to Kashmir. C. Asia. 2100–3600 m. Stony slopes, drier areas; common in Ladakh. Jun.–Sep.
A low much-branched greyish shrubby perennial to 60 cm, with many slender stiff stems with narrow leaves, and with few pink flowers in widely-spaced whorls. Flowers 2–2.5 cm, hairy; corolla-tube short, upper lip oblong notched, longer than the tube, lower lip 3–lobed with a longer mid-lobe and short lateral lobes; calyx to 9 mm, with triangular long-tipped spreading or erect lobes. Leaves lanceolate to narrow-elliptic 1.3–2.5 cm, entire or shallowly lobed, narrowed to a short stalk, or stalkless. **Pl. 104**.

13 ERIOPHYTON Corolla 2–lipped, the upper lip broad, hooded, arching over the smaller 3–lobed lower lip; calyx bell-shaped, 10–veined. 1 sp.

1130 E. wallichii Benth.
W. Nepal to S.W. China. 4300–5400 m. Stony slopes, screes. Jul.–Sep.
A dwarf very densely long-woolly-haired plant, rather like a ball of cotton-wool, with rounded overlapping leaves, and with whorls of few wine-red to pale purple densely woolly-haired flowers. Corolla 2.5–3 cm, with a broad very hairy upper lip arching over the smaller lower lip; calyx papery, bell-shaped, with 5 fine-pointed lobes; bracts wedge-shaped papery, deeply toothed. Leaves 4–5 cm broad, covered with very soft long white woolly hairs, like the bracts and the calyx; stem mosty below surface, slender, with scales below, 5–20 cm. **Pl. 103**.

14 LEUCAS Corolla 2–lipped, the lower lip longer than the upper. Calyx 10–veined, usually 10–toothed. 6 spp.

LABIATEAE

1131 **L. lanata** Benth.
Kashmir to C. Nepal. India. China. 700–3000 m. Stony slopes, cultivated areas. Apr.–Sep.

A softly densely woolly-haired silvery-leaved perennial, with small ovate or oblong leaves, and with numerous dense axillary whorls of small white 2–lipped flowers. Corolla *c.* 1 cm, the tube included in the calyx, the upper lip hooded, very hairy, the lower 3–lobed, longer, with the mid-lobe rounded and larger than the 2 blunt side-lobes; calyx *c.* 8 mm, tubular, with 10 short teeth, densely woolly-haired. Leaves short-stalked, 2–3.5 cm, coarsely saw-toothed, with impressed veins above; stems several, often woody below, stout, bluntly 4–angled, 30–45 cm. **Pl. 104.**

15 ROYLEA Corolla 2–lipped, the upper lip erect, entire, the lower lip spreading 3–lobed. Calyx 10–veined, with 5 large broad lobes. 1 sp.

1132 **R. cinerea** (D. Don) Baillon (*R. elegans* Wallich ex Benth.)
Kashmir to W. Nepal. 600–2400 m. Open slopes. Mar.–Apr.

An erect much-branched shrub 1–2 m, with pale brown finely woolly branches with small leaves, and with axillary whorls of small white flowers tinged with pink and with conspicuous green leafy calyx largely concealing the corolla. Corolla hairy, 1.3 cm, scarcely longer than the calyx, 2–lipped, the upper lip hood-like erect, the lower 3–lobed, with the mid-lobe longest; calyx pale green, 10–veined, with 5 blunt broadly spathulate lobes as long as the tube. Leaves 1.5–3 cm, ovate, coarsely toothed, woolly beneath, shortly stalked. **Page 497.**

16 COLQUHOUNIA Flowers orange or scarlet; corolla-tube incurved and inflated at the throat, upper lip short, lower lip 3–lobed. Calyx obscurely 10–veined, with 5 lobes. Nutlets winged above. 1 sp.

1133 **C. coccinea** Wallich
Uttar Pradesh to S.W. China. Burma. S.E. Asia. 1200–3000 m. Shrubberies; common in Nepal. Sep.–Oct.; also occasionally in summer.

A tall rambling shrub 1.3–3 m, with branchlets and usually the undersides of leaves and leaf-stalks covered with thick woolly star-shaped hairs, and with large tubular red flowers striped with orange within. Flowers in dense axillary whorls, or forming dense terminal spikes to 8 cm or more. Corolla 1.5–5 cm long, tube funnel-shaped, 2–lipped, the upper lip elliptic entire, the lower deeply 3–lobed; calyx woolly-haired, 10–veined and 5–lobed. Leaves 8–15 cm, opposite, broadly lanceolate long-pointed, toothed, rugose above, stalked and sometimes white-woolly beneath. **Pl. 103.**

17 NEPETA CATMINT Corolla with a slender and strongly curved tube, the upper lip erect, 2–lobed, the lower 3–lobed with the mid-lobe much larger and concave, and lateral lobes usually deflexed; calyx with 5 nearly equal lobes, 15–veined. *c.* 31 spp.

328

Whorls of flowers in dense unbranched spikes, usually without distant whorls below.

a *Leaves stalkless, or almost so*

  i *Leaves linear to linear-lanceolate*

1134 N. linearis Royle ex Benth.
Pakistan to Uttar Pradesh. 1500–4000 m. Open slopes. May–Aug.
A hairy perennial with linear-lanceolate entire leaves, and with compact whorls of purplish-blue to yellowish flowers crowded into small heads or terminal spikes to 5 cm long, often with interrupted whorls below. Flowers *c*. 1.5 cm long; corolla-tube curved, twice as long as the calyx which is hairy and has linear-lanceolate, ciliate, spine-tipped lobes; bracts elliptic, spine-tipped. Leaves 2.5–6 cm, the upper stalkless, blunt or acute; stems often several, with few leaves, ascending, 15–45 cm.

1135 N. connata Royle ex Benth.   ✓ Manali 7.9.98
Pakistan to Himachal Pradesh. 2100–3600 m. Shrubberies, open slopes. Aug.–Sep.
Differs from 1134 in having stalkless leathery linear-lanceolate long-pointed entire leaves with heart-shaped bases, 8–20 cm. Flowers larger, purplish-violet, *c*. 2 cm, with corolla-tube longer than the calyx, in a dense cylindrical spike 2.5–12 cm, with large ovate fine-pointed lower bracts, and narrow upper bracts. Calyx *c*. 1.5 cm, with very long fine-pointed bristly-hairy lobes as long as the calyx-tube, or longer. Stem unbranched, 30–60 cm. **Pl. 101**.

1136 N. nervosa Royle ex Benth.
Pakistan to Kashmir. 2100–3600 m. Meadows, open slopes. Jul.–Sep.
A showy species with a stout cylindrical spike 3–15 cm long of pale blue or purple flowers. Flowers *c*. 1.5 cm, twice as long as calyx; calyx papery, with long slender lobes and spreading hairs; bracts ovate to lanceolate fine-pointed. Leaves linear-lanceolate fine-pointed, 5–10 cm, toothed, with rounded or heart-shaped bases; stems little-branched, hairless, 30–60 cm.

  ii *Leaves oblong to ovate-heart-shaped* (see also 1140)

1137 N. podostachys Benth.
Afghanistan to Himachal Pradesh. 2700–4300 m. Stony slopes; common in Ladakh and Lahul. Jul.–Sep.
A slender plant with many spreading stems, with small leaves, and with terminal cylindrical spikes formed of whorls of small white flowers. Corolla *c*. 8 mm, upper lip narrow, entire, hooded, lower lip with the two lateral lobes reflexed. Leaves few, oblong-elliptic, 1–3 cm, finely toothed, short-stalked; stems slender, many, to 30 cm, **Pl. 100**.

1138 N. coerulescens Maxim. (*N. thomsonii* Benth. ex Hook.f.)
Kashmir to Sikkim. Tibet. 3600–5300 m. Open slopes; Tibetan border areas. Jul.–Aug.

Flowers purple or blue, in dense ovoid or conical leafy spikes usually 2.5 cm long, with large green leafy bracts much longer than the calyx. Corolla *c*. 1 cm; calyx with ovate-lanceolate lobes much shorter than the calyx-tube. Leaves oblong with heart-shaped base, 2.5–4 cm, blunt or acute, with rounded teeth, many closely arranged up stem, stalkless or the lower short-stalked; stems several, rather stout, erect to 30 cm. An aromatic plant.

b *Leaves distinctly stalked* (see also 1138)

1139 **N. discolor** Royle ex Benth.
Afghanistan to W. Nepal. 2700–4800 m. Stony slopes; common in Ladakh and Lahul. Jun.–Aug.

A low spreading perennial, with erect flowering stems with white or pale purple flowers, in ovoid or cylindric spikes 2–4 cm long, with narrow-ovate pointed often purplish bracts. Corolla to *c*. 1 cm, lower lip often paler; calyx with awl-shaped spiny-pointed lobes as long as the calyx-tube, shaggy-haired. Leaves small, mostly 1–2 cm, broadly ovate-heart-shaped, rounded-toothed, the upper short-stalked, often densely white-woolly beneath; stems weak trailing, sometimes reddish, to 30 cm.

1140 **N. lamiopsis** Benth. ex Hook.f.
W. Nepal to Bhutan. Tibet. 3600–4500 m. Open slopes, drier areas only. Jul.–Aug.

Flowers deep purple, in dense whorls forming ovoid spikes with distinctive large rounded, toothed, green bracts. Spikes 2–4 cm long, whorls few, much shorter than the subtending leaves; corolla 12–15 mm; calyx mauve, with awl-shaped teeth. Leaves few, heart-shaped blunt with coarsely rounded-toothed blade 6–30 mm, usually short-stalked, often paler beneath; stems several, slender unbranched, 15–45 cm.

1141 **N. laevigata** (D. Don) Hand.–Mazz. (*N. spicata* Wallich ex Benth.)
Afghanistan to S.W. China. 3000–4500 m. Open slopes; drier areas only. Jul.–Sep.

Flowers blue-purple, in dense whorls crowded into long terminal spikes 5–10 cm long, sometimes with one or a few widely spaced whorls at the base. Flowers 8–12 mm; corolla-tube much longer than calyx, upper lip deeply 2–lobed, lower lip short, rounded; calyx with linear-lanceolate long-pointed, bristly-hairy lobes as long as the calyx-tube; lower bracts ovate, the upper lanceolate purplish. Leaves usually stalked, 2.5–5 cm, ovate acute to triangular-lanceolate, coarsely saw-toothed; stems erect, 30–90 cm. **Pl. 105**.

1142 **N. longibracteata** Benth.
Pakistan to Himachal Pradesh. 4400–4800 m. Screes, stony slopes of Tibetan borderlands. Jun.–Aug.

A dwarf spreading, strongly aromatic lemon-scented, densely woolly perennial, with lobed grey-woolly leaves, and with globular woolly heads of violet-blue darker-spotted flowers with long curved corolla-tubes. Flower-heads dense *c*. 2.5 cm across; corolla to twice as long as silky-haired calyx, lower lip of corolla broad, upper lip 2–lobed, stamens included; calyx-tube slender, with 5 lanceolate acute lobes, *c*. 9 mm; inner bracts linear, very hairy, often purple. Leaves with a wedge-shaped, rounded-toothed or lobed blade *c*. 8 mm long and broad; stems several 8–15 cm, erect or spreading. Rootstock fibrous, creeping. **Page 497**.

Whorls of flowers in branched clusters, or whorls distant
a Flowers 12 mm or less long

1143 **N. floccosa** Benth.
Pakistan to Uttar Pradesh. 2700–4400 m. Stony slopes; prominent in Ladakh. Jun.–Sep.
Readily distinguished by its greyish, lemon-scented leaves densely covered with matted woolly-white hairs, and its dense rounded widely spaced woolly whorls of pinkish-mauve to blue flowers. Inflorescence sparsely branched at base; whorls 1.5–2 cm across; flowers *c*. 8 mm,with slender corolla-tube twice as long as the calyx which is densely hairy, purple, and has linear pointed lobes; bracts linear pointed, hairy. Leaves with rounded-heart-shaped blade 1–4 cm, obscurely toothed; stems many, spreading from the base, to 60 cm.

1144 **N. leucophylla** Benth.
Himachal Pradesh to C. Nepal. 2400–3600 m. Open slopes; drier areas. Jul.–Aug.
A hairy slender-stemmed perennial, with small lilac or blue flowers in an interrupted spike, with at least the lower whorls distinctly stalked and distant. Flowers 10–12 mm, hairy, twice as long as the hairy calyx which has linear-lanceolate lobes which are shorter than the calyx-tube. Leaves heart-shaped, 2–4 cm, blunt, with rounded teeth, wrinkled above, silvery-haired beneath, stalked; stem branched, 60–90 cm.

b Flowers more than 12 mm

1145 **N. govaniana** (Benth.) Benth.
Pakistan to Uttar Pradesh. 2400–3300 m. Forests. Aug.–Sep.
Distinguished by its large yellow or pale mauve flowers 2.5–3 cm long, in distant whorls, borne in branched clusters, the lower stalked. Corolla-tube curved and conspicuously widened towards the mouth, upper lip 2–lobed, concave, lower lip broader, shallowly 2–lobed, purple spotted within; calyx *c*. 8 mm, with triangular lobes shorter than the straight tube. Leaves 8–12 cm, stalked, ovate-oblong, toothed; stems hairy, erect, 60–120 cm. An aromatic and sticky plant.

1146 **N. clarkei** Hook.f.
Pakistan to Kashmir. 2700–3300 m. Open slopes; in clumps. Jul.–Sep.
Flowers blue with paler lower lip, and with a slender corolla-tube to three times as long as the calyx, borne in dense but widely spaced whorls

and forming narrow terminal spikes 8–15 cm long, with the lower whorls shortly stalked. Corolla *c*. 15 mm, upper lip short with 2 reflexed lobes, lower lip with 2 rounded lateral lobes and a broader rounded central lobe streaked and spotted with purple. Calyx 7 mm, with triangular lobes shorter than the tube, hairy; bracts lanceolate, the lower leafy. Leaves ovate-lanceolate to lanceolate, 2.5–6 cm, toothed, shortly stalked; stem erect to 60 cm. **Pl. 105.**

1147　**N. erecta** (Benth.) Benth.
Pakistan to Uttar Pradesh. 2100–3600 m. Forests, open slopes. Jun.–Aug.

Flowers large, blue to blue-purple, with a curved corolla-tube much enlarged at the mouth and with a larger lower lip, borne in distant whorls subtended by longer leaves. Flowers 2–2.5 cm; calyx curved 10 mm, with triangular lobes much shorter than the tube. Leaves triangular to oblong-heart-shaped with acute or blunt teeth, blade usually 3–6 cm, stalked; stems erect, softly hairy, 30–90 cm. **Pl. 105.**

1148　**N. royleana** R. R. Stewart (*N. salviaefolia* Royle ex Benth.)
Pakistan to Kashmir. 2400–3600 m. Stony slopes. Jul.–Sep.

A slender branched densely white-woolly perennial, with either long narrow interrupted spikes up to 30 cm, or with branched clusters, of pale blue or white flowers *c*. 1.5 cm long. Corolla-tube very slender more than twice as long as calyx, corolla-lobes small *c*. 6 mm across; calyx 6 mm, with triangular lobes much shorter than the tube; bracts linear-lanceolate. Leaves shortly stalked, 1.5–4 cm, oblong to narrow-ovate, thick, wrinkled above and densely and closely white-woolly haired beneath, sparsely scattered up stem; stems usually many, erect, 30–60 cm.

1149　**N. glutinosa** Benth.
Pakistan to Kashmir. 3300–4400 m. Stony places; prominent in Ladakh. Jul.–Aug.

Flowers violet, borne in a long lax leafy spike in loose axillary whorls subtended by coarsely toothed very glandular-sticky leaves. Corolla slender to *c*. 2 cm, with short lips and exserted style; calyx slender *c*. 1 cm, with narrow triangular lobes, glandular-hairy; bracts lanceolate, toothed. Leaves broadly ovate, with shallow triangular lobes, coarsely saw-toothed, stalkless, 1.5–3 cm; stems many, stout, to 60 cm, forming big clumps to 60 cm or more across.

18　GLECHOMA　Like *Nepeta*, but corolla-tube straight, and hairy within; anther-cells at right-angles, each opening by a separate slit (anther-cells divergent and opening by a single slit in *Nepeta*). 3 spp.

1150　**G. nivalis** (Benth.) Press (*Nepeta n*. Benth.)
Kashmir to S.W. China. Tibet. 4300–5500 m. Screes of high altitudes, in Tibetan borderlands. Jun.–Jul.

A dwarf hairy perennial, with crowded rounded grey-green wrinkled aromatic leaves, and with axillary whorls of up-curved pale mauve flowers.

Corolla to 2 cm, the lips small; calyx curved, 2–lipped, the lobes lanceolate, unequal. Upper leaves with distinctive rounded blades to 2 cm, lower leaves smaller, ovate, obscurely toothed; stems erect 8–16 cm, arising from spreading fibrous underground stems. **Page 497.**

19 DRACOCEPHALUM Calyx 5–toothed more or less 2–lipped, the upper lip entire or 3–toothed with the middle tooth largest; calyx-tube 15–veined. 5 spp.

Upper lip of calyx entire, ovate pointed; flowers blue to mauve

1151 **D. nutans** L.
Pakistan to Kashmir. N. & C. Asia. 2700–4500 m. Alpine slopes. Jun.–Aug.

Flowers blue to blue-lilac, in many-flowered whorls forming a long terminal spike 10–15 cm. Flowers *c.* 1.5 cm long, horizontal or drooping, with a slender corolla-tube; upper lip of corolla oblong, deeply notched, lower lip longer with a rounded terminal lobe and small lateral lobes at base; calyx often purple, with the upper lip broad and the lowest bristle-like; bracts leafy, ovate entire. Lower leaves short-stalked, with elliptic to ovate-heart-shaped, coarsely toothed blade 1–2 cm, the upper leaves stalkless; stems 4–angled, many, spreading, shortly hairy, 15–30 cm. **Pl. 100**.

1152 **D. wallichii** Sealy (*D. speciosum* Wallich ex Benth.)
Himachal Pradesh to S.W. China. 3600–4500 m. Open slopes. Jun.–Aug.

Flowers large blue-purple with darker spots, in whorls forming a dense head surrounded by broad, toothed, upper leaves. Flowers 2.5–3 cm, tubular 2–lipped, shortly hairy; calyx *c.* 16 mm, leathery, the lips unequal, the upper lip ovate, the lower narrower; bracts small, ovate acute, often purple-flushed like the calyx. Leaves with rounded broadly heart-shaped, toothed blade 5–6 cm across, the lower very long-stalked to 20 cm, the upper stalkless and smaller, undersides of leaves sometimes purplish; stems unbranched, 4–angled, 15–45 cm. **Page 497**.

Upper lip of calyx broad, with 3 teeth; flowers white

1153 **D. heterophyllum** Benth.
Pakistan to S.E. Tibet. 3000–5000 m. Open slopes; only in dry country along the Tibetan border. Jun.–Aug.

Flowers white, sometimes tinged mauve or pink, in large dense leafy cylindrical spikes to 10 cm long and 2.5–4 cm across, (often larger than the rest of the plant) and with green elliptic, lobed bracts. Flowers 2–3 cm long, hoary, with corolla-tube inflated above; calyx to 2 cm, but variable in size, hairless, with upper 3 teeth ovate, with bristle-like tips. Leaves leathery, stalked, oblong-ovate with cut-off or heart-shaped base, rounded-toothed, 2–3 cm; stems stout, ascending 15–20 cm. An aromatic plant.

LABIATEAE

20 PRUNELLA Corolla 2–lipped, the upper lip strongly hooded, the lower 3–lobed. Calyx 2–lipped, the upper lip broad with 3 short teeth, the lower lip with 2 long narrow lobes, calyx tube 10–veined. 1 sp.

1154 **P. vulgaris** L. COMMON SELF-HEAL
Afghanistan to Bhutan. N. Temperate Zone. 1500–3600 m. Meadows, open slopes. May–Sep.

Flowers bright blue-violet, rarely white or pink, in dense cylindrical terminal heads 2.5–5 cm long, usually with broad ovate, purplish, overlapping bracts, with purplish calyx, and with a pair of leaves immediately below and spreading beyond the inflorescence. Flowers *c.* 1.5 cm long, in whorls of 6; corolla 2–lipped; calyx 2–lipped, the throat closed after flowering. A thinly hairy erect perennial, with elliptic to lanceolate, entire or toothed, stalked leaves, 2.5–5 cm; stems many, spreading or ascending, 10–30 cm.

21 CLINOPODIUM Corolla-tube longer than calyx, 2–lipped, the upper lip entire the lower lip 3–lobed. Calyx tubular curved, 13–veined, the upper lip 3–lobed, the lower 2–lobed. 3 spp.

1155 **C. piperitum** (D. Don) Press (*Calamintha longicaule* (Wallich ex Benth.) Benth.)
Pakistan to E. Nepal. 1500–2400 m. Shrubberies. Feb.–May.

Flowers violet or reddish-purple, with a slender corolla-tube twice as long as the calyx, borne in lax rather one-sided whorls at the ends of the leafy branches. Corolla 1.5–2 cm; calyx-lobes triangular long-pointed, one third as long as the calyx-tube; bracts shorter than calyx. An erect or spreading, softly hairy perennial, with mostly entire elliptic leaves 1–3 cm; stems 20–90 cm, slender, spreading, woody below. **Pl. 101**.

1156 **C. umbrosum** (M. Bieb.) K. Koch (*Calamintha u.* (Bieb.) Fischer & C. Meyer)
Afghanistan to S.W. China. Iran. Japan. S.E. Asia. 1000–3400 m. Forests, weed of cultivated areas. Apr.–Sep.

Flowers small, pink or purple, in lax few-flowered whorls, with few short slender bracts (not forming an involucre); corolla *c.* 8 mm; calyx *c.* 6 mm with unequal lobes, the lower 2 longer and narrower, with bristly hairs. A softly hairy perennial, with spreading or erect stems 30–90 cm, and with short-stalked, ovate, sharply toothed leaves 2–4 cm.

1157 **C. vulgare** L. WILD BASIL
Pakistan to C. Nepal. Temperate Eurasia. 1800–3300 m. Open slopes; common in Kashmir. Jun.–Aug.

Differs from 1156 in having dense compact globular whorls of many pink or purple flowers, each whorl surrounded by an involucre of numerous linear bristly-haired bracts. Corolla *c.* 1 cm, with tube little longer than calyx, and with a flat blunt upper lip, and a slightly longer lower lip with 3 rounded lobes, and with a pale centre; calyx *c.* 9 mm. Leaves triangular-ovate, mostly 2–3 cm; stems many, erect, 10–40 cm.

22 MICROMERIA Corolla-tube straight, 2–lipped, the upper lip erect and entire or notched, the lower 3–lobed. Calyx with 13–15 veins, and 5 equal pointed lobes. 2 spp.

1158 **M. biflora** (Buch.–Ham. ex D. Don) Benth.
Afghanistan to Bhutan. India. Burma. 1000–3000 m. Sandy banks, dry slopes. Mar.–Sep.
A small usually tufted thyme-like shrublet 15–30 cm, with numerous tiny leaves and with axillary whorls of 1–4 small, stalked, pink, pale purple or nearly white flowers. Corolla *c.* 6 mm, corolla-tube slightly longer than calyx, 2–lipped, the upper lip erect and nearly flat, the lower spreading and 3–lobed; calyx *c.* 3 mm hairy, prominently 13–veined, with 5 equal awl-shaped lobes. Leaves 4–6 mm, elliptic to ovate acute, gland-dotted, stalkless. **Page 498**.

23 ORIGANUM Corolla 2–lipped, the upper lip erect and notched, the lower lip 3–lobed. Calyx hairy within, bell-shaped, with 13 veins, and with 5 equal lobes. Stamens diverging. 2 spp.

1159 **O. vulgare** L. MARJORAM
Pakistan to Bhutan. Temperate Eurasia. 1500–3600 m. Open slopes. Jun.–Sep.
A very aromatic perennial, with small pale pink flowers crowded into a branched domed inflorescence, often with calyx and bracts flushed purple, or less commonly flowers in short spike-like clusters. Flowers 5–8 mm, 2–lipped, the upper lip notched; calyx glandular-hairy, lobes equal, enlarging in fruit; bracts green or purple, ovate. An erect perennial 20–80 cm, with ovate entire, stalked leaves 1–4 cm.
The plant is used medicinally. **Page 498**.

24 THYMUS THYME Corolla weakly 2–lipped, the upper lip entire or notched, the lower 3–lobed. Calyx 10–13–veined, 2–lipped, the upper 3–lobed, the lower with 2 narrow bristles, throat closed with a ring of hairs. Stamens 4, 2 usually longer than corolla. 1 sp.

1160 **T. linearis** Benth. ex Benth.
Afghanistan to C. Nepal. Tibet. India. N. China. Japan. 1500–4300 m. Rocky slopes. Apr.–Sep.
A small spreading very aromatic often tufted shrublet, with many tiny elliptic-oblong leaves, and with small whorls of purple flowers crowded into short dense terminal clusters. Flowers *c.* 6 mm or more, weakly 2–lipped; calyx 2–lipped with ciliate lobes, veined, with shaggy hairs in throat; bracts minute. Leaves *c.* 8 mm, nearly stalkless, entire blunt, gland-dotted; stems spreading 10–20 cm.
The plant is used medicinally. **Page 498**.

25 MENTHA MINT Corolla tubular with 4 equal lobes. Calyx 10–13–

LABIATEAE

veined, and usually with 5 equal lobes. Stamens diverging, exserted or not.
3 spp.

1161 **M. longifolia** (L.) Hudson (*M. sylvestris* L.) HORSE-MINT
Pakistan to C. Nepal. Tibet. N. & W. Asia. Europe. 1500–3800 m. Bogs,
streamsides, irrigation channels. Jul.–Aug.

A very aromatic hairy perennial, with tiny lilac flowers in whorls
forming slender spikes often interrupted below, borne at the ends of
branches and forming a lax densely hairy inflorescence. Flowering spikes
2.5–8 cm long by 8–12 mm broad. Flowers *c.* 3 mm across, hairy outside,
stamens exserted; calyx with 5 triangular lobes; bracts lanceolate. Leaves
oblong-obovate to lanceolate 2.5–8 cm, saw-toothed, with rounded base,
densely hoary, paler beneath; stems 30–100 cm.
The plant is used medicinally. **Page 498**.

26 SALVIA SAGE Stamens 2, each with 1 fertile anther, hinged. Corolla
2–lipped, the upper usually hooded, the lower 3–lobed, larger. Calyx
2–lipped, the upper lobe entire or 3–lobed, the lower 2–lobed. 11 spp.

Leaves thick, wrinkled; calyx spiny-toothed; bracts large

1162 **S. lanata** Roxb.
Pakistan to W. Nepal. 1500–3000 m. Open slopes, drier areas. Apr.–Jun.

A white-woolly perennial, with mostly basal oblong-lanceolate leaves,
and with blue flowers in many distant whorls in a terminal spike, with
conspicuous rounded sticky bracts. Flowers 2.5 cm long; corolla-tube not
longer than calyx; calyx sticky-glandular, ribbed, 2–lipped with spine-
tipped teeth; bracts rounded, broader than long, abruptly pointed. Leaves
obscurely toothed, 8–15 cm, stalkless, woolly or nearly hairless above,
paler and densely woolly and closely wrinkled beneath; stems 15–25 cm.
Pl. 101.

1163 **S. moorcroftiana** Wallich ex Benth.
Pakistan to W. Nepal. 1500–2700 m. Open slopes, wasteland; common in
Kashmir valley. May–Jun.

Differing from 1162, in being a more robust plant with long-stalked ovate
to elliptic, toothed leaves, and with pale blue, lilac or nearly white flowers
in a lax branched cluster. Flowers *c.* 2.5 cm long; corolla-tube distinctly
longer than calyx, lower lip with lateral lobes reflexed; calyx bristly; bracts
conspicuous, pale, green-veined, nearly rounded and with an abrupt
point. Leaves mostly 15–25 cm; stem branched above, 45–90 cm. **Pl. 101**.
The plant is used medicinally.

Leaves thin, not wrinkled; calyx-lobes shallow, not spiny; bracts small
a *Flowers purple or violet*

1164 **S. hians** Royle ex Benth.
Pakistan to Bhutan. 2400–4000 m. Forests, shrubberies, open slopes; a
common Kashmir alpine. Jun.–Sep.

336

Readily distinguished by its large deep blue flowers *c.* 4 cm long, with a broad inflated corolla-tube, and much shorter lips which are often marked with white. Whorls few-flowered, distant, in a spike-like and little-branched cluster; calyx 1.5 cm, conspicuous, dark, broadly funnel-shaped with 2 broad shallow lips. A robust sticky perennial, with large ovate-heart-shaped or arrow-shaped, rounded-toothed leaves to 25 cm or more including the long leaf-stalk; stem 30–80 cm. **Pl. 102**.

b *Flowers yellow*

1165 **S. campanulata** Wallich ex Benth.

Uttar Pradesh to S.W. China. 2700–4000 m. Forests, shrubberies. Jun.–Jul.

Like 1164, but flowers large yellow, with a short inflated corolla-tube which is often little longer than the funnel-shaped very shallow-lobed calyx. Flowers 2–3 cm, 2–lipped, the lower lip darker yellow and with reflexed triangular lobes; corolla sometimes streaked with brownish purple. A robust glandular-sticky hairy perennial, with ovate-heart-shaped, toothed, long-stalked leaves. **Pl. 101**.

1166 **S. nubicola** Wallich ex Sweet (*S. glutinosa* auct. non L.)
Afghanistan to Bhutan. 2100–4300 m. Open slopes, forest clearings. Jun.–Aug.

A tall robust strong-smelling, sticky glandular-hairy perennial, with large yellow flowers with orange upper lips, borne in large branched spreading terminal inflorescences. Whorls few-flowered, lax; corolla 2–3 cm, the tube little longer than the calyx, and upper lip long curved and with an in-curved margin, lower lip 2–lobed; calyx *c.* 1 cm, shallowly 2–lipped, the upper lip ovate acute entire. Leaves glandular-hairy, ovate-oblong with triangular spreading basal lobes, toothed, stalked, 10–18 cm; stem 60–120 cm. **Pl. 105**.

27 **PEROVSKIA** Flowers small; corolla 2–lipped, with a ring of hairs within, upper lip erect 4–lobed, lower lip elliptic-ovate entire; calyx 2–lipped. Fertile stamens 2. 2 spp.

1167 **P. abrotanoides** Karelin
Afghanistan to Kashmir. C. Asia. 2700–3600 m. Stony slopes; prominent, forming large clumps in Ladakh. Jun.–Sep.

A lavender-like white- or greyish-hairy, twiggy, strong-smelling shrub, with linear-oblong, lobed leaves, and with usually a branched inflorescence of small distant whorls of small blue-violet flowers. Corolla *c.* 1 cm, 2–lipped, with the tube longer than the calyx, the upper lip 4–lobed, the lower entire smaller; calyx with long cottony hairs and triangular lobes. Leaves 2–5 cm, with oblong blunt widely spaced lobes, wrinkled and covered with small bright brown scales, upper leaves nearly entire; stems 60–120 cm.

Plant contains a reddish pigment used in dyeing. **Page 498**.

28 ELSHOLTZIA Flowers small; corolla-tube longer than calyx, 4–lobed, the upper lobe erect, the others spreading, stamens exserted; calyx 5–toothed, enlarging in fruit. *c.* 10 spp.

Shrubby plants to 1 m or more

1168 E. flava (Benth.) Benth.
Uttar Pradesh to S.W. China. 1500–2400 m. Shrubberies. Sep.–Oct.
A shrub with ovate pointed toothed leaves, and with many dense slender terminal and axillary spikes of numerous tiny yellow flowers with protruding stamens. Spikes 5–10 cm long; corolla *c.* 5 mm long, twice as long as the calyx, which enlarges in fruit. Leaves glandular, 10–20 cm, with rounded or heart-shaped bases, with conspicuously impressed lateral veins above, long-stalked; stems to 2 m, branches rough, 4–angled.
Pl. 100.

1169 E. fruticosa (D. Don) Rehder (*E. polystachya* Benth.)
Pakistan to S.W. China. 1800–3300 m. Open slopes, shrubberies; common in Nepal. Sep.–Oct.
Like 1168, but with narrower elliptic-lanceolate leaves which are narrowed below to a very short leaf-stalk, and with longer narrower spikes of tiny white hairy flowers. Spikes strongly aromatic, 10–25 cm long and to 8 mm broad, with lanceolate bracts; corolla 3–4 mm long, densely hairy like the much shorter calyx. Leaves 8–16 cm, saw-toothed, glandular, finely hairy; stems mostly 1–3 m, with grey fibrous bark on older stems which peels off in strips. **Pl. 100**.

Annuals less than 1 m

1170 E. eriostachya (Benth.) Benth.
Pakistan to S.W. China. 3000–4500 m. Open slopes, damp places in drier areas. Jul.–Aug.
An aromatic mint-like slender erect annual, with oblong to lanceolate leaves and with terminal stout cylindrical shaggy-haired spikes of numerous very tiny yellow flowers. Spikes 2.5–6 cm by 4 mm at flowering, enlarging to 12 cm in fruit; corolla 2 mm; calyx 1 mm, with linear hairy lobes, enlarging to 4 mm. Leaves 2.5–4 cm, toothed, softly hairy, or sometimes woolly beneath, shortly stalked; stems 4–angled, often unbranched, or sometimes branched below, 15–45 cm. **Pl. 105**.

NYCTAGINACEAE   Four O'Clock Family

A family of mostly tropical herbs, trees, and shrubs, with petal-like tubular perianth (there are no true petals) with 5 lobes, and usually 5 stamens, free or fused at the base. Ovary superior, of one carpel with one ovule. Fruit not splitting, enclosed in persistent perianth.

MIRABILIS 1 native sp.

1171 **M. jalapa** L. MARVEL OF PERU
Native of Peru. Semi-naturalized in the foothills, from Pakistan to Bhutan; often planted in gardens. 600–1800 m. Jul.–Oct.

A leafy herbaceous plant, with fragrant long-tubular flowers which open in the evening, varying in colour from white, yellow to crimson, and often striped or blotched. Flowers with slender tube 4–5 cm, widening abruptly to a wide limb with rounded petal-like lobes which spread to 3–4 cm across, and with an involucre of 5 calyx-like bracts; flowers borne in clusters among the leaves at the ends of the branches. Leaves large entire, ovate-lanceolate long-pointed, with rounded or cut-off bases, short-stalked, 6–10 cm; stem to 120 cm. Fruit elliptic, *c.* 9 mm.

Plant used medicinally.

Bougainvillea glabra CHOISY Commonly grown in gardens; native of S. America. A scrambling shrub, with alternate leaves, and large conspicuous bright magenta bracts surrounding the yellow cylindrical flowers.

AMARANTHACEAE   Cockscomb Family

A large family of herbs and shrubs of the tropical, sub-tropical and temperate regions. Leaves entire, opposite or alternate, without stipules. Inflorescence in our species arranged in a spike-like cluster or head, with chaffy scales subtending the flowers. Flowers regular, with 4–5 usually dry and papery perianth-segments, which are sometimes fused below; stamens 1–5, often united into a tube at base; ovary superior, of 2–3 fused carpels, which are free from or united with the perianth. Fruit a capsule or berry enclosed by the persistent perianth.

AMARANTHUS

1172 **A. caudatus** L. LOVE-LIES-BLEEDING
Country of origin unknown; cultivated in the Himalaya to 2400 m. May–Aug.

Very prominent in the autumn, with its bright red long pendulous clusters of numerous tiny flowers borne on robust erect leafy stems. Inflorescence branched, with numerous slender spikes; flowers 1–2 mm long; perianth-segments papery, obovate fine-pointed; bracts awl-shaped longer. Leaves entire, elliptic to oblong and usually blunt, long-stalked, to 12 cm; a robust annual to 1 m.

Another common cultivated plant is **A. hybridus** L. subsp. **hypochondriacus** L. PRINCE'S FEATHER, which has deep crimson erect spikes of flowers, oblong-lanceolate acute purple or greenish-purple leaves, and stem to 2 m.

339

This and other hybrids are grown as pot-herbs, and for their edible seeds. **Pl. 106.**

CHENOPODIACEAE Goosefoot Family

A widely distributed family, particularly of salt-rich habitats: by the sea, in steppes, and alkaline prairies. Usually annual or perennial herbs, or fleshy shrubby plants, with small lobed or spiny leaves, often mealy or covered with hairs. Flowers inconspicuous, arranged in spikes or branched clusters, usually regular; perianth-segments similar, brownish or greenish, 2, 3 or 5, more or less fused; stamens similar in number. Ovary superior, with a single chamber with 1 ovule. Fruit a nutlet.

CHENOPODIUM Annuals. Perianth-segments 2–5, persistent in fruit; stamens 1–5. *c.* 10 spp.

1173 **C. foliosum** (Moench) Asch. (*C. blitum* Hook.f.)
Pakistan to C. Nepal. Temperate Eurasia. N. Africa. 1800–3600 m. Wastelands, cultivated areas; largely in drier regions; common in W. Himalaya. Jun.–Sep.
Readily distinguished by its bright red fleshy fruit borne in dense stalkless clusters in the axils of triangular and coarsely and irregularly toothed leaves. Flower-clusters 6–8 mm across, stalkless, axillary, green, forming a terminal leafy spike. A hairless spreading or ascending annual, with bright green leaves 2.5–6 cm, with a long leaf-stalk shorter or longer than the blade, leaves progressively smaller up stem; stem rather stout, white, 30–100 cm. Individual fruit 1–2 mm.

1174 **C. album** L. Fat Hen
N. Temperate Zone. Throughout Himalaya. 1500–3600 m. Field weed, around shepherds' encampments; common. May–Aug.
Usually a mealy-white annual to 1 m, with often purple-tinged leaves and stem and with tiny green flowers in rounded clusters, borne in slender spikes or forming a lax leafy branched terminal inflorescence. Leaves 2.5–6 cm, the lower ovate or oblong, with wavy or toothed margins, stalked, the upper often narrow entire. Perianth-segments keeled, covering the fruit; seeds smooth.
Sometimes cultivated for its edible leaves and seeds; cultivated plants may grown to 3 m, with leaves 10–15 cm.

HALOGETON Annuals, with fleshy cylindrical leaves. Perianth-segments 5, developing a transverse wing in fruit; stamens 3–5. 1 sp.

1175 **H. glomeratus** (M. Bieb.) C. Meyer
Afghanistan to Kashmir. C. Asia. 2700–4000 m. Sandy places in Ladakh. Jul.–Sep.

A much-branched, glaucous or hoary, erect or ascending annual, with clusters of tiny fleshy cylindrical green leaves tipped with a bristle. Flowers minute, solitary or clustered, perianth-segments 4–5, enlarging in fruit with distinctive broad fan-shaped or rounded transparent apices, with pinkish or brownish veins, and spreading to 6–8 mm across. Leaves spathulate, to 5 mm; stems spreading, to 20 cm.

KRASCHENINNIKOVIA Perennial herbs or small shrubs, with star-shaped hairs; leaves flat. Flowers 1–sexed; male flowers with 4 perianth-segments, and 4 stamens; female flowers without perianth, but with 2 persistent bracteoles. 1 sp.

1176 **K. ceratoides** (L.) Gueldenst. (*Eurotia c.* (L.) C. Meyer)
Afghanistan to C. Nepal. Tibet. W. & N. Asia. S. & E. Europe. 3600–4500 m. Stony slopes in dry country. Jul.–Aug.
A shrubby hoary plant with small linear-oblong to narrow-elliptic leaves, and with conspicuous fruiting bracteoles covered with long dense silky red-brown hairs. Flowers yellowish, minute; the male in a terminal spike, 8–15 mm; the females below, axillary, with 2 fused bracteoles veined, hairy, with free tips with 2 beaks which at length close round the fruit and become leathery. Leaves variable, 1–3.5 cm; stems with many slender leafy branches, to 1m, but often dwarfed at high altitudes. **Page 499.**

PHYTOLACCACEAE   Pokeweed Family

A largely tropical family of herbs, shrubs, trees and woody climbers, with small clustered flowers. Leaves alternate, simple, entire; stipules absent or minute. Flowers regular, usually with one whorl of 4–5 perianth-segments which persist in fruit. Stamens 4–5, or more numerous as a result of branching. Ovary superior, of one to many separated or united carpels, each with a single ovule. Fruit a fleshy berry, or a dry nut.

PHYTOLACCA 3 spp.

1177 **P. acinosa** Roxb.
Kashmir to S.W. China. S.E. Asia. 1500–3000 m. Forests, shrubberies; cultivated areas. Jun.–Sep.
A nearly hairless herbaceous plant with robust succulent stems 90–150 cm, with large lanceolate leaves, and with long dense erect cylindrical clusters of greenish-white flowers borne opposite the leaves. Flower clusters 5–15 cm; corolla *c.* 7 mm across, with 5 obovate spreading perianth-segments; stamens 8–10. Leaves mostly 15–25 cm, long-pointed, entire, narrowed to a short stalk. Each fruit with about 8 fleshy dark purple carpels, the fruits crowded into stout cylindrical clusters 10–20 cm long.

Plant said to have narcotic properties, and produces a bitter toxic substance. The leaves make an excellent pot-herb if well boiled. **Pl. 106**.

POLYGONACEAE   Dock Family

A medium-sized cosmopolitan family of herbs and some shrubs, largely of the northern temperate regions. Flowers small and not showy, usually numerous and grouped into branched inflorescences. Leaves alternate, simple, with a characteristic basal sheath, uniting the stipules. Flowers regular, with 3–6 perianth-segments, often enlarged in fruit; stamens 6–9. Ovary superior, of 2–4 fused carpels with one chamber and a single ovule; styles 2–4. Fruit a triangular nutlet.

The genus *Polygonum* has recently been divided into the following genera: *Aconogonum, Bistorta, Persicaria* and *Polygonum*.

Perianth-segments 5
  a Flowers in terminal branched clusters   1 ACONOGONUM
  b Flowers in a spike-like cluster or globular head   2 BISTORTA
                                                       3 PERSICARIA
  c Flowers in axillary clusters   4 POLYGONUM   5 FAGOPYRUM

Perianth-segments 6
  a Perianth-segments not enlarging in fruit   6 RHEUM
  b Perianth-segments enlarging in fruit   7 RUMEX

Perianth-segments 4   8 OXYRIA

1 ACONOGONUM Inflorescence branched. Perianth coloured, petal-like; stamens 8; styles 3; nut 3–angled. *c*. 7 spp.

Shrubby plants, at least at base

1178 **A. molle** (D. Don) Hara (*Polygonum m*. D. Don)
Uttar Pradesh to S.W. China. 1200–2400 m. Forests, shrubberies, damp places; often gregarious. May–Oct.

A robust shrubby perennial to 2.5 m or more, with usually hairy branches and large lanceolate-elliptic silky-hairy leaves, and with large terminal, much-branched, hairy clusters of numerous tiny white flowers. Flower-clusters usually 10–40 cm long; flowers 3–5 mm across, with oblong spreading perianth-segments. Leaves 10–18 cm, short-stalked, either silky-haired, or hairless; stipules variable, to 7 cm; stem rounded in section. Fruit black, rounded, 3 mm. **Pl. 106**.

1179 **A. tortuosum** (D.Don) Hara (*Polygonum t*. D.Don)
Afghanistan to Bhutan. Tibet. 3300–5600 m. Stony slopes; Tibetan borderlands. Jun.–Aug.

A low dense shrub-like plant with spreading branches, with entire leathery stalkless leaves, and with short dense cylindrical clusters of

cream-coloured flowers borne at the ends of the branches. Clusters
1.3–2.5 cm long; flowers *c.* 5 mm across, with ovate erect perianth-
segments. Leaves 1–4 cm, elliptic to broadly ovate, blunt or acute, pale
green, becoming crimson in autumn; stems to 45 cm, smooth, shining
red-brown. **Page 499**.

Tall herbaceous plants
a *Flowers pink or white*

1180 **A. alpinum** (All.) Schur (*Polygonum angustifolium* auct. non Pallas)
Afghanistan to Himachal Pradesh. C. & N. Asia. Europe. 1500–3000 m.
Open slopes, shrubberies; common in Kashmir and Lahul. Jun.–Aug.

A tall herbaceous plant to 2 m, with very numerous tiny pale pink or
white flowers, in much-branched rather narrow pyramidal clusters to 30 cm
or more, and with narrow lanceolate long-pointed leaves. Flowers 1–2 mm
long. Leaves 8–12 cm long, narrowed to a short stalk, hairless or finely
hairy beneath; stipules long, lax; stems grooved, pale reddish or white.

The stems are edible.

1181 **A. campanulatum** (Hook.f.) Hara (*Polygonum c.* Hook.f.)
Uttar Pradesh to Bhutan. Burma. 2100–4000 m. Forests, shrubberies,
damp places. Jul.–Sep.

An erect perennial 60–120 cm, often distinguished by the pale pinkish-
brown or greyish hairs on the rather leathery leaves, and by its spreading
or nodding hairy branched clusters of pink or white flowers. Flowers
bell-shaped, 3–5 mm long, with perianth fused at the base and with
rounded spreading lobes. Leaves elliptic to lanceolate, pointed, 8–16 cm,
short-stalked; stipules large, lax and with long bristly hairs. **Pl. 106.**

b *Flowers green*

1182 **A. rumicifolium** (Royle ex Bab.) Hara (*Polygonum r.* Royle ex Bab.)
Afghanistan to C. Nepal. Tibet. 2700-4300 m. Shrubberies, alpine slopes.
Jun.–Aug.

A robust very leafy perennial, with large dock-like leaves, and with
usually dense axillary and terminal clusters of tiny green flowers. Flowers
4–6 mm across, with rounded spreading perianth-segments. Leaves green
and rather fleshy, broadly ovate to ovate-heart-shaped, 8–13 cm, blunt or
almost acute, leaf-stalk short, stout, to 2.5 cm; stipules large, lax, hairless;
stem stout, unbranched, pale, 15–120 cm.

The young shoots are edible. **Page 499**.

2 **BISTORTA** Flowers usually in dense terminal spikes; perianth coloured,
petal-like. Stamens 6–8; styles 3, long slender; nut 3–angled. 8 spp.

Spreading mat-forming plants, shrubby at least at base

1183 **B. affinis** (D. Don) Greene (*Polygonum affine* D. Don)
Afghanistan to E. Nepal. 3000–4800 m. Open slopes, screes; common.
Jun.–Sep.

A low creeping densely tufted mat-forming alpine plant, with narrow-elliptic leaves which are glaucous beneath, and with cylindrical spikes of very many pale or deep pink flowers, borne at the ends of short erect stems. Flower-spikes 5–7.5 cm; flowers densely crowded round stem, each 4–6 mm long; stamens shortly exserted. Leaves mostly basal, 3–8 cm, narrowed to a short stalk, entire or finely toothed, with a conspicuous mid-vein and with inrolled margin; stipules brown-papery conspicuous, to 2.5 cm. Flowering stems several, 5–25 cm, with few smaller leaves; rootstock woody, branched. **Pl. 106**.

1184 **B. emodi** (Meissner) Hara (*Polygonum e.* Meissner)
Kashmir to S.W. China. 2500–4000 m. Rocks in forests. Jul.–Sep.

A low trailing perennial, readily distinguished by its numerous narrow pointed leaves with inrolled margins, and by its short-stemmed slender cylindrical spikes of red flowers. Spikes 2.5–3.5 cm long; flowers *c.* 4 mm long, short-stalked. Leaves 3–8 cm by 2–4 mm, linear-lanceolate long-pointed; stipules conspicuous, long slender, brown-papery; branches rather woody, creeping, often with widely spaced internodes and with spine-like remains of old leaves. Flowering stems slender, erect, nearly leafless, 5–10 cm.

1185 **B. vaccinifolia** (Wallich ex Meissner) Greene (*Polygonum v.* Wallich ex Meissner)
Kashmir to S.E. Tibet. 3000–4500 m. Rocks, open slopes; common in wetter parts of the Nepal alpine zone. Aug.–Sep.

A low trailing woody-stemmed plant, often carpeting the ground or covering rocks, distinguished by its small ovate leaves, and its erect cylindrical clusters of pink flowers. Clusters 1.5–4 cm long, dense; flowers *c.* 4 mm long, very short-stalked. Leaves 1–2 cm, ovate entire, short-stalked, rather glaucous beneath; stipules brown, rigid, cut into fine teeth. Flowering stems mostly axillary, leafy, mostly 5–10 cm. **Page 500**.

Erect herbaceous plants

1186 **B. amplexicaulis** (D. Don) Greene (*Polygonum a.* D. Don)
Afghanistan to S.W. China. 2100–4800 m. Shrubberies, open slopes. Jul.–Aug.

A slender erect perennial, with ovate-heart-shaped clasping upper leaves, tapering to a long point, and with slender terminal spikes of pink, deep red or white flowers. Spikes 5–15 cm long, usually solitary, sometimes branched; flowers 3–6 mm long, numerous crowded; perianth-segments 5; bracts papery, ovate pointed. Lower leaves long-stalked, the blade 5–15 cm, minutely toothed; stipules tubular, papery, 2.5–6 cm; stem to 1 m or more; rootstock stout, horizontal. Flowering spikes usually erect, but in C. and E. Nepal they are pendulous (var. **pendula** Hara) **Pl. 106. Page 499**.

344

1187　**B. vivipara** (L.) Gray (*Polygonum* v. L.)
Pakistan to S.W. China. N. Temperate Zone. 3300–5000 m. Open slopes, damp places. Jun.–Jul.

Flowers pale pink to white, in a solitary slender spike, with the lower flowers usually replaced by brown or purple bulbils. Spikes 2.5–10 cm long; flowers *c.* 3 mm, but variable, more or less erect; stamens exserted, anthers dark; bracts ovate long-pointed. Leaves leathery, variable, linear to oblong, to 10 cm, acute or blunt, margins inrolled, lower leaves stalked, the upper stalkless; stipules papery, to 4 cm; stem erect simple, 5–40 cm, arising from a thick rootstock covered with old fibrous leaf-bases. Sometimes a very slender plant to 10 cm, with linear leaves, and a rather globular rootstock.

1188　**B. macrophylla** (D. Don) Soják (*Polygonum m.* D. Don)
Uttar Pradesh to S.W. China. 1700–4500 m. Open slopes; in Nepal in drier areas only. Jun.–Aug.

A slender plant like 1187, but with a rounded or oval terminal cluster of pink or red flowers, usually *c.* 1 cm long, borne on a slender erect, nearly leafless stem. Flowers drooping, *c.* 2 mm. Leaves oblong acute to broadly linear, parallel-sided, to 2 cm broad but often much narrower, blade 3–12 cm long, with a rounded base; lower leaves long-stalked; stem mostly 5–15 cm, but occasionally to 30 cm. **Pl. 107. Pl. 57.**

1189　**B. milletii** Léveillé (*Polygonum sphaerostachyum* auct. non Meissner)
W. Nepal to S.W. China. 3000–4000 m. Shrubberies, open slopes. Jun.–Sep.

Flowers crimson, in a very dense broad-cylindrical or rounded head, usually 1.5–4 cm long. Flowers *c.* 4 mm. Leaves oblong blunt with a distinctive parallel-sided blade narrowed to a winged leaf-stalk below, to 30 cm long by 2–4 cm broad, the upper leaves sometimes enlarged and clasping at the base; stem 20–50 cm; rootstock stout, covered with old leaf-bases.

3　PERSICARIA　Flowers in globular heads or spikes, bracts few or inconspicuous; perianth usually petal-like. Stamens 5–8; styles usually 2; nut not 3-angled. *c.* 25 spp.

1190　**P. capitata** (Buch.–Ham. ex D. Don) Gross (*Polygonum c.* Buch.–Ham. ex D. Don)
Pakistan to S.W. China. 600–2400 m. Cultivated areas, banks, rocks; common. Mar.–Nov.

A low trailing and rooting perennial, with broadly ovate leaves, and with erect, sparingly branched stems, with terminal globular pink flower-heads 6–13 mm across. Flower-stalks glandular-hairy; flowers *c.* 2 mm; perianth 5-lobed, the lobes blunt; stamens 8. Leaves in 2 rows, short-stalked, mostly 2–3 cm, more or less downy on both sides, margins fringed with bristly hairs; stipules lax, brown-papery, *c.* 5 mm; stems many, trailing, 15–25 cm. **Pl. 107.**

1191 **P. polystachya** (Wallich ex Meissner) Gross (*Polygonum p.* Wallich ex Meissner)
Afghanistan to S.W. China. 2000–4000 m. Forests, shrubberies, open slopes; often gregarious. Jul.–Sep.

A shrubby erect perennial to 2 m, with white or pale pink fragrant flowers, in usually erect much-branched terminal clusters 15–45 cm long. Flowers *c.* 6 mm across, with unequal perianth-segments, the two outer narrower than the 3 inner. Leaves oblong-lanceolate long-pointed, 10–25 cm, hairless or nearly so above, usually softly and densely hairy beneath; stipules tubular below, very long, hairy; stems angled, often hairy above. Easily mistaken for 1181 *Aconogonum campanulatum.*
Leaves used as a pot-herb.

4 POLYGONUM Flowers axillary, much shorter than the leaves; stipules silvery-white. *c.* 4 spp.

1192 **P. paronychioides** *C.* Meyer ex Hohen.
Afghanistan to Himachal Pradesh. 2700–4300 m. Stony slopes, drier areas; prominent in Lahul and Ladakh. Jun.–Aug.

A low spreading rather matted plant, with small linear leaves, with conspicuous silvery papery stipules almost as long, and with stalkless bright pink flowers. Flowers *c.* 3 mm across, hidden amongst the stipules and leaves. Leaves fleshy and with a fine point, grooved above, mostly 5–8 mm, hairless; silvery stipules of young shoots concealing both leaves and stem; stem much branched, white or red-brown, spreading 2.5–10 cm.

1193 **P. plebeium** R. Br.
India, W. Asia. Africa. China. S.E. Asia. Australia. To 1800 m. Weed of dry paddy fields throughout the Himalaya. Dec.–May.

A small prostrate branched plant, with small linear to narrow-obovate leaves, and with minute white or pale pink stalkless flowers, borne in axillary clusters and half-concealed by the short papery fringed stipules. Flowers *c.* 2 mm, short-stalked, 4–5–lobed; stamens 4–5. Leaves alternate, rarely more than 12 mm; stipules short, transparent; stems spreading 15–40 cm, leafy and flowering throughout their length and forming pink carpets. **Page 500**.

5 FAGOPYRUM BUCKWHEAT Distinguished from *Polygonum* by its fruit which is much longer than the persistent perianth. Leaves about as wide as long. 3 spp.

1194 **F. dibotrys** (D. Don) Hara (*F. cymosum* (Trev.) Meissner)
Pakistan to S.W. China. 1500–3400 m. Cultivated areas, forests. Jun.–Oct.

A hairy annual to 1 m or more, with broadly triangular leaves, and with terminal branched often elongated clusters of small white flowers. Clusters 5–12 cm across; perianth 5–lobed, *c.* 5 mm across; flower-stalks jointed near the middle. Leaves with a triangular blade 6–15 cm long and with a heart-shaped base and triangular basal lobes, the lower leaves long-stalked, the uppermost narrower and clasping stem. Nutlet 3–angled, to 8 mm, more than twice as long as persistent perianth.

1195 **F. esculentum** Moench
Pakistan to Bhutan. 2000–4400 m. Grown for grain throughout Temperate Eurasia, widely cultivated in the Himalaya. Jun.–Sep.

A slender erect annual like 1194, but a smaller plant with smaller leaves and smaller dense inflorescence. Flowers pink, in terminal branched clusters, usually 3–5 cm across; perianth 4–5 mm across. Leaves with a triangular heart-shaped blade, mostly as long as broad, 2–5 cm, and with rounded or triangular basal lobes, the lower leaves stalked, the upper stalkless; stem unbranched, often tinged red at maturity, to 60 cm. Nutlet c. 6 mm, with smooth faces and angles.

1196 **F. tataricum** (L.) Gaertn.
Pakistan to Bhutan. 1400–4400 m. Cultivated in Temperate Eurasia, though less commonly than 1195. Jul.–Aug.

Differs from 1195 in having less conspicuous flower-clusters with smaller usually greenish flowers c. 2 mm long, and nutlets with rough faces and toothed wavy angles. Often taller, hairless, but seldom red-tinged like 1195; leaves triangular-heart-shaped, usually wider than long, light green.

Both 1195 and 1196 are widely cultivated in the Himalaya for their grain; when young the leaves are used as a pot-herb.

6 RHEUM Rhubarb Robust perennials with woody rhizomes. Flowers with 6 free perianth-segments; stamens 9. Fruit with 3 papery wings. 7 spp.

Flowering-spikes stemless or almost so, borne direct from the rootstock

1197 **R. spiciforme** Royle
Afghanistan to Bhutan. Tibet. 3600–4800 m. Shrubberies, open slopes; in drier areas only. Jun.–Jul.

A robust perennial with leaves all basal, with large rounded blades, and with 1–3 erect cylindrical spikes of greenish flowers turning to reddish as the fruit matures. Spikes 5–30 cm; flowers 3 mm across, borne on short slender stalks; bracts minute, papery. Leaves with a very stout leaf-stalk 5–15 cm, blade rather thick and leathery, with rounded or heart-shaped base, 15–30 cm across when mature, becoming red-brown; rootstock very stout, covered with old leaf-bases. Nutlets 8–10 mm, with broad papery wings.

1198 **R. moorcroftianum** Royle
Himachal Pradesh to C. Nepal. 3600–4700 m. Open slopes. Jun.–Jul.

Very like 1197, and sometimes included in it, but differing in having much longer and often more numerous hairy spikes (spikes hairless in 1197), which are 25–60 cm long. Nutlets ovoid, with narrow wings.
Pl. 107.

POLYGONACEAE

Flowering spikes borne on stems 30–100 cm
a *Flowers pale yellow*

1199 **R. webbianum** Royle
Pakistan to W. Nepal. 2400–4300 m. Shrubberies, open rocky slopes; very
common in Chenab valley. Jun.–Jul.

A stout erect perennial with many axillary spikes of tiny pale yellowish
flowers clustered in a dense much-branched terminal inflorescence which
elongates in fruit. Flowers *c*. 2 mm across, slender-stalked. Lower leaves
long-stalked, with rounded-heart-shaped or kidney-shaped blades 10–60
cm across; stipules of upper leaves very large and partially encircling the
young inflorescence; stems very variable, 30–200 cm. Nutlets 11 mm long,
with broad wings.

b *Flowers purple or red*

1200 **R. australe** D. Don (*R. emodi* Wallich ex Meissner)
Himachal Pradesh to E. Nepal. 3000–4200 m. Open slopes. Jun.–Jul.

Stem stout, streaked green and brown, 1.5–2 m, with rounded leaves,
and with dense branched clusters of small dark reddish-purple flowers
borne in an inflorescence 20–30 cm long, which enlarges greatly in fruit.
Flowers *c*. 3 mm across. Leaves with a very stout leaf-stalk 30–45 cm, and
with rounded to broadly ovate blade with a heart-shaped base, hairy
beneath, the basal leaves up to 60 cm across; rootstock very stout. Nutlets
purple, *c*. 12 mm long, with narrow wings, and with a rounded heart-
shaped base and notched apex.

The roots are used medicinally. **Pl. 107.**

1201 **R. acuminatum** Hook.f. & Thoms. ex Hook.
C. Nepal to S.E. Tibet. 3600–4300 m. Open slopes. Jun.–Aug.

Distinguished from 1200 by its triangular to rounded heart-shaped long-
pointed leaves which are nearly hairless beneath, and by its dark red flowers,
4 mm across, in a more lax inflorescence. A considerably smaller and less
robust plant to 1 m, with leaf-blades 8–20 cm across. Nutlets *c*. 8 mm.

The roots are used medicinally.

c *Flowers concealed by inflated overlapping bracts*

1202 **R. nobile** Hook.f. & Thoms.
E. Nepal. to S.E. Tibet. 3600–4500 m. Open slopes. Jun.–Jul.

A very striking and distinctive-looking plant, with a stout erect stem
bearing a slender conical spike of large pale cream-coloured, rounded and
bladder-like, drooping and overlapping bracts which conceal the short
flower-clusters. Bracts progressively smaller up stem, the lowest to 15 cm,
the uppermost 1–2 cm. Flower-clusters branched, to 6 cm; flowers green,
very numerous, *c*. 2 mm across. Leaves leathery, to 30 cm across, rounded
with a wedge-shaped or rounded base, margin usually edged with red, and
with a stout stalk; leaves gradually changing upwards into the cream-
coloured bracts; stem very stout, to 1.5 m. Nutlets broadly 2–4–winged.

348

Rhizomes and roots used medicinally. Stems edible, either raw or cooked. **Pl. 107.**

7 RUMEX Perianth-segments in 2 whorls of 3, the outer remaining small and thin, the inner enlarging and becoming wing-like and often hardened in fruit; stamens 6. Nutlets 3–angled. *c.* 8 spp.

Fruit with entire papery wings

1203 **R. acetosa** L. SORREL
Afghanistan to C. Nepal. Temperate Eurasia. 2100–4300 m. Open slopes; shrubberies. Jun.–Aug.

Flowers numerous, reddish-green, in slender branched terminal and axillary clusters, and with distinctive oblong-oval leaves with pointed triangular basal lobes. Flowers *c.* 3 mm across, one-sexed; perianth 3–4 mm in fruit, the outer lobes reflexed after flowering. Leaves with blade 4–11 cm long; the lower with a slender stalk, the upper stalkless; stem erect, little branched, deeply grooved, 30–80 cm.

1204 **R. hastatus** D. Don
Afghanistan to S.W. China. 1000–2700 m. Stony slopes, banks; common on walls of cultivated terraces; in Nepal in drier areas only. Apr.–Aug.

A rather bushy plant with many ascending stems, with narrow arrow-shaped leaves with a pair of narrow spreading basal lobes, and with numerous thin branches with terminal very slender clusters of distant whorls of tiny greenish-white flowers. Flowers very small, flower-stalk lengthening in fruit. Leaves mostly 2–4 cm, some broadly triangular, stalked; stems woody at base, 20–60 cm. Fruit pinkish, wings papery and delicately net-veined, to 5 mm across. **Page 500.**

Fruit with conspicuously toothed wings

1205 **R. nepalensis** Sprengel
Afghanistan to S.W. China. W. Asia. S.E. Europe. 1200–4300 m. Culti-vated areas, grazed ground. Apr.–Jul.

A robust perennial to 120 cm, readily distinguished by its fruits which have broad wings which are fringed with comb-like hooked teeth, and one wing with an oblong swelling. Flowers bisexual, in whorls, forming long interrupted leafless spikes. Leaves entire, the lower with oblong-ovate blade often with heart-shaped base, to 15 cm, and leaf-stalk as long, the uppermost leaves stalkless, lanceolate; stem robust, with spreading branches.

Roots used medicinally. **Page 500.**

8 OXYRIA Perianth-segments in 2 whorls of 2, the inner enlarging in fruit, without swelling; stamens 6. Fruit lens-shaped, broadly winged. 1 sp.

1206 **O. digyna** (L.) Hill MOUNTAIN SORREL
Pakistan to S.W. China. N. Temperate Zone. 2400–5000 m. Open slopes, grazing grounds; common. May–Jul.

Distinguished by its rather fleshy, pale green, rounded to kidney-shaped leaves, and its long slender sparsely branched spikes of many whorls of tiny greenish flowers tinged with pink. Flowers 2 mm across, with the outer two perianth-segments spreading or reflexed, the inner two spathulate, erect. Leaves mostly basal, with blade usually 2.5–5 cm across, long-stalked; stems often several, 10–80 cm; stems and leaves often turning reddish. Fruit at length red, with papery wings to 6 mm across.

The leaves are edible, both raw and cooked. **Pl. 106**.

ARISTOLOCHIACEAE   Birthwort Family

A family of herbs and woody climbers, mostly of the Tropics and warmer Temperate Forest Zone. Leaves alternate, entire. Flowers solitary, or in clusters, bisexual, with bell-shaped or irregular trumpet-shaped petal-like perianth; stamens usually 6, or more, free or fused. Ovary inferior, with 4–6 chambers with numerous ovules, and 4–6 stigmas. Fruit usually a many-seeded capsule.

ARISTOLOCHIA 5 spp.

1207 **A. griffithii** Hook.f. & Thoms. ex Duchartre
C. Nepal to S.E. Tibet. 1800–2900 m. Climber on shrubs. Apr.–May.

A tall climber with large heart-shaped stalked leaves, and with very distinctive brown-purple, spotted, solitary axillary flowers, with curved trumpet-shaped corolla with a large rounded mouth. Flowers long-stalked, foetid; corolla inflated below and ribbed and veined, narrowed above and ultimately enlarged to a wide mouth 7.5–10 cm across which is spotted with yellow within. Leaves 10–20 cm long and wide, woolly-haired beneath; young shoots covered with rust-coloured woolly hairs. Capsule to 18 cm long by 1.5 cm wide, twisted at base, with 6 strong ribs. **Pl. 109**.

SAURURACEAE

A small family of temperate and sub-tropical herbaceous plants, with alternate simple leaves and with stipules partly fused to the leaf-stalk. Flowers in dense racemes or spikes, with coloured bracts surrounding the base of the inflorescence. Flowers bisexual; perianth absent; stamens 3, 6 or 8, free or more or less fused to the carpels. Carpels 3–5, superior or inferior, free or fused; styles similar in number, free. Fruit of 3–5 splitting capsule-like units; or if fused, a thick capsule opening at the top.

HOUTTUYNIA 1 sp.

1208 **H. cordata** Thunb.
Himachal Pradesh to S.W. China. Burma. S.E. Asia. 1500–2400 m.

Cultivated areas, shrubberies, damp places. Jun.–Jul.

A hairy erect herbaceous plant, with broadly ovate-heart-shaped leaves, and with distinctive erect cylindrical spikes of tiny green flowers with an involucre of 4–6 large spreading white elliptic petal-like bracts at their bases. Flower-spikes axillary 1–2 cm, stalked, with densely clustered flowers with 3 stamens and 3 carpels each with a curved style; coloured bracts 8–15 mm. Spikes elongating in fruit to 5 cm. Leaves with blade narrowed to a slender point, usually 3–7 cm, stalked, gland-dotted; stipules to 2 cm, oblong blunt; stems leafy, erect to *c.* 30 cm, spreading and rooting at the nodes below. Fruit globular, of 3 fused carpels. **Pl. 108**.

LAURACEAE   Laurel Family

Mostly trees and shrubs of the tropical and sub-tropical regions, with the main concentrations in S.E. Asia and tropical America, and with a few genera in the temperate regions. Leaves evergreen, leathery, usually with oil glands and aromatic, alternate or opposite. Flowers in branched clusters, or umbels, and usually with an involucre of leafy bracts surrounding the inflorescence. Flowers regular, unisexual or bisexual and borne on the same plant. Parts of flowers in multiples of 3, often not well differentiated from each other. Ovary usually superior, surrounded by a cup-shaped base, one-chambered and with a single ovule. Fruit a berry, or drupe, often enclosed by the cup-shaped base which may become fleshy.

Flowers in branched clusters
  a Fruit supported on enlarged perianth   1 CINNAMOMUM
                                                       2 DODECADENIA
  b Fruit supported by reflexed perianth   3 PERSEA

Flowers in compact umbels
  a Leaves with several pairs of lateral veins   5 LITSEA
                                                        6 NEOLITSEA
  b Leaves with 3 main veins from base   4 LINDERA
                                                        6 NEOLITSEA

1 CINNAMOMUM Leaves 3–veined, often opposite. Flowers in large axillary stalked clusters: stamens 9. *c.* 6 spp.

1209 **C. tamala** (Buch.–Ham.) Nees & Eberm.
    Kashmir to Bhutan. Burma. 450–2100 m. Forests. Apr.–May.
    A small evergreen tree, with ovate-lanceolate long-pointed leathery leaves, and with pale yellowish flowers in terminal and axillary branched clusters about as long as the leaves. Perianth 7 mm long, with oblong lobes and a short tube, silky-haired; fertile stamens 9; ovary hairy, with slender style. Leaves short-stalked, entire 10–15 cm, glaucous beneath,

with 3 conspicuous nearly parallel veins arising from near the base, the leaf-tip often curved. Leaves bright pink when young in spring; aromatic when crushed. Fruit black, succulent, ovoid 12 mm, seated on the somewhat enlarged perianth-tube.

The thin dark brown bark is used as a substitute for cinnamon. The leaves are used medicinally. **Page 501**.

1210 **C. camphora** (L.) J. S. Presl CAMPHOR TREE
Native of Japan. Cultivated in the Himalaya to 1500 m, and sometimes naturalized; common in Kathmandu. Apr.–May.

A medium-sized evergreen tree, with elliptic long-pointed 3–veined leaves, which are bright green above and pale and glaucous beneath, and with tiny greenish-yellow flowers in small spreading branched clusters each to 5 cm across. Flowers *c.* 3 mm across, with 6 lobes, hairy within. Leaves 5–9 cm, with a short slender leaf-stalk about a third as long as the blade. Fruit globular *c.* 9 mm across, with a thickened club-shaped base.

Grown for a white crystalline substance known as Japan Camphor which is obtained from the leaves and twigs; used in medicine, disinfectants, and for other purposes.

2 DODECADENIA Leaves with many lateral veins. Perianth with 6–9 lobes; fertile stamens 10–18. Fruit seated on the thick flat perianth-tube. 1 spp.

1211 **D. grandiflora** Nees
Uttar Pradesh to E. Nepal. S.E. Tibet. Burma. 2400–2700 m. Forests. Feb.–Apr.

An evergreen tree with silky-haired young leaves and branchlets, with elliptic-lanceolate long-pointed entire leaves, and with solitary or few pale yellow flowers in axillary clusters 1.5–2 cm. Flowers to *c.* 15 mm across, with very short flower-stalks covered with bracts; perianth-lobes oblong blunt, hoary within; stamens exserted, hairy. Leaves usually 7.5–12 cm, minutely netted on both sides, with 6–8 pairs of conspicuous lateral veins; leaf-stalk short, slender. Fruit *c.* 1.3 cm long, ellipsoid, seated on a dilated base. **Pl. 109**.

3 PERSEA Leaves alternate, with many pairs of lateral veins. Flowers in apparently terminal clusters. Perianth 6–lobed, persistent in fruit. 6 spp.

1212 **P. odoratissima** (Nees) Kosterm. (*Machilus o.* Nees)
Pakistan to Bhutan. 1500–2100 m. Forests. Mar.–Apr.

A small or medium-sized evergreen tree, with leaves which are bright green shining above and glaucous beneath, and with small yellowish-green flowers in lax branched clusters which are at first terminal but become lateral as the shoot grows. Flowers 6–10 mm across, with oblong-linear acute lobes, and with hairs towards the base within; filaments of the stamens hairy below. Leaves 7.5–18 cm long by 3–7.5 cm broad, variable oblong to obovate, acute or long-pointed and narrowed to the short leaf-stalk; twigs hairless. Fruit purple, ellipsoid 1–1.6 cm, supported by the persistent reflexed perianth.

1213 **P. duthiei** (King ex Hook.f.) Kosterm.
Pakistan to E. Nepal. 1500–2700 m. Forests. Apr.–May.

Differs from 1212 in having finely hairy inflorescences, and globular black fruit (inflorescence hairless, and fruit ellipsoid in 1212). A medium-sized evergreen tree, with oblong to oblong-lanceolate pointed leaves which are glaucous beneath, and with terminal branched clusters of greenish-yellow flowers. **Pl. 109.**

4 LINDERA Flowers numerous, in umbels; anthers 2–celled. Leaves with 3 veins in our species. 6 spp.

1214 **L. pulcherrima** (Nees) Benth. ex Hook.f.
Uttar Pradesh to S.W. China. Burma. 1200–2700 m. Oak forests, common in understory. Mar.–Apr.

A large evergreen tree or often a shrub, with distinctive ovate-lanceolate leaves with a long tail-like tip often as long as 2.5 cm, and with 3 conspicuous pale parallel raised veins. Flowers 5–6, greenish-yellow, with hairy stalks, in globular axillary clusters *c.* 1–1.5 cm across, borne in axils of the leaves and encircled by 4–6 large pale overlapping silky-hairy bracts which fall when the flowers open; perianth *c.* 5 mm across, silky-haired. Leaves 5–15 cm, short-stalked, leathery, paler and often glaucous beneath, hairless; branchlets hoary; buds silky-haired. Fruit black, ellipsoid 8 mm. **Pl. 109.**

5 LITSEA Differs from *Lindera* in having anthers with 4 cells. Leaves with many pairs of lateral veins in our species. 8 spp.

1215 **L. doshia** (Buch.-Ham. ex D. Don) Kosterm. (*L. oblonga* (Wallich ex Nees) Hook.f.)
C. Nepal to Sikkim. Burma. 1500–2100 m. Oak forests, common in understory. Oct.–Nov.

A quite hairless evergreen shrub or small tree, with rather large elliptic to oblong-lanceolate leaves which are pale brown and somewhat glaucous beneath, and with short-stalked globular axillary clusters of yellow flowers. Flowers 3–6 in each cluster, *c.* 4 mm, hairy; bracts 4–6, ovate and enclosing flower-bud. Leaves 10–18 cm, leathery, shining above, with 8–10 pairs of slender lateral veins. Fruit cylindrical to 1.6 cm, dark purple, with a funnel- or club-shaped base. **Page 501.**

6 NEOLITSEA Perianth-segments 4, tube not enlarged in fruit; stamens usually 6. 2 spp.

1216 **N. pallens** (D. Don) Momiyama & Hara (*Litsea umbrosa* var *consimilis* (Nees) Hook.f.)
Pakistan to Sikkim. 1500–2700 m. Forests, in understory. Mar.–May.

A small or medium-sized evergreen tree, with oblong-elliptic long-pointed leaves which are pale and glaucous beneath. Flowers yellowish-white, in dense stalkless axillary clusters 1–2 cm across, with funnel-shaped perianth-tube with 4 elliptic blunt lobes, densely hairy outside. Leaves 6–10 cm, alternate, with often 2 stronger basal lateral veins (thus appearing 3–veined) and alternate lateral veins above, hairless, short-stalked. Fruit globular 10 mm, black when ripe, stalked. **Page 501.**

THYMELAEACEAE   Daphne Family

A medium-sized family mainly of shrubs, of both temperate and tropical regions. Leaves entire, alternate, without stipules. Flowers regular, with parts in fours or fives. Receptacle tubular and with the petal-like lobes borne on the rim and appearing continuous with it; stamens inserted in the receptacle-tube, half to twice as many as the lobes; nectar-disk often present. Ovary at the base of the receptacle-tube usually of one or two fused carpels, each with one ovule. Fruit an achene, drupe, or berry.

DAPHNE   Receptacle-tube cylindrical, lobes 4, acute; style absent; nectar-disk absent. Leaves thick, hairless, alternate. 5 spp.

1217   **D. bholua** Buch.–Ham. ex D. Don
Uttar Pradesh to S.W. China. 1800–3100 m. Forests, shrubberies. Oct.–Apr.

An erect or spreading evergreen or deciduous shrub, with elliptic to oblanceolate entire dull green leathery hairless leaves, and with very sweet-scented white flowers flushed externally pink or purplish, borne in terminal rounded stalkless clusters. Flowers with slender silky-haired tube 6–12 mm long, and with 4 broad or narrow ovate acute spreading lobes *c.* 6–8 mm long. Leaves alternate, 5–10 cm or more, very short-stalked; stems usually 60 cm. Fruit ellipsoid, black when ripe.

Var. **glacialis** (W. W. Smith & Cave) Burtt is a deciduous shrub with pink to purple, very sweet-scented flowers, appearing in stalkless clusters on bare branches in spring, and occasionally in winter; from W. Nepal to Sikkim; 2000–3000 m.

The inner bark is used to make paper. **Pl. 108.**

1218   **D. papyracea** Wallich ex Steud. (*D. cannabina* Lour. ex Wallich)
Pakistan to C. Nepal. 1500–2100 m. Forests. Nov.–Apr.

A much-branched erect evergreen shrub with smooth grey bark, with dull green narrow-lanceolate to oblanceolate leathery leaves, and with scented white or greenish-white flowers borne in terminal clusters with persistent hairy bracts. Flowers with tube 10–13 mm long, very downy outside, and 4 acute spreading lobes *c.* 8 mm long. Leaves 5–12 cm, hairless, with veins impressed above, stalk short and grooved; stems to

1.25 m long. Fruit *c*. 1 cm, fleshy, at first orange then deep red when fully ripe.

Paper is made out of the inner fibrous bark. **Pl. 109**.

1219 **D. mucronata** Royle (*D. cachemireana* Meissner, *D. angustifolia* C. Koch)

Pakistan to Himachal Pradesh. W. Asia. 1500–2700 m. Shrubberies, open slopes; common in Kashmir and Chenab valleys. Apr.–May.

An erect much-branched evergreen wiry-stemmed shrub, with narrowly oblanceolate, grey leathery leaves, and with creamy-white or yellowish fragrant flowers borne in dense terminal clusters. Flowers with tube 10 mm long, densely woolly-haired outside, lobes spreading, ovate blunt. Leaves 2–4 cm, stalkless, alternate; stems to 2 m or more; young branches hairy, reddish-brown, older branches greyish. Fruit ellipsoid, reddish-orange, somewhat hairy, enclosed in the enlarged calyx. **Pl. 108**.

Poisonous to animals and hence not grazed.

1220 **D. retusa** Hemsl.

Kashmir to S.W. China. 3300–3700 m. Alpine shrubberies, open slopes; not common. May–Jun.

A dense much-branched evergreen shrub 20–60 cm, with small shiny green, broadly oblanceolate leaves, and with pinkish-purple to white very fragrant flowers in terminal clusters. Flowers with slender tube 8–12 mm, pinkish outside, not hairy, lobes 7–10 mm, ovate. Leaves 2.5–4 cm long, with margins inrolled, clustered towards ends of branches; young branches green, then light brown. Fruit orange to red, nearly globular 8–10 mm. **Pl. 110**.

EDGEWORTHIA Receptacle-tube cylindrical, lobes 4, blunt; style long, stigma linear; nectar-disk a ring, or lobed. Leaves thin, silky-haired beneath, alternate. 1 sp.

1221 **E. gardneri** (Wallich) Meissner

Uttar Pradesh to S.W. China. 1500–3000 m. Forests, shrubberies. Nov.–Apr.

A large much-branched shrub with long scrambling branches, with elliptic-lanceolate long-pointed leaves, and with dense pendulous globular clusters of numerous small golden-yellow, sweet-scented, tubular flowers, fading to white. Flower clusters 2.5–5 cm across, very hairy; receptacle-tube 1–2 cm long, densely silky-haired, lobes short rounded, spreading to 1 cm; bracts linear-oblong, conspicuous in bud. Leaves 7.5–12.5 cm, hairless above, hairy or silky-haired beneath, narrowed to a short stalk; stems brownish, to 2 m or more.

Bark used for paper-making. **Pl. 109**.

WIKSTROEMIA Receptacle-tube cylindrical, lobes 4, blunt; style short, stigma large globular; nectar-disk of 4 erect scales. Leaves becoming hairless above. 1 sp.

1222 **W. canescens** Meissner
Afghanistan to Sikkim. Sri Lanka. China. 1800–3000 m. Shrubberies, light forests; often gregarious. May–Sep.

A small shrub with many slender branches, with narrow-elliptic papery leaves, and with small yellowish tubular flowers in rounded or sometimes elongate short-stalked clusters. Flowers with a slender tube 8–12 mm long, silky-haired outside, with 4 short blunt spreading lobes, and with 4 linear nectar-scales. Leaves 3–6 cm, silky when young, but becoming hairless except on the mid-vein beneath; young shoots hairy; stems to 2 m. Fruit narrow ovoid, black when ripe, enclosed at first in the hairy tube, which splits and falls. **Pl. 108**.

STELLERA Herbaceous perennials. Upper part of receptacle-tube splitting off after flowering. 1 sp.

1223 **S. chamaejasme** L.
Uttar Pradesh to C. Nepal. Bhutan. Tibet. N. China. 2700–4300 m. Stony slopes in Tibetan borderlands; abundant on abandoned terrace-fields. May–Jul.

A very attractive herbaceous plant with usually many simple very leafy stems arising from a thick rhizome, and with stalkless dome-shaped heads of sweet-scented white flowers with pinkish tubes. Flower with a hairless receptacle-tube 1–1.5 cm long, and 5 short oblong lobes. Leaves numerous, overlapping, 1.5–2 cm long, elliptic-lanceolate long-pointed, stalkless, the upper forming an involucre round the flower-heads; stems 20–30 cm. Fruit dry, included in the persistent base of the receptacle-tube.

The roots are used in Tibet for paper-making. **Pl. 110**.

ELAEAGNACEAE   Oleaster Family

A small family of shrubs, commonly of coastal and steppe regions of N. America, Europe, S. Asia and Australia. Stems and leaves covered with distinctive silvery, brown, or golden scaly or flat-topped hairs. Leaves leathery, entire, without stipules. Flowers regular, bisexual or unisexual; perianth-lobes 2–8, fused in a single whorl; petals absent; stamens 4–8; receptacle flat or tubular. Ovary superior, with one carpel with one ovule; style long. Fruit a one-seeded unit, often enclosed by the thickened lower part of the persistent perianth.

ELAEAGNUS OLEASTER Flowers bisexual, borne on present year's shoots; perianth 4–lobed. 6 spp.

1224 **E. parvifolia** Wallich ex Royle (*E. umbellata* auct. non Thunb.)
Afghanistan to S. W. China. 1500–3000 m. Shrubberies, cultivated areas; common throughout the Himalaya; sometimes planted. Mar.–Jun.

A deciduous, somewhat spiny shrub, with oblong-elliptic leaves which are silvery-scaly beneath, and with small tubular dull yellowish-white fragrant flowers appearing with the young leaves. Flowers few, in axillary clusters; perianth-tube 8 mm, with 4 triangular acute lobes, silvery-scaly outside; stamens 4, slightly exserted. Leaves alternate, 2.5–5 cm, stalked; shoots and young branches covered with silvery scales, lateral shoots ending in a straight thorn; shrub to 2 m or more. Fruit ovoid 7 mm long, fleshy, edible, red and covered with silvery scales; nut bony. **Pl. 110**.

HIPPOPHAE BUCKTHORN Flowers one-sexed, occurring on different plants, borne on previous year's shoots. Flower with perianth 2–lobed. 3 spp.
Leaves 6 mm or more broad, with star-shaped hairs
1225 **H. salicifolia** D. Don
Himachal Pradesh to S.E. Tibet. 2000–3500 m. Colonizing alluvial gravel, wet landslips, riversides. Apr.–May.

A deciduous thorny willow-like shrub or small tree to *c.* 5 m, with rusty-scaly shoots, and oblong-lanceolate leaves densely white-downy beneath, and with dense stalkless clusters of yellowish-brown male flowers, the female flowers solitary appearing on leafless stems. Male flowers *c.* 3 mm, with 2 scaly lobes, stamens usually 4; female flowers short-stalked, 2–lobed, stigma exserted. Leaves 3–7 cm by 6–8 mm broad, with star-shaped hairs beneath, and above when young, and with a scaly midrib, very short-stalked. Fruit *c.* 7 mm globular, orange or red, succulent, ranged along the stems.
Leaves to 4 mm broad, with scales
1226 **H. rhamnoides** L. subsp. **turkestanica** Rousi
Pakistan to Himachal Pradesh. C. Asia. 2100–3600 m. Riversides; common in Ladakh and Lahul; planted as a hedge in irrigated areas. Flowering in spring.

Usually dwarf much-branched very thorny shrubs, with scaly young twigs, and with a silvery-waxy covering to the older shoots; distinguished from 1225 by the leaves which are smaller and narrower and clothed beneath with silvery or rusty scales (not star-shaped hairs). Leaves variable, oblong blunt to 4 cm by 2–4 mm broad, green above or silvery-scaly on both sides. Fruit *c.* 6 mm, orange or red.
1227 **H. tibetana** Schltr.
Himachal Pradesh to S.E. Tibet. N.W. China. 3300–4500 m. Riversides, stony moraines; gregarious; in Nepal in the inner drier valleys only. Flowering in spring.

A very dense much-branched shrublet, with long stout terminal spines formed from the tips of old branches, and with small narrow-elliptic leaves covered with rust-coloured scales. Flowers stalkless, yellowish, *c.* 4 mm across, in clusters appearing on leafless stems. Leaves blunt, 1.5–2 cm by 2–4 mm broad, numerous; stems to 30 cm. Ripe fruit orange-red, ranged up stem. **Pl. 110**.

LORANTHACEAE   Mistletoe Family

A family of parasitic plants with green leaves, attached by means of modified roots to the host plant, widely distributed in tropical forests, and to a lesser extent in temperate forests of the world. Leaves evergreen, opposite, without stipules. Flowers usually in branched clusters, regular, with a cup-shaped receptacle bearing on the rim green or coloured petal-like perianth-segments; stamens equal in number to perianth-segments and fused to them. Ovary sunk in the receptacle, 1–celled, with numerous ovules; style simple or absent. Fruit a berry or drupe, with a characteristic layer of sticky juice which surrounds the seeds and sticks to the beaks of birds, thus bringing about the dispersal of seed to other host-plants.

HELIXANTHERA   Flowers bisexual, red; corolla-tube short with 6 spreading lobes. Bracts boat-shaped or cup-shaped. Fruit ovoid-ellipsoid. 2 spp.

1228   **H. parasitica** Lour.
E. Nepal to Sikkim. India. S.E. Asia. To 1400 m. Forest parasite. Apr.–Jun.
Flowers small, bright red, numerous in long slender spikes, borne from the leaf-axils, often on leafless stems. Flower-spikes to 10 cm; axis of inflorescence and short flower-stalks red; corolla-tube short *c.* 5 mm with 6 equal spreading lobes; anthers yellow, exserted. Leaves elliptic entire, 8–13 cm, stalked; a shrubby parasitic plant to 3 m. **Pl. 110**.

SCURRULA   Flowers bisexual, showy, with a long coloured 4–lobed corolla-tube, borne in axillary clusters. 4 spp.

1229   **S. elata** (Edgew.) Danser
Himachal Pradesh to S. E. Tibet. 1500–3000 m. Parasitic mostly on oaks; quite common. Apr.–Jun.
A rather robust parasitic shrub growing on the branches of trees, with ovate leaves, and with axillary clusters of curved tubular flowers which are red at the base and green in the upper half. Corolla-tube slender 2.5–3.5 cm long, with 4 long oblong reflexed lobes; calyx short, rusty-haired. Leaves opposite, stalked, blade thick, hairless 7.5–15 cm; youngest shoots and young leaves brown-hairy; stems to 3 m. Berry *c.* 8 mm, broadly top-shaped. **Pl. 110**.

VISCUM   Flowers unisexual, tiny, greenish corolla-tube very short or absent, 3–4–lobed; anthers opening by many pores. 2 spp.

1230 **V. album** L. Mistletoe
Afghanistan to C. Nepal. Temperate Eurasia. 1000–2700 m. Common on walnut trees in Kashmir. Mar.–May.

A much-branched evergreen parasitic shrub with yellow-green branches and with similar-coloured narrow-elliptic blunt opposite leaves. Flowers yellowish-green, in stalkless clusters in the axils of the branches, with cup-shaped ciliate bracts; perianth-segments triangular, thick. Leaves variable, usually 3–5 cm long, thick fleshy; stems forked, 60–90 cm. Berry globular, white translucent, c. 7 mm.

Fruit and plant used medicinally.

1231 **V. articulatum** Burm. f.
Himachal Pradesh to S.W. China. 300–2100 m. Parasitic on trees; quite common on oaks in Nepal. Jun.–Oct.

A green leafless parasitic shrub forming pendulous tufts of long flattened and distinctively jointed green stems 15–100 cm long. Flowers 1–2 mm, yellow-green, clustered at the tops of the joints, the male and female flowers in the same cluster. Joints of stem 1.5–7 cm long, longitudinally ribbed; stems often branched. Berry globular, yellow, 4 mm. **Page 502.**

ARCEUTHOBIUM Whole plant minute, leafless; flowers 1–sexed. 1 sp.

1232 **A. minutissimum** Hook. f.
Pakistan to W. Nepal. 2400–3300 m. Common on *Pinus wallichiana* in the W. Himalaya. Jul.–Sep.

The smallest flowering plant of the Himalaya, looking more like a fungus, consisting of small green swellings rarely up to 5 mm long, breaking through the bark of branches of the host tree. Flowers appear as minute 2–lipped cups. After flowering the 2–jointed stem elongates to 6 mm at most, bearing the tiny berries which are ovoid c. 2 mm long, with persistent perianth above.

Does considerable damage to its host and causes witches-brooms.

BALANOPHORACEAE

A family of plants parasitic on the roots of trees of the upland forests of the tropical regions. Plants fleshy, pale yellow or brown, with a club-shaped inflorescence appearing above ground and looking very like a fungus; green leaves are absent. The inflorescence is formed inside a tube which ruptures as the inflorescence develops and remains as a cup round the base. Inflorescence bearing stalkless flowers over its surface; male and female flowers often occur on separate inflorescences. Fruit minute.

BALANOPHORA 3 spp.

1233 **B. involucrata** Hook. f.
Pakistan to S.W China. 2100–3400 m. In dense forests; rarely collected

but probably mistaken by many for a fungus. Jul.–Oct.

A short reddish-brown or yellowish fleshy plant, with a club-shaped or globular flower-head borne on a stout stem and with a large involucre of scales half way up the stem, and a cup-shaped involucre below. Flower-head 2–2.5 cm, borne on a stem 5–10 cm; involucre *c.* 3 cm long. Male flowers with the tube of the perianth sunk in the head, usually 3–lobed; anthers 3, stalkless; female flowers with an ovary and a slender style, but flowers sometimes bisexual. Nut 1–seeded. **Page 502**.

EUPHORBIACEAE   Spurge Family

A large and important family of trees, shrubs and herbaceous plants, largely of the Tropics. Leaves simple or palmate. Inflorescence very variable and complex; flowers regular, unisexual, with usually 5 perianth segments; stamens 1 to very numerous. Ovary superior, of 3 fused carpels, 3–chambered, each chamber with 1–2 ovules. Fruit a capsule, or a drupe.

EUPHORBIA SPURGE Soft-wooded shrubs, or herbs, with milky juice. Inflorescence unique: flowers borne in heads consisting of a cup-shaped involucre which encircles many male flowers each with a single stamen, and which surround a single central female flower with a 3–chambered, stalked ovary; the involucre usually bears 4–5 shining glands from the rim. Fruit a 3–valved capsule. *c.* 22 spp.

Soft-wooded fleshy-stemmed shrubs, with or without spines

1234 **E. royleana** Boiss.
Pakistan to E. Nepal. To 1800 m. Rocks, dry river valleys, cultivated areas. Mar.–May.

A large cactus-like shrub to 5 m, with a stout trunk and many thick fleshy branches with 5–7 thick undulate winged ridges armed with pairs of spines on their margins. Leaves fleshy, spathulate 10–15 cm, but stems becoming leafless during the hot and cold seasons. Flower-heads yellow-green, 3–4, in an axillary, almost stalkless cluster. Fruit 3–lobed, *c.* 1 cm across. **Pl. 112**.

Grows readily from cuttings and is often used as a hedge plant.

1235 **E. milii** Des Moul. (*E. splendens* Bojer ex Hook.) CROWN OF THORNS.
Native of Madagascar; often used to protect the tops of walls in the Nepal valley and elsewhere. 900–1500 m. Mar.–Apr. and throughout the summer.

Distinguished by its very spiny stems, its dense cluster of scarlet flower-heads with pairs of scarlet bracts surrounding each cluster, and by its sticky red flower-stalks. Stems brown, with 4 ranks of many long tapering spines; leaves obovate, thin, persisting at the ends of the branches. **Pl. 111**.

1236 **E. pulcherrima** Willd. ex Klotzsch POINSETTIA.
Native of C. America; commonly cultivated in the tropics; a common

hedge plant in the Himalaya, to 2000 m. Nov.–Jan.

A deciduous spineless branched soft-wooded shrub to 4 m, distinguished by its large vermilion bracts surrounding the yellow flower-heads. Individual flower-heads *c.* 1 cm across; bracts leaf-like to 10 cm or more. Leaves alternate, to 15 cm, ovate to lanceolate, entire, toothed or lobed, leaf-stalks red.

Herbaceous plants, without spines
a *Erect or spreading perennials*

1237 **E. wallichii** Hook.f.
Afghanistan to S.W. China. 2300–3600 m. Open slopes, grazing grounds. Apr.–May.

An erect perennial with very leafy stems terminating in several yellow flower-heads in a more or less flat-topped cluster, each flower-head with 3–4 large rounded or ovate acute golden-yellow bracts. Involucre with large kidney-shaped glands and with hairs within; bracts 1–3 cm. Leaves 5–10 cm, linear to narrow-elliptic; stems several, unbranched 30–60 cm, with ovate to elliptic bracts below. Fruit *c.* 8 mm, finely hairy; seeds blue-grey, very smooth. **Pl. 111.**

1238 **E. cognata** (Klotzsch & Garcke) Boiss. (*E. pilosa* L. var. *cognata* Hook.)
Afghanistan to Uttar Pradesh. 2100–3600 m. Open slopes. May–Jun.

A slender perennial with a branched spreading flat-topped compound inflorescence to 15 cm across, of many flower-heads each surrounded by yellow-green bracts. Distinguished from 1237 by its rounded entire (not kidney-shaped) glands of the involucre and with many long hairs within, and by its rounded fruit which is covered with small swellings. A variable perennial with leaves mostly 3–6 cm, oblong acute to narrow-elliptic; stems leafy, to 60 cm. **Pl. 111.**

1238a **E. longifolia** D. Don
W. to E. Nepal. 1700–2900 m. Wasteland, cultivated areas, grazed slopes; common around Jumla. Mar.–Jun.

A tall nearly hairless perennial with leathery linear-oblong leaves, and an inflorescence with few short rays each with 3–4 rounded or broadly ovate fine-pointed bracts. Involucre bell-shaped, hairy within, with rounded lobes. Leaves 2.5–10 cm; stem *c.* 70 cm. Fruit with obscure sparse conical swellings.

1239 **E. stracheyi** Boiss.
Kashmir to S.W. China. Tibet. 3300–4700 m. Meadows, stony slopes. May–Jul.

A small perennial with spreading stems often in a rosette, with small elliptic to ovate leaves, and with dark reddish flower-heads *c.* 8 mm across. Involucre 5–lobed; glands kidney-shaped, dark purple; bracts obovate, reddish. Leaves mostly 5–8 mm, stalkless, many and overlapping;

stems spreading, mostly 2–5 cm; underground stems long; rootstock stout. Fruit smooth, hairless.

b *Erect annuals*

1240 **E. helioscopia** L. Sᴜɴ Sᴘᴜʀɢᴇ

Afghanistan to Himachal Pradesh. W. Asia. Europe. 300–1800 m. Fields, wasteland. Apr.–May.

An erect hairless annual 15–40 cm, with obovate to spathulate finely toothed leaves, and with yellow flower-heads in a flat-topped cluster, with oval finely toothed bracts like the upper leaves but smaller and often yellowish. Glands of involucre yellow, transversely oval, entire. Leaves 1.5–3 cm, finely toothed towards apex, the lower leaves smaller. Fruit 3–lobed, smooth; seeds with a network of ridges.

Plant used medicinally.

RICINUS Flowers in branched clusters. Perianth-segments 3–5; stamens very numerous. Leaves palmately lobed. 1 sp.

1241 **R. communis** L. Cᴀsᴛᴏʀ Oɪʟ Pʟᴀɴᴛ

Native of the Tropics; common in the hills in cultivated areas and wasteland; to 2400 m. Flowering intermittently throughout year.

A herbaceous plant or a soft-wooded shrub, readily distinguished by its large green or reddish, long-stalked, palmately 5–9–lobed leaves 20–50 cm across. Flowers reddish or yellowish, in terminal spikes to 15 cm; the upper flowers female, the lower male. Leaf-blade with triangular-lanceolate toothed lobes, cut to two-thirds the width of the blade; leaf-stalk as long as the blade. Fruit distinctive, in a cylindrical cluster of many capsules, each 2.5 cm, usually covered with long stiff prickles, splitting into 3 valves, and with large smooth shiny mottled grey, white or brown seeds, each with a large swelling (aril) at its base.

The seeds are poisonous, but yield the well known castor oil, which is purgative; other parts of the plant are used medicinally. **Page 502.**

BUXACEAE   Box Family

A small family of evergreen shrubs and small trees scattered through the Temperate, Sub-tropical and Tropical Zones. Leaves entire, leathery, dark green, opposite or alternate. Flowers tiny, regular, unisexual, borne in small axillary clusters, or spikes. Sepals usually 4, fused at the base; petals absent; stamens usually 4. Female flowers few; ovary superior, 3–chambered each with 1–2 ovules; styles 3. Fruit a capsule, or a drupe, seeds shining black.

BUXUS Box Leaves opposite. Flower-clusters with female flowers at the apex. Fruit a capsule. 2 spp.

1242 **B. wallichiana** Baillon
Afghanistan to C. Nepal. 1800–2700 m. In shady rocky ravines; often gregarious. Mar.–May.

A dense leafy evergreen shrub to 2 m or a small tree to 7 m, with 4–sided branchlets, and with soft corky yellowish-grey bark fissured into 'crocodile-scales' on old stems. Flowers yellowish, in dense rounded axillary clusters 6–8 mm. Leaves entire leathery, linear-lanceolate to elliptic-oblong, 1.5–6 cm, with a blunt or notched apex, shining above and hairless. Capsule ovoid 7–12 mm, with 3 diverging styles, 3–valved.

Wood hard, fine-grained, and durable, used for fine carving, engraving, turning, geometrical instruments, and combs. Extracts from the leaves, bark, and wood are used medicinally. **Page 502**.

SARCOCOCCA Leaves alternate. Flower clusters with female flowers at the base, or clusters 1–sexed. Fruit a drupe. 4 spp.

1243 **S. saligna** (D. Don) Muell. Arg. (*S. pruniformis* auct. non Lindley)
Afghanistan to W. Nepal. 1200–2400 m. Moist and shady places, forests; common, gregarious. Sep.–May.

A small evergreen shrub to 1 m, with green stems, and with shining leathery lanceolate entire leaves. Flowers white, in short dense axillary clusters 6–10 mm; stamens with stout white filaments much longer than the minute rounded sepals. Leaves hairless, variable, narrowly to broadly lanceolate long-pointed, 5–10 cm. Fruit purple, ovoid 7–8 mm long, tipped with the 3 style-bases. **Pl. 111**.

1244 **S. coriacea** (Hook.) Sweet (*S. pruniformis* Lindley)
Himachal Pradesh to E. Nepal. Burma. 600–1800 m. Forests, shrubberies. Oct.–Nov.

Like 1243, but leaves broader, elliptic long-pointed, and flower greenish-yellow, in axillary elongate cylindrical clusters to 2 cm. Stamens much exserted to 6 mm, filaments pale green. Leaves with 3 parallel veins; stems to 180 cm.

1245 **S. hookerana** Baillon
W. Nepal to S.W. China. 2100–3600 m. Forests; very common. Feb.–Apr.

Like 1243 with lanceolate leaves, but female flowers solitary, borne on short bract-covered stalks. Male flowers white, sweet-scented, in rounded or elongate clusters; filaments stout, twice as long as sepals. Leaves *c.* 8 cm, mostly 1 cm or less broad; stems branched, 50–100 cm. Fruit globular.

URTICACEAE Nettle Family

A medium-sized family of the tropical and temperate regions, usually of herbaceous plants or small shrubs, less commonly of trees. Leaves alternate or opposite, with stipules. Inflorescence of branched clusters of small green-ish flowers which are unisexual, though borne on the same inflorescence. Perianth of 4–5 sepal-like lobes or segments; stamens 4–5, anthers exploding when ripe. Female flowers with a superior ovary, 1–chambered and with 1 ovule; style 1. Fruit an achene, often enclosed in the persistent perianth.

Trees or shrubs	1 BOEHMERIA 2 DEBREGEASIA

Herbaceous plants
  a Plants with stinging hairs   3 GIRARDINIA 4 LAPORTEA
                                                5 URTICA

  b Plants without stinging hairs
    i Leaves alternate   6 ELATOSTEMA
    ii Leaves opposite   7 LECANTHUS 8 PILEA

1 BOEHMERIA Shrubs or small trees. Flowers in spikes; fruit enclosed in dry persistent perianth. 8 spp.

1246 **B. platyphylla** D. Don
Kashmir to S.W. China. India. Burma. 800–2700 m. Wasteland, second-ary forest; very common. Jul.–Sep.
    A variable deciduous shrub (or sometimes a perennial herb) covered with rough or sometimes smooth hairs, with large ovate to rounded, toothed leaves, and with small greenish-white to pinkish flowers in long interrupted tail-like spikes composed of widely-spaced flower clusters. Male spikes erect, often branched, 7–15 cm; perianth 4–lobed; stamens 4; female spikes pendulous 15–30 cm; flowers with a tubular 4–lobed perianth and with a long thread-like hairy style. Leaves variable, 7–23 cm, mostly opposite, rough and wrinkled or nearly smooth above, sometimes with a slender tail-like point, saw-toothed, stalked; stems and branches 4–sided, to 2 m, or sometimes more. Achenes enclosed in the dry bristly-haired perianth. **Page 503.**

2 DEBREGEASIA Evergreen shrubs. Flowers in rounded stalkless axil-lary clusters. Fruit a cluster of many achenes with fleshy perianths. 3 spp.

1247 **D. salicifolia** (D. Don) Rendle (*D. hypoleuca* (Hochst.) Wedd.)
Afghanistan to E. Nepal. W. Asia. Abyssinia. 1500–2400 m. Forests, shrubberies; common. Apr.
    A softly hairy evergreen shrub, with broad to narrow lanceolate sharply toothed alternate leaves which are densely white-woolly beneath, and with densely woolly young branches and leaf-stalks, conspicuous in the autumn with its yellow to orange raspberry-like fruits. Flowers 1-sexed, males and females on different plants, in rounded stalkless axillary clusters *c.* 1 cm across, with numerous bracteoles. Male flowers with 4-lobed hairy peri-anth, and 4 stamens; female flowers with a tubular 4-lobed perianth constricted at the mouth and enclosing the ovary; stigma resembling a tuft

of hairs. Leaves alternate, long-pointed and short-stalked, with rough upper surface, 7–15 cm long and usually 1.5–2.5 cm broad; stems to 5 m. Fruit yellow, of many achenes with fleshy perianths.

Twine and rope is made from the bark. The fruit is edible.

1248 **D. longifolia** (Burm.f.) Wedd. (*D. velutina* Gaudich.)
Uttar Pradesh to S.W. China. India. Burma. S.E. Asia. 1000–2000 m. Forests, shrubberies. Jul.–Sep.

Very like 1247, but differing in its larger broadly lanceolate leaves which are white beneath with very fine dense white hairs, and its orange fruits borne on short branched stalks (stalkless in 1247) which are prominent in autumn. **Pl. 111**.

3 **GIRARDINIA** Herbaceous perennial, with stinging hairs. Leaves alternate, with stipules fused below. Achene flattened, seated on the perianth. 1 sp.

1249 **G. diversifolia** (Link) Friis (*G. heterophylla* Decne.)
Pakistan to Bhutan. India. Burma. China. S.E. Asia. 1200–3000 m. Wasteland, shrubberies, edges of cultivation. Jul.–Aug.

A robust nettle-like plant to 2 m or more, with deeply lobed saw-toothed leaves with bristles and stinging hairs, and with axillary and terminal branched spikes of greenish flowers. Female spikes very long, to 40 cm in fruit; male spikes much-branched. Leaves variable, usually with 5 lanceolate long-pointed coarsely saw-toothed lobes; blade and leaf-stalk with long awl-shaped stinging hairs (leaves sometimes ovate and shallow-lobed). Fruiting spikes with many bristles; achenes *c*. 3 mm.

This plant can give a very severe sting. **Page 503**.

4 **LAPORTEA** Herbaceous perennials with stinging hairs; leaves alternate. Flowers in branched clusters. Achenes flattened, seated on perianth. 1 sp.

1250 **L. terminalis** Wight
Uttar Pradesh to S.E. Tibet. India. Burma. China. S.E. Asia. 1900–3300 m. Forest clearings, shrubberies. Jul.–Sep.

A perennial to 120 cm, with few or many stinging hairs which sting severely, with broadly to narrowly elliptic long-pointed, coarsely saw-toothed leaves, and short axillary branched clusters of male flowers, and larger and long-stalked branched spreading clusters of female flowers. Male flowers with 4–5 equal, hairy perianth-segments; stamens 4–5; female flowers with broadly winged flower-stalks and 4 very unequal perianth-segments. Leaves 10–25 cm, leaf-stalk 3–5 cm. Achenes 3 mm, flattened. **Page 503**.

5 **URTICA** Herbaceous annuals or perennials, with opposite leaves and with stinging hairs. Flowers in branched clusters. Achenes encircled by the papery perianth. 4 spp.

365

1251 U. dioica L. STINGING NETTLE

Pakistan to S.W. China. N.& S.Temperate Zones. 1000–2500 m. Forests, shrubberies, edges of cultivation. Aug.–Sep.

A stinging herbaceous perennial, with opposite ovate to lanceolate, coarsely toothed leaves, and tiny green flowers in long axillary tassel-like hanging or spreading clusters not longer than the leaves. Flowers 1-sexed, on different plants; male flowers with 4 ovate perianth-segments and 4 stamens; female flowers with unequal segments; stigma like a tuft of hairs. Leaves 5–10 cm, with rounded or shallow heart-shaped base, stalked and with a pair of elliptic stipules at the base. Stem rather robust, grooved, 80–175 cm. Achenes flattened, encircled by the persistent perianth.

The young shoots are eaten as vegetables.

1252 U. hyperborea Jacquem. ex Wedd.

Pakistan to Bhutan. Tibet. Mongolia. 4100–5100 m. Stony high altitude steppes. Aug.

A densely tufted hairy perennial with many stout stems 15–35 cm, with small crowded coarsely toothed leaves, and with green flowers in short dense axillary clusters mostly 1 cm, but lengthening to 5 cm in fruit. Leaves 3–5 cm, ovate heart-shaped, with soft hairs between the small stinging hairs, nearly stalkless. Achenes *c.* 5 mm, encircled by the perianth, the outer segments bristly-haired.

Young shoots eaten as vegetables. **Page 503**.

6 ELATOSTEMA Herbaceous plants or undershrubs, with alternate leaves without stinging hairs. Flowers tiny, borne on a fleshy disk. Achenes minute and with persistent perianth. *c.* 14 spp.

1253 E. platyphyllum Wedd.

Uttar Pradesh to S.W. China. 1200–1800 m. Shrubberies, forests. Apr.–Jun.

The largest Indian species of this genus growing to nearly 2 m, with large leathery rhomboid to oblong-lanceolate very long-pointed, toothed leaves, and with minute yellowish flowers crowded on the surface of fleshy saucer-like receptacles, and usually bordered with fringed bracts. Flower-clusters borne in axils of upper leaves; male clusters 0.8–2 cm across, stalkless or short-stalked, with lobed involucral bracts; perianth 4–5-lobed; stamens 4–5; female clusters small, woolly-haired. Leaves 15–25 cm, variable, with an unequal wedge-shaped, rounded or eared base, and long pointed stipules at the base. Achenes ovoid, dotted with red. **Pl. 111**.

7 LECANTHUS Herbaceous perennials without stinging hairs. Like *Elatostema* with flowers borne on a disk-like receptacle, but leaves opposite. 1 sp.

1254 **L. peduncularis** (Royle) Wedd.

Pakistan to S.W. China. India. Burma. S.E. Asia. Africa. 1200–3200 m.
Forests. Jul.–Sep.

A fleshy spreading plant with creeping rooting stems, with ovate saw-toothed leaves, and with erect often leafless stems bearing a globular cluster of minute pink flowers borne on a variable saucer-like receptacle 0.3–7.5 cm across, bordered with involucral bracts. Male flowers with 4–5 perianth-segments; female flowers with 3 minute unequal lobes, the larger hooded at the top; male and female flowers on the same or different plants. Leaves opposite, 1.5–5 cm, unequal, 3–veined, borne on erect leafy stems, sometimes also with axillary long-stalked flower-clusters; flowering stems 10–20 cm. Some plants much smaller. Achenes narrow-oblong, red. **Page 503**.

8 PILEA Herbaceous perennials without stinging hairs; leaves opposite, 3–veined at base. Flowers minute, in long dichotomously branched clusters. Achenes ovoid. *c.* 12 spp.

1255 **P. scripta** (Buch.–Ham. ex D. Don) Wedd.

Pakistan to S.W. China. 1000–2500 m. Forests, shrubberies. Aug.–Sep.

A hairless erect perennial, with elliptic long-pointed leaves with 3 conspicuous parallel veins, and with minute white to pinkish flowers in much-branched slender spreading axillary clusters shorter than the leaves. Male and female flowers on the same or different plants; male with 4–lobed perianth; stamens 4, with white anthers; female flowers with 3 unequal lobes; stigma a tuft of hairs. Leaves 7–20 cm, with 3 fine transverse veins between the parallel veins, teeth very small, long-stalked; stem 30–150 cm. Achenes rough, ridged, *c.* 1 mm. **Pl. 111**.

ULMACEAE Elm Family

A family of trees and shrubs of tropical and temperate regions. Leaves alternate, simple, with stipules which are shed when the leaves unfold. Flowers green and inconspicuous, borne in clusters. Perianth usually 5–lobed, sepal-like; stamens usually 5. Ovary superior, usually 1–chambered with 1 ovule; styles 2. Fruit a winged nutlet or a drupe.

ULMUS ELM *c.* 3 spp.

1256 **U. wallichiana** Planchon

Afghanistan to W. Nepal. 1800–3000 m. In broad-leaved forests and moist ravines. Mar.–Apr.

A large tree with rough grey bark, with elliptic to obovate long-pointed, double-toothed leaves, and with green flowers in globular clusters appearing before the leaves. Flowers *c.* 3 mm long, short-stalked; perianth 5–6–lobed, the lobes ciliate, blunt, about as long as the tube. Mature leaves rough above, with an unequal oblique base, and with 15–20 pairs of lateral veins each ending in a large tooth, usually 8–10 cm; young shoots finely hairy. Fruit a nutlet surrounded by a wide papery wing spreading to 1.5 cm and contracted to a short stalk longer than the persistent perianth. Branches much lopped for fodder. Wood used for making furniture; it has a handsome grain. Bark contains a strong fibre used for rope, sandals, and matches. **Page 504**.

CANNABACEAE   Hemp Family

Close to the Moraceae and often included in it, but differing in having the parts of the flowers in fives instead of fours. Herbaceous plants with leaves usually lobed, and with stipules. Fruit a nutlet enclosed in the persistent perianth. A small widely distributed family.

CANNABIS HEMP 1 sp.

1257 **C. sativa** L. HEMP
Probably native of C. Asia; cultivated throughout the temperate and tropical regions. 2100–2700 m. Wasteland, edges of fields; very common. Jun.–Sep.

An erect strong-smelling annual, with digitate leaves with narrow coarsely toothed leaflets, and with short axillary clusters of yellow-green male flowers, and on different plants stalkless axillary clusters of green female flowers. Flowers c. 4 mm long; male flowers with 5–lobed perianth, and 5 stamens with yellow anthers with thread-like filaments. Female flowers with entire perianth enclosing the ovary, and with 2 protruding styles. Upper leaves with 1–5 linear-lanceolate long-pointed leaflets, the lower with 5–11 leaflets, all leaves long-stalked, gland-dotted, pale, very finely downy beneath; leaflets variable in size 3–10 cm. Stem grooved, finely hairy, with slender branches, to 3 m or more in cultivation. Fruit c. 4 mm, glandular hairy.

The intoxicating drugs 'ganja' and 'charas' are obtained from the resinous exudations of the stem, young leaves and flowers. 'Bhang' or 'pot' consists of the larger leaves mixed with some fruits.

The fibre hemp is obtained from the stems of the plant; it is used for weaving cloth. **Page 504**.

MORACEAE   Mulberry or Fig Family

An important family of trees and shrubs, mainly of the tropical and subtropical regions. Distinguished from related families by having a milky sap containing latex. Leaves alternate or opposite, entire or lobed; stipules present. Flowers small, unisexual, borne on the same or different plants, in heads, or on a flattened or hollowed receptacle, or in catkins. Male flowers with 4 stamens; female flowers with 2 carpels, one of which is aborted; styles 2. Ovary with 1 ovule. Fruit very variable in structure, often fleshy and edible.

FICUS FIG Distinguished by its inflorescence which is composed of a hollow flask-like receptacle with the minute flowers lining the inner surface, and which is almost completely closed at the mouth by numerous overlapping scales. Fruit comprising a fleshy receptacle, lined with carpels. *c.* 40 spp.

Fruit paired and stalkless, or solitary and stalked

1258 **F. benghalensis** L. BANYAN

India. In the Himalaya to 1400 m. Commonly planted as a shade tree at wayside resting places. Fruit ripening Oct.–Nov., also in Apr.

A large spreading tree, with aerial roots descending from the spreading branches which root in the ground and thicken and become trunk-like, thus extending the diameter of the tree. Distinguished by its leathery ovate blunt leaves, with entire margins, with blades 8–20 cm and with thick leaf-stalks 1–5 cm. Figs in pairs, globular, hairy, stalkless 1.3–2 cm, red when ripe. Trees become massive, sometimes spreading to 80–100 m; bark greyish-white, smooth; young trees epiphytic.

Sacred to Hindus. Wood of aerial roots elastic and can be used for carrying-poles, tent-poles and cart-yokes. Leaves and twigs good cattle fodder. Bird-lime is made from the milky juice. Fruits edible in times of scarcity; the leaves used as plates. **Page 504**.

1259 **F. religiosa** L. PIPAL

Pakistan to Bhutan. India. Widely cultivated in the Himalaya to 1400 m. Commonly planted with 1258. Fruit ripening Oct.–Nov., and in Apr.

A large tree with grey bark, distinguished by its broad ovate leaves with a long tail-like point one third to half as long as the blade, shining above, reddish when young beneath, and with white swellings when mature. Leaf-blade including apex 11–18 cm long, with undulate margin and rounded or shallow heart-shaped base, and with 6–8 pairs of lateral veins; leaf-stalk 7–10 cm. Fruit globular *c.* 1.3 cm, in stalkless pairs, dark purple when ripe. Tree leafless or nearly so for a short time during the hot weather.

Sacred to Hindus. Wood can be used for charcoal and packing cases, and shoots used for cattle fodder. An epiphytic tree, destructive to forest trees, buildings, and walls. **Page 504**.

1260 **F. palmata** Forsskal

Afghanistan to W. Nepal. Africa. To 2300 m. Rocky slopes in drier areas. Fruiting Jun.–Oct.

A shrub or small tree with grey-hairy young branches, smooth grey bark, and with broadly ovate, toothed leaves which are rough above and shortly woolly-haired beneath. Leaves variable, often deeply lobed, 7–13 cm including the hairy leaf-stalk 2.5–5 cm. Fruit usually solitary, stalked, rounded or top-shaped 1.2–2.5 cm, generally hairy, yellow when ripe.

Fruit edible; leaves used as cattle-fodder. Some varieties of cultivated figs possibly belong to this species.

Fruit clustered on short leafless branches

1261 **F. auriculata** Lour. (*F. roxburghii* Wallich ex Miq.)
Pakistan to Bhutan. India. Burma. China. To 1600 m. Forests; common in Nepal villages. Fruit ripening Apr.–Jun.

A spreading tree with grey warty bark, and with large alternate broadly ovate heart-shaped, obscurely toothed or wavy-margined stalked leaves, usually 15–30 cm but sometimes larger, softly hairy beneath. Fruit short-stalked, in clusters of 6–20 borne on short thick leafless branches arising from the trunk, sometimes low down and close to the ground. Fruit top-shaped, strongly ribbed, hairy when young but becoming hairless, russet-brown and tinged with red or purple when ripe, 2.5–5 cm long by 5–7.5 cm broad.

Fruit edible; leaves lopped for cattle-fodder.

1262 **F. oligodon** Miq. (*F. pomifera* Wallich ex King, *F. hamiltoniana* Wallich)
W. Nepal to Bhutan. Burma. S. China. S.E. Asia. To 1800 m. Forests and cultivated in Nepal villages. Fruit ripening Mar.–Jun.

A small tree with white bark, and with elliptic to oblong-elliptic blunt leaves with rounded or heart-shaped bases, irregularly toothed, hairy beneath. Leaf-blade variable in size 12–35 cm by 8–25 cm broad, long-stalked. Fruit in stalked clusters borne on shortened leafless branches from the trunk and main branches, depressed-globular 2.5–9 cm, red when ripe, edible.

1263 **F. semicordata** Buch.–Ham. ex Smith (*F. cunia* Buch.–Ham. ex Roxb.)
Pakistan to Bhutan. India. S. China. Burma. S.E. Asia. To 2000 m. Forests; common Nepal village tree. Fruit ripening May–Sep.

A small spreading tree with dark grey bark, and with rough oblong-lanceolate to ovate leaves, which are very unequally lobed at the base, the lower lobe rounded and much longer. Leaves long-pointed, toothed, short-stalked, 10–30 cm. Fruit reddish-brown, to 2 cm, short-stalked, rough, wrinkled and often scaly, numerous and borne on long leafless branches arising from the trunk and older branches.

Fruit edible; leaves lopped for fodder. Fibres from the bark used to make coarse rope.

MORUS MULBERRY Distinguished by its fleshy fruit composed of many fleshy perianths encircling the nutlets which are clustered over the fleshy axis. 4 spp.

1264 **M. serrata** Roxb.
Pakistan to C. Nepal. 1200–2700 m. Forests, shrubberies. Apr.–May.
A large deciduous tree with reddish or greyish-brown bark, and with broadly ovate-heart-shaped, coarsely-toothed, often deeply 3–lobed leaves. Male flowers in stalked spikes 2.5–7 cm long, with sepals 3–4, and 4 stamens. Female flowers in short-stalked spikes 1–2 cm; styles long, very hairy. Leaves 5–20 cm, mostly with a long-pointed apex, sometimes paler woolly-haired beneath when young. Fruit purple, 1.6–2.5 cm, borne on a stout hairy stalk.
Fruit sweet and edible. Wood used for cabinet-making and agricultural implements. Leaves cut for cattle-fodder. **Page 504**.

1265 **M. alba** L. WHITE MULBERRY
Probably native of China; cultivated extensively in the drier areas of the W. Himalaya to 3300 m. Mar.–Apr.
Distinguished by its white or purple fruit, and by the female flowers which have a perianth of 4 unequal sepals, the outer 2 keeled (sepals 3–4, all similar in 1264). A medium-sized tree with hairy shoots. Leaves 5–7.5 cm, ovate to lanceolate, toothed, sometimes lobed, hairless except on the veins beneath. Male spikes 1.2–4 cm long; female spikes ovoid.
Fruit edible; silk-worms are fed on the leaves.

PLATANACEAE   Plane Family

A small family of deciduous trees almost exclusively of N. America, distinguished by its scaling bark, its large palmately lobed leaves, and its pendulous clusters of globular flowers and fruits. Flowers 1–sexed; sepals small, hairy, 3–8; petals spathulate, many, scaly. Male flowers with 3–8 stamens with their connectives fused above to form a shield-shaped cap; female flowers with 3–9 carpels each with tapering styles. Fruit a globular head of many 1–seeded carpels each with persistent style and with a ring of long bristly hairs.

PLATANUS 1 sp.

1266 **P. orientalis** L. ORIENTAL PLANE
Native of S.E. Europe, W. Asia and possibly Pakistan west of the Indus; frequently cultivated as a shade tree in W. Himalaya. To 2400 m. Apr.–May.
A large deciduous tree with characteristic grey scaling bark, particularly on younger trunks and branches, with broad deeply cut leaves, and with long-stalked pendulous globular clusters of numerous small greenish flowers Male flower-clusters 2–3; female clusters 2–5, larger 1–1.5 cm. Leaves broadly heart-shaped in outline, 8–24 cm by 12–30 cm broad, deeply cut into 5–7 narrow triangular-pointed, coarsely toothed lobes; leaf-stalk long, with an enlarged base which encircles the axillary bud.

Fruiting heads globular 2–3 cm, of many carpels each with a long persistent style, and with long yellow hairs at the base.

The wood is of relatively little value; used for boxes and cabinet-making. The bark is used medicinally. **Page 505.**

JUGLANDACEAE   Walnut Family

A small family of deciduous trees, largely of north temperate and sub-tropical regions, with characteristic pinnate leaves. Flowers small one-sexed, both male and female borne on the same plant. Male flowers in pendulous catkin-like inflorescences borne on the previous year's branches; female flowers few, in small erect spikes borne on present year's branches. Male flowers with many stamens, and usually a 4–lobed perianth; female flowers with an inferior ovary of 2 fused one-chambered carpels, each with a single ovule; stigmas 2. Fruit a nut, or a drupe.

JUGLANS   Fruit a drupe with a hard shell. Leaves pinnate, with paired leaflets and a terminal leaflet. 1 sp.

1267   **J. regia** L. var. **kamaonia** C. DC. Himalayan Walnut
Kashmir to S. E. Tibet. 1500–3000 m. Forests; widely cultivated in the W. Himalaya. Feb.–Apr.

A large deciduous tree with grey vertically fissured bark, distinguished by its pinnate leaves, and by its large green drupes containing wrinkled nuts. Male catkins pendulous, green, 6–12 cm long; female flowers very small, in a short spike. Leaflets 5–13, elliptic to ovate pointed, entire, leathery 7–20 cm.

The wood is hard, strongly grained and polishes well, excellent for furniture, carved work, gun-stocks, and veneers. The fruit is an important article of diet, and in some regions pressed for oil; the rind of the unripe fruit is used to intoxicate fish, and for tanning and dyeing. The bark, leaves, and fruit are used medicinally.

ENGELHARDIA   Distinguished from *Juglans* by its fruit which has a large 3–lobed membraneous wing, and by its pinnate leaves with paired leaflets but without a terminal leaflet. 1 sp.

1268   **E. spicata** Leschen. ex Blume
Himachal Pradesh to S.W. China. S.E. Asia. To 1600 m. Forests. Mar.–May.

A large deciduous tree with smooth grey bark, with pinnate leaves, and with long pendulous fruiting clusters with characteristically papery-winged fruits. Male spikes solitary or 3–5, to 8 cm long, hairy, borne on leafless branches; female spikes long pendulous. Leaflets 6–10, elliptic blunt, rather leathery, usually woolly-haired and dotted with minute yellow glands beneath. Fruiting clusters long dense, with numerous overlapping fruits, to 30 cm; fruit a small globular nut with a 3–lobed papery wing to 5 cm, densely hairy at base. **Page 505**.

MYRICACEAE

A small more or less cosmopolitan family of aromatic trees and shrubs, with usually simple alternate leaves. Flowers unisexual, borne in dense catkin-like axillary spikes. Male flowers with usually 4 bracteoles and usually 4 stamens; female flowers with 2–4 bracteoles, and with an ovary of 2 fused carpels with 1 chamber and 1 ovule. Fruit a succulent, resinous or waxy, wrinkled drupe.

MYRICA 1sp.

1269 **M. esculenta** Buch.–Ham. ex D. Don (*M. nagi* auct. non Thunb.)
Kashmir to Bhutan. India. Burma. China. S.E. Asia. 1000–2300 m. Cultivated areas. Sep.–Dec.

A small evergreen tree or large shrub, with narrow oblong to oblong-lanceolate entire leathery dark green leaves, dotted beneath with minute resinous glands. Male flower-spikes reddish, in branched axillary clusters; female flowers in slender spikes, or occasionally at end of male spikes. Leaves short-stalked, pale or rust-coloured beneath, mostly 8–12 cm; bark greyish-brown, rough, vertically fissured; shoots greyish. Fruit reddish, with red flesh and a rough stone, to 1.5 cm.

Fruit edible; it makes a refreshing drink. Bark aromatic; it is used medicinally and for fish-poisoning, and gives a good yellow dye. **Page 505**.

BETULACEAE Birch Family

A family of trees and shrubs of the north temperate regions and tropical mountains. Leaves simple, alternate, deciduous, with stipules. Male and female flowers borne in separate spikes or catkins on the same tree. Male flowers in pendulous cylindrical catkins; female flowers in often erect cone-like spikes. Male flowers with a 4–lobed perianth, and 2–12 stamens; female flowers with a naked ovary of 2 fused carpels with 2 long thread-like styles. Fruiting cones cylindrical or ovoid, with deciduous or persistent bracts. Nutlet 1–seeded, winged.

ALNUS ALDER Distinguished by its fruiting cones which are woody and persistent. 2 spp.

1270 **A. nepalensis** D. Don
Himachal Pradesh to S.W. China. Burma. S.E. Asia. 1000–3000 m. Riversides, wet gullies, damp forests; common, and often planted to contain landslips. Oct.–Dec.

A large deciduous tree with a dark green bark which becomes silvery-grey in the open. Male catkins slender to 12 cm; female clusters cone-like, becoming woody, ovoid to 1.5 cm long. Leaves elliptic to ovate with rounded or shortly pointed apex, entire or with an undulate margin, finely hairy on the veins beneath when young. Nutlet with a narrow papery wing.

Wood soft and light and easily worked. Bark used for dyeing and tanning. **Page 505**.

1271 **A. nitida** (Spach) Endl.
Kashmir to W. Nepal. 1000–2700 m. Riversides; common. Sep.–Oct.

Very like 1270, but with ovate to elliptic leaves which are narrowed from about the middle to the apex, usually toothed, and with lateral veins running to the margin, smooth beneath (lateral veins not reaching the margin, and leaves with minute swellings (lens required) in 1270). Female cones mostly solitary, axillary; nutlet with a narrow leathery margin. A large tree with smooth dark brown shining bark, becoming rough and furrowed on old trunks; twigs usually hairless.

BETULA BIRCH Distinguished from *Alnus* by its fruiting cones which are not woody and which fragment when mature into 3–lobed scales and small winged nutlets. 3 spp.

1272 **B. utilis** D. Don
Pakistan to S.W. China. 2700–4300 m. Forms forests at upper limit of tree growth. Apr.–May.

Distinguished by its white to brownish bark, its leaves which are woolly-haired below, and its solitary female spikes. A moderate-sized tree, with very distinctive bark which peels off in very thin pale almost transparent horizontal strips. Male catkins mostly 5–10 cm, reddish, appearing on bare branches, or with the young leaves. Leaves 5–8 cm, ovate fine-pointed, irregularly saw-toothed, finely hairy when young. Fruiting catkins 2.5–4 cm by 1.2 cm broad. Nutlets *c.* 2 mm, conspicuously winged, with a 3–lobed much longer bract.

Wood used for building in the inner drier regions. Bark used as paper, for water-proofing and roofing houses. **Page 506**.

B. jacquemontii Spach Very like 1272 and by some considered to be a subspecies; found in Afghanistan and W. Himalaya. Distinguished by its minutely downy, non-glandular branchlets, and by its leaves with wedge-shaped or rounded-ovate base, gland-dotted beneath, with 7–9 pairs of lateral veins. Fruiting catkin solitary, stalked. There are many intermediate forms.

1273 **B. alnoides** Buch.–Ham. ex D. Don
Himachal Pradesh to S.W. China. 1500–2700 m. Forests, ravines, stream-sides. Nov.–Dec.

Distinguished from 1272 by its greyish-brown to reddish papery bark, and by its long slender female catkins which are usually in clusters of 2–4.

A large tree with ovate to ovate-lanceolate acute leaves 8–15 cm, very irregularly doubly saw-toothed, and which are hairy beneath when young and gland-dotted when older. Nutlet with a wing broader than itself; bracts with 2 small lobes or teeth. **Page 506**.

CORYLACEAE   Hazel Family

A small family of deciduous trees and shrubs, similar to the Betulaceae and often included in it. Distinguished by its nuts which are partly encircled by a greenish bract or involucre. Male flowers solitary, perianth absent, many borne in catkins; female flowers with perianth clustered or in spikes. (Male flowers in groups of 3, with a perianth; female flowers lacking perianth in Betulaceae.)

CARPINUS Hornbeam Nuts borne on drooping spikes, each nut with a large papery bract-like involucre which aids in its dispersal. 2 spp.

1274 **C. viminea** Lindley
Kashmir to S.W. China. Burma. S.E. Asia. 1500–2700 m. Forests, shady ravines, watercourses. Mar.–Apr.
    A medium-sized tree with compact grey bark with darker streaks, with ovate-lanceolate long-pointed, doubly saw-toothed leaves, and with pendulous fruiting clusters 5–10 cm long. Male catkins pendulous to 8 cm, borne on leafless branches; bracts triangular-ovate; stamens hairy. Leaves 5–12 cm; stipules oblong, soon falling. Nut *c.* 3 mm, ribbed, with lanceolate toothed involucre *c.* 2.5 cm, with small triangular basal lobes.
    Wood used for fuel. **Page 506**.

1275 **C. faginea** Lindley
Kashmir to W. Nepal. 1800–2400 m. Forests. Mar.–Apr.
    Like 1274, but differing in its softly hairy young branches, and leaves which remain silky-haired on the veins beneath. Bark dark brown, wrinkled. Involucre surrounding nut broader, narrow-elliptic, unequal-sided, toothed and usually not lobed, to 4 cm.

CORYLUS Hazel Distinguished by its large nuts encircled by a large lobed or incised green involucre. 2 spp.

1276 **C. ferox** Wallich
C. Nepal. to S.W. China. 2400–3000 m. Forests. Oct.–Nov.
    Readily distinguished when in fruit by its dense globular clusters of nuts, with involucres with narrow lobes ending in long spines. A small or medium-sized deciduous tree, with silky-haired elongated buds, and with ovate-lanceolate long-pointed leaves, with sharply saw-toothed margins, 6–15 cm. Male catkins many, dense, to 8 cm; bracts silky-haired. Nuts in clusters of 3–6 *c.* 4 cm across; nut *c.* 1 cm or more; involucre thick almost fleshy, hairy. **Page 506**.

1277 **C. jacquemontii** Decne.

Afghanistan to W. Nepal. 1800–3000 m. Forests, shrubberies; often gregarious. Mar.–Apr.

A medium-sized deciduous tree, with thin dark grey bark. Leaves broadly ovate long-pointed, 8–15 cm, often slightly lobed, unequally and doubly saw-toothed. Nuts in clusters of 2–3, each with a leathery glandular-hairy involucre to 4 cm, much longer than the nut, and cut into narrow lanceolate lobes.

Trees are cultivated for the nuts, which are an important article of food. **Page 506.**

FAGACEAE   Beech or Oak Family

An important family of hardwood trees which often dominate large areas of the temperate and sub-tropical regions of the Northern Hemisphere, and to a lesser extent in the Southern Hemisphere. Leaves evergreen or deciduous, alternate, entire or deeply lobed, usually with stipules. Flowers 1–sexed; male flowers in erect spikes or drooping catkins; female flowers in separate spikes, or at base of male spike. Distinguished by its fruit which consists of a nut encircled or enclosed in a hard woody involucre of many often spine-tipped bracts, or bracts with small swellings.

CASTANOPSIS Nuts wholly enclosed in a spiny involucre. Evergreen trees. 5 spp.

1278 **C. indica** (Roxb.) Miq.

C. Nepal to S.W. China. S.E. Asia. 1300–2900 m. Forests; often gregarious in association with *Schima wallichii*. Oct.–Nov.

A moderate-sized or large evergreen tree, distinguished by its large oblong to oblong-elliptic leaves 12–28 cm, with 14–16 pairs of prominent lateral veins, with saw-toothed margins, and finely hairy undersides. Flowering spikes cream-coloured, erect, clustered, to 15 cm. Young shoots rusty-haired; bark silvery-grey, warted and fissured; tree to 25 m. Fruit in spike-like clusters 10–15 cm, often branched; individual fruits 2.5–4 cm across, with long unequal straight slender crowded hairy spines covering the involucre.

Nuts edible. Wood used for roof-shingles. **Page 507.**

1279 **C. hystrix** Miq.

C. Nepal to S.W. China. Burma. S.E. Asia. 1000–2400 m. Forming forests, often mixed with 1280, but not so common. Apr.–May.

A large evergreen tree, with usually entire lanceolate fine-pointed leaves 7.5–10 cm, with 7–9 pairs of lateral veins, and with distinctive fine reddish woolly hairs on the underside of the blades, but leaves variable in shape and sometimes toothed. Male spikes stout, spreading or somewhat drooping; female flowers in threes. Fruit more or less 4–angled, *c.* 5 cm, thickly covered with numerous simple or branched hairy spines with hairless tips.

Nuts edible. **Page 507**.

1280 **C. tribuloides** (Smith) A. DC.

Uttar Pradesh to S.W. China. S.E. Asia. 450–2300 m. Forests; gregarious and common; often in association with *Schima wallichii*. Mar.–May, and sometimes Aug.–Sep.

Like 1279, but often distinguishable by its fruits in which the flat surface of the involucre is visible between the tufts of spines (some forms have the involucre completely covered with spines). Flowering spikes cream-coloured, male spikes densely clustered to 20 cm; female spikes solitary. Fruiting spikes to 25 cm, fruits widely spaced. Leaves mostly entire to 25 cm, usually lanceolate long-pointed, but variable and sometimes broader and toothed, often paler or reddish beneath; leaf-stalks thickened below. A medium-sized or rarely large evergreen tree, with finely hairy shoots. **Page 507**.

LITHOCARPUS Involucre of fruit thick woody, several involucres coalescing in fruit and forming clusters on the fruiting spike. 2 spp.

1281 **L. pachyphylla** (Kurz) Rehder

E. Nepal to Burma. 1800–2700 m. Locally dominant in forests in extreme E. of Nepal. Jun.–Aug.

A large evergreen tree, distinguished by its long fruiting clusters with the fruits coalesced by their involucres into groups, and the involucral cups almost completely encircling each nut. Flowering spikes very slender to 15 cm, longer than the leaves, solitary or clustered; female flowers in groups of 3. Leaves elliptic-lanceolate long-pointed, entire, 13–20 cm, hairless above, with minute star-shaped hairs beneath. Nut depressed-globular, *c.* 2.5 cm.

Used for building; wood excellent for firewood. **Page 507**.

1282 **L. elegans** (Blume) Hatus ex Soep. (*L. spicata* Rehder and Wilson)

W. Nepal to S.W. China. S.E. Asia. 600–2100 m. Forests; rarely gregarious. Sep.–May.

Like 1281 with fruits in confluent clusters, but distinguished by its smaller shining nuts c. 1.5 cm, which are encircled by the involucral cup to one third at most. Male spikes stout, hairy, in dense terminal branched clusters 5–20 cm. Leaves lanceolate to oblanceolate 13–30 cm, leathery shining, very variable. A large or medium-sized evergreen tree. In Nepal there are two distinct forms; a high-altitude form with long-stalked lanceolate leaves with acute bases; a low-altitude form with much larger broader leaves with eared bases. **Page 507**.

QUERCUS OAK Distinguished by its fruits which have a conical or globular nut (acorn) partly encircled by an involucral cup of fused or overlapping bracts. Fruits usually borne in clusters, but involucral cups separate (not fused as in *Lithocarpus*). Himalayan oaks are evergreen; they are very important in the temperate hill forests, particularly in the hill-village and cultivated zone, where these forests are often much lopped for cattle-fodder and firewood. For certain identification it is essential to obtain fruit; these can often be found on the ground beneath the trees. 10 spp.

Bracts of involucral cups fused in concentric rings
a *Involucral cup 1.2–2 cm across; leaves to c. 15 cm long*

1283 **Q. glauca** Thunb.
Pakistan to S.W. China. Japan. S.E. Asia. 800–3000 m. Moist forests, ravines; not very gregarious. Apr.–May; fruiting the following year.

A large handsome evergreen tree, distinguished by its conical nuts *c.* 1.5 cm, with a distinctive involucral cup of 5–8 concentric velvety belts of scales, covering less than half the nut when mature. Male catkins solitary, or in a lax cluster, *c.* 8 cm; female spikes 2–4–flowered. Leaves oblong to oblong-lanceolate, long-pointed, 8–15 cm, saw-toothed, leathery and shining above, becoming glaucous beneath. Bark rough, fissured.
Wood used largely for fuel; foliage cut for fodder. **Page 508**.

1284 **Q. oxyodon** Miq. (*Q. lobbii* var. *oxyodon* (Miq.) Wenzig)
C. Nepal to S.W. China. S.E. Asia. 1800–2700 m. Moist forests; gregarious, often occurring with 1285. Apr.–May.

Distinguished from 1283 by its hemispherical nuts *c.* 2 cm, half-encircled by the involucral cups; nuts solitary, stalkless. Leaves lanceolate, glaucous beneath, with conspicuous slender spike-like marginal teeth; a tree to 20 m. **Page 508**.

b *Involucral cup large, to 6.5 cm across; leaves 15 cm long or more*

1285 **Q. lamellosa** Smith
C. Nepal to S.W. China. 1600–2800 m. Moist forests; gregarious and common in wet areas. Apr.–May.

A very large evergreen tree to 60 m, distinguished by its large dark green leaves and the contrasting glaucous undersides of mature leaves, and by its very large stalkless fruits borne on short spikes with concentricly ringed involucral cups almost encircling the nuts. Leaves oblong-elliptic, conspicuously toothed, with many pairs of strong lateral veins, short-stalked, 15–30 cm; young leaves silvery-hairy or buff-hairy beneath. Nut globular or top-shaped and with a fine point; involucral cup 5–6.5 cm across.
Wood very hard, and durable under cover, and used for building; gives excellent fuel; foliage cut for fodder. **Page 508**.

Bracts of involucral cup distinct, overlapping, and with free tips
a *Mature leaves saw-toothed, never spiny*

1286 **Q. lanata** Smith (*Q. lanuginosa* D. Don)
Uttar Pradesh to Arunachal Pradesh. Burma. S.E. Asia. 1400–2400 m. Forests; gregarious and common. Apr.–May.

378

A large or medium-sized evergreen tree, distinguished by its mature leaves which are dark shiny green above and contrastingly rust-coloured and woolly-haired beneath, and by its young shoots which are covered in rust-coloured hairs. Male catkins woolly-haired. Leaves leathery, elliptic-lanceolate, conspicuously toothed, with 9–14 pairs of lateral veins, 8–15 cm. Fruit solitary or paired, stalkless, axillary; nut oblong, *c.* 1.5 cm, half encircled in the woody involucral cup which has numerous oblong adpressed finely hairy bracts.

Wood used for fuel; foliage cut for fodder. **Page 508.**

1287 **Q. leucotrichophora** A. Camus (*Q. incana* Roxb.)
Pakistan to C. Nepal. Ceylon. Burma? 1200–2400 m. Forests; gregarious and common; often associated with *Rhododendron arboreum* and *Lyonia ovalifolia.* Apr.–May.

Similar to 1286, but distinguished by its leaves which have dense white-woolly hairs beneath, and its white-woolly young shoots. A large or medium-sized tree, with leathery dull-green, sharply toothed leaves, 6–16 cm. Nuts ovoid, to 1.5 cm long, half encircled by the involucral cup when mature, with involucral bracts as broad as long, rough hoary, involucral cup *c.* 1.5 cm across.

Wood used locally for building; gives good fuel. Foliage cut for fodder. Acorns used medicinally. **Page 508.**

b *Some mature leaves entire, some spiny, particularly on lower branches*

1288 **Q. semecarpifolia** Smith
Afghanistan to S.W. China. 2100–3800 m. Forming forests; gregarious and common, sometimes dominant up to the upper tree-line. May–Jun.; fruiting 15 months later.

A large evergreen tree distinguished by its globular nuts and its thin concave involucral cups which only cover the extreme base of the nuts. Leaves elliptic to oblong,mostly 5–10 cm, usually with spiny marginal teeth on young trees, but often with entire margins on old trees, dark green glossy above and generally with rust-coloured hairs beneath, but often nearly hairless on old leaves. Nut solitary, to 2.5 cm across, black when ripe; involucre with narrow overlapping bracts.

Wood used locally for building; gives good fuel. Foliage cut for cattle-fodder. The bark contains much tannin. **Page 509.**

1289 **Q. baloot** Griffith (*Q. ilex* auct. non L.)
Afghanistan to Himachal Pradesh. 1000–2600 m. In inner drier valleys only; gregarious and often with *Pinus gerardiana.* Apr.–May.

A small or medium-sized evergreen tree, or a bush, distinguished by its cylindrical or top-shaped nut, and by its dark green leathery leaves which are softly white-haired beneath. Leaves oblong to rounded 2–6 cm, either entire or with spiny-toothed margins. Nuts solitary or paired, two and a half to three times as long as the involucral cup. Similar to the European HOLM OAK, *Q. ilex* L. and considered by some to be conspecific. **Page 509.**

379

SALICACEAE

1290 **Q. floribunda** Lindley ex A. Camus (*Q. dilatata* Lindley)
Afghanistan to C. Nepal. S.W. China. 1800–2700 m. Gregarious and
often mixed with other oaks in damp cool situations. Apr.–May.

Like 1289 with small leathery evergreen leaves with spiny margins, or
with entire margins, but leaves hairless beneath, and lateral veins branched
short of the margin (veins not conspicuously branched in 1289). A large
evergreen tree, with grey or black bark, peeling in strips. Leaves mostly to
8 cm, variable, lanceolate to elliptic. Nut ovoid with a fine point, to 2 cm,
twice as long as the involucral cup, which has finely hairy bracts.

Wood strong, used locally for many purposes. Foliage cut for fodder.
Page 509.

SALICACEAE   Willow Family

An important and common family throughout the N. Temperate Zone, with a
few species occurring in the S. Hemisphere. Deciduous trees and shrubs, with
simple usually alternate leaves. Flowers unisexual, on different plants, numerous and borne in a *catkin*, which has a central axis bearing numerous tiny
densely clustered flowers each subtended by papery bracts. Male flowers with
2 to several, free or united stamens; female flowers of 2 carpels, 1–chambered,
with many ovules. Fruit a capsule, splitting into 2–4 valves; seeds with a tuft
of white silky hairs which aids their dispersal.

SALIX WILLOW Flowers insect-pollinated, with stiff erect 1–sexed catkins. Buds with a single covering scale; leaves narrow and short-stalked. *c.*
35 spp.

Trees, or erect shrubs to 1 m or more

1291 **S. disperma** Roxb. ex D. Don (*S. wallichiana* Andersson)
Afghanistan to S.W. China. 1500–3600 m. Streamsides, shrubberies;
common. May–Jun.

A shrub or a small tree to 5 m, with silky young leaves and branches,
and with greenish-grey bark. Catkins usually appearing before the leaves;
male catkins to 4 cm, stalks and bracts very silky-haired; female catkins to
6 cm, ovaries silky-haired, and bracts ovate-oblong, dark brown, silky-
haired. Fruiting catkins to 10 cm; capsules 6–8 mm, densely silky–haired.
Leaves oblong to ovate-lanceolate, usually narrowed to a long slender
point, mostly 5–10 cm, with entire margins or with obscure glandular
teeth, silky-haired when young, becoming hairless and glaucous beneath,
or with persistent silky hairs. A very variable plant.

Branches used for basket-making, the twigs for tooth-brushes.

1292 **S. sikkimensis** Andersson
E. Nepal to S.W. China. 3300–4400 m. Rhododendron shrubberies,
streamsides; gregarious. May.

A small tree, or a shrub at high elevations, with lanceolate leaves which are silky-haired and copper-coloured beneath, and with stout angled, shining branchlets. Male catkins borne on leafless branches almost stalk-less, stout curved, 2.5–5 cm long, very silvery-haired. Female catkins to 10 cm, silvery-haired, with obovate notched bracts; capsules densely silky-haired. **Pl. 112**.

1293 **S. daltoniana** Andersson
C. Nepal to Arunachal Pradesh. 3600–4400 m. Open slopes, streamsides; common and gregarious in Nepal in the lower alpine zone. May–Jun.

A shrub 2–3 m, or sometimes a small tree to 5 m, with large oblong to narrow-elliptic leaves which are densely rusty-haired beneath, and with finely saw-toothed margins. Male catkins slender, occurring with or before the young leaves, 3.5–6 cm, brownish-green; bracts spathulate, with long woolly hairs; stamen-filament hairy; female catkins axillary. Fruiting catkins to 12 cm. Mature leaves mostly 7–10 cm, stalked; young leaves densely silvery-haired beneath; branchlets dark, hairless.

1294 **S. karelinii** Turcz.
Afghanistan to C. Nepal. C. Asia. 2100–4500 m. Open slopes, stream-sides; common and gregarious in the drier parts of Kashmir. May–Jul.

A shrub to 120 cm, but sometimes a small tree to 5 m, with elliptic to ovate finely toothed leaves 4–5 cm, becoming paler and hairless beneath when mature; young leaves woolly-haired. Catkins borne on leafless stems; male catkins stout, very silky-haired, 3–4 cm; female catkins stout, to 4 cm, very silky-haired, with spathulate bracts which are blackish. Fruiting catkins to 9 cm; capsules *c.* 5 mm.

1295 **S. denticulata** Andersson (*S. elegans* Wallich ex Andersson)
Afghanistan to C. Nepal. 1800–3700 m. Forests, shrubberies, open slopes; common and gregarious. Mar.–May.

A shrub to 3 m, or a small tree to 5 m, with narrow-elliptic to elliptic, finely toothed leaves 4–5 cm, bright green above, becoming hairless and glaucous or pale beneath, and with darker brown twigs. Catkins appearing with the leaves; male catkins on short leafy shoots, *c.* 4 cm, with ovate truncate bracts which are hairy on the upper surface; female catkins green, borne on leafy shoots, slender flexuous, 2–5 cm long by 4 mm wide; bracts truncate; axis hairy. Fruiting catkins to 8 cm; capsules *c.* 3 mm, hairless.

Cultivated species

Larger arboreal species of willow are very important in the economy of villages in the dry inner valleys of W. Himalaya; they are extensively planted along irrigation channels. They are used for fuel, baskets and building. The commonest species at lower altitudes is **Salix acmophylla** Boiss. Other species include: **S. alba** L. White Willow; **S. excelsa** S. Gmelin (*S. fragilis* auct. non L.) Crack Willow; **S. babylonica** L. Weeping Willow; this latter species is also often planted in Nepal. (These names are current in India and do not conform with European usage.)

Dwarf alpine shrublets, usually less than 1 m
a *Leaves tiny, usually less than 1 cm*

1296 **S. lindleyana** Wallich ex Andersson
Pakistan to S.W. China. 3600–4500 m. Alpine slopes, screes, rocks; gregarious. Jun.–Jul.

A small creeping tight mat-forming shrublet, with glossy hairless elliptic-lanceolate leaves mostly 6 mm but variable. Male catkins ovoid 6–10 mm, pale green; stamens 2, with dark brown anthers and hairless filaments; bracts oblong, hairless. Female catkins hidden amongst the leaves, to 1.5 cm in robust plants. Capsules ovoid *c.* 5 mm, hairless.

1297 **S. hylematica** Schneider (*S. fruticulosa* Andersson)
Uttar Pradesh to Bhutan. 3000–4300 m. Alpine slopes; common in Nepal. May–Jun.

A laxer more spreading shrublet than 1296, with young shoots, axis of catkins, and stamen-filaments with long hairs, and with leaves which are glaucous beneath. Male catkins ovoid to 1.5 cm, with hairless dark red-purple bracts, borne on short lateral leafy branches. Female catkins 7–20 mm, axis densely hairy; style deeply 2–lobed. Fruiting catkins red, stalked, cylindrical 2–5 cm; capsules 2–3 mm, hairless. Leaves elliptic-lanceolate to oblanceolate, obscurely toothed, shining, hairless; stems spreading, sometimes to 20 cm.

b *Leaves usually more than 1 cm*

1298 **S. calyculata** Hook.f. ex Andersson
Uttar Pradesh to S.E. Tibet. Burma. 3600–4300 m. Alpine slopes; common in Nepal. Jun.–Jul.

A low creeping lax mat-forming shrublet, often with ascending stems to 20 cm. Male spikes to 2 cm, ovoid, hairless, borne on short leafy stems. Fruiting catkins ovoid *c.* 1.5 cm; capsules *c.* 7 mm. Leaves narrow-elliptic, narrowed to the base, short-stalked, mostly 1–3 cm, toothed, glossy above; young shoots and young leaves silky-haired. **Pl. 112.**

1299 **S. flabellaris** Andersson
Afghanistan to Uttar Pradesh. 3000–4000 m. Alpine slopes; common and gregarious in W. Himalaya. Jun.–Jul.

A dwarf shrublet creeping over rocks with stout spreading branches, or sometimes erect to 60 cm, with elliptic to obovate, shiny hairless leaves mostly 2 cm long, which are paler or glaucous beneath and with toothed or almost entire margins. Male catkins cylindrical 2 cm, yellowish, hairless, short-stalked and borne above the leaves; bracts obovate, brown, hairless; stamen-filaments hairless. Female catkins *c.* 2.5 cm, similar to male catkins, but stigmas entire. Fruiting catkins rather lax; capsules 3.5 mm, reddish, hairless. **Pl. 112.**

POPULUS POPLAR Flowers wind-pollinated, numerous in pendulous 1–sexed catkins. Buds with several scales; leaves with a broad blade and slender leaf-stalk. *c.* 8 spp.

Leaves lobed; buds dry, hairy

1300 **P. caspica** Bornm. (*P. alba* auct. non L.)
Afghanistan to Himachal Pradesh. W. & C. Asia. 1200–3000 m. Native, and also planted. Rarely flowering in India.

A medium-sized tree with smooth greenish-grey bark, and with distinctive 3–5–lobed upper leaves with conspicuous white-woolly hairy undersides. Catkins pendulous, hairy, slender, to 20 cm; flowers with oblanceolate coarsely toothed brown ciliate bracts; males with 6–10 stamens; females with 4 stigmas. Leaves ovate, with acute or blunt triangular lobes, the lobes often irregularly toothed, blade mostly 4–8 cm long, leaf-stalk 2.5–5 cm; young shoots, leaf-stalks and leaves densely covered with white-woolly hairs. Bark becoming very rough and dark on old trunks. Capsule *c.* 7 mm, oval-conical, smooth, shortly stalked.

Leaves toothed (not lobed); buds sticky

1301 **P. ciliata** Wallich ex Royle   HIMALAYAN POPLAR
Pakistan to S.W. China. 2100–3600 m. Streamsides; coniferous forests in the drier areas. Mar.–Apr.

A large tree with smooth greenish-grey bark, becoming brown and deeply vertically fissured on old trunks, and with hairless twigs. Leaves large with blade 8–18 cm, broadly ovate-heart-shaped long-pointed, toothed, pale and usually minutely hairy beneath; leaf-stalk 5–12 cm long. Catkins compact in flower; male catkins 7.5–10 cm, bracts oblanceolate, fringed with long hairs; female flowers with 1–4 very large stigmas. Fruiting catkins lax, 15–30 cm; capsule broadly ovoid, 8–10 mm long, 3–4–valved; seeds enveloped in long silky hairs. **Page 505.**

1302 **P. pamirica** Komarov (*P. balsamifera* auct. non L.)
Afghanistan to Pakistan. C. Asia. 2400–4000 m. Cultivated in W. Himalaya in the drier areas, particularly in Kashmir. May.

A large tree with long flexous angular yellowish-brown or grey hairy branchlets, and with sticky buds. Leaves with ovate long-pointed blade mostly 5–8 cm, rather leathery, hairless, usually paler beneath; leaf-stalk 2.5–7.5 cm. Male catkins with a slightly winged axis; stamens 18–25, with slender filaments. Fruiting catkins to 13 cm; capsule with usually 3 thick valves.

1303 **P. jacquemontiana** Dode var. **glauca** (Haines) Kimura.
E. Nepal to Bhutan. 2600–2900 m. Forest clearings; quite common around villages. Mar.–Apr.

Like 1301, but with hairy capsules. Leaves with broadly ovate acute blade with small marginal teeth, to 18 cm; leaf-stalk as long. Male catkins crimson, to 12 cm; female catkins greenish-yellow. Fruiting catkins *c.* 15 cm; capsule ovate, *c.* 6 mm.

1304 **P. nigra** L. cultivar **italica** LOMBARDY POPLAR is commonly cultivated in Kashmir and Ladakh; it is distinguished by its narrow columnar
form. Rarely flowering in India; always propagated by cuttings.

EPHEDRACEAE   Joint-Pine Family

A small family with sporadic distribution in arid regions of the Tropics and
Sub-Tropics of both hemispheres. Shrublets with closely clustered green
stems bearing deciduous brown scales (all that remain of the leaves). Cones
in small axillary clusters; male cones in stalked clusters with anthers borne
in the axils of bracts; female cone a solitary naked ovule, with bracts at the
base.

EPHEDRA   3 spp.

1305 **E. gerardiana** Wallich ex Stapf
Afghanistan to Bhutan. 2400–5000 m. Stony slopes, gravel terraces, in
drier areas. May–Jun.
A low-growing rigid tufted shrub 30–60 cm, with numerous densely
clustered erect slender smooth green jointed branches, arising from a
branched woody base. Branches with scales at the joints. Male cones ovate
6–8 mm, solitary or 2–3, with 4–8 flowers each with 5–8 anthers with
fused filaments, and rounded fused bracts. Female cones usually solitary.
Fruit ovoid 7–10 mm, with fleshy red succulent bracts enclosing the 1–2
seeds.
Goats and yaks feed on the branches during winter. Plant used medicinally; it yields the alkaloid ephedrine. **Page 510**.

PINACEAE   Pine Family

An important family of resinous woody plants, dominating large areas in the
temperate regions. Trees with needle-like or linear leaves arranged spirally on
the stem, either solitary, or clustered on short lateral shoots, usually persistent, but sometimes deciduous as in *Larix*. Male and female organs borne on
seperate cones, each cone consisting of an axis with spirally arranged scales
bearing the sexual organs. Male cones deciduous, each scale bearing 2–6
anthers; female cones persistent, woody, erect or pendulous, remaining intact
when ripe, or fragmenting into scales as in *Cedrus* and *Abies*. Scales of
female cones bearing 2 naked seeds, which are usually winged and dispersed
by wind.

Long shoots only present; leaves not clustered
   a Leaves needle-like; leaf-scars raised on peg-like extensions   1 PICEA

b Leaves flattened; leaf-scars flat or raised on ridges
  i Cones elliptic, less than 4 cm .... 2 TSUGA
  ii Cones cylindrical, more than 8 cm .... 3 ABIES

Long shoots and short shoots present; leaves on short shoots clustered
  a Clusters many-leaved .... 4 CEDRUS 5 LARIX
  b Clusters with 3 or 5 leaves .... 6 PINUS

1 PICEA SPRUCE Leaves needle-like, quadrangular in section, spiny-pointed, arranged spirally; leaf-scars on peg-like projections. Cones long-cylindrical, pendulous, not fragmenting. 1 sp.

1306 **P. smithiana** (Wallich) Boiss. (*P. morinda* Link) W. HIMALAYAN SPRUCE
Afghanistan to C. Nepal. 2100–3600 m. Locally common; often associated with *Cedrus deodara, Pinus wallichiana, Abies pindrow*, and sometimes with evergreen oaks. Apr.–May.

A conical tree to 40 m, but sometimes taller, with spreading whorled branches which are pendulous at the ends, and with stiff dark green needle-leaves. Bark greyish-brown, shallowly creviced into rounded or four-sided scales. Leaves mostly 2.5–4 cm, arranged at an angle of about 60 degrees to the twigs, branchlets pale grey. Ripe cones pendulous, dark shining brown, 10–15 cm by 2.5–5 cm in diameter.

Timber not of good quality; used for planks, boxes, shingles, and pulp for paper-making. Gives indifferent fuel but fairly good charcoal. **Page 510**.

P. spinulosa (Griffith) Henry, EAST HIMALAYAN SPRUCE, occurs in Sikkim but has not been recorded for Nepal.

2 TSUGA HEMLOCK Leaves solitary, short-stalked, leaving scars on raised ridges. Cones small, pendulous, with thin cone-scales. 1 sp.

1307 **T. dumosa** (D. Don) Eichler (*T. brunoniana* (Wallich) Carrière) HIMALAYAN HEMLOCK.
Uttar Pradesh to Bhutan. Burma. 2100–3500 m. Forming pure forests in more humid valleys and mountain slopes, absent from drier areas. May–Jun.

A beautiful pyramidal evergreen tree to 40 m, with spreading branches, and pendulous finely hairy branchlets, with leaves in two ranks. Leaves linear flattened, 1.5–2.5 cm by 2 mm broad, mealy-white beneath, with recurved margins. Readily distinguished by its small ovoid cones *c.* 1.5–2.5 cm long, with thin rounded scales.

Produces timber of little value, but logs often used to bridge small rivers. **Page 511**.

3 ABIES SILVER FIR Leaves solitary, flattened, leaving flat circular leaf-scars. Cones erect, cylindrical; cone-scales separating from central axis when ripe and releasing the seeds. 3 spp.

385

PINACEAE

1308 **A. spectabilis** (D. Don) Mirbel (*A. webbiana* Lindley) HIMALAYAN
SILVER FIR.
Afghanistan to Bhutan. 2800–4000 m. Forming forests over large areas in
the upper tree zone. Apr.–May.

Sometimes a slender pyramidal tree to 50 m or more, but often a gnarled
and much smaller tree with the upper branches horizontal, and branchlets
horizontal and flattened, and with dark grey, deeply longitudinally grooved
bark. Distinguished by the young branchlets which are hairy in the grooves
but soon become hairless, and by the leaves which are distinctly parted on
the upper sides of the branchlets and point sideways, and with the apex of
leaf rounded and notched. Leaves flattened, to 4 cm long, pale beneath
with incurved margins. Cones dark purple, erect, cylindrical, 10–20 cm by
4–7.5 cm; male catkins cylindrical *c*. 5 cm.
Timber used for building and roof-shingles. **Page 510**.

1309 **A. pindrow** Royle WEST HIMALAYAN SILVER FIR.
Afghanistan to W. Nepal. 2100–3600 m. Often forming mixed forests with
Picea smithiana. Apr.–May.

Very like 1308, and considered by some not to be a distinct species.
Generally growing at lower altitudes and distinguished by the branchlets
which are hairless in the grooves, and by the leaves borne on the upper
side of the branchlets, erect or directed forwards and loosely overlapping
(not parted), and by apex of leaf with two slender sharp tapering points.
Leaves 4–7 cm long, shining dark green above, with two faint silvery lines
beneath.

The wood is soft, light, and not very durable, but is used for internal
building work, and rough furniture, tea-chests and general carpentry; the
pulp is used to make paper.

1310 **A. densa** Griffith ex R. Parker EAST HIMALAYAN SILVER FIR.
E. Nepal to S.E. Tibet. 3000–4000 m. Only once positively recorded from
E. Nepal but possibly quite common there. Apr.

Very like 1308, and often confused with it, but distinguished by its bark
which soon becomes scaly, by its brownish (not yellowish) twigs, and by its
leaves mostly with incurved margins. Cones with bracts which protrude
between the scales and widen into a spathulate blade ending in a long
point. A tall tree to 60 m, with a broad flattened crown.

4 **CEDRUS** Leaves evergreen in dense clusters on short shoots. Female
cones ovoid-cylindrical, scales numerous, separating from the central axis
when ripe and releasing the winged seeds. 1 sp.

1311 **C. deodara** (Roxb. ex D. Don) G. Don DEODAR, HIMALAYAN CEDAR.
Afghanistan to W. Nepal. 1800–3000 m. Forming forests. Oct.

A large robust tree to 80 m, with a girth of up to 10–12 m in very old
trees, and with spreading branches with drooping branchlets, and with
drooping leading shoots. Leaves either solitary or in dense clusters, dark

green or sometimes bluish-green, rigid, leathery, three-sided, sharp-pointed, 2.5–4 cm long. Bark smooth and with vertical fissures. Male cones cylindrical 5–12 cm, erect; female cones ovoid-cylindrical, erect, large 10–13 cm by 8–10 cm with numerous thin scales; young cones bluish-purple.

A valuable timber tree with very durable aromatic wood, extensively used for building, furniture, vehicles and boats. Oil extracted by distillation is used medicinally, and as a seal for skin boats. Gives poor fuel. **Page 510**.

5 **LARIX** Larch Leaves deciduous, needle-like, borne in dense clusters on short shoots. Cones erect, not fragmenting, and bracts projecting between scales at least when young. 2 spp.

1312 **L. griffithiana** Carrière East Himalayan Larch.
E. Nepal to S.E. Tibet. 2800–4000 m. Forming forests in Kambachen valley north-west of Kanchenjunga. May.

A small pyramidal tree to 20 m, with pale green foliage and pendulous branches with downy young shoots. Bark thick, brown. Leaves on short lateral shoots in clusters of 30–50, deciduous, 2.5–3.1 cm long, soft flexible. Cones erect on pendulous branches, cylindrical 5–8 cm by 2.5–2.9 cm broad; bracts persisting at least when young, longer than the scales and with reflexed tips.

1313 **L. himalaica** W. C. Cheng & L. K. Fu
Endemic to C. Nepal and Tibet. 2400–4000 m. In mixed forests; sometimes dominant. May.

Distinguished from 1312 by its yellowish-grey branchlets, by its shorter and broader cones 4–5 cm long, which are light brown when ripe, and with broad, minutely irregularly toothed cone-scales, and erect bracts about as long as the scales and abruptly narrowed into a rigid tooth. Leaves to 3 cm. A tree closely related to *L. potaninii* Batalin of W. China. **Page 510**.

6 **PINUS** Pine Leaves needle-like, on short shoots borne in clusters of 3 or 5, solitary on long shoots. Cones not fragmenting, with scales thickened at their apices. 3 spp.

Leaves on short shoots in clusters of 5; cones cylindrical

1314 **P. wallichiana** A. B. Jackson (*P. excelsa* Wallich ex D. Don, *P. griffithii* McClell., *P. chylla* Lodd.) Himalayan Blue Pine.
Afghanistan to S.E. Tibet. 1800–4300 m. Forming forests in drier areas over a wide range of altitudes, also found in wet areas in secondary forest. Apr.–Jun.

Distinguished by its clusters of long cylindrical pendulous cones, and its slender drooping grey- or blue-green needle-like leaves borne on short shoots in clusters of 5. A tall symmetrical pyramidal tree to 50 m, with smooth slate-grey bark which becomes rough and shallowly fissured on mature trees. Leaves 15–20 cm long. Cones 15–25 cm long, in clusters of 2–3.

Wood highly resinous, used for local construction, carpentry, matches, tea-chests. Yields turpentine and tar. Honey-dew from aphid-infested leaves is eaten locally. Good for firewood, but with a pungent resinous smoke which makes the eyes smart and blackens the face. **Page 511**.

Leaves in clusters of 3; cones ovoid-conical

1315 **P. roxburghii** Sarg. (*P. longifolia* Roxb. ex Lambert) LONG-LEAVED or CHIR PINE.
Afghanistan to Bhutan. Usually 1000–2000 m, exceptionally 500–2700 m. Forming extensive forests; gregarious. Feb.–Apr.

A large tree to 40 m, with very thick and deeply fissured rough bark. Leaves dark or bright green, needle-like 23–38 cm, with greyish basal sheaths, borne in clusters of 3 on short shoots. Cones solitary or clustered, ovoid-conical, 10–18 cm by 8–13 cm broad.

Wood not durable, but used for general construction purposes. Sapwood yields resin, used in the manufacture of bangles; also for charcoal. Bark yields tannins, used in dyeing; oil of turpentine is obtained from the wood, it is used medicinally. Seeds edible. **Page 511**.

1316 **P. gerardiana** Wallich ex Lambert
Afghanistan to Uttar Pradesh. 1800–3000 m. Dry inner valleys; commonly associated with *Quercus baloot*. Jun.–Jul.

Differs from 1315 in having much shorter stiff, dark green leaves only 5–10 cm long, and thin smooth bark peeling off in large flakes. Mature cones glaucous, 12–20 cm by 7–11 cm broad, with very thick woody scales with stout recurved apices. A tree to 20 m.

The seeds, known as 'chilghoza', are edible and considered a luxury. **Page 511**.

TAXODIACEAE   Swamp Cypress Family

A small family of ancient trees, all that remain in the living state of a once very important fossil family. Distinguished by its small round woody female cones with scales bearing 2–9 seeds each. Male cones with spirally arranged scales each bearing 2–8 pollen sacs.

CRYPTOMERIA   1 sp.

1317 **C. japonica** (L.f.) D. Don   JAPANESE RED CEDAR
Native of Japan; planted extensively round Darjeeling and occasionally in E. Nepal and Kathmandu. 1300–2600 m. Mar.; cones ripening. Jul.–Aug.

A dense dark green, evergreen tree to 40 m, with a narrow conical crown, and brown bark which peels off in long strips. Distinguished by its small curved awl-shaped leaves which are 4–angled with thickened bases, 7–20 mm long, and are arranged spirally on the twigs. Female cones brown, globular 1.5–2 cm, composed of 20–30 overlapping cone-scales, each with 4–6 spiny tips.

A fast-growing tree, producing wood of little value, but used for tea-chests.

CUPRESSACEAE   Cypress Family

Trees and shrubs of cosmopolitan distribution, with evergreen needle-like or scale-like leaves, distinguished by their cones which have scales arranged in opposite pairs at right angles with each other. In *Cupressus* the scales are dry and distinct. In *Juniperus* the scales are fused together to form a somewhat fleshy berry-like fruit containing one or several seeds. Male cones small solitary, at tip of branchlets. Leaves of two kinds; either 'juvenile' needle-shaped and spreading, or 'adult' scale-like and closely pressed to the branchlets and arranged in ranks.

CUPRESSUS   Distinguished by its globular fruiting cones with flat-topped scales which separate at their margins, releasing the 4–6 seeds from each scale. 1 native sp.

1318   **C. torulosa** D. Don   HIMALAYAN CYPRESS.
Himachal Pradesh to C. Nepal. 1800–3300 m. Drier areas, usually on limestone. Feb.

A broadly pyramidal tree to 35 m, with whorled spreading branches with drooping branchlets. Leaves scale-like *c.* 1 mm, ovate-triangular and with white margins, closely overlapping; bark peeling off in long strips. Cones loosely clustered, bluish, globular *c.* 1.5 cm, composed of woody flat-topped rounded or angular scales, which separate when dry.

Timber very durable; used for building; burnt in temples for incense. **Page 511.**

C. corneyana Hort. ex Carrière (*C. funebris* auct. non Endl.)
A native of China; sometimes grown near temples, to 2000 m. It is distinguished by its wide-spreading branches, its pendulous bright green fan-shaped branchlets, and its smaller cones 1 cm across.

JUNIPERUS   Readily distinguished from other conifers by its fruit which is berry-like, somewhat fleshy, and contains 1–5 seeds. Leaves of two kinds: 'juvenile' and 'adult'. 5 spp.

Leaves all 'juvenile', linear spiny-pointed

1319   **J. communis** L. COMMON JUNIPER
Afghanistan to C. Nepal. N. Temperate Zone. 1800–3600 m. Drier areas. Apr.–May.

A dense erect shrub to 1.5 m, or at high altitudes a prostrate shrub. Leaves needle-like sharp-pointed, 6–13 mm long, with a broad bluish-white band above, in whorls of 3. Male cones ovoid, recalling leaf-buds. Female cone blue-black when ripe, 6–8 mm; seeds usually 3. Fruit used medicinally. **Page 512**.

1320 **J. recurva** Buch.–Ham. ex D. Don   DROOPING JUNIPER
Pakistan to S.W. China. 2500–4600 m. Gregarious and often covering large areas as a shrub, rarely dominant as a forest tree. Jun.–Jul.

A low spreading shrub, or a tree to 10 m or occasionally taller, with awl-shaped leaves 6–8 mm long, in whorls of 3, more or less adpressed to the branchlets and loosely overlapping. Growth rather lax; stems often brown, with ultimate branches tail-like and curving separately in various directions. Fruit purplish-brown to black, shining when ripe, ovoid 8–13 mm, 1–seeded.

The common dwarf Juniper in the wetter parts of the Nepal alpine zone. **Page 512**.

1321 **J. squamata** Buch.–Ham. ex D. Don
Afghanistan to S.W. China. 3000–4500 m. Inner valleys. Jun.–Jul.

A dwarf spreading shrub with a compact growth, with only the leading shoots separating from the mass. Leaves awl-shaped, incurved but not adpressed, 3–5 mm, in whorls of 3. Fruit reddish-brown to black, 6–8 mm, 1–seeded.

Used for fuel, and burnt for incense. **Page 512**.

Leaves of two kinds: awl-like and spreading, and scale-like and pressed to stem

1322 **J. indica** Bertol. (*J. wallichiana* Hook.f. & Thoms. ex Brandis, *J. pseudosabina* auct. non Fischer & Meyer)   BLACK JUNIPER
Pakistan to S.W. China. 2800–4500 m. Shrubberies in drier areas; sometimes forming low open forests, or scattered on alpine slopes in clumps to 1 m high. May.

A large gregarious shrub, or small tree to 20 m, with a stout trunk, and with two kinds of leaves; those on lower branches awl-shaped 3–6 mm, spreading, those on terminal branches scale-like *c.* 1.5 mm, adpressed overlapping in four ranks giving a smooth cord-like appearance to the branches. Fruit 1–seeded, at first brown then shining blue, 6–13 mm.

The common Juniper of drier parts of the alpine zone of Nepal & W. Himalaya. Used as incense in Buddhist temples. **Page 512**.

1323 **J. macropoda** Boiss.   PENCIL CEDAR.
Pakistan to Uttar Pradesh. 2400–4300 m. Dry river valleys, gregarious; forming open forests in Lahul. May.

A tree to 18 m, very like 1322, but with fruit with 2 or more seeds, and with light open foliage with spreading sharp-pointed leaves on lower branches, and with scale-like leaves on the upper branches and branchlets.

Lower leaves linear *c.* 8 mm; scale leaves *c.* 1.5 mm, closely pressed to stem and with a large resinous gland on the back. Fruit bluish-black, resinous.

Wood most valuable for furniture, building and pencil-making; used for fuel and charcoal. **Page 512**.

TAXACEAE   Yew Family

A small family of 3 genera. *Taxus* occurs sporadically in Europe, Asia, N. Africa and N. & C. America. Distinguished by its fruit which consists of a solitary seed encircled by a fleshy red aril. Male cones small globular, with flat-topped scales bearing the anthers beneath; female cones consisting of a solitary ovule, with fleshy lobes beneath. Leaves linear flat, two-ranked and alternate, with pale lines beneath.

TAXUS 1 sp.

1324 **T. baccata** L. subsp. **wallichiana** (Zucc.) Pilger HIMALAYAN YEW.
Afghanistan to S.W. China. Burma. S.E. Asia. 2100–3400 m. Forests, shady ravines; usually in the understory. Mar.–May.

A tree to nearly 30 m, but usually not more than 10 m, with dark green foliage, and thin dark reddish-grey flaking bark. Leaves linear flattened, curved, spiny-tipped, leathery and dark glossy green, 2–3.5 cm long by 3 mm broad but often narrower, 2–ranked, borne on hairless branchlets. Fruit red fleshy, *c.* 8 mm, often almost concealing the single green seed.

The wood is used for cabinet-making, furniture, poles and axles of carts. The fleshy fruit is eaten. The foliage is used as litter and fed to cattle in Pakistan. The leaves contain an alkaloid poisonous to live-stock, but the alkaloid content varies from area to area.

ORCHIDACEAE   Orchid Family

A very large family of herbaceous plants ranging from the high mountains of the world to tropical rain forests. Either *terrestrial* and growing in the soil, or *epiphytic* and growing on trees or sometimes on damp rocks. The latter develop special aerial roots able to absorb moisture and nutrients. Both terrestrial and epiphytic orchids may have *pseudobulbs*, borne on or above the soil, which are swollen stems which store water and nutrients, and bear leaves. Some orchids are *saprophytic* and have leaves reduced to brownish scales. Leaves alternate, often in two ranks, often fleshy and sheathing at the base. Flowers unique in structure. Sepals 3, often coloured; petals 3, 2 often similar and coloured like the sepals, and the third forming a lower lip which is often the distinctive feature of the orchid flower. Lip usually much larger than the other parts, entire, or 2–3–4–lobed and often contrasting conspicuously in shape, colour, and markings from the other petals and sepals, and often

ORCHIDACEAE

with a spur. Stamens and ovary united above to form a *column*, on which the anthers and stigmatic patches are borne, and which are separated by a flap of tissue, the beak or *rostellum*. The pollen is borne in sacs or *pollinia*. Innumerable variations in the structure of the column have been evolved, giving a great variety of pollination mechanisms. Ovary inferior, 1–chambered. Fruit a capsule; seeds minute and requiring the appropriate symbiotic fungus for germination.

This world-wide family comprises about 18,000 species; 453 have been recorded for Sikkim, and 314 for Nepal; only the most frequently encountered can be described here.

Terrestrial orchids
  a Flowers with a spur or spurs
    i Flowers predominantly pink or purple   7 EULOPHIA
        2 CALANTHE   5 DACTYLORHIZA   8 GALEARIS
            17 PONERORCHIS   18 SATYRIUM
    ii Flowers predominantly white, green or yellow   2 CALANTHE
        8 GALEARIS   11 HABENARIA   15 PECTEILIS
          16 PLATANTHERA   17 PONERORCHIS
  b Flowers without a spur
    i Flowers predominantly pink, purple or red-brown   1 ARUNDINA
        4 CYPRIPEDIUM   6 EPIPACTIS   14 OREORCHIS
                     20 SPIRANTHES
    ii Flowers predominantly white   3 CEPHALANTHERA
        4 CYPRIPEDIUM   10 GOODYERA
                     20 SPIRANTHES
    iii Flowers predominantly yellow   2 CALANTHE   9 GALEOLA
        13 MALAXIS   19 SPATHOGLOTTIS
    iv Flowers green   6 EPIPACTIS   12 HERMINIUM
        13 MALAXIS   4 CYPRIPEDIUM

Epiphytic orchids growing on trees, or on rocks
  a Flowers with a spur
    i Flowers predominantly pink or purple   28 RHYNCHOSTYLIS
    ii Flowers predominantly green or yellow   29 VANDA
    iii Flowers predominantly white   21 AERIDES
                  28 RHYNCHOSTYLIS
  b Flowers not spurred
    i Flowers predominantly pink, purple or brown   23 COELOGYNE
        27 PLEIONE   28 RHYNCHOSTYLIS
    ii Flowers predominantly green or yellow   22 BULBOPHYLLUM
        24 CRYPTOCHILUS   25 CYMBIDIUM
                  26 DENDROBIUM
    iii Flowers predominantly white   23 COELOGYNE
    25 DENDROBIUM   27 PLEIONE   28 RHYNCHOSTYLIS
                  30 VANDOPSIS

TERRESTRIAL ORCHIDS

1 ARUNDINA Petals and sepals similar; lip curved into a tube at the base and encircling the column, two-lobed at apex. 1 sp.

1325 **A. graminifolia** (D. Don) Hochr. (*A. bambusifolia* Lindley)
C. Nepal to Bhutan. S. China. India. S. E. Asia. 450–1500 m. Open grassy slopes; quite common. Jun.–Sep.

A tall reed-like plant 120–220 cm, with linear pointed curved leaves, and with a lax terminal branched cluster of a few large pink-purple, or brilliant rose-purple flowers, each 7–8 cm across. Petals and sepals similar-coloured; lip curved into a tube below, with an oblique mouth and a squarish, deeply-cleft apex. Leaves 20–30 cm by 1.8–2.5 cm broad, in two ranks. **Pl. 113**.

2 CALANTHE Terrestrial orchids usually with pseudobulbs. Flowers in erect racemes; sepals and petals similar; lip attached to column, 3–lobed, the apical lobe often bi-lobed. Pollinia 8. Capsule drooping. 12 spp.

Spur absent, or less than half as long as ovary

1326 **C. tricarinata** Lindley
Pakistan to S.W. China. Japan. 1500–3000 m. Forests; common. Apr.–Jul.

Flowers *c*. 2 cm across, in a terminal spike-like cluster, with greenish-yellow petals and sepals, and a reddish-brown 3–lobed lip, the mid-lobe rounded and with distinctive large purple fleshy ridges, and an undulate greenish margin; basal lobes of lip short, rounded, erect. Leaves 2–3, broadly oblanceolate, 20–35 cm; stem 30–60 cm. **Pl. 115**.

1327 **C. brevicornu** Lindley
C. Nepal to Sikkim. 1800–2500 m. Forests. May–Jun.

Flowers in a spike-like cluster, dull purple to brick-red and with white lines to the sepals and petals and to the margins of the lip, and flushed white outside. Flowers 3–4 cm across, with a short straight conical spur; lip 3–lobed, basal lobes oblong blunt, diverging, mid-lobe obovate with a notched apex and irregularly toothed margin, and with 3 fleshy ribs; bracts lanceolate, half as long as ovary. Leaves usually 3, oblong, 4–6 cm broad; flowering stem 30 cm, longer than leaves; pseudobulbs ovoid. **Pl. 114**.

Spur as long as ovary or longer

1328 **C. plantaginea** Lindley
Pakistan to S.W. China. 1500–2100 m. Forests. Mar.–Apr.

Flowers pale pink, lilac or white, sweet-scented, many borne on a stout stem longer than the leaves. Flowers *c*. 3 cm across; spur long and very slender, horizontal; lip 3–lobed, the lateral lobes broadly obovate, the mid-lobe broadly wedge-shaped, with three small orange ridges near the base; sepals and petals lanceolate pointed. Leaves several, elliptic-lanceolate, 20–30 cm, enlarging in fruit. **Pl. 114**.

3 CEPHALANTHERA Sepals and petals similar, grouped together and not spreading widely. Lip with a basal *hypochile* which clasps the column, and an apical *epichile* with longitudinal ribs or crests and with a recurved tip; spur absent. Pollinia 2. 1 sp.

1329 **C. longifolia** (L.) Fritsch NARROW-LEAVED HELLEBORINE
Afghanistan to S.E. Tibet. W. & N. Asia. Japan. Europe. N. Africa. 1800–3000 m. Forests in drier country. May–Aug.

Flowers white, 8–20 in a terminal spike-like cluster, the lip small with a few yellow markings and concealed by the longer encircling sepals and petals. Flowers 9–15 mm long; sepals lanceolate acute, longer than the blunt petals; epichile oblong, 3–4–ribbed, concave towards the apex. Leaves alternate, in two ranks, curved, lanceolate and clasping stem, 5–10 cm; stem 15–45 cm. **Pl. 113**.

4 CYPRIPEDIUM LADY'S-SLIPPER ORCHID Flowers large, usually solitary. Dorsal sepal larger than lateral sepals; petals often long; lip with narrow or ear-like lateral lobes and with a large inflated sac-like mid-lobe; spur absent. Anthers 2. 3 spp.

Flowers white

1330 **C. cordigerum** D. Don
Pakistan to Bhutan. 2100–4000 m. Light forests, shrubberies, open slopes; common in drier areas. May–Jun.

Flowers solitary, with green, yellow or white sepals, green or white spreading petals, and with a large oblong bag-shaped white lip which is open-mouthed. Sepals 4–5 cm, ovate-lanceolate long-pointed; petals lanceolate; lip with a few purple spots outside, 2.5–3 cm long; bract large, leaf-like. Leaves several, ovate to lanceolate acute, to 15 cm; stem 25–60 cm. **Pl. 116**.

Flowers greenish to brownish-purple

1331 **C. elegans** Reichb.f.
C. Nepal to S.E. Tibet. 3000–3800 m. Light forests, shrubberies, open slopes; not common. Jun.

Flowers 2.5 cm across, with greenish sepals and petals, and a bag-like lip which is brown with darker brown lines, and which is shorter than the sepals; bract large, as long as flower. Leaves 2, stalkless, broadly ovate, 3–3.7 cm; stem very hairy, to 10 cm or less with the basal sheath short.

1332 **C. himalaicum** Rolfe
W. Nepal. to S.E. Tibet. 3000–4300 m. Light forests, shrubberies, open slopes; quite common in drier areas. Jun.–Jul.

Flowers *c.* 5 cm across, with a broadly ovoid pendent bag-like lip with a wavy to crenate mouth, streaked with purple, longer than the petals; bract leafy, larger than the flower. Petals and sepals green and with red veins; upper sepal broadly ovate, the lateral sepals narrower; petals spreading, narrow-oblong. Leaves 3–4, elliptic to lanceolate, 7–9 cm; stem with several sheaths at base, 20–30 cm. **Pl. 115**.

5 DACTYLORHIZA Differs from *Orchis* in having palmately lobed or divided root-tubers. Bracts leaf-like; petals and sepals erect or spreading, not all forming a hood over the column. 1 sp.

1333 **D. hatagirea** (D. Don) Soó (*O. latifolia* auct. non L.)
Pakistan to S. E. Tibet. 2800–4000 m. Shrubberies, open slopes, marshes; common. Jun.–Jul.

Flowers spotted rosy-purple, in a many-flowered dense cylindrical terminal spike, borne on a robust leafy stem. Flowers *c.* 1.8 cm long, including the stout curved cylindrical spur; sepals and petals nearly equal, three coming together in a hood, the two lateral sepals spreading; lip rounded, shallowly three-lobed, spotted dark purple; lower bracts longer than the flowers. Leaves many, oblong-lanceolate to 30 cm; flowering stem 30–90 cm.

Tubers used medicinally. **Pl. 114**.

6 EPIPACTIS Terrestrial plants with fibrous roots. Flowers in spike-like clusters; sepals and petals similar; lip divided into a hard basal cup-like deflexed *hypochile* and a terminal entire *epichile*. 3 spp.

1334 **E. helleborine** (L.) Crantz (*E. latifolia* (L.) All.) BROAD-LEAVED HELLEBORINE
Pakistan to S.E. Tibet. Temperate Eurasia. N. Africa. 2000–4000 m. Forests, shrubberies; mostly in drier areas. Jul.–Aug.

Flowers greenish or dull purple, numerous in a rather dense one-sided spike. Flowers *c.* 1.3 cm across; lip shorter than the elliptic-ovate sepals, hypochile purplish, pink or greenish-white, epichile heart-shaped with 2 rounded swellings at base. Leaves many, the lower elliptic, the upper narrower and grading into the linear bracts above; stem to 30 cm, minutely hairy above.

1335 **E. royleana** Lindley
Pakistan to S.E. Tibet. 1600–3500 m. Grassy slopes, damp places, shrubberies; quite common. Jun.–Jul.

Flowers 2.5 cm across, green and veined with red, with a yellowish or reddish lip, borne in a lax spike. Distinguished from 1334 by the lip which has a large bag-like hypochile, with many veins, and which is very much broader than the ovate-lanceolate epichile (epichile broader than hypochile in 1334). Leaves ovate-lanceolate to linear-lanceolate 1–1.5 cm long, clasping stem; flowering stem 30–60 cm, with sheathing bracts at base. **Page 513**.

7 EULOPHIA Terrestrial plants with fleshy tubers or rhizomes. Leaves appearing with or after the flowering spikes. Sepals and petals similar; lip with a short spur, 3–lobed, with a ridged or crested mid-lobe. Pollinia 2. 6 spp.

1336 **E. dabia** (D. Don) Hochr. (*E. campestris* Wallich ex Lindley)
Himachal Pradesh to C. Nepal. 1500–2400 m. Dry hillsides, pine forests. Apr.–May.

Flowers appearing before the leaves, pale pink with darker lines, many in a one-sided spike 6–9 cm long. Flowers to 2 cm long; petals blunt, shorter than the acute sepals, lip short three-lobed, with short side-lobes and with transversely oblong, orange, toothed mid-lobe, which is three-ridged at base and with a patch of densely crowded short bristles at apex. Spur short, conical. Leaves 2, linear pointed, 25–40 cm by 1 cm broad; stem 15–40 cm, sheathed with papery bracts. **Pl. 113.**

8 GALEARIS Differs from *Orchis* and *Dactylorhiza* by the absence of tubers, and in having a rhizome with fleshy roots; and with 2 pouch-like expansions of the stigma and divergent anther cells. 1 sp.

1337 **G. spathulata** (Lindley) P. F. Hunt (*Orchis s.* (Lindley) Reichb.f. ex Benth.)
Uttar Pradesh to Bhutan. 3000–4300 m. Shrubberies, open slopes. Jun.–Jul.

Flowers white or dark purple, borne on a short stem with usually a solitary basal leaf. Flowers *c.* 12 mm across; lip broadly elliptic, entire or obscurely three-lobed, as long as the sepals and with many shallow longitudinal grooves. Spur stout cylindrical, about half as long as the ovary; sepals unequal, the dorsal ovate, forming a hood with the petals, the lateral sepals spreading. Bracts leafy, as long or longer than the flowers. Leaf blunt, fleshy, elliptic-oblong, stalked, 2.5–5.7 cm; stem angled, 5–20 cm.

1338 **G. stracheyi** (Hook.f.) P. F. Hunt (*Orchis s.* Hook.f.)
Uttar Pradesh to Bhutan. 3000–4800 m. Shrubberies, open slopes. Jun.–Jul.

Like 1337, but differing in its lip which is longer than the sepals, and broadly wedge-shaped and deeply three-lobed, and its spur which is stout blunt, curved, as long as or longer than the ovary. Bracts often purple and leaves purple-spotted. **Page 514.**

9 GALEOLA Saprophytic plants without green leaves. Flowers in branched clusters; lip with a large broad apical lobe, with or without lateral lobes; column enlarged into a broad apex. 2 spp.

1339 **G. lindleyana** (Hook.f. & Thoms.) Reichb.f.
Himachal Pradesh to S.W. China. 1200–2400 m. Forests; not uncommon in extreme E. Nepal. Jun.–Jul.

A strange-looking plant with a tall inflorescence with short, widely spaced branched clusters of few drooping yellow globular flowers, borne

on a robust brownish-purple stem with brown scale-like sheaths. Flowers
c. 3.5 cm, with broadly ovate, strongly ribbed thick spongy petals, and a
nearly hemispherical cup-shaped lip with a very narrow three-lobed tip,
the upper surface covered with small swellings, and 2 red patches. Stem
120–150 cm; rootstock stout. Capsule 15–18 cm long. **Pl. 116**.

10 GOODYERA Flowers small, numerous, in a spike. Leaves mostly
basal. Sepals and petals forming a hood over the column. Lip entire,
bag-like with an acute down-curved apex; spur absent. 8 spp.

1340 **G. repens** (L.) R. Br. CREEPING LADY'S-TRESSES
Pakistan to S.E. Tibet. North Temperate Zone. 2400–4000 m. Forests;
widespread and common. Aug.–Oct.

A small creeping orchid with an erect twisted spike of tiny whitish rather
globular flowers flushed with brownish-pink. Petals and dorsal sepal
forming a hood *c.* 4 mm, lip as long as hood, with ovate acute apex; bracts
linear-lanceolate, longer than the ovary. Leaves all basal, in a lax rosette,
spear-shaped 1.2–2.5 cm, with conspicuous pale netted veins. Flowering
stem erect, with bracts, 10–20 cm. **Page 513**.

1341 **G. fusca** (Lindley) Hook.f.
W. Nepal to S.E. Tibet. 3200–4100 m. Forests, shrubberies; common.
Aug.–Oct.

Like 1340, but a much more robust plant with uniformly green leaves,
and with twisted spikes of numerous close-set white flowers. Flowers *c.* 6
mm; lip flushed pinkish-yellow, rounded sac-like at base and suddenly
contracted into a down-curved green beak; bracts ovate-oblong. Leaves
thick, broadly ovate 2.5–4 cm, with in-curved margins; stem 15–30 cm.

11 HABENARIA Flowers in spikes or racemes; dorsal sepal and petals
forming a hood, the lateral sepals spreading or reflexed. Lip 3–lobed,
spurred. 17 spp.

1342 **H. pectinata** D. Don
Pakistan to S.E. Tibet. 900–3000 m. Grassy slopes. Jul.–Aug.

A striking terrestrial orchid, with green and white flowers with a very
distinctive 3–lobed white lip with side-lobes deeply cut on the outer
margin into numerous long slender segments (the whole lip recalling a
stag's antler), the mid-lobe linear. Flowering spike dense 8–15 cm, with
many flowers, each 3–4 cm across; bracts leafy, longer than the ovary.
Petals and dorsal sepal forming a greenish-white hood, lateral sepals erect
green, spreading and with recurved apex. Spur 2 cm, longer than the
ovary. Leaves linear-lanceolate long-pointed; stem leafy above, 30–60 cm.
Pl. 114.

1343 **H. intermedia** D. Don
Pakistan to C. Nepal. Tibet. 2000–3000 m. Grassy slopes. Jul.–Aug.

Differs from 1342 in having fewer, larger greenish-white flowers each *c.*

5 cm across, with a similar 3–lobed 'stag's antler' lip, but lip green, larger and longer than the sepals (lip shorter than sepals in 1342), and with a very stout spur 6.5 cm long, up to twice as long as the ovary. Flowers 2–6; hood white, lateral sepals green. Leaves ovate long-pointed, 8–18 cm; stem 20–50 cm.

12 HERMINIUM Flowers green, in a slender terminal spike; sepals and petals usually forming a hood, lip usually longer than petals, narrow, 3–lobed; spur absent. 9 spp.

1344 **H. lanceum** (Thunb.) Vuijk (*H. angustifolium* (Lindley) Benth. ex C. B. Clarke)
Pakistan to S.W. China. India. S.E. Asia. 1500–3300 m. Open slopes, shrubberies; quite common. Jul.–Aug.

Flowers green, tiny, with green bracts as long as the ovaries, crowded into a narrow cylindrical terminal spike 7–25 cm, borne on a leafy stem. Flowers *c*. 6 mm; upper sepal and petals forming a hood, the two lateral sepals spreading; lip oblong, much longer than sepals and with two long narrow curved lobes and a central tooth; spur absent. Leaves narrow-linear, 10–20 cm; stem 25–75 cm. **Page 513**.

1345 **H. macrophyllum** (D. Don) Dandy (*H. congestum* Lindley)
C. Nepal to S.E. Tibet. 3800–4600 m. Shrubberies, alpine slopes. Jun.– Aug.

Flowers green, very tiny, in a spike 5–10 cm long; like 1344 but flowers drooping, only 2–3 mm long, and lip fleshy, ovate-lanceolate entire, or very shallowly 3–lobed; bracts ovate, minute. Leaves obovate to oblanceolate; stem 10–20 cm.

1346 **H. monorchis** (L.) R. Br. MUSK ORCHID
Pakistan to C. Nepal. Tibet. China. N. Asia. Europe. 3000–4300 m. Forests, marshy ground in drier areas. Jun.–Aug.

A slender perennial with an erect usually lax spike of tiny greenish-yellow, honey-scented flowers, each 2–3 mm across. Petals and sepals somewhat converging but not forming a hood; lip 3–lobed with 2 lateral lobes projecting at right angles or curved forwards, mid-lobe shorter. Lower leaves elliptic-oblong to linear-lanceolate 2–7 cm, upper leaves bract-like; stem 7–15 cm; tubers globular. **Pl. 116**.

13 MALAXIS Like *Herminium*, but lip 2–lobed, inverted, the lobes facing upwards; column very short; pollinia 4. Pseudobulbs present. 6 spp.

1347 **M. muscifera** (Lindley) Kuntze (*Microstylis m.* (Lindley) Ridley)
Afghanistan to S.E. Tibet. 2500–4000 m. Forests, shrubberies, grassy slopes. Jul.–Aug.

Flowers yellowish-green, tiny, very numerous in a very slender rather lax spike 8–25 cm, and leaves broad stalkless, paired unequal, arising towards the base of the stem. Flowers c. 3 mm long, with lateral sepals curved upwards; petals linear; lip erect, nearly circular, with thick obscure basal lobes and narrow lanceolate tip. Leaves unequal, elliptic, 2.5–6.5 cm; stem 15–38 cm; pseudobulbs ovoid. **Pl. 115**.

1348 **M. acuminata** D. Don (*Microstylis wallichii* Lindley)
Himachal Pradesh to Arunachal Pradesh. Burma. S.E. Asia. 600–3000 m. Shrubberies, shady rocks; common. Jun.–Aug.
   Flowers pale yellow-green and tinged with purple especially near the centre, in a lax many-flowered spike-like cluster. Distinguished by the lip which is erect, shield-like and broadly ovate (the whole flower is inverted) and with erect basal lobes with a notch between them exposing the strongly recurved sepal, which looks superficially like a spur. Sepals oblong; petals linear; bracts linear. Leaves 2–4, ovate-lanceolate acute, with undulate margin, 7.5–11 cm; flowering stem 20–35 cm.

14 OREORCHIS Terrestrial plants with pseudobulbs; leaves basal, 1–2. Flowers in spike-like clusters. Petals and sepals similar; lip 3–lobed, the mid-lobe broad; spur absent. Pollinia 4. 3 spp.

1349 **O. foliosa** (Lindley) Lindley
Himachal Pradesh to S.W. China. 2700–3800 m. Forests. May–Jul.
   Flowers few, red-brown, borne in the axils of tiny lanceolate bracts, in a long cluster 7.5–20 cm, on a slender leafless stem. Flowers *c.* 1.3 cm long; sepals ovate-lanceolate, the lateral sepals curved and pointed; petals broad, blunt; lip longer, deflexed from the middle, the mid-lobe broadly obovate, notched, strongly 5–veined, the lateral lobes shorter, ovate. Spur absent. Leaf solitary, narrow-lanceolate, strongly 3–veined, 10–22 cm, arising from the rounded pseudobulb; stem 25–30 cm, with 3 tubular sheaths. **Pl. 117. Page 513**.

15 PECTEILIS Flowers showy; upper sepal and petals forming a hood. lateral sepals spreading; lip 3–lobed, the lateral lobes finely cut; spur prominent; column short, basally extended into distinct canals. 2 spp.

1350 **P. susannae** (L.) Raf. (*Habenaria s.* (L.) Blume)
Himachal Pradesh to S.W. China. India. Burma. S.E. Asia. 1000–2000 m. Grassy slopes. Aug.–Sep.
   A tall robust plant with 3–5 very large, fragrant, white flowers tinged with yellowish-green, each 7–10 cm across. Sepals very broad, spreading, the dorsal rounded *c.* 4 cm wide and long; petals curved, linear acute; lip fleshy, about as long as the sepals, with side-lobes cut into long slender thread-like segments, and with a linear mid-lobe. Spur more than twice as long as the ovary. Leaves 5–15 cm, overlapping up stem; stem 60–80 cm. **Pl. 114**.

16 PLATANTHERA Lip strap-shaped entire; spur long slender. Column short; pollinia 2. 10 spp.

1351 **P. latilabris** Lindley (*Habenaria l.* (Lindley) Hook.f.)
Kashmir to Bhutan. Tibet. India. 1000–4000 m. Open slopes, shrubberies; common. Jun.–Sep.

Flowers greenish-yellow, in a moderately dense narrowly cylindrical spike 7–25 cm, borne on a leafy stem. Flowers *c.* 1.3 cm across; sepals unequal, the dorsal broadly ovate, the laterals narrow; petals spreading, lanceolate, fleshy; lip linear or lanceolate, entire with recurved margin, longer than the sepals. Spur curved, longer than the curved ovary. Leaves about 5, ovate to narrow-elliptic, the middle leaves 5–10 cm long; stem 20–45 cm.

17 PONERORCHIS   Like *Orchis* but differing in its beak and pollinia. Sepals and dorsal petal forming a hood; lateral petals reflexed; lip rhomboid 3–lobed; spur cylindrical. 1 sp.

1352 **P. chusua** (D. Don) Soó (*Orchis c.* D. Don)
Uttar Pradesh to S.E. Tibet. Burma. 2400–4200 m. Shrubberies, open slopes. Sep.–Oct.

Flowers few, borne in a short spike, dark purple or sometimes white, with a lip with three equal oblong-rounded toothed lobes, and a stout spur as long as, and pressed to, the ovary. Lateral sepals reflexed, the dorsal sepal smaller, broader; petals ovoid, projecting forward. Leaves 1–3, usually linear-lanceolate 3.5–7.5 cm; stem 10–25 cm. **Pl. 113.**

18 SATYRIUM   Readily distinguished by the paired spurs to each flower which are as long or up to twice as long as the ovary; lip inverted, broad, hood-shaped; column arched. 2 spp.

1353 **S. nepalense** D. Don
Pakistan to S.W. China. 1500–4000 m. Open grassy slopes; quite common. Jul.–Sep.

Flowers pink, fragrant, in a dense terminal spike and with reflexed bracts tinged pink and much longer than the flowers. Flowers *c.* 1.3 cm across, in a spike 2.5–15 cm long; lip erect projecting upwards, hood-shaped, curved and with two down-projecting spurs; petals recurved, smaller than the spreading sepals. Leaves narrow-elliptic with sheathing bases, 10–25 cm; stem 50–75 cm. **Pl. 113.**

1354 **S. ciliatum** Lindley
W. Nepal to S.E. Tibet. 2000–3500 m. Shrubberies, open slopes. Sep.–Oct.

Similar to 1353, but a smaller plant less than 25 cm, and with sepals conspicuously ciliate. Petals pink, irregularly toothed; spurs shorter than the sepals.

19 SPATHOGLOTTIS   Terrestrial plants with pseudobulbs; leaves all basal. Sepals and petals equal, spreading; lip 3–lobed, with a ridged or swollen disk at the base. Pollinia 8. 1 sp.

1355 **S. ixioides** (D. Don) Lindley
C. Nepal to Arunachal Pradesh. 2000–3500 m. Rocks, steep banks.
Jul.–Aug.

Flowers pale yellow, 1–2 borne on slender leafless stems each from a
small pseudobulb. Flowers 2.5–4 cm across; lip 3–lobed, lateral lobes
conical, terminal lobe obcordate, minutely spotted with red on the lower
half and with basal swelling and a median ridge; sepals and petals equal,
elliptic, spreading. Leaves 2–3, linear-lanceolate 10–20 cm; flowering
stems erect, 10–15 cm. Pseudobulbs globular, 1.5–1.8 cm. **Page 513**.

20 SPIRANTHES Flowers small, arranged spirally in a terminal spike.
Lip entire or 3–lobed, without a spur, about as long as the sepals and
petals which are usually free and sometimes form a hood. 1 sp.

1356 **S. sinensis** (Pers.) Ames (*S. australis* (R. Br.) Lindley)
Afghanistan to China. N. Asia. India. S. E. Asia. Australia. 100–4500 m.
Open slopes, cultivated areas. Apr.–Aug.

Flowers very small, pink or sometimes white, arranged spirally in a
slender but dense spike 8–15 cm long, and borne on a hairy stem. Flowers
3–4 mm long, hooded and with spreading lateral sepals; lip oblong, the
apical part with crisped margin and dilated recurved tip. Spur absent.
Leaves 4–5, linear-lanceolate, 2–6 cm; flowering stem 15–45 cm, leafy
only near the base and with sheathing bracts above. Ranging from the
tropics to the alpine zone; one of the widest ranging species. **Pl. 113**.

EPIPHYTIC ORCHIDS

21 AERIDES Epiphytic plants with linear leathery leaves arising from a
stout stem. Flowers many, in drooping spikes; lip shortly spurred, usually
with small side-lobes and a much larger mid-lobe. 4 spp.

1357 **A. multiflora** Roxb.
Uttar Pradesh to Sikkim. India. Burma. 200–1100 m. Forest epiphyte.
May–Jun.

Flowers showy, white and conspicuously marked with pink or purple, in
a long cylindrical many-flowered drooping spike. Flowers *c.* 2 cm across;
sepals and petals equal, oblong; lip oblong, twice as long as sepals, the
basal part pale, thick, the apical part dark rosy-purple. Spur short, conical.
Leaves folded longitudinally, broadly linear, recurved, 1.8–3 cm broad;
stem stout, erect or ascending. **Pl. 117**.

22 BULBOPHYLLUM Epiphytic plants with creeping rhizomes and
pseudobulbs. Leaves solitary, thick. Flowers solitary or many, lip jointed
on the foot of the column, mobile. Pollinia 4. *c.* 24 spp.

1358 **B. affine** Lindley
C. Nepal to Sikkim. 800–1500 m. Forest epiphyte. Jun.

A creeping orchid with yellowish-green flowers streaked conspicuously with brown, arising individually or in pairs, on short stalks from the creeping rhizome, and much shorter than the leaves. Flowers 1.8–2.4 cm long; lip lanceolate, shortly stalked, edged with dark purple, shorter than the elliptic petals and sepals. Spur absent. Leaves fleshy, linear-oblong blunt, 7–18 cm, narrowed to a short stalk; rhizome unbranched, densely rooting along its length; pseudobulbs arising about 8 cm apart, ovoid-oblong to ovoid-cylindrical 3.6–10 cm long. **Pl. 116**.

23 COELOGYNE Epiphytes with pseudobulbs. Inflorescence spike-like, borne from the apex of the pseudobulb. Floral bracts large, deciduous. Lip oblong, 3–lobed, or rounded and unlobed; column often hooded; pollinia 4. 12 spp.

Pseudobulbs 5–10 cm long; flowers usually more than 4 in each cluster
a *Flowers white*

1359 **C. ochracea** Lindley (*C. nitida* Lindley)
Uttar Pradesh to Bhutan. Burma. 1500–2500 m. Forest epiphyte, and growing on rocks; common. Apr.–May.
Flowers 6–8 in erect clusters, white, with a lip with rounded incurved lateral lobes with 2 yellow red-margined blotches, and with a rounded white mid-lobe with 2 orange-red spots at the base and with 2 ridges. Flowers 4 cm across, with white sepals and petals; bracts oblong, sheathing the slender stalk of the ovary. Leaves narrow-oblong, narrowed to the channelled leaf-stalk. Pseudobulbs cylindrical, furrowed, 7–10 cm. **Pl. 117**.

1360 **C. cristata** Lindley
Uttar Pradesh to Sikkim. 1000–2000 m. Forest epiphyte, and growing on rocks; very common. Mar.–Apr.
Flowers white in pendulous clusters, with a white lip with 4 yellow ridges at the base between the lateral lobes, and with 2 broad crenulate yellow plates on the mid-lobe. Flowers larger than those of 1359, 5–9 cm across, 3–10 borne in clusters 15–30 cm long; sepals and petals 4–5 cm, oblong blunt with undulate margins; bracts oblong, persistent. Spur absent. Leaves paired, linear-lanceolate 15–30 cm by 2–3 cm broad; pseudobulbs oblong-ovoid, polished, 5–8 cm, arising from a long stout rhizome.

b *Flowers pale brown*

1361 **C. flaccida** Lindley
C. Nepal to Sikkim. 1000–2100 m. Forest epiphyte. Apr.
Distinguished by its drooping cluster of many pale brown unpleasant-smelling flowers, with lips with 3 flexuous ridges. Flowers 3.5 cm across; sepals and petals pale brown, the former oblong-lanceolate spreading, the latter narrower and shorter. Lip darker brown, 3–lobed, the lateral lobes acute, the mid-lobe oblong, reflexed, white with yellow patch near the

base, and with yellow ridges. Leaves narrow-oblong pointed, 10–15 cm; pseudobulbs ovoid-cylindrical 5–15 cm, with papery sheaths at base.

Pseudobulbs to 4 cm; flowers usually 4 or less in each cluster

1362 **C. corymbosa** Lindley
C. Nepal to S.W. China. 2200–3300 m. Epiphytic. Apr.–Jun.
Distinguished by its small pseudobulbs, and its erect flower-clusters of only 2–4 white fragrant flowers each to 5 cm across. Sepals and petals white, broadly lanceolate; lip oblong with broad blunt erect lateral lobes, the terminal lobe triangular-ovate, brown at base with 2 yellow blotches and with a white terminal part. Leaves oblong-elliptic 10–18 cm; pseudobulbs 2.5–4 cm, borne on a thick rhizome. **Pl. 117.**

24 CRYPTOCHILUS Flowers small, in a 2–ranked spike shorter than the bracts. Sepals fused into a tube below and encircling the petals and lip. Pollinia 8. 2 spp.

1363 **C. luteus** Lindley
C. Nepal to Bhutan. 1200–2300 m. Epiphytic; not common. Jun.
Flowers yellow, nearly globular and shorter than the linear-lanceolate bracts, numerous, in 2 ranks in an erect spike. Flowers *c.* 8 mm; sepals fused and forming a flask-shaped tube with 3 triangular apices; petals included in tube, narrowly and obliquely rhomboid; lip oblong, slightly exserted from the sepals. Leaves 1–2, linear-lanceolate 7–12 cm; pseudobulbs crowded, cylindrical. **Pl. 114.**

25 CYMBIDIUM Usually epiphytic plants with a short stout stem-like pseudobulb. Flowers many, or few, in erect or drooping clusters, with numerous sheaths. Lip 3–lobed, with the middle lobe recurved and with undulate edges, usually with 2 ribs; column long; pollinia 2. 9 spp.

1364 **C. hookeranum** Reichb.f. (*C. grandiflorum* Griffith)
E. Nepal to S.E. Tibet. 2000–2500 m. Forest epiphyte; not common. Apr.–May.
Flowers large, sweet-scented, with sepals and petals apple-green, and with an ochre lip speckled with purple, the lip and column turning red after fertilization. Flowers 7.5–10 cm across; lip 3–lobed, with 2 parallel hairy ridges, the large terminal lobe with an irregular wavy margin. Leaves linear-oblong 40–60 cm; flowering stem very robust, arched and pendant, 60–150 cm. **Page 514.**

1365 **C. longifolium** D. Don. (*Cyperorchis elegans* (Lindley) Blume)
C. Nepal to Arunachal Pradesh. 1800–2700 m. Forest epiphyte. Sep.–Oct.
Flowers appearing funnel-shaped, pale yellow, numerous, borne in long-stalked drooping or pendulous clusters. Flowers 3.8–4.5 cm long with sepals, petals and lip nearly equal; lip narrow-oblong, with a small rounded incurved mid-lobe and with small narrow side-lobes near the apex, and

403

with 2 darker yellow parallel ridges, and with a hairy swelling at the base. Leaves narrow-linear long-pointed, 45–60 cm; pseudobulb short. **Pl. 117.**

26 DENDROBIUM   Epiphytes with stem-like pseudobulbs. Lateral sepals fused to the foot of the column and forming a sac; lip with side-lobes encircling column, or spreading, terminal lobe variable; disk with ridges. Column short; pollinia 4. A difficult genus. *c.* 26 spp.

Lip rounded, funnel-shaped

1366 **D. densiflorum** Lindley
C. Nepal to Bhutan. Burma. 600–1600 m. Forest epiphyte, and growing on rocks; not uncommon. Apr.–May.
   Flowers yellow with a rich orange lip which is funnel-shaped below and expands outwards into a rounded hairy and toothed apex. Flowers 3.5–4 cm across, densely crowded into large dense pendulous clusters; flower-buds encircled by overlapping yellow bracts, the inflorescence appearing cone-like in bud. Leaves leathery, borne towards the apex of the pseudo-bulbs, oblong-elliptic 10–15 cm; pseudobulbs club-shaped 30–45 cm, with distinct internodes, narrowest near the base and increasing markedly in thickness above, deeply grooved between the coarse ribs. **Pl. 117.**

Lip longer than broad, entire

1367 **D. amoenum** Wallich ex Lindley
Uttar Pradesh to Bhutan. Burma. 1000–2000 m. Forest epiphyte; common. May–Jun.
   Flowers borne in pairs from the nodes of the pendulous leafless stem-like pseudobulbs, with sepals and petals white with violet tips, and with a white lip which is yellow towards the middle and with a purple spot near the tip. Flowers smelling of violets, *c.* 4 cm across; lip funnel-shaped with broad inrolled side-lobes, minutely downy on both sides. Leaves linear-lanceolate 8–10 cm; pseudobulbs slender 30–45 cm. **Pl. 116.**

1368 **D. transparens** Wallich ex Lindley
Uttar Pradesh to Bhutan. 700–1800 m. Forest epiphyte. May.
   Flowers borne in pairs along the erect leafless stem-like pseudobulbs, with sepals and petals white and sometimes tinged with pink towards the tips, and with the lip with a patch of deep purple encircled by a yellow zone. Differs from 1367 in its acute sepals, and large bracts (sepals blunt, bracts small in 1367). Lip with an inrolled base without distinct side-lobes, and with a rounded undulate mid-lobe. Leaves linear-lanceolate 7–10 cm; pseudobulb slender 30–60 cm.

Lip with a finely fringed margin

1369 **D. fimbriatum** Hook.
C. Nepal to Bhutan. Burma. 500–2400 m. Forest epiphyte. Apr.–May.
   Flowers showy, brilliant yellow with a rounded wavy shortly fringed lip, sometimes with a large reddish-brown central patch. Flowers 5–7.5 cm

across, in pendulous clusters of 7–12, borne towards the ends of the stem-like pseudobulbs; flowering stem with sheathing bracts at base. Leaves oblong-lanceolate long-pointed, 9–15 cm; stem-like pseudobulbs tapering upwards, grooved, 60–150 cm.

27 PLEIONE Flowers 1–2, borne on short stalks from the base of the pseudobulbs. Lip rounded or kidney-shaped and without side-lobes. Leaves solitary or paired. 4 spp.

Pseudobulbs flask-shaped, gradually narrowed into a beak

1370 **P. hookerana** (Lindley) J. Moore
C. Nepal to S.E. Tibet. 2000–3700 m. Epiphytic, often on oaks or conifers; common. May–Jun.
Flowers white, with sepals and petals flushed with violet, and with a rounded lip mottled with reddish-brown and with 5–7 parallel hairy ridges. Flowers solitary, 5–6 cm across, appearing with the leaves; bract blunt. Leaves solitary, stalked, blade narrow-elliptic 5–7 cm; flowering stem 2.5–4 cm; pseudobulbs ovoid to elliptic, less than 2.5 cm long. The highest growing epiphytic orchid in our area.

1371 **P. humilis** (Smith) D. Don.
C. Nepal to S.E. Tibet. Burma. 2000–3000 m. Forest epiphyte; also growing on rocks and steep banks. Oct.–Nov.
Flowers with petals and sepals white, and with an elliptic funnel-shaped lip with a central yellow part with brown spots and with the margin irregularly toothed or fringed. Flowers 5–6 cm across, appearing before the leaves; bract pale purple. Leaves elliptic 7.5–10 cm; pseudobulbs ovoid-oblong, 3–4.3 cm, encircled by old bracts. **Pl. 117.**

Pseudobulbs barrel-shaped, abruptly contracted into a beak

1372 **P. praecox** (Smith) D. Don
Uttar Pradesh to S.W. China. S.E. Asia. 1800–2500 m. Forest epiphyte, also growing on rocks and steep mossy banks; common. Oct.–Nov.
Flowers rose-pink; petals and sepals lanceolate, spreading, the lip funnel-shaped and folded round the column, with toothed margin and 5 central yellow toothed ridges, white within. Flowers 7.5–10 cm across, 1–2 borne on a short stalk arising from the base of the pseudobulb; bract oval, longitudinally inrolled. Leaves paired, elliptic, appearing after the flowers; pseudobulbs barrel-shaped with a short conical beak, 2.5–4 cm across, sparsely sheathed with loose fibres. **Pl. 115.**

28 RHYNCHOSTYLIS Flowers many, in pendulous clusters; sepals and petals spreading; lip with a sac-like hypochile (not a true spur), and a concave epichile. Column short, stout; pollinia 2. Pseudobulbs absent. 1 sp.

1373 **R. retusa** (L.) Blume

Uttar Pradesh to Bhutan. India. Burma. Sri Lanka. 1200–1500 m. Epiphytic; not common. May–Jun.

Flowers fragrant, very numerous in a long cylindrical pendulous cluster, white or purple and variably marked with purple or violet. Flowers *c.* 2 cm across; lateral sepals ovate, broader than the dorsal sepal; lip with a wide and deep sac-like hypochile and an obovate acute epichile with an erect margin. Leaves 15–50 cm, crowded on the stem above, fleshy, curved, channelled, broadly linear *c.* 2.5 cm broad; stem stout, creeping, clothed with the sheaths of old leaves below. **Pl. 116.**

29 VANDA Flowers large, spurred, in simple clusters. Lip large, with a sac-like hypochile and 2 short side-lobes, the mid-lobe fleshy, with a ridged disk. Column short, stout; pollinia 2. 5 spp.

1374 **V. cristata** Lindley

Uttar Pradesh to Arunachal Pradesh. 1000–2000 m. Forest epiphyte; common. May–Jun.

Flowers with yellow-green oblong spreading sepals and petals, and a very distinctive buff-coloured lip conspicuously blotched with rich purple. Flowers 3.5–5 cm across, usually few in short erect clusters; lip with 5 raised ridges, and with two spreading spindle-shaped lobes and a shorter fleshy down-curved horn-like mid-lobe. Spur very short, conical. Leaves numerous, long and narrow, channelled, 8–10 cm; stem stout, 8–15 cm long. **Pl. 117.**

30 VANDOPSIS Like *Vanda*, but flowers not spurred. Epiphyte, with short rigid leafy stems. Lip shorter than sepals and petals, continuous with the column, spreading, concave, the lateral lobes short, the mid-lobe longer tongue-like. Pollinia 2. 1 sp.

1375 **V. undulata** (Lindley) J. J. Smith (*Vanda u.* Lindley)

C. Nepal to Arunachal Pradesh. To 2100 m. Forests; not common. Apr.–May.

Flowers white, flushed with pink and with a yellowish-green lip striped with pink, few borne at the ends of stems much longer than the leaves. Flowers *c.* 4 cm across; sepals and petals fleshy, wavy-margined, reflexed; lip with rounded base and a tapering concave, ridged tongue-shaped apex. Leaves narrow-oblong notched at apex, 7–10 cm by less than 2 cm broad, in two ranks; flowering stem and young shoots spotted with purplish-brown; stem 20–35 cm.

ZINGIBERACEAE   Ginger Family

A mainly tropical family of aromatic herbaceous forest plants, with fleshy branched underground rhizomes, and with aerial stems which are naked or leafy. Leaves alternate, usually 2–ranked, entire, stalked or with a basal sheath and usually with a long *ligule* at the junction of the blade with the sheath or leaf-stalk. Flowers solitary, or in dense heads, with individual flowers or groups of flowers, often subtended by sheathing bracts. Flower structure unique: calyx tubular, 2–3–toothed; corolla-tube short or long with 3 petal-like lobes, the upper often larger, the two lateral spreading. Stamens potentially in 2 whorls, but only one fertile stamen present and opposite it lies a petal-like *lip* which is commonly 2–3–lobed (formed from 2 fused staminodes), the outer stamen-whorl sometimes absent, or of 2 petal-like staminodes. Ovary inferior, 3–chambered; many-seeded. Fruit brightly coloured, often fleshy; seeds often encircled by a distinctive red aril.

(The petal-like lobes of the corolla are described as petals in the following descriptions.)

Leaves spirally arranged, the leaf-sheath tubular, closed   1 COSTUS
Leaves in 2 ranks, the leaf-sheath open on one side
  a Primary bracts fused for about ⅓ their length forming a basal pouch
                                         2 CURCUMA
  b Primary bracts boat-shaped, not fused below
    i Flowers purple (rarely white); dorsal petal broad, more or less hooded
                                       3 ROSCOEA
    ii Flowers yellow, orange or white; dorsal petal narrow
                    4 CAUTLEYA   5 HEDYCHIUM

1 COSTUS Leaves distinctive in being spirally arranged. Lip of flower bell-shaped; fertile stamen with a petal-like filament much longer than the anther; lateral staminodes absent. 2 spp.

1376 **C. speciosus** (König) Smith
Uttar Pradesh to Sikkim. India. Burma. To 1500 m. Forests, shrubberies. Jun.–Aug.
    Flowers white, funnel-shaped, with contrasting large bright red bracts, numerous in a very dense cylindrical spike 5–10 cm long, borne on a stout red leafy stem 1–2 m. Flowers *c*. 4 cm; bracts ovate, 1.5–4 cm long. Calyx with 3 oval lobes; corolla-tube shorter than calyx, with unequal petals 2.5–4 cm; lip white with an orange-red centre, rounded, 5–8 cm, forming a funnel with margins incurved and meeting. Leaves oblong acute, sheathing at base, 15–30 cm. Capsule red, crowned with persistent calyx.
    The root is used medicinally. **Pl. 118.**

2 CURCUMA Inflorescence cone-like, with overlapping papery bracts, and with the uppermost bracts forming a brightly coloured tuft. 2 spp.

1377 **C. aromatica** Salisb. WILD TURMERIC
C. Nepal to Bhutan. India. Sri Lanka. 150–1600 m. Light forests, open slopes. Apr.–May.
    Flowers pinkish-white with an orange lip, in a dense leafless spike

crowned with enlarged coloured bracts tipped with pink, and borne on a short stem with papery bracts, usually appearing before the leaves. Floral bracts papery, the lower green, funnel-shaped and encircling several flowers which open in succession. Calyx short-cylindrical; corolla-tube funnel-shaped, the petals unequal, the upper concave and broader; staminodes oblong; lip rounded with a reflexed swollen apex. Leaves elliptic, 90–120 cm and up to 20 cm wide, leaf-stalk as long as blade which is finely hairy beneath. Rootstock tuberous, tubers yellow and aromatic inside.

Tubers used medicinally. **Pl. 118.**

3 **ROSCOEA** Flowers usually purple, or sometimes pale lilac or white. Corolla-tube long, with 3 petals, the upper erect, broad, incurved, hooded, the lateral spreading; staminodes erect, more or less united round the anther; lip 2–lobed. 4 spp.

Corolla-tube much longer than calyx

1378 **R. alpina** Royle
Pakistan to S.W. China. 2500–4000 m. Forests, open slopes; the only alpine species of this family. Jun.–Jul.

Flowers several, but appearing one at a time, dark purple (rarely white) and with a paler corolla-tube much longer than the calyx, borne on a short stem with or without the leaves. Flowers with a rounded hooded upper petal and narrower reflexed lateral petals, and with the staminodes shorter than upper petal; lip 2–lobed, 12–18 mm long. Leaves 2–4, narrow-elliptic to lanceolate 6–10 cm long; stem 6–20 cm with papery bracts below and few leaves above. **Pl. 119.**

Corolla-tube shorter or little longer than calyx

1379 **R. purpurea** Smith
Uttar Pradesh to S.E. Tibet. 1500–3000 m. Forest clearings, shrubberies, open slopes. Jun.–Aug.

Flowers purple, rarely pale lilac or white, few in a terminal spike, borne on leafy stems 15–30 cm. Flowers *c.* 5 cm long, the upper petal narrow hooded, the lip 2–3–lobed. Leaves often 6, to 20 cm long at flowering; leaf-sheaths broad, overlapping. A variable plant.

Roots used in veterinary medicine. **Pl. 119.**

1380 **R. capitata** Smith
Endemic to C. Nepal. 1500–2500 m. Open slopes; quite common. Jun.–Jul.

Distinguished by its dense compact oblong head of flowers borne at the end of a slender stem which is leafy only below. Flowers pink-purple, many, 4–5 cm long, upper petal hooded, the lateral broader, reflexed, the lip oblong-spathulate, 2–lobed at apex; bracts large, lanceolate to 4 cm, encircling flower-buds. Leaves many, linear 1–2 cm broad; stem to 45 cm. **Pl. 119.**

4 CAUTLEYA Flowers yellow or orange. Like *Roscoea*, but upper petal not wider than lateral petals. 2 spp.

1381 **C. gracilis** (Smith) Dandy (*C. lutea* Royle)
Kashmir to S.W. China. 1200–2800 m. Forest epiphyte, and growing on rocks. May–Aug.

Flowers yellow to orange, few in a lax drooping spike 5–10 cm, with green bracts much shorter than the reddish calyx. Corolla-tube longer than calyx, the upper longer petal narrow concave, to 2 cm, the lateral petals broader reflexed; staminodes forming a hood over the anther; lip deeply 2-lobed, reflexed. Leaves linear long-pointed 13–25 cm by 2.5–4 cm, blade green above and tinged red beneath; stem 30–45 cm. Capsule globular, red; seeds glossy-black, partly encircled in a white aril.

1382 **C. spicata** (Smith) Baker
Himachal Pradesh to Sikkim. 1800–2800 m. Shrubberies, rocks; sometimes epiphytic. Jul.–Sep.

Flowers yellow, few or many, crowded into a terminal erect spike with red bracts as long as the calyx; spike 13–23 cm long; corolla-tube hardly as long as calyx. A larger plant than 1381, with broader leaves and longer coloured bracts. Leaves narrow-elliptic, 5–8 cm broad; stem 30–60 cm. Capsule waxy-grey; seeds black and encircled in a white aril. **Page 514**.

5 HEDYCHIUM Flowers yellow or orange. Corolla-tube long narrow, petals narrow; stamen distinctive with a long slender filament and with anthers which are not spurred as in the 3 preceding genera. 12 spp.

Stamen longer than petals

1383 **H. aurantiacum** Roscoe
C. Nepal to Bhutan. 450–2000 m. Shrubberies, open slopes. Jul.–Aug.

Flowers very showy, bright orange-red, many in a long spike to 30 cm, each with a long-projecting stamen and linear petals. Stamen with a bright red filament twice as long as the petals; staminodes lanceolate; lip obovate, deeply lobed; bracts oblong, green, nearly as long as corolla-tube. Leaves sheathing at base, with lanceolate blade 5–8 cm broad and up to 45 cm long; stem to 2 m. Capsule as in 1385. **Pl. 118**.

1384 **H. ellipticum** Buch.–Ham. ex Smith
Uttar Pradesh to Bhutan. 300–3000 m. Forests, shrubberies, rocks; common. Jun.–Aug.

Flowers white or orange-yellow in a very dense spike 8–10 cm long, with bright green ovate densely overlapping bracts. Flowers to 7 cm, with narrow white petals and linear orange staminodes; fertile stamen with stout orange filament much longer than the petals; lip white, oblong, notched at apex; corolla-tube twice as long as bracts. Leaves oblong pointed 15–30 cm by 8–12 cm broad; stem to 1 m. **Pl. 118**.

Stamen shorter than petals

1385 H. spicatum Smith
Himachal Pradesh to Arunachal Pradesh. 1800–2800 m. Forest clearings, shrubberies. Jul.–Aug.

Flowers fragrant, white with an orange-red base, in a dense terminal spike 15–25 cm, borne on a robust leafy stem 90–150 cm. Floral bracts large green, 1–flowered; calyx papery, 3–lobed, shorter than bracts. Corolla-tube 5–6.5 cm and much longer than calyx, petals white, linear, spreading; lip white with two elliptic lobes with an orange base; filament of stamen red. Leaves oblong, to 30 cm by 4–12 cm broad. Capsule globular, 3–valved, with an orange-red lining; seeds black, with a red aril.

Rootstock used medicinally. A perfume *abir* is obtained from the rootstock. **Pl. 119**.

Amomum subulatum Roxb. CARDAMOM is cultivated on shady slopes in extreme E. Nepal and Sikkim. The ripe fruits are dried and give a strongly flavoured spice.

Several species of **Canna**, a native American genus, are grown in gardens in the hills.

IRIDACEAE   Iris Family

A family of world-wide distribution, of herbaceous plants usually with rhizomes or corms. Leaves alternate, usually narrow-linear, often two-ranked. Distinguished from the Liliaceae by its inferior ovary, and 3 stamens. Petals 6, coloured, either all similar as in *Crocus*, or the outer and inner whorls of 3 differing as in *Iris*. Ovary inferior, 3–chambered, with many ovules; styles 3, coloured and petal-like in *Iris*. Fruit a capsule splitting longitudinally.

BELAMCANDA Differing from *Iris* in having all petals equal, and stamens alternating with the styles. 1 sp.

1386 B. chinensis (L.) Redouté
Native of China; widespread but doubtfully native in the Himalaya. 1000–2300 m. Shrubberies, banks at edge of cultivation; common. Jul.–Aug.

An iris-like plant with spear-shaped leaves, and a branched cluster of orange flowers spotted with brownish-purple. Flowers *c.* 5 cm across, several borne at the ends of the branches in the axils of papery bracts. Petals all similar, elliptic and narrowed to a short stalk, *c.* 2 cm; corolla-tube very short. Leaves linear-lanceolate, to 3 cm broad; stem leafy below, to 1 m. Capsule 4 cm, splitting and leaving a central axis; seeds black. **Page 514**.

CROCUS Flowers stalkless, with a long slender corolla-tube and 6 more or less equal petals; stamens 3; style 3–lobed; ovary below ground. Corm with fibrous scales. 1 sp.

1387 **C. sativus** L. SAFFRON CROCUS
Native of S. Europe and W. Asia; cultivated in the Kashmir valley, especially in Pampur. 1600 m. Oct.

Flowers deep blue-violet, funnel-shaped, stalkless, and flowering in autumn with the leaves. Petals narrow-elliptic, equal, fused below into a long slender corolla-tube; styles orange, stigmas brick-red. Leaves linear, channelled.

Cultivated for its styles, which when dried form saffron, which is much-prized for flavouring food, for perfume, and for dyeing. Saffron is also used medicinally.

IRIS Distinguished by its showy flowers, with 3 outer spreading or recurved petal-like *falls*, and 3 inner erect petal-like *standards* which are fused below into a corolla-tube. Falls either smooth above, or with a central *crest*, or patch of hairs, the *beard*; each fall is overlaid by a petal-like style which covers each stamen. Flowers borne in the axils of 2 sheaths or *spathes*. Rhizomes creeping. 11 spp.

Falls smooth above, without a beard or crest

1388 **I. lactea** Pallas (*I. ensata* Thunb.)
Pakistan to Himachal Pradesh. Tibet. C. Asia. 1500–3300 m. Edge of fields, irrigation channels; common in Kashmir and Ladakh. Apr.–Jun.

Flowers 1–2, pale mauve; falls 4–5 cm, with narrow-elliptic blade shorter than its stalk; standards oblanceolate, blunt, erect, *c.* 5 mm broad. Spathes narrow, green, papery-margined, 6–10 cm. Leaves rigid, glaucous, linear, to 6 mm broad; stem 15–30 cm. Fruit to 5 cm, 6–ribbed, beaked. **Pl. 120**.

1389 **I. spuria** L.
Native of W. Asia and Europe. Grown on graves in Kashmir valley. 1300 m. Jun.

Flowers mauve; falls elliptic or rounded, 3–8 cm long; standards obovate, 3–6 cm long. Spathes green, leafy. Leaves spear-shaped, 6–10 mm broad; stems 30–90 cm. **I. crocea** Jacquem. ex R. C. Foster (*I. aurea* Lindley) is also grown in Kashmir. It is probably no more than a yellow-flowered form of 1389. **Pl. 120**.

1390 **I. clarkei** Baker ex Hook.f.
C. Nepal to Bhutan. 3000–4000 m. Marshes, wet meadows. Jun.–Jul.

Flowers large, *c.* 8 cm across, bright lilac blotched with violet and with a yellow throat; falls oblong-wedge-shaped narrowed to a bearded stalk. Standards with small oblong blades and long stalks; corolla-tube *c.* 12 mm. Spathes green, to 10 cm. Leaves variable, glossy above, glaucous beneath, linear long-pointed, usually 6–10 mm broad; stems stout, 30–60 cm. Rhizome stout, creeping. Fruit cylindrical to 6 cm.

411

IRIDACEAE

Falls with a raised ridge or crest

1391 **I. decora** Wallich (*I. nepalensis* D. Don)
Pakistan to S.W. China. 1800–4000 m. Drier inner valleys. Jun.–Jul.
Flowers solitary or few, pale lilac, in a branched sometimes long-stalked cluster; falls to 2.5 cm broad, stalked, with a central yellow ridge-like crest; standards narrower and smaller; corolla-tube 4 cm. Spathes narrow, 4–6 cm. Leaves linear, to 6 mm broad; stems slender 10–30 cm. Fruit cylindrical 2.5–3.5 cm, beaked.
The root is used medicinally.

1392 **I. milesii** R. C. Foster
Kashmir to Uttar Pradesh. 1600–2700 m. Coniferous forests and clearings. Jun.
Easily distinguished by its stout branched inflorescence with large flowers to 8 cm across, pale mauve with dark veins and blotches, and with obovate falls with a pale much-cut crest. Standards large obovate, bluish-purple, spreading, undulate. Spathes many-flowered, blunt, to 2.5 cm. Leaves pale green, curved, gradually tapering to apex, 2.5–8 cm broad; stems to 1 m. **Pl. 120.**

Falls with a beard of orange or yellow hairs
a *Dwarf plants with stems less than 25 cm*

1393 **I. goniocarpa** Baker
W. Nepal to S.W. China. 3600–4400 m. Dry scrub, alpine slopes, drier inner valleys. Jun.–Jul.
Flowers lilac, usually solitary; falls obovate blunt, shortly stalked, *c.* 13 mm broad, and with a conspicuous and copious yellow beard; standards spreading, with an oblong blade narrowed to a long stalk. Spathes papery, narrow-elliptic. Leaves few, narrow *c.* 3 mm broad; stem usually to 20 cm. Fruit narrow-elliptic, with a slender beak. **Pl. 120.**

1394 **I. kemaonensis** D. Don ex Royle
Pakistan to Arunachal Pradesh. 2800–4000 m. Grazing grounds, alpine slopes; common, often in large clumps. Apr.–Jul.
Flowers solitary, stemless or nearly so, appearing with the young leaves, bright lilac to purple with darker spots and blotches, and with falls with conspicuous yellow beards. Falls obovate, shortly stalked, to 5 cm, corolla-tube slender, longer than the spathes; standards lilac, elliptic, erect and incurved. Spathes broader than the leaves, to 6 cm long. Leaves 3–9 mm broad, soon overtopping the flowers. Fruit 3–4 cm, long-beaked, often nearly stalkless. **Pl. 120.**

1395 **I. hookerana** R. C. Foster
Pakistan to Kashmir. 2400–3300 m. Open grassy slopes, grazing grounds; forming large clumps. Apr.–Jul.
Differs from 1394 in having one or several purple blotched flowers borne on short leafy stems 5–10 cm. Falls obovate, stalked, with a white

412

AMARYLLIDACEAE

beard, to 5 cm; corolla-tube short *c.* 1.3 cm. Spathes to 6 cm, broader than the leaves which are up to 1 cm broad. Capsule long-stalked. **Pl. 120.**

b *Plants with stems more than 25 cm*

1396 I. germanica L.
Origin unknown; widely cultivated in Temperate Eurasia; common on graves in the Kashmir valley. Apr.–May.
Flowers large, usually mauvish-blue; falls 5.5–9 cm by 4–6 cm broad with yellow beard; standards broadly elliptic, 4.5–6 cm broad. Spathes with a green base, and a papery apex. Leaves 2–3.5 cm broad; stems 40–90 cm.
Oil extracted from the rhizome is used medicinally.
A fragrant white-flowered iris, **I. kashmiriana** Baker, is also common on graves in Kashmir. Apr.–May. Apart from the colour it is very similar to 1396.

AMARYLLIDACEAE   Daffodil Family

Bulbous or tuberous herbaceous plants, with linear entire deciduous leaves, mainly of the warm temperate, sub-tropical and tropical regions. Distinguished from Iridaceae by its 6 stamens, and from Liliaceae by its inferior ovary (except *Allium* which has a superior ovary, but which because of its inflorescence is now placed in the Amaryllidaceae). Flowers usually in an umbel, which is surrounded in bud by a papery *spathe*, or flowers borne in branched clusters. Petals 6, free, or fused into a corolla-tube at the base. Ovary of 3 fused carpels, 3–chambered, each with many ovules; style 1. Fruit a capsule, or rarely a fleshy berry.

CRINUM Flowers large, funnel-shaped, borne in an umbel. Fruit splitting irregularly; seeds few, large. 1 sp.

1397 C. amoenum Roxb. ex Ker-Gawler
C. Nepal to Bhutan. Burma. 700–1700 m. Cliffs, rocky slopes. May–Jun.
Flowers large, white, in a dense umbel, with slender spreading petals and a long green corolla-tube, and with conspicuous reddish stamen-filaments, borne on a slender stem 30–60 cm. Flowers 6–12, almost stalkless; petals linear-lanceolate 5–9 cm; corolla-tube very slender 8–10 cm long; style long purple; spathe papery, 2–7 cm. Leaves 10–12, spear-shaped, 2.5–4 cm broad. Bulb large, 5–7.5 cm in diameter. **Pl. 122.**

ALLIUM ONION, GARLIC. Usually distinguished by its dense solitary umbel of many small, stalked flowers, borne on an unbranched stem. Umbel with a papery spathe at its base which encircles the flower-buds and later splits into lobes. Petals usually free, not fused below; ovary superior. Leaves usually smelling of garlic when crushed. *c.* 18 spp.

Leaves linear
a *Leaves cylindrical in section, often hollow*

1398 **A. schoenoprasum** L. CHIVES
Pakistan to Uttar Pradesh. Temperate Eurasia. 3300–4000 m. Rocks, alpine slopes. Jul.–Aug.

Flowers pink or purple, numerous in a dense globular umbel 2.5–5 cm across, with stamens shorter than the linear to lanceolate pointed petals. Corolla bell-shaped, petals 7–14 mm; flower-stalks mostly shorter than petals. Spathe-lobes usually 2, broad and shorter than the umbel. Leaves 1–2, grooved above, 10–25 cm long; stem hollow, 15–35 cm, leafless or with one leaf.

1399 **A. semenovii** Regel
Pakistan to Himachal Pradesh. 3000–4000 m. Open slopes, wet places. Jun.–Jul.

Flowers pale yellow, in a dense globular umbel to 3 cm across, with a broad persistent papery spathe. Petals oblong-lanceolate long-pointed, 13–18 mm; stamens shorter than petals. Leaves usually 2, stout, hollow, acute, 6–13 mm broad, with a long sheath; the leaves about as long as the stout stem which is 15–40 cm. Bulb cylindrical, covered with old fibrous sheaths. Leaves eaten as a vegetable. **Page 515**.

1400 **A. przewalskianum** Regel (*A. jacquemontii* Regel, *A. stoliczkii* Regel)
Pakistan to C. Nepal. Tibet. S.W. China. 2700–4300 m. Dry stony slopes, steppes. Jul.–Sep.

Flowers usually lilac, in a dense umbel 1.5–7.5 cm across, with 2 papery spathe-lobes, borne on a stem to 30 cm. Petals *c.* 5 mm, oblong-lanceolate, the stamens much longer; filaments of inner stamens with broad bases, toothed at each side, outer filaments thread-like. Leaves 3–6, round in section, hollow, 1 mm broad. Bulb cylindrical, covered with rusty-brown, finely-netted scales.

b *Leaves flattened*
 i *Flowers white or greenish*

1401 **A. humile** Kunth (*A. govanianum* Wallich)
Pakistan to W. Nepal. 3000–4000 m. Open alpine slopes. Jun.–Aug.

Flowers white, star-shaped, in a rather lax umbel 2.5–4 cm across, borne on a leafy stem. Petals narrow-elliptic acute to *c.* 10 mm, spreading, at length reflexed, much longer than the stamens; outer flower-stalks usually longer than the flowers; spathe-lobes broadly ovate, persistent. Leaves several, flat, 2–5 mm broad, blunt, usually shorter at flowering than the acute-angled stem, which is 7–25 cm. Bulbs clustered, cylindrical and covered with fibrous leaf-bases. **Pl. 121**.

1402 **A. oreoprasum** Schrenk
Pakistan to W. Nepal. C. Asia. 2700–5000 m. Stony slopes in Tibetan borderlands; common in Ladakh. Jun.–Jul.

Flowers white with purple mid-veins, or less commonly rose-purple, borne in a dense umbel with equal flower-stalks mostly twice as long as the flowers, and with shorter broadly ovate spathe-lobes. Flowers bell-shaped; petals 5–7 mm, elliptic fine-pointed, longer than the stamens. Leaves several, basal, narrowly linear, usually 3–4 mm broad; flowering stem 10–20 cm; base of stem with numerous dense netted brown fibres. Bulbs tufted, cylindrical. **Pl. 121**.

1403 A. fasciculatum Rendle
W. Nepal to Bhutan. Tibet. 2800–4500 m. Stony slopes in Tibetan borderlands. Jul.–Aug.

Flowers white or pale green, sweet-scented, in a lax rounded umbel, borne on a stem as long or shorter than the leaves; distinguished by the absence of a bulb and by its cluster of many stout tuberous roots. Flowers *c*. 5 mm long; petals lanceolate pointed, shorter than the stamens. Leaves several, 3–4 mm wide; flowering stem usually 12–20 cm, sometimes more.

Leaves eaten as garlic.

ii *Flowers pink or purple* (see also 1402)

1404 A. carolinianum DC.
Afghanistan to C. Nepal. 3300–4800 m. Stony slopes. Jul.–Aug.

A rather stout onion with a very dense globular umbel of pink flowers, borne on stout stems usually 10–30 cm, with usually shorter broad flat and curved, glaucous leaves. Umbels 2–3.5 cm across; flowers cylindrical; petals elliptic pointed, to *c*. 6 mm, much shorter than the stamens. Spathe-lobes broadly ovate, shorter than umbel. Leaves several, blunt, 4–12 mm broad. Bulbs relatively large, oblong-cylindric, covered with conspicuous leathery scales. **Pl. 121**.

1405 A. wallichii Kunth
Pakistan to S.W. China. 2800–4300 m. Forest clearings, shrubberies; the only species found in those parts of Nepal fully open to the monsoon rains. Aug.–Sep.

Flowers purple, numerous, long-stalked, in a lax rounded umbel 5–7 cm across. Petals broadly linear blunt, spreading in a star, at length reflexed, longer than the purple stamens and ovary. Leaves many, linear or spear-shaped, flat and keeled, to 2 cm broad, often as long as the acutely 3–angled flowering stem which is 30–90 cm. Bulb absent, but numerous fibrous roots present. **Pl. 121. Page 515**.

iii *Flowers blue*

1406 A. caesium Schrenk
Himachal Pradesh. C. Asia. 2400–3300 m. Stony slopes; quite common in Lahul. Jun.–Jul.

Distinguished by its dark blue flowers, in a lax umbel with slender spreading flower-stalks twice as long as the flowers. Umbel 3–4 cm across; flowers *c*. 5 mm long; petals narrow-elliptic acute, longer than the stamens which have broad-elliptic bases to the filaments; spathe with elliptic lobes shorter than the flower-stalks. Leaves 1–2 mm broad, shorter than the flowering stem which is 15–30 cm. **Pl. 121**.

1407 **A. sikkimense** Baker
C. Nepal to Bhutan. S.W. China. 3000–4800 m. Open slopes, rocks. Jul.–Sep.

A rather slender plant with a dense globular umbel of pale blue flowers, borne on a slender stem longer than the narrow basal leaves. Flowers short-stalked, cup-shaped; petals elliptic 5–9 mm; filaments blue, those of the inner stamens markedly broader at base; spathe yellowish, persistent, papery, shorter than umbel. Leaves 1–4 mm broad; stems sometimes tufted, covered with fibrous sheaths at base, 10–30 cm. Bulb absent.

Leaves elliptic, narrowed to a short stalk

1408 **A. victorialis** L. ALPINE LEEK
Pakistan to Uttar Pradesh. W. Asia. Europe. 3000–3600 m. Forests. Jun.–Aug.

Distinguished by its lax head of greenish-white to yellowish star-shaped flowers, and its broad flat leaves which are narrowed to a short leaf-stalk below. Umbels 2.5–4 cm across; petals lanceolate to narrow-ovate *c.* 5 mm, spreading and later reflexed, shorter than the stamens; spathe shorter than the unequal flower-stalks. Leaves usually 3–5, elliptic, 2.5–6 cm broad, shorter than the flowering stem which is 30–60 cm.

1409 **A. prattii** C. H. Wright
W. Nepal to S.W. China. 2400–4300 m. Forest clearings, shrubberies, open slopes. Jul.–Aug.

Differs from 1408 in having rose or pink flowers, and narrow-elliptic, usually paired leaves which are narrowed to the leaf-stalk below. Umbel many-flowered, *c.* 4 cm across; flowers *c.* 6–7 mm, with stamens much longer than the elliptic blunt petals; flower-stalks about as long as the flowers or a little longer; spathe elliptic. Leaves to 30 cm; flowering stems much longer, to 45 cm, with fibrous sheaths surrounding base of stem.

Leaves used for seasoning.

IXIOLIRION Flowers blue or violet, borne in a raceme; petals 6, equal, free to their base. 1 sp.

1410 **I. karateginum** Lipsky (*I. montanum* (Labill.) Herbert)
Native of C. Asia and Pakistan; found in the saffron beds in the Kashmir valley. 1500–1800 m. Apr.–May.

Flowers bright blue-lilac, funnel-shaped, several in a long spike-like cluster or umbel, borne on a slender stem 30–45 cm. Petals oblanceolate fine-pointed, spreading, 2–3.5 cm long; stamens shorter. Leaves linear to *c.* 4 mm broad, the upper shorter, all shorter than the flowering stem. Capsule cylindric, *c.* 2 cm, splitting at the apex. **Page 515**.

ZEPHYRANTHES Flowers solitary, short-stalked, with a spathe; petals fused below; stamens 6, 3 long and 3 short; stigma 3–lobed. 1 sp.

1411 **Z. carinata** Herbert

Native of Mexico; quite common in cultivated areas of the Himalaya. 600–2500 m. Apr.–Jul.

Flowers large solitary, funnel-shaped, usually pink, borne on a leafless stem 13–20 cm. Flowers 7–9 cm long and broad; petals lanceolate, spreading and narrowed below, with a long green corolla-tube which is encircled at the base by a purplish 2–lobed spathe. Leaves several, flat grass-like, usually 4–6 mm broad. Bulb ellipsoid, with many papery scales. **Pl. 121**.

HYPOXIDACEAE

Closely related to the Amaryllidaceae and often included in it. Distinguished by its solitary or few flowers, not borne in umbels, and its tuberous, not bulbous rootstock. A small more or less cosmopolitan family, absent in Europe and N. Asia.

HYPOXIS 1 sp.

1412 **H. aurea** Lour.

Pakistan to Bhutan. Burma. China. Japan. S.E. Asia. 1500–2800 m. Grassy slopes, grazed ground. May–Jul.

A small hairy tuberous plant, with slender grass-like leaves, and solitary or paired yellow flowers, with spreading lanceolate petals which are green and hairy on the undersides. Flowers 1.3–2 cm across, with conspicuous cylindrical inferior ovary, and borne on a slender stem much shorter than the leaves. Leaves 2–5 mm broad, strongly veined, with long sparse hairs; flowering stem 3–5 cm. Fruit cylindrical and topped by the persistent petals. **Pl. 121**.

AGAVACEAE Sisal Family

A family of woody rhizomatous plants of the Tropics and Subtropics, particularly in arid and semi-arid regions. Leaves usually stiff, fleshy, narrow and sharp-pointed, and crowded at the base of the stem. Flowers borne in large branched clusters; perianth of 2 petal-like whorls, each whorl fused at the base in a long or short tube; stamens 6. Ovary of three fused carpels, either superior or inferior, with three chambers; style single, slender. Fruit a splitting capsule, or a berry. An important fibre-producing family.

AGAVE CENTURY PLANTS Leaves in a very large basal rosette, narrow-lanceolate, spine-tipped and often with a spiny margin. Infloresence very large, branched, with usually greenish flowers, occurring only once and followed by the death of the plant. Fruit a capsule, with flattened black seeds. Natives of America; commonly cultivated to 2000 m in the Himalaya. Used as hedge-plants, and the leaves are soaked and pounded to yield a fibre. **Pl. 120.**

The commonest species seen in the Himalaya are: **A. angustifolia** Haworth with grey-green leaves with a black terminal spine, and marginal black-brown teeth; **A. cantula** Roxb. with light or dark green leaves with a curved tip, and large terminal spine and marginal hooked teeth; **A. sisalana** Perrine SISAL or HEMP with leaves spreading, mostly with smooth margins and a terminal black-brown spine.

YUCCA   Leaves stiff, sword-shaped; plants often with woody stems. Inflorescence annual or periodic. Fruit fleshy in our species.

1414   Y. aloifolia L.

Native of N. & C. America; cultivated and planted by roadsides in Chamba and the Kulu valley in Himachal Pradesh. To 2000 m. Jun.–Aug.

Flowers cream or white, tinged with purple, in long branched clusters 60–100 cm, borne on slender flowering stems. Flowers *c*. 10 cm across. Leaves narrow spine-tipped, and with minutely toothed margins, glaucous green, 45–50 cm long; stem simple or branched, to 7 m.

LILIACEAE   Lily Family

A very large and important cosmopolitan family of herbaceous plants usually possessing swollen corms, rhizomes, or bulbs. Leaves usually entire, narrow, and parallel-veined. Flowers usually in racemes, regular, with 6 coloured usually similar petals (perianth-segments) in two whorls, free to the base, or fused below into a tube. Stamens 6. Ovary superior, 3–chambered, with a single style, or style 3–lobed; ovules numerous. Fruit usually a dry splitting capsule, or less commonly a fleshy berry.

Flowers in umbels
   a Umbels several, axillary and borne on lateral branches
                                    8 DISPORUM   20 SMILAX
   b Umbels solitary, terminating stem   6 CLINTONIA
                                    8 DISPORUM   11 GAGEA

Flowers not in umbels
   a Flowers large; petals 3–25 cm long
      i Flowers several   4 CARDIOCRINUM   10 FRITILLARIA
            12 HEMEROCALLIS   13 LILIUM   15 NOTHOLIRION
      ii Flowers solitary or two, rarely more   10 FRITILLARIA
            13 LILIUM   15 NOTHOLIRION   17 PARIS
                        23 TRILLIDIUM   24 TULIPA

b Flowers small; petals less than 3 cm long
  i Flowers many in a terminal spike-like cluster   1 ALETRIS
                             3 CAMPYLANDRA 5 CHLOROPHYTUM
                   6 CLINTONIA 9 EREMURUS 16 OPHIOPOGON
                          19 SMILACINA 22 THEROPOGON
  ii Flowers few, or in small axillary clusters
    x Flowers white   14 LLOYDIA 18 POLYGONATUM
                                   21 STREPTOPUS
    xx Flowers green   2 ASPARAGUS
    xxx Flowers yellow   7 COLCHICUM 11 GAGEA
                                   14 LLOYDIA
    xxxx Flowers pink, violet or purple   1 ALETRIS
               6 CLINTONIA 13 LILIUM 18 POLYGONATUM
                    19 SMILACINA 22 THEROPOGON

1 ALETRIS Flowers small, in a spike or raceme borne on a leafless stem; petals shortly fused below; style 3–lobed. 4 spp.

1415 A. pauciflora (Klotzsch) Hand.–Mazz.
Kashmir to S.W. China. 3000–4300 m. Shrubberies, alpine slopes; common. Jun.–Aug.

A small inconspicuous alpine plant, with a slender spike of tiny whitish to pinkish bell-shaped flowers, and grass-like leaves. Flowers 2–4 mm long, few or many in a spike 2–5 cm long, either stalked or stalkless, and borne in the axils of linear bracts which are longer than the flowers; petals fused below, out-curved. Leaves 5–12, shorter than the flowering stem which is 3–15 cm and woolly-haired or glandular-hairy above. **Page 515.**

2 ASPARAGUS Distinguished by the absence of leaves; in their place are clusters of needle-like *cladodes* (stems performing the function of leaves), and the leaves are reduced to scales or spines. Flowers tiny; fruit a berry. 4 spp.

1416 A. racemosus Willd.
Pakistan to Sikkim. India. S.E. Asia. Australia. Africa. 600–2100 m. Shrubberies. Jun.–Aug.

A tall woody, much-branched climbing plant, with straight or curved spines on the stems below, and with slender pointed, often curved cladodes in clusters of 2–6, mostly 1–2 cm long. Flowers tiny, scented, 2–3 mm across, green and with purplish anthers, in simple or branched clusters 2.5–5 cm long. Fruit globular, 4–7 mm.

1417 A. filicinus Buch.–Ham. ex D. Don
Pakistan to S.W. China. Burma. S.E. Asia. 2100–3000 m. Forests, shrubberies. May–Jun.

A tall erect perennial, or a twining plant, without spines and with flat curved cladodes in clusters of 2–6. Flowers solitary or paired, on slender stalks, greenish to reddish-green, 2–3 mm long; anthers white. Cladodes 6–9 mm; stems variable, 30–100 cm. Roots tuberous. Fruit black, c. 6 mm. Roots used medicinally. **Page 515.**

3 CAMPYLANDRA Flowers in a dense spike borne on a very short leafless stem; petals fused into a broad tube below. Leaves all basal, arising from a thick rootstock. 1 sp.

1418 **C. aurantiaca** Baker (*Tupistra a.* ((Baker) Wallich ex Hook. f.)
C. Nepal to S.W. China. 2100–2800 m. Forests, shrubberies; quite common at village level. Mar.–Apr.

Flowers yellow to grey-green, in a dense almost ovoid cluster, borne on a short stout stem surrounded by many long linear spreading leaves. Flowering clusters 3–5 cm long; flowers tubular c. 1 cm long, with rounded wavy-margined lobes, waxy, unpleasant-smelling; bracts lanceolate, at least twice as long as flowers. Leaves pointed, to 45 cm, all basal; rootstock thick tuberous, branching above. Fruit a red berry; seeds white. **Page 515.**

4 CARDIOCRINUM Flowers very large, white, in a terminal cluster. Leaves large, heart-shaped. Bulb with narrow overlapping scales. 1 sp.

1419 **C. giganteum** (Wallich) Makino
Kashmir to S.W. China. Burma. 1800–3000 m. Forests; quite common. Jun.

A very distinctive plant with a tall stout stem 2–4 m, with large heart-shaped glossy leaves, and with a terminal cluster of very large, white, drooping funnel-shaped flowers. Flowers very fragrant, 13–18 cm long. Leaf-blades 10–25 cm broad and long, the lower long-stalked. Capsule 5–7.5 cm; seeds flattened and with a broad papery wing. **Pl. 122.**

5 CHLOROPHYTUM Flowers borne in axils of bracts in a spike-like terminal cluster. Leaves all basal. Capsule acutely 3–angled. 3 spp.

1420 **C. nepalense** (Lindley) Baker
C. Nepal to Arunachal Pradesh. 1500–3000 m. Rocks, open slopes. Jul.–Sep.

Flowers white, drooping short-stalked, in a long terminal spike-like cluster borne on a leafless stem 30–90 cm. Flowers solitary or several, in the axils of small lanceolate sheathing bracts; petals narrow-elliptic c. 9 mm, spreading; anthers bright yellow in a cone-like long-pointed cluster. Leaves all basal, linear to c. 1 cm broad, usually shorter than the stem. Fruit a 3–lobed capsule c. 1 cm across. **Page 516.**

6 CLINTONIA Flowers in a terminal umbel, or a spike-like cluster, borne on leafless stems. Flowers funnel-shaped. Fruit a fleshy berry. Rootstock creeping. 1 sp.

1421 **C. udensis** Trautv. & Meyer (*C. alpina* Kunth ex Baker)
Uttar Pradesh to S.W. China. 3000–4000 m. Birch forests, alpine shrubberies; quite common. May–Jun.

Flowers white or pale violet, in a dense terminal drooping cluster, borne
on leafless stems arising from a basal rosette of broad leaves. Flowers
several or many, funnel-shaped 6–10 mm long; petals oblanceolate blunt;
bracts as long as the short flower-stalks. Leaves 3–4, elliptic to obovate
mostly to *c*. 12 cm long; flowering-stem silvery-hairy, 15–50 cm. Fruits a
lax cluster of black berries, borne on up-curved stalks. **Pl. 123**.

7 COLCHICUM Flowers very short-stalked, with 6 equal petals and a
slender corolla-tube; styles 3, free. Ovary below ground. 1 sp.

1422 **C. luteum** Baker
Pakistan to Himachal Pradesh. 1000–2700 m. Open slopes recently cleared
of snow. Feb.–Jun.

Flowers solitary or 2, with golden-yellow petals, appearing from the
ground with the young developing leaves. Petals *c*. 2 cm long, oblong to
oblanceolate blunt, fused below into a long corolla-tube 8–10 cm which is
partly below ground. Leaves narrow-oblong blunt, lengthening to 15–30
cm in fruit. Corms cylindrical, covered with numerous brown scales. One
of the earliest plants to flower in spring.

The corms are used medicinally; they contain the drug colchicine.
Pl. 122.

8 DISPORUM Flowers bell-shaped, in terminal or axillary umbels; stem
leafy. Fruit a berry. Rootstock creeping. 2 spp.

1423 **D. cantoniense** (Lour.) Merr.
Uttar Pradesh to Arunachal Pradesh. Burma. S.E. Asia. 1500–2500 m.
Forests, shrubberies; quite common. May–Jun.

An erect simple or branched herbaceous plant, with many broadly
lanceolate leaves, and with terminal and axillary umbels of few pendulous
white, or sometimes greenish or purplish, funnel-shaped flowers. Flowers
1–2 cm long, stalked or stalkless; petals oblanceolate, not spreading. Leaves
5–10 cm long, pointed; stems 60–120 cm, with large sheathing papery
bracts arising from the lower nodes. Fruit a black fleshy berry. **Page 516**.

9 EREMURUS DESERT CANDLE Flowers very numerous in a long cylindrical spike-like cluster with bracts, borne on a long leafless stem. Leaves
all basal. Rootstock fibrous. 1 sp.

1424 **E. himalaicus** Baker
Afghanistan to Himachal Pradesh. C. Asia. 2100–3300 m. Rocky slopes in
drier areas; often locally abundant. May–Jun.

Unmistakable with its tall stout spike-like cluster of hundreds of white
flowers with protruding orange anthers, and with a basal tuft of long

narrow leaves. Flower-clusters 30–40 cm long; flowers *c.* 2.5 cm across; stalked, borne in the axils of awn-shaped papery bracts; petals oblong blunt, with brown line outside, 1–1.5 cm. Leaves erect, 30–90 cm by 2–4 cm wide; stem 60–175 cm. Capsule *c.* 1.5 cm. The only Himalayan representative of this C. & W. Asian genus. **Pl. 124**.

10 FRITILLARIA Flowers bell-shaped, nodding; petals with a conspicuous nectary at base. Capsule 6–angled. 3 spp.

Flowers usually solitary, usually chequered

1425 **F. roylei** Hook.
Pakistan to Uttar Pradesh. 2700–4000 m. Alpine slopes, shrubberies; quite common. Jun.–Jul.
Flowers yellowish-green to brownish-purple and usually chequered with dull purple, broadly bell-shaped, pendulous and usually solitary but sometimes 2–4. Petals narrow-ovate, 4–5 cm long. Leaves linear-lanceolate often long-pointed, 5–10 cm, opposite or 3–6 in a whorl; stem 15–60 cm.
The bulbs are used medicinally. **Pl. 125**.

1426 **F. cirrhosa** D. Don
W. Nepal to S.W. China. 3000–4300 m. Alpine slopes, shrubberies; quite common. May–Jun.
Very like 1425, but a plant of the E. Himalaya, and more slender and often distinguishable by the coiled tips of the upper leaves. Flowers variable in colour; maroon, yellow, green or purple, usually chequered; petals 3.5–5 cm, narrow-elliptic blunt.

Flowers several, not chequered

1427 **F. imperialis** L. Crown Imperial
Afghanistan to Kashmir. W. Asia. 2100–2700 m. Rare in our area, but grows on the Banihal pass leading to the Kashmir valley. Apr.–May.
A rather robust leafy plant, with a terminal cluster of pendulous usually yellow bell-shaped flowers, topped by a crown of narrow leafy bracts. Flowers 4–5 cm, not chequered. Leaves lanceolate to 12 cm, the upper in whorls; stem robust, 1–1.3 m.

11 GAGEA Yellow-Star-of-Bethlehem Bulbous perennials, with 1 or several yellow flowers. Probably several species occur in the extreme west of our area but their identification is uncertain.

1428 **G. elegans** Wallich ex D. Don (*G. lutea* auct. non Schultes f.)
Pakistan to C. Nepal. 2000–4300 m. Alpine slopes, shrubberies; locally abundant around herdsmens' camps, flowering before the herds arrive. Apr.–May.
A delicate bulbous plant, with grass-like leaves, and with small yellow star-shaped flowers which are greenish or reddish-brown on the outside of the petals. Flowers long-stalked; petals oblanceolate acute, 1.3–7 cm,

spreading; stamens shorter. Basal leaf linear to lanceolate, stem-leaves 2, bract-like, the lower broader but smaller than the basal leaf; stem 5–13 cm. Bulb 5–10 mm. **Pl. 125**.

12 HEMEROCALLIS Day Lily Rhizomatous perennials. Flowers funnel-shaped, usually yellowish, in a terminal shortly branched cluster. Petals fused below; stamens and style curved upwards. Capsule 3–seeded. 1 sp.

1429 **H. fulva** (L.) L.

Native of China; cultivated and naturalized in the Himalaya. 2400–3500 m. Common in the Kulu valley, and in some parts of W. & C. Nepal. May–Jun.

Flowers large funnel-shaped, pink to orange-yellow, several borne on stems which are shortly branched above. Flowers erect, with out-curved petals to 8 cm, with undulate margins, the inner petals distinctly broader than the outer; corolla-tube slender c. 3 cm; filaments of stamens yellow. Leaves mostly basal, narrow-linear to 2 cm broad, 30–60 cm long, keeled, somewhat glaucous; flowering stem 60–100 cm. **Pl. 122**.

13 LILIUM Lily Usually tall leafy plants. Flowers usually large funnel-shaped, solitary or in spike-like clusters. Bulbs with fleshy overlapping scales without an encircling tunic. 7 spp.

Flowers small, less than 6 cm long

1430 **L. nanum** Klotzsch (*Nomocharis n.* (Klotzsch) E. H. Wilson)
Himachal Pradesh to S.W. China. 3300–4300 m. Shrubberies, alpine slopes; quite common. Jun.–Jul.

A small plant with usually a solitary drooping dull-purple bell-shaped flower, with elliptic petals 1.8–2.5 cm long. Leaves 10–12, linear, the uppermost much overtopping the flower; stem usually 10–30 cm. Bulb with lanceolate scales. **Page 517**.

1431 **L. oxypetalum** (D. Don) Baker (*Fritillaria o.* D. Don, *Nomocharis o.* (Royle) E. H. Wilson)
Himachal Pradesh to W. Nepal. 3300–4000 m. Rocks, open slopes. Jun.–Jul.

Flowers pale yellow or cream, solitary, cup-shaped with petals spreading outwards from the base not curved, borne almost stalkless amongst the upper leaves. Petals ovate acute, 4–5 cm, with hairs above the nectaries; stigma 3–lobed. Leaves many, elliptic-lanceolate, 4–8 cm long; flowering stem stout, leafy, 30–45 cm. **Pl. 125**.

Flowers large, 6–25 cm long

1432 **L. nepalense** D. Don
Uttar Pradesh to Arunachal Pradesh. 2300–3500 m. Rocks, open slopes. Jun.–Jul.

Flowers predominantly yellow and usually with a large or small brown-purple zone within, with spreading petals which are recurved above the middle, and with protruding stamens with reddish anthers. Flowers to 15 cm long, sweet-scented, solitary or few; corolla-tube greenish on the outside. Leaves many, broadly lanceolate, to 8 cm; stem leafy, 60–100 cm. **Pl. 122.**

1433 **L. polyphyllum** D. Don
Afghanistan to Uttar Pradesh. 2100–3300 m. Forests, shrubberies. Jun.– Aug.

Flowers yellowish or greenish without, white and speckled with pink within, with petals strongly curved outwards from the middle, and with protruding red stamens. Flowers many, long-stalked, pendulous, sweet-scented, 6–8 cm long; petals narrow-elliptic. Leaves linear to oblanceolate, 10–13 cm; stem leafy to 1 m. **Page 517.**

The only species of the W. Asian group of lilies; the others in our area are related to the Chinese group.

1434 **L. wallichianum** Schultes & Schultes f.
Uttar Pradesh to Arunachal Pradesh. S. India. 1200–2000 m. Open slopes, grasslands. Jun.–Aug.

Flowers very large, white, very sweet-scented, usually solitary, narrow funnel-shaped with petals outcurved in the upper third only. Flowers 12–25 cm long; petals with a long narrow stalk and elliptic pointed blade; stamens shorter. Leaves linear, 15–30 cm by 3–6 mm broad; stem 1.5–2 m. **Pl. 122.**

1435 **L. sherriffiae** Stearn
E. Nepal. Bhutan. 2700–4000 m. Open slopes, alpine shrubberies, rocks. May–Jun.

A rare lily recorded only once in E. Nepal, with reddish-brown flowers with gold chequering inside. Flowers 1–2; petals c. 6 cm. Leaves linear pointed; stem 30–60 cm.

14 **LLOYDIA** Small slender bulbous plants, with solitary or few, white or yellow funnel-shaped flowers. Stigmas usually entire; capsule 3–ribbed. 4 spp.

Flowers white, marked with orange or brown

1436 **L. serotina** (L.) Reichb.
Pakistan to S.W. China. Europe. 3500–4800 m. Alpine slopes, rocks. Jun.–Aug.

A small delicate bulbous plant with solitary or few white flowers with blunt spreading petals, and with thread-like leaves. Petals elliptic, 8–17 mm, usually with a small transverse nectary-fold situated above the base; stamens shorter than petals. Leaves 5–15 cm; flowering stem slender, similar in length to basal leaves and with 1–2 much smaller stem-leaves; base of stem with old scaly leaf-bases; bulb ovoid, with papery tunic.

1437 **L. longiscapa** Hook.
Kashmir to S.W. China. 3600–4500 m. Alpine slopes. Jun.–Jul.
Similar to 1436 with white flowers with brown-purple or orange towards the base, but with the inner petals somewhat acute and generally hairy within towards the base. Filaments of stamens shaggy-haired at least below (hairless in 1436). **Page 517**.

Flowers yellow

1438 **L. flavonutans** Hara
C. Nepal to S.E. Tibet. 3600–4500 m. Open slopes. Jun.–Jul.
A delicate plant with a usually solitary nodding yellow flower with a reddish-orange patch at the base. Flower *c.* 2 cm long; petals elliptic blunt, with green veins, usually hairless. Plant to 15 cm.

15 NOTHOLIRION Like *Lilium*, but stigmas with 3 short narrow recurved lobes. Bulb with a dry brown outer tunic (not with exposed scales). Basal leaves longer than stem leaves. 3 spp.

1439 **N. thomsonianum** (Royle) Stapf
Afghanistan to Uttar Pradesh. 800–1800 m. Fields, rocky slopes. Mar.–Apr.
Flowers pale rose to rose-purple, sweet-scented, funnel-shaped, 5–6 cm long, many in a dense terminal spike-like cluster, borne on a stout stem 30–60 cm. Petals narrow-spathulate, outcurved at the apex only. Stem-leaves 7–13, linear 6–8 mm wide, narrowed to a very fine point. Bulbs with dark brown papery tunic. **Pl. 122**.

1440 **N. macrophyllum** (D. Don) Boiss.
W. Nepal to S.E. Tibet. 2800–4000 m. Shrubberies, open slopes. Jun.–Jul.
A much smaller plant than 1439 and found at higher altitudes, with usually 1–5 deeper purple funnel-shaped flowers, borne on a stem 30–60 cm. Flowers 3–5 cm long; petals oblanceolate blunt. Stem-leaves 3–7, linear. **Pl. 124**.

16 OPHIOPOGON Flowers in a terminal spike-like cluster, with spreading petals (thus differing from *Theropogon*); bracts papery. Ovary inferior; ovules 2 in each chamber. 3 spp.

1441 **O. intermedius** D. Don
Afghanistan to S.W. China. S.E. Asia. 1500–3000 m. Forests, shrubberies. May–Jul.
A small tufted plant of shady forest banks, with many grass-like leaves, and with spike-like clusters of small white drooping flowers borne on leafless stems. Spikes mostly 2–5 cm; flowers cup-shaped, with spreading petals 4–6 mm, short-stalked, and subtended by papery bracts longer than the flower-stalks. Leaves linear 2–5 mm broad, tufted; flowering stems 10–15 cm. Rootstock short, covered with old fibres. **Page 517**.

17 PARIS HERB PARIS   Similar to *Trillidium* but differing in having 4–9 leaves in a whorl; flowers with 8–12 perianth-segments in 2 differing whorls. 1 sp.

1442 **P. polyphylla** Smith
Pakistan to S.W. China. 2000–3000 m. Forests. Apr.–May.

Flower solitary, terminal, short-stalked, greenish and relatively inconspicuous, with 4–6 lanceolate long-pointed green leaf-like perianth-segments 5–10 cm, and with an inner whorl of thread-like yellow or purple segments, as long or shorter than the outer. Stamens 10, short; stigmas lobed. Leaves 4–9 in a whorl, elliptic short-stalked, to 10 cm; plant to 40 cm. Rhizome stout, creeping. Fruit globular; seeds scarlet. Plant shows wide variation in different parts of its range. **Pl. 124**.

18 POLYGONATUM SOLOMON'S SEAL   Flowers borne in axils of leaves; corolla tubular and with 6 lobes; anthers included. Fruit a berry. Rootstock creeping. 9 spp.

Flowers mauve or pinkish, solitary or paired in leaf-axils

1443 **P. hookeri** Baker
Uttar Pradesh to S.W. China. 3300–5000 m. Shrubberies, open slopes. May–Jun.

A dwarf plant only 2.5–5 cm high, with usually solitary, or paired, nearly erect pink flowers arising from the axils of the lower leaves. Corolla-tube slender, about as long as the out-curved narrow-elliptic lobes, which spread 1.3–2 cm across. Leaves opposite, linear blunt, 2–9 cm long by up to 5 mm broad, glaucous beneath; stem with papery scales below. Rhizome creeping, branched. **Pl. 123**.

Flowers white, two or more in axils of upper leaves
a *Leaves whorled*

1444 **P. verticillatum** (L.) All. WHORLED SOLOMON'S SEAL
Pakistan to S.E. Tibet. W. Asia. Europe. 1500–3700 m. Forests, shrubberies, open slopes. May–Jul.

An erect rather robust plant, with many whorls of narrow lanceolate leaves, bearing in their axils branched clusters of 2–3 small pendulous tubular white flowers with green tips. Flowers 8–12 mm long, fused into a broad tube below and with short triangular spreading lobes. Leaves in whorls of 4–8, linear or lanceolate, 9–20 cm; stem angled and grooved, 60–120 cm. Rootstock thick, creeping. Fruit a berry, at first bright red, becoming dark purple.

Shoots eaten when young.

1445 **P. cirrhifolium** (Wallich) Royle
Himachal Pradesh to S.W. China. 1500–3700 m. Forests, shrubberies, open slopes. May–Jul.

Differs from 1444 in that it usually has coiled tendril-like tips to the leaves. Flowers white tinged with purple or green, in short-stalked some-times paired clusters of 2–4 from the axils of the whorled leaves. Leaves in whorls of 3–6, linear to narrow-lanceolate, usually 6–15 cm, with margins often inrolled; stems weak, 60–120 cm. Rhizome stout, creeping. **Pl. 123**.

b *Leaves alternate*

1446 **P. multiflorum** (L.) All.
Pakistan to Uttar Pradesh. Temperate Eurasia. 1500–2700 m. Forests; quite common. May–Jun.

Distinguished by its large alternate elliptic leaves from the axils of which are borne a stalked pendulous cluster of 2–4 tubular white flowers. Corolla to 2 cm long, the tube much longer than the short out-curved lobes. Leaves short-stalked, 8–20 cm; stem slender *c.* 30 cm. **Pl. 123**.

19 SMILACINA Flowers small, with spreading petals, in terminal spike-like or branched clusters. Leaves alternate; stem simple, leafy. Fruit a berry. 3 spp.

1447 **S. purpurea** Wallich (*S. pallida* auct. non Royle)
Himachal Pradesh to S.W. China. 2400–4200 m. Forests. Apr.–Jun.

A leafy erect plant, with alternate broadly elliptic leaves in two ranks and with a terminal often pendulous narrow spike-like cluster of small dark purple, or less commonly white flowers. Petals oblong or elliptic 4–6 mm, stamens shorter; inflorescence sometimes with one or two branches below. Leaves 3–9, usually 5–12 cm, distinctly ciliate on margin, clasping stem at the base; stem often hairy, unbranched, 20–40 cm. Rhizome creeping.

Young plants make good boiled vegetables. **Pl. 123**.

1448 **S. oleracea** (Baker) Hook. f.
E. Nepal to S.W. China. 2400–3600 m. Forests. May–Jun.

Distinguished from 1447 by its broad spreading branched terminal inflorescence composed of many spike-like clusters of pure white or pinkish-purple flowers. Flowers to 1 cm across; petals ovate 5–7 mm. Leaves 7–13, elliptic long-pointed, short-stalked, to 25 cm; stem slender, to 1 m. **Pl. 123**.

20 SMILAX Usually woody climbers, with tendrils developed from the stipules, often spiny. Flowers inconspicuous, in axillary umbels. Fruit a berry. 15 spp.

1449 **S. aspera** L.
Pakistan to Bhutan. India. Europe. E. Africa. 1200–2500 m. Climbing in shrubberies. Sep.–Nov.

A tendril-climber with flexuous usually prickly stems, linear-lanceolate to rounded leaves, and with axillary clusters of small fragrant white flowers. Umbels numerous, forming long spike-like clusters 2.5–15 cm;

flowers one-sexed, *c.* 5 mm across; petals linear. Leaves glossy, very variable, to 10 cm or more, margins with or without prickly teeth, and leaf-bases rounded, heart-shaped, or lobed, usually with prickly leaf-stalks; tendrils paired, arising from base of the leaf-stalk. Fruit a red berry 6–8 mm, turning blue-black when ripe.

Roots used medicinally.

1450 **S. ferox** Wallich ex Kunth
C. Nepal to S.W. China. 1100–2500 m. Climber in forests and shrubberies. Apr.–May.
Differs from 1449 in having greenish flowers in umbels which are borne on the young shoots only, and leaf-stalks which are swollen at the base and encircle the stem and become hard and persistent after the blade has fallen. Tendrils few or absent. **Page 516.**

21 STREPTOPUS Flowers bell-shaped, axillary. Leaves alternate, stalkless and clasping. Fruit a berry. Rootstock creeping. 2 spp.

1451 **S. simplex** D. Don
Uttar Pradesh to S.W. China. 2500–3700 m. Forests, shrubberies. Jun.–Aug.
An erect plant with flexuous forked leafy stems, and with small white solitary, pendulous, long-stalked, bell-shaped flowers arising from the axils of most of the leaves. Flowers to 2.5 cm across; petals elliptic acute; stamens much shorter than petals. Leaves elliptic acute with heart-shaped clasping base, 5–12 cm, glaucous beneath; stem 60–120 cm. Fruit a berry, orange when ripe. **Pl. 122.**

22 THEROPOGON Flowers in a terminal spike-like cluster, nodding, borne on a leafless stem; corolla globular-bell-shaped; petals free. Fruit a berry. Rootstock branched. 1 sp.

1452 **T. pallidus** (Kunth) Maxim.
Uttar Pradesh to S.W. China, 1800–2700 m. Rocks, open slopes. Jun.–Aug.
A tufted plant, often forming clumps, with grass-like leaves, and with erect leafless stems bearing spike-like clusters of many stalked drooping white or pink flowers which are borne from the axils of shorter or longer bracts. Resembling *Ophiopogon*, but flowers rounded bell-shaped with erect petals (not spreading in a star), 7 mm long, in clusters 5–8 cm long. Leaves linear, 15–25 cm by 2–6 mm broad, outcurved, sheathing at the base; flowering stems mostly shorter than the leaves; roots tuberous. Fruit green, mottled. **Pl. 123.**

23 TRILLIDIUM Flower solitary; perianth with 6 persistent spreading petals. Leaves 3, short-stalked. Fruit a berry. 1 sp.

1453 **T. govanianum** (D. Don) Kunth (*Trillium g.* Wallich ex D. Don)

Pakistan to Bhutan. 2700–4000 m. Forests. May–Jun.

Resembling *Paris* with its single central rather inconspicuous flower borne at the apex of the stem and surrounded by leaves, but leaves 3, broadly ovate acute and conspicuously stalked. Flower brown-purple, with narrow spreading petals, the outer 3 narrowly lanceolate, the inner 3 linear; anthers large, yellow; styles 3, long and conspicuous; ovary purple-brown. Leaf-blade 3.5–10 cm, leaf-stalk 3–16 mm; stem to *c*. 30 cm. Fruit a globular red berry, 1–2 cm. **Pl. 125**.

24 TULIPA TULIP Flowers large, erect, solitary, at first bell-shaped, petals later spreading; stem leafy only at base. Fruit a capsule. Bulb coated with thin papery tunic. 1 sp.

1454 **T. stellata** Hook.f.
Pakistan to Uttar Pradesh. 1500–3300 m. A weed of cornfields, rocky slopes. Apr.–May.

Flowers solitary, white with a broad red band on the outer side of the outer petals, or flowers yellow and red, borne on a long stem with 4–6 leaves towards the base. Flowers erect, at first bell-shaped, but petals later spreading widely; petals elliptic pointed, 3–5 cm long. Leaves linear acute, 23–30 cm, channelled, sheathing below; flowering stem 15–40 cm. Bulb-tunic with woolly hairs on the inner side. The form with yellow and red flowers, var. **chrysantha** Boiss., grows at higher altitudes on rocky slopes. **Pl. 124**.

COMMELINACEAE Spiderwort Family

A moderate-sized family of annuals and perennials of the Tropical and Subtropical Zones, with rather fleshy stems with swollen nodes, and with alternate entire leaves with basal sheaths encircling the stem. Flowers with 3 green sepals, alternating with 3 often blue petals. Stamens 6, all fertile, or 3 sterile. Ovary superior, 3–chambered; style 1 with stigma entire or branched. Fruit a dry splitting capsule.

COMMELINA Fertile stamens 3, sterile stamens 2–3. Petals unequal, free below. 6 spp.

1455 **C. paludosa** Blume
Pakistan to S.W. China. India. Burma. S.E. Asia. 300–3500 m. Waste ground, edges of cultivation, shrubberies. Jun.–Aug.

An erect or ascending herbaceous plant, with broad lanceolate leaves with sheathing bases, and with flowers with 3 conspicuous pale blue to whitish petals, the flowers borne in the axils of obliquely funnel-shaped leafy bracts. Flowers *c*. 8 mm across; petals ovate, stalked; filaments of stamens hairless; bracts 2–3.5 cm. Leaves usually 5–10 cm, with basal

sheaths fringed with hairs; stem sparsely branched, 20–60 cm, rooting at the lower nodes.
Used in the past medicinally. **Page 518**.

CYANOTIS All 6 stamens fertile. Petals unequal, united below into a short tube. 2 spp.

1456 **C. vaga** (Lour.) Schultes & Schultes f. (*C. barbata* D. Don)
Pakistan to S.W. China. India. Burma. S.E. Asia. Africa. 800–2700 m. Grasslands, cultivated areas. Jul.–Sep.

A slender sparsely-branched plant, with narrow-lanceolate leaves often with long woolly hairs on the leaf-sheaths, and with small dark blue flowers subtended by long straight or curved bracts. Flowers *c.* 1 cm across, in terminal or axillary clusters; filaments of stamens with long blue or white hairs; bracts hairy, with enlarged rounded bases. Leaves 3–7.5 cm, variably hairy; stem usually 8–30 cm, often tufted and rooting at the nodes below. **Page 518**.

JUNCACEAE Rush Family

A rather small family of the cooler temperate and montane regions, of tufted grass-like plants with small inconspicuous flowers which are wind-pollinated. Flowers usually in a dense head, or in a branched cluster, regular, with 6 papery scale-like green, brown, or whitish petals (perianth-segments). Stamens 6; ovary 3–celled, many-seeded; styles 1 or 3, stigmas feathery. Fruit a capsule.

JUNCUS RUSH Stem-leaves sheathing at the base, usually cylindrical in section. *c.* 32 spp., many of which are quite common in the temperate and alpine regions of the Himalaya.

Flowers in a solitary head

1457 **J. thomsonii** Buchenau (*J. leucomelas* auct. non Royle)
Pakistan to S.W. China. C. & N. Asia. 3000–5200 m. Open slopes, damp ground; a common alpine species. Jun.–Sep.

Flowers in solitary dense terminal heads, usually brown or sometimes whitish, borne at the ends of hollow stems, and subtended by brown bracts, the upper bracts pale and papery. Petals *c.* 4 mm, the inner petals with papery margins, stamens much longer, conspicuous. Stems tufted, 5–15 cm, each with 1–3 short thread-like leaves. Capsule 1–celled, glossy, with a conspicuous beak. **Pl. 126**.

1458 **J. leucanthus** Royle ex D. Don.
Uttar Pradesh to Bhutan. 3000–5000 m. Open slopes, damp places; quite common. Jul.–Sep.

Very like 1457 with solitary terminal pale yellow or white heads of flowers 6–10 mm across, and with short brown bracts, the lowermost equalling the flowers. Differing in having one or more very slender leaves borne above the middle of the stem (all leaves near the base and much shorter in 1457), and capsule 3–celled. Petals papery, *c.* 6 mm; stamens longer. Stems tufted, with brown basal sheaths, 6–23 cm.

Flowers in several heads or clusters on each stem

1459 **J. himalensis** Klotzsch
Pakistan to S.E. Tibet. 3000–5000 m. Forests, open slopes, marshes; common. Jun.–Aug.

Flowers in two or more dense dark brown clusters subtended by longer leafy bracts. Inflorescence irregularly branched; bracts surrounding flower-clusters papery; outer petals 9 mm long, blunt, inner petals with papery margins, stamens shorter; styles exserted. Bracts broad at base, long-pointed. Leaves cylindrical, hollow, channelled above, shorter than the stem which is stout rigid, leafy below, 20–50 cm. Capsule dark shining brown, much longer than persistent petals. **Page 518**.

PALMAE   Palm Family

A very important tropical family occupying all types of habitat from mangrove swamps, tropical rain-forests, to mountains. Trees, shrubs, or woody climbers, usually with unbranched trunks, and with massive feather-like leaves in a terminal crown. Leaves palmate or pinnate, rarely simple. Flowers numerous, in axillary branched clusters arising from the crown of leaves, at first enclosed in one or more large bracts or *spathes*. Flowers usually bi-sexual; sepals 3; petals 3; stamens 6. Ovary usually one. Fruit a berry, drupe, or nut.

PHOENIX FEATHER PALM Leaves lanceolate in outline, pinnately divided into many pairs of leaflets. Spathe 1. 3 spp.

1460 **P. sylvestris** Roxb. WILD DATE PALM
Himachal Pradesh to W. Nepal. India. Burma. 150–1500 m. Open hills, rocky slopes; prominent in the Beas gorge on road to Kulu, Himachal Pradesh. May–Jun.

Trunk stout 7–13 m, covered with persistent leaf-bases and topped by a crown of large feathery pinnate leaves each 3–5 m long. Flower-clusters inclined, and with spreading branches; male clusters white, scented, 30–60 cm; female flowers terminating the branches, in large bunches or spikes; spathe 30–40 cm. Leaflets 30–60 cm; leaf-stalk spiny. Fruit 2.5–3 cm, yellowish to reddish-brown.

Mats and baskets are made from the leaves. A sugary juice, rich in vitamins, is obtained from the trunk and is used as a beverage, or for the manufacture of sugar.

TRACHYCARPUS FAN PALM Leaves rounded or kidney-shaped in outline and palmately divided into narrow lobes. Spathes many. 2 spp.

1461 **T. takil** Becc.
Uttar Pradesh to W. Nepal. 1500–2400 m. Oak forests; quite common; possibly cultivated elsewhere in the hills.

Trunk 3–7 m, with a crown of leaves with rounded blades *c.* 1 m across, cut to about half-way into 30–40 linear lobes, and with a spiny leaf-stalk about as long as the blade. Flowering stems much-branched, shorter than the leaf-stalks, recurved in fruit, brown hairy; petals ovate, whitish. Fruit yellow, then dark glossy-blue, kidney-shaped, *c.* 1.3 cm across. Young trunks covered with old leaf-bases and brown fibres; old trunks smooth, ringed.

The fan-leaved palm prominent on precipitous rocks in the Marsiandi gorge, C. Nepal is probably **T. martianus** (Wallich) Wendl.

PANDANACEAE   Screw-Pine Family

A large tropical family of trees, shrubs and climbers. Stems tall slender, branched, and ringed with leaf-scars, and with terminal crowns of long narrow stiff sword-like leaves arising spirally on the stem and arranged in three ranks. Aerial roots supporting the stems are often present. Flowers one-sexed, on separate plants; sepals and petals absent; male flowers with numerous stamens; female flowers with many carpels in a ring. Fruit cone-like, of many berries, or drupes.

PANDANUS 1 sp.

1462 **P. nepalensis** H. St. John (*P. furcatus* auct. non. Roxb.)
C. & E. Nepal eastwards. To 1000 m. Damp shady places; common.

Stems 3–17 m and up to 15 cm in diameter, sparingly branched above, with branched aerial roots arising from the lower trunk. Leaves dark green, entire, 3–5 m by 10 cm broad, with stout curved spines on the margin and midrib. Spathes several, leathery, golden-yellow, the lowest up to 1 m long. Fruit of numerous drupes, cone-like, orange-red when ripe, 15–25 cm long; drupes fleshy, obconical 5–6–angled. **Pl. 126.**

ARACEAE   Arum Family

A large family of pan-tropical herbaceous plants with a few temperate species, with aerial stems or with underground tubers or rhizomes, and with a very distinctive unique, often foetid inflorescence which attracts flies and other insects. Inflorescence comprising a large *spathe* which is often conspicuous and coloured, subtending, or encircling and often fused below round a fleshy central axis, the *spadix*, which bears numerous tiny inconspicuous flowers, either covering the spadix, or borne only round the basal part. The sterile apical part of the spadix, the *appendage*, is often well developed. Flowers bisexual, or unisexual in which case the male flowers are usually borne in a zone above the female flowers, facilitating a unique method of pollination. Perianth-segments 4–6; stamens 1–6, often fused together. Ovary superior, or imbedded in the spadix, of 1 to many carpels. Fruit a cluster of berries.

Spadix with a flowerless apex – the appendage
  a Male and female flowers on different plants   1 ARISAEMA
  b Male and female flowers on same spadix
    i Base of spathe fused below into a tube   2 SAUROMATUM
    ii Base of spathe not fused below   3 TYPHONIUM
                          4 THOMSONIA 5 COLOCASIA

Spadix covered with flowers to the apex, appendage absent
  a Leaves undivided, peltate; not climbing   5 COLOCASIA
                    6 GONATANTHUS 7 REMUSATIA
  b Leaves pinnately lobed; large climbers   8 RHAPHIDOPHORA

1 ARISAEMA Male and female flowers borne on separate plants. Leaves either with 3 leaflets, *trifoliate*; or with 5 or more equal leaflets spreading finger-like from the tip of the leaf-stalk, with mid-veins all meeting at a common point, *digitate*; or with 5 or more leaflets with the two outer smaller leaflets, with their mid-veins not meeting at a common point, *pedate*; or with 7–11 equal leaflets spreading or radiating round the apex of the leaf-stalk, *radiate*. Fruit a cylindrical cluster of red berries.

    The tuberous roots of many species can be ground into a flour and eaten. Care must be taken as the tubers contain minute sharp particles which can damage the digestive tract. 17 spp.

Leaves trifoliate, usually solitary; spadix with a whip-like appendage

  a *Spathe conspicuously striped, or netted with green or brown*

1463 **A. costatum** (Wallich) Martius ex Schott
Endemic to C. & E. Nepal. 2000–2600 m. Shrubberies. May–Jun.
    Spathe dark purple with longitudinal white stripes, 8–12 cm long, the blade down-curved and with a tail-like tip 1–4 cm long. Spadix appendage very long, 15–45 cm. Distinguished by its leaflets which have very numerous parallel lateral veins which are conspicuously raised beneath; leaflets 3, elliptic to ovate 10–20 cm, the outer leaflets unequal with an angled base on the outer side and a narrow wedge-shaped base on the inner side; flowering stem greenish, shorter than the leaves, to 40 cm. **Pl. 127**.

1464 **A. propinquum** Schott (*A. wallichianum* Hook.f.)
Kashmir to S.E. Tibet. 2400–3600 m. Forests, shrubberies, open slopes. May–Jun.
    Spathe 10–15 cm long, dark purple or green, with white or purple

stripes and longitudinally ribbed inside, often netted with pale veins towards the apex, and with an oblong-ovate blade narrowed to a tail-like tip 1–4 cm long. Spadix appendage long, thread-like, 8–20 cm, somewhat thicker at its base. Leaflets 3, rhombic-ovate, 8–20 cm by 4–15 cm wide; leaf-stalk often brown-spotted 15–70 cm. **Pl. 127**.

1465 **A. utile** Hook. f. ex Schott
Himachal Pradesh to Bhutan. 2400–4300 m. Forests, open slopes. May–Jul.

Spathe dark purple with whitish stripes, 10–15 cm long, but differing from 1464 in having a broadly obovate blade which is rounded or notched at the apex and with a short tail-like tip 1–3 cm long, and margin of blade conspicuously netted with transparent veins. Spadix with a thread-like appendage about as long as the spathe. Leaflets 3, rhombic-ovate, nearly equal, to 25 cm, reddish near margin; leaf-stalk 20–50 cm, dark-spotted.

1466 **A. speciosum** (Wallich) Martius ex Schott
C. Nepal to S.W. China. 1800–2800 m. Forests, shrubberies. May–Jun.

Spathe to 20 cm, dark black-purple and longitudinally striped towards the base with white, with a broadly lanceolate curved blade narrowed to the tip. Spadix appendage dark purple, thickened curved and white at the base, and with very long thread-like tail 20–80 cm. Distinguished also by its stout elongated horizontal rhizome, and its distinctly stalked unequal elliptic leaflets to 30 cm, with red margins, and its marbled dark purple leaf-stalk to 40 cm, or more.

1467 **A. griffithii** Schott
C. Nepal to Bhutan. 2400–3000 m. Forests. May–Jun.

A very striking aroid with a very large broad spathe 10–20 cm long which is curved back on itself, and with large rounded ear-like flaps 10–15 cm across which are conspicuously netted with green over dark purple, and with a slender purple apex. Tube of spathe pale-ribbed, to 8 cm. Spadix abruptly narrowed to a purple tail-like appendage 20–80 cm long. Leaves usually 2, each with 3 large rhombic-ovate leaflets 10–40 cm; leaf-stalk stout, 15–60 cm, green or purple-spotted. Basal bracts very large. **Pl. 126**.

b *Spathe not conspicuously striped, or netted*

1468 **A. intermedium** Blume
Kashmir to Sikkim. 2100–3000 m. Forests. May–Jun.

Spathe pale yellowish-green, rarely striped dark purple, with an ovate-lanceolate blade narrowed to a short tail-like tip 2–3 cm. Spadix appendage very long, slender, 15–45 cm, curved upwards from its broader purple base then hanging downwards and white. Leaves 1–2; leaflets 3, ovate long-pointed, the 2 lateral leaflets very unequal-sided, minutely net-veined beneath, 9–20 cm; leaf-stalk slender 15–50 cm; flowering stem much shorter than leaf-stalk. **Pl. 127**.

Leaves with 5–20 leaflets; spadix usually without long whip-like appendage (except 1470)

a *Leaves digitate or pedate, with 5 or more leaflets*
 i *Spathe small 2–4 cm; yellowish*

1469 **A. flavum** (Forsskal) Schott
Afghanistan to S.W. China. 1800–4500 m. Open forests, stony slopes; in Nepal only in drier areas. May–Jun.

Distinguished from all other species by its very small spathe which has a yellowish-green ovoid tube, and a small yellowish or greenish triangular blade 1.5–4 cm, which is usually dark purple inside at least in the lower part. Spadix appendage very short, ellipsoid, greenish or yellowish. Leaves pedate, with 5–11 oblong-lanceolate pointed leaflets 2.5–12 cm long; flowering stem 10–40 cm. **Pl. 128.**

 ii *Spathe more than 6 cm, not yellowish*
 x *Spadix appendage long slender, curved upwards*

1470 **A. tortuosum** (Wallich) Schott
Kashmir to S.W. China. 1500–3000 m. Forests, shrubberies, open slopes. May–Jun.

Spathe green, somewhat glaucous, rarely purple, blade ovate acute 4–12 cm, curved forward. Spadix distinctive with a rather thick up-curved appendage gradually tapering to a long green or sometimes dark purple erect tail-like tip 8–12 cm. A tall plant, often to 150 cm, with usually 2 pedate leaves with 5–7 variable, broadly ovate to oblanceolate abruptly-pointed leaflets, and a leaf-stalk 5–20 cm. **Pl. 128.**

 xx *Spadix appendage short*

1471 **A. jacquemontii** Blume
Afghanistan to S.E. Tibet. 2400–4000 m. Shrubberies, rocky slopes; in Nepal in drier areas only; common in upper forest and lower alpine zones. Jun.–Aug.

Distinguished from 1470 by its green sometimes white-striped spathe which has a long up-curved green or dark purple tail-like tip, while in contrast the appendage is short-cylindrical, dark purple, and projects forwards only a short distance from the mouth of the spathe-tube. Spathe *c.* 15 cm. A smaller plant 10–70 m, usually with 1 leaf which is digitate with 5–9 narrow-elliptic to ovate long-pointed leaflets. **Pl. 127.**

1472 **A. nepenthoides** (Wallich) Martius ex Schott COBRA PLANT
C. Nepal to S.W. China. Burma. 2000–3300 m. Forests, shrubberies. May–Jun.

Quite distinctive with its mottled greenish-brown to reddish-brown spathe, resembling a cobra about to strike, with white stripes on the back, and with a short fat thick pale green spadix scarcely longer than the spathe-tube. Spathe triangular-ovate, curved forward, and with conspicuous rounded spreading lobes at the base of the blade at its junction with the spathe-tube. Leaves usually 2, digitate, with 5 thick glossy narrow-elliptic leaflets 6–12 cm; leaf-stalk and stem spotted with greenish- to reddish-brown like the inflorescence. **Pl. 128**.

b *Leaves radiate, with 7–20 nearly equal spreading leaflets*

 i *Leaflets 2–5 cm broad*

1473 **A. concinnum** Schott
Himachal Pradesh to S.E. Tibet. Burma. 1700–2500 m. Forests, shrubberies, open slopes. May–Jun.

Spathe green and longitudinally striped with white, ovate-lanceolate, and with a long green or dark purple, down-curved tail-like tip 2–7 cm; spadix slender, green or purple, slightly longer than the cylindrical spathe-tube. Distinguished by its 7–11 rather broad, oblanceolate long-pointed more or less equal radiating leaflets usually 2–5 cm broad, borne on a stout greenish leaf-stalk 30–50 cm; flowering stem green, shorter than the leaf-stalk.

 ii *Leaflets 1.5 cm or less broad*

1474 **A. erubescens** (Wallich) Schott
C. Nepal to Sikkim. 2000–2600 m. Forests, shrubberies. May–Jun.

Spathe reddish-brown, pink and white, or violet and white, with white stripes inside in the lower part, with an ovate long-pointed down-curved blade. Spadix blunt, slightly longer than the cylindric spathe-tube, brownish to greenish. Leaf 1, with 7–14 radiating narrow-lanceolate long-pointed leaflets which are dark green above and somewhat glaucous beneath; leaf-stalk variegated with brown, 12–30 cm; flowering stem usually shorter than leaf-stalk. **Pl. 126**.

1475 **A. consanguineum** Schott
Uttar Pradesh to S.W. China. 1800–3000 m. Forest, shrubberies. May–Jun.

Spathe green, with green-striped oblong-ovate curved blade, to 12 cm, with a long thread-like apical tail 5–15 cm; spadix with a stout green cylindrical appendage little longer than the spathe-tube. Readily distinguished by its numerous (11–20) radiating, narrow linear-lanceolate leaflets with long thread-like tips, borne on a dark variegated leaf-stalk. Fruiting stalk recurved.

2 SAUROMATUM Spathe with a short cylindrical tube and long reflexed blade; spadix with very long appendage; male and female flowers in widely separated zones. Leaf 1, pedate. 2 spp.

1476 **S. venosum** (Aiton) Kunth (*S. guttatum* Schott)
Himachal Pradesh to C. Nepal. S.E. Tibet. India. 1000–2300 m. Stony slopes, hot river valleys. Mar.–May.

Inflorescence short-stemmed, foetid, with a large lanceolate conspicuously dark purple-blotched spathe, and a greenish-yellow to dark purple, erect fleshy tapering spadix not quite as long. Spathe 32–50 cm, including the spathe-tube which is 8–10 cm, spathe-blade narrowed to a long slender curved tip, greenish-purple outside, strongly blotched within. Leaf solitary, appearing in the autumn after flowering, pedate with 9–11 large unequal lanceolate leaflets; leaf-stalk mottled with black, to 2 m. Corm large, flattened globular, *c*. 8 cm across. **Pl. 128**.

3 TYPHONIUM Spathe with a short tube and long blade which is deciduous; spadix with elongate smooth appendage; male and female flowers in separate zones. Leaves variable. 2 spp.

1477 **T. diversifolium** Wallich ex Schott
Himachal Pradesh to S.E. Tibet. 2500–4300 m. Shrubberies, open slopes. Jun.–Jul.

As its name implies, the leaves are very variable, being either entire, heart-shaped, arrow-shaped, spear-shaped, or with 5–7 lanceolate to linear lobes. Spathe green outside, variously striped, spotted or netted with dark reddish-purple within, variable in size 5–20 cm, with a short green tube and lanceolate finely-pointed blade. Spadix ending in a blunt cylindrical dark purple appendage to *c*. 8 cm long, which protrudes from, or is enclosed in the longer spathe. Leaves usually solitary, appearing with the inflorescence, 8–15 cm broad and long. Corm to 2 cm in diameter. **Pl. 126**.

4 THOMSONIA Spathe large; spadix stout exposed, apex covered with swellings. Leaf appearing after flowering, with many leaflets; tuber large. 1 sp.

1478 **T. napalensis** Wallich
C. Nepal to Sikkim. 600–1800 m. Forests, shrubberies. Apr.–May.
Inflorescence appearing before the leaves, with a strong and nasty smell. Spathe large, green, or pale yellow to brown, oblong-boat-shaped 30–46 cm; spadix exposed, very stout, with a rounded apex covered with swellings, green changing to yellow, 13–25 cm long by 2–3 cm broad. Male flowers in a zone 5–13 cm wide; female flowers below, in a zone 1–2.5 cm wide; appendage 8–10 cm. Leaf-blade large 30–40 cm across, three times divided into ovate to oblong-lanceolate long-pointed leaflets, 8–13 cm; leaf-stalk very stout, to 40 cm. Tuber 10–13 cm across. **Pl. 128**.

5 COLOCASIA Tall stout plants with stout-stalked, peltate leaves. Spathe with a thick persistent tube and erect deciduous blade; spadix shorter, appendage variable or absent. Leaves and inflorescence appearing together. 3 spp.

1479 **C. esculenta** (L.) Schott (*C. antiquorum* Schott) ELEPHANT'S EAR, TARO, DASHEEN

Semi-naturalized in places in Nepal; widely cultivated in tropical countries. To 1500 m. Aug.–Sep.

A tall coarse herbaceous plant with long-stalked, dark green leaves with ovate-triangular blades to 66 cm across, attached in the middle to the leaf-stalk (peltate). Leaf-stalks 90–120 cm, stout, green or violet. Inflorescence short-stemmed, solitary or clustered; spathe pale yellow, 20–40 cm, with a thick persistent tube constricted at the mouth, and later enclosing the fruits, and with a deciduous heart-shaped pointed blade; spadix shorter, club-shaped. Rootstock tuberous.

The tubers and leaf-stalks are edible. Extracted juice is used medicinally.

6 **GONATANTHUS** Spathe with an ovoid tube which is constricted above and abruptly bent at the junction with the conical blade; spadix included. Habit and foliage like *Remusatia*. 1 sp.

1480 **G. pumilus** (D. Don) Engler & Krause
Himachal Pradesh to S.W. China. 1200–2500 m. Rocks, shady banks; sometimes epiphytic. Jun.–Aug.

Spathe 15–25 cm, with slender conical golden-yellow blade and with a green inflated tube 2 cm, which gapes above to expose the tip of the purple spadix. Leaves 1–3, with ovate-oblong heart-shaped often mottled blade 8–15 cm by 5–10 cm broad, attached to the stalk in the middle (peltate); leaf-stalk 10–30 cm. Fruit a head of small yellow berries enclosed in the persisting spathe-tube. Rootstock tuberous, bearing slender branches with small bulbils with long slender curved apices to the scales. **Page 518.**

7 **REMUSATIA** Spathe with ovoid tube, not constricted or bent, and with broad or narrow, erect or spreading and reflexed blade; spadix short, included. Leaf solitary, peltate. 2 spp.

1481 **R. hookerana** Schott
Himachal Pradesh to Sikkim. 1200–2500 m. Forests, mossy boulders; sometimes epiphytic. May–Jun.

Inflorescence fragrant; spathe with an erect yellow ovate-lanceolate blade and green ovoid tube, and with the spadix included within the spathe. Spathe 4–6 cm, blade with margins overlapping at apex, often twisted and gaping below and exposing the spadix which has male flowers to the tip. Leaf solitary, appearing with or before the inflorescence; blade ovate-heart-shaped, to 15 cm by 10 cm broad, the upper side often variegated pale and dark green; leaf-stalk *c.* 13 cm, attached to the middle of the blade. Tuber *c.* 2.5 cm across, bearing extensively branched, spreading or pendulous shoots bearing bulbils with long curled hairs. **Pl. 126.**

1482 **R. vivipara** (Roxb.) Schott
Uttar Pradesh to Bhutan. India. Burma. S.E. Asia. 1000–2000 m. Forests;

sometimes epiphytic on trees, and growing on rocks. Apr.

Differing from 1481 in having a larger broader reflexed spathe-blade 8–13 cm, with a cylindrical tube and with the tip of the spadix exposed. Inflorescence generally appearing before the leaf, which has a rounded or heart-shaped blade 13–30 cm long, and a leaf-stalk to 30 cm. Long erect or ascending very distinctive bulbil-bearing shoots to 35 cm, arise direct from the tuber; bulbils ovoid, with hooked spines. Inflorescence with large sheathing bracts below. Tuber to 5 cm across. Fruit bright red.

8 RHAPHIDOPHORA Stout climbers with pinnately-lobed leaves. 2 spp.

1483 **R. glauca** (Wallich) Schott
C. Nepal to Bhutan. 600–2100 m. Forests. Aug.–Oct.

A large climber with thick green stems climbing up and rooting on the trunks and branches of trees, and with 2 ranks of large dark green, pinnately-lobed leaves. Leaf-blade broadly ovate 15–30 cm, divided almost to the base into 2–4 pairs of irregular oblong to elliptic long-pointed lobes, glaucous-blue beneath; leaf-stalk 15–25 cm. Spathe variable in size 8–15 cm, yellow; spadix pale yellow to 6 cm, without a sterile appendage. Fruit white.

ALISMATACEAE   Water Plantain Family

A small family of aquatic and marsh plants widely distributed in the world, with narrow underwater leaves and robust stalked aerial leaves with rounded or narrow blades. Flowers in branched inflorescences. Sepals 3, green; petals 3, coloured; stamens usually 6; ovary superior, carpels many. Fruit a cluster of 1–seeded carpels.

ALISMA Flowers bisexual; stamens 6. Carpels numerous, in a single whorl. 2 spp.

1484 **A. lanceolatum** With. NARROW-LEAVED WATER PLANTAIN
Kashmir. W. Asia. Europe. N. Africa. 1400 m. Shallow water-channels; common in Kashmir valley. Jun.–Jul.

Flowers white to pale lilac, in a widely branched lax pyramidal cluster, borne above water on erect leafless stems 20–100 cm. Flowers c. 1 cm across; sepals oval; petals acute. Leaves arising from the underwater rhizome, borne above water, with lanceolate blades 6–15 cm, narrowed gradually into long stalks. Fruit of numerous carpels in one whorl.

SAGITTARIA Flowers mostly one-sexed; stamens usually numerous. Carpels numerous, spirally arranged. 2 spp.

1485 **S. trifolia** L. (*S. sagittifolia* auct. non L.) ARROWHEAD
Kashmir. Temperate and sub-tropical regions of the Northern Hemis-

phere. 1400 m. Shallow water-channels; common in Kashmir valley. Jun.–Oct.

Flowers conspicuous, with white petals and yellow anthers, c. 2 cm across, short-stalked, borne in whorls of 3–5, on erect leafless stems carried above water 30–90 cm. Leaves all basal, arising from the underwater rhizome, the aerial leaves with arrow-shaped blades 8–25 cm with long basal lobes, the floating leaves with oval-lanceolate blades, and the submerged leaves strap-shaped. Fruiting head globular, to 1.5 cm.

BUTOMACEAE   Flowering Rush Family

A family of one genus and one species (according to some authorities), of Temperate Eurasia.

BUTOMUS 1 sp.

1486 **B. umbellatus** L. FLOWERING RUSH
Kashmir. Temperate Asia. Europe. Naturalized in N. America. 1400 m. Shallow water-channels; common in Kashmir valley. Jun.–Jul.

Quite distinctive with its large conspicuous terminal umbel-like cluster of numerous pink flowers borne above water on a leafless stem, and its long triangular-sectioned twisted rush-like leaves 2–5 mm wide, borne above water. Flowers 2–2.5 cm across, long-stalked; petals 6, ovate, persistent; stamens 9, conspicuous; bracts triangular, papery. Flowering stem 50–100 cm above water, usually longer than the leaves. Carpels pink, 6, fused by their bases, in one whorl.

GRAMINEAE   Grass Family

One of the most important families in the world, though not the largest; cosmopolitan and dominating large areas. Structurally it shows many distinctive vegetative features such as fibrous roots, hollow stems with conspicuous nodes, and narrow entire leaves with an encircling basal sheath and a terminal blade. Flowers borne in spikelets composed of two rows of scales; the lower 2, or *glumes*, are flowerless, the upper scales, or *lemmas*, are fertile, which with an opposite scale the *palea* encircles the 3 stamens, and the single ovary. Spikelets variously arranged in spikes, racemes, or branched panicles. Ovary superior, 1–seeded, with 2 feathery styles. A large number of species occur in the Himalaya; Nepal for example, has over 350 species. Only a few conspicuous bamboo species are described here.

DENDROCALAMUS Stems woody, over 10 m, more than 5 cm in diameter. Stamens 6. c. 5 spp.

1487 **D. hamiltonii** Nees & Arn. ex Munro
Uttar Pradesh to Arunachal Pradesh. Burma. S.E. Asia. Native in the lower jungles; grown round villages in Nepal and W. Himalaya, to 2000 m. Flowering sporadically.

Forms impenetrable thickets of stems up to 26 m, with over-hanging often horizontal branching. Stems 10–13 cm in diameter; internodes 30–50 cm long; stem-sheaths triangular, persisting, to 30 cm long. Branches on lower part of stem seated on woody knobs the size of a fist, and on these may be half-developed swollen buds with brown sheaths. Leaves narrow-lanceolate long-pointed, to 45 cm long by 13 cm wide; leaf-sheaths with stiff hairs. Spikelets blunt, in dense rounded axillary clusters 1.5–4 cm across, on stout unbranched stems; bracts with ciliate margins.

Used for making baskets, mats, screens. Stems rather soft for building purposes.

1488 **D. strictus** (Roxb.) Nees   MALE BAMBOO
Pakistan to Burma. India. To 1500 m. Common around villages in Nepal. Flowering sporadically.

Forms densely packed clumps 7–13 m high, of greyish-green often blotched stems, with stiff spreading branches. Stem 5–8 cm in diameter; internodes 10–25 cm long; stem-sheaths hairless or with greyish-brown hairs, little shorter than the internodes. Leaves narrow-lanceolate, finely hairy on both sides, 10–25 cm long. Flower-spikelets spiny-tipped, in dense globular heads to 2.5 cm, borne along slender somewhat branched stems; bracts hairy and with distinctive spiny tips.

Stems are used for many purposes such as scaffolding, poles, masts, bamboo bridges, implements, mats, and swings. Valuable in paper manufacture; the pulp is used in the rayon industry. Bamboo charcoal is valuable for metal-smiths' work.

1489 **D. hookeri** Munro
Native and cultivated in the hills, from Uttar Pradesh to Sikkim; quite common around Nepali villages, to 1500 m. Flowering sporadically.

Stems tufted, 16–20 m and 10–15 cm in diameter; internodes 45–50 cm long; branches tufted, drooping; stem-sheaths usually very hairy, 20–30 cm with blade 7.5–17.5 cm. Leaves lanceolate, 30–38 cm. Spikelets acute, numerous, in large soft dense usually globular heads.

Used for many purposes.

1490 **D. sikkimensis** Gamble
Probably E. Nepal to Bhutan. 1300–2000 m. Cultivated round E. Nepal villages.

Stems tufted, few, very tall 17–24 m and 12–18 cm in diameter; internodes 45 cm long; stem-sheaths densely covered with dark golden-brown matted hairs, and blade decurrent on each side into a long fringed auricle. Leaves 15–30 cm, with long adpressed white hairs beneath. Spikelets in dense globular red-brown heads *c.* 4 cm across.

441

GRAMINEAE

Used for making milk- and water-containers, also for butter-churns.

BAMBUSA Distinguished from *Dendrocalamus* by the fruit in which the pericarp is fused to the seed (separable from the seed in *Dendrocalamus*). *c.* 2 spp.

1491 **B. arundinacea** (Retz.) Willd.
Native of S. India, Burma, Ceylon; widely cultivated in N. India and in the hills to 1000 m. Flowering sporadically.

Distinguished by its tall crowded bright green stems with spiny lower branches; stems 28–34 m by 15–18 cm in diameter. Internodes prominent and with zig-zag nearly horizontal spiny branches; stem-sheaths thickly covered with golden hairs when young, the blade hairy without, matted with dark bristles within. Leaves linear or linear-lanceolate to 20 cm, but often smaller. Inflorescence an enormous branched cluster.

Its great strength and size make it valuable for building purposes.

1492 **B. nutans** Wallich ex Munro
Himachal Pradesh to Sikkim; commonly cultivated to 1700 m. Flowering sporadically.

Stems solitary, not crowded, arising from a creeping rhizome, and without spiny branches, 7–17 m and 3.5–7.5 cm in diameter, graceful, nodding. Internodes 38–45 cm; stem-sheaths 15–23 cm, with adpressed black hairs within, and with 2 large wavy densely bristly auricles. Leaves linear-lanceolate 15–30 cm.

ARUNDINARIA Stem usually less than 10 m and less than 4 cm in diameter, rounded in section. Stamens 3. *c.* 8 spp.

1493 **A. falcata** Nees
Pakistan to E. Nepal. 1200–2100 m, rarely to 3600 m. Evergreen oak forests; often forming impenetrable underwood thickets. The commonest low-altitude bamboo in the W. Himalaya. Flowering sporadically.

Stems erect, densely tufted, slender, often glaucous and waxy, 2–3.5 m. Internodes 15–30 cm long; stem-sheaths thin, as long or longer than internodes and with pointed blades 1.5–5 cm. Leaves thin, linear long-pointed, 8–13 cm to 8 mm wide. Flowers occurring on leafless stems in large branched clusters; spikelets 1–1.5 cm, some stalkless, others long-stalked, in a densely whorled large branched cluster. Glumes conspicuously ribbed, without awns.

Used to make mats and baskets.

1494 **A. maling** Gamble (*A. racemosa* Munro)
C. Nepal to Sikkim. 1800–3600 m. Forming dense thickets in forests; very common in E. Nepal and Sikkim. Flowering sporadically.

Stems 3–10 m, clustered, with prominently ribbed stem-sheaths 18–25 cm. Stems slender, 1.6–4 cm in diameter; internodes 30–38 cm long. Leaves narrow-lanceolate 10–18 cm by 8–16 mm wide. Flowers in branched clusters; spikelets *c.* 1 cm, awned, solitary, short-stalked.

Used for matting, roofing, fencing, and fodder for animals. Very young shoots good eating for a short period in summer.

THAMNOCALAMUS Distinguished from *Arundinaria* by the bracts of the inflorescence which are large and sheathing. *c.* 3 spp.

1495 **T. spathiflora** (Trin.) Munro (*Arundinaria s.* Trin.)

Himachal Pradesh to E. Nepal. 2100–3600 m. In cedar, fir, and oak forests; gregarious; the common high altitude bamboo of the W. Himalaya.

Forming dense thickets with stems 4–7 m tall and 1.3–2 cm in diameter, with oblong stem-sheaths with rounded apex and blade 5–10 cm. Leaves net-veined, 7.5–12.5 cm by 8–13 mm wide, the leaf-sheath 5–7.5 cm with a few purple bristles at the mouth, ligule long, dark. Bracts of inflorescence 7.5 cm, oblong, clasping stem and usually bearing a rudimentary blade; spikelets crowded into numerous drooping spike-like branches 25–45 cm long. (Distinguished from 1493 by the prominent transverse veinlets of the leaves.)

The stems are used for pipes and baskets.

DRAWINGS

By Ann Farrer

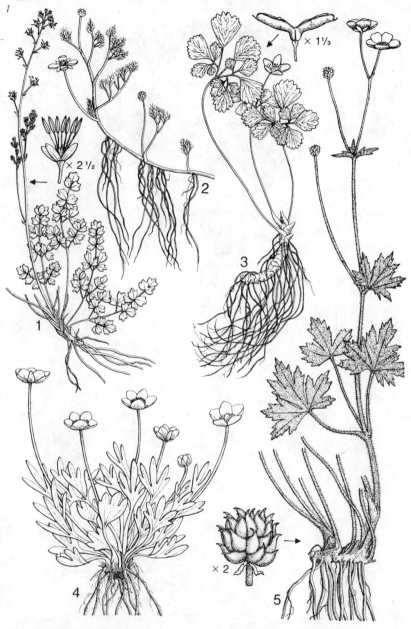

× 1⅓

× 2½

× 2

1 *Thalictrum alpinum* 56
2 *Ranunculus trichophyllus* 41
3 *Isopyrum adiantifolium* 11

4 *Ranunculus pulchellus* 37
5 *Ranunculus laetus* 33

All × ²/₅ except where shown

× ²/₃

× 1¹/₃

× ²/₃

1 *Anemone vitifolia* 47   3 *Anemone biflora* 45
2 *Anemone rivularis* 49   4 *Magnolia campbelli* 72

All × ¹/₃ except where shown

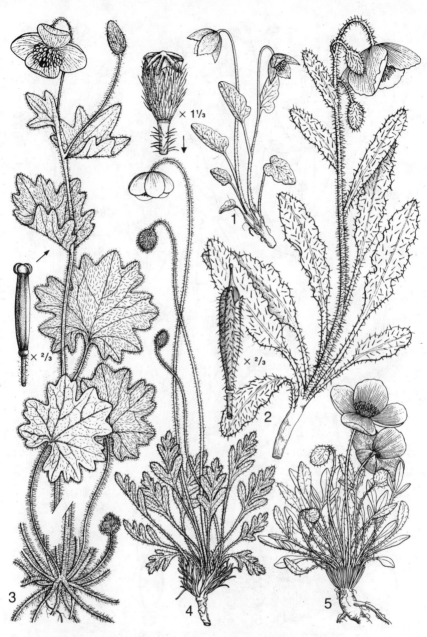

1 *Meconopsis lyrata* 105
2 *Meconopsis sinuata* 104
3 *Meconopsis villosa* 96
4 *Papaver nudicaule* 109
5 *Meconopsis bella* 99

All × ²/₅ except where shown

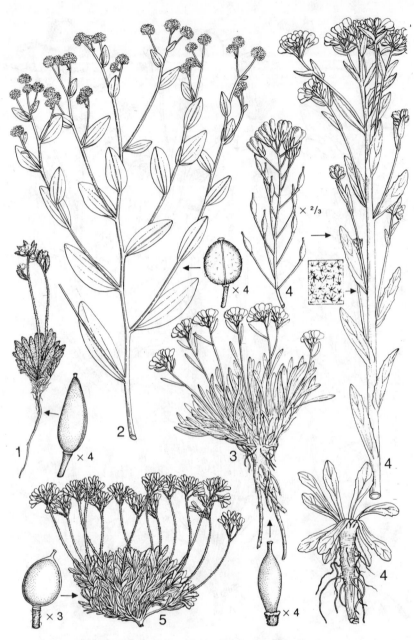

1 *Draba altaica* 127
2 *Lepidium latifolium* 134
3 *Braya oxycarpa* 124

4 *Draba amoena* 126
5 *Draba oreades* 131

All × ²/₅ except where shown

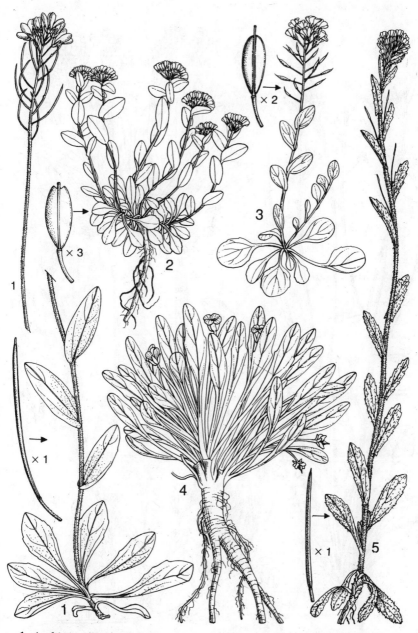

1 *Arabis amplexicaulis* 140
2 *Thlaspi griffithianum* 137
3 *Thlaspi andersonii* 136
4 *Pegaeophyton scapiflorum* 135
5 *Arabidopsis himalaica* 139

All × ²/₃ except where shown

× 2/3

× 1 1/3

× 3

× 2/3

1 *Matthiola flavida* 145
2 *Phaeonychium parryoides* 146

3 *Ermania himalayensis* 144
4 *Christolea crassifolia* 143

All × 2/5 except where shown

1 *Barbarea intermedia* 158

2 *Cardamine violacea* 152

3 *Cardamine macrophylla* 151

4 *Lignariella hobsonii* 154

All × ²/₅ except where shown

1 *Viola canescens* 165

2 *Polygala arillata* 171

3 *Cerastium dahuricum* 174

4 *Viola kunawarensis* 170

5 *Viola betonicifolia* 169

All × ²/₃ except where shown

1 *Geranium refractum* 243   3 *Oxalis corniculata* 246
2 *Eurya acuminata* 216   4 *Hypericum japonicum* 206

All × ⁴/₉ except where shown

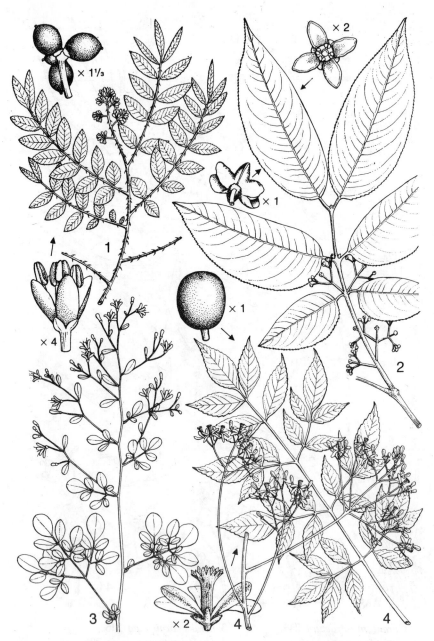

× 1⅓

× 2

× 1

× 4

× 1

× 2

1 *Zanthoxylum oxyphyllum* 263   3 *Boenninghausenia albiflora* 257
2 *Euonymus hamiltonianus* 271   4 *Melia azedarach* 266

All × ⅓ except where shown

1 *Rhamnus purpureus* 274   3 *Dodonaea viscosa* 279
2 *Maytenus rufa* 273   4 *Parthenocissus himalayana* 276

All × ⅓ except where shown

1 *Acer oblongum* 281   3 *Acer campbellii* 284
2 *Acer cappadocicum* 282   4 *Acer caesium* 285

All × ¹/₃ except where shown

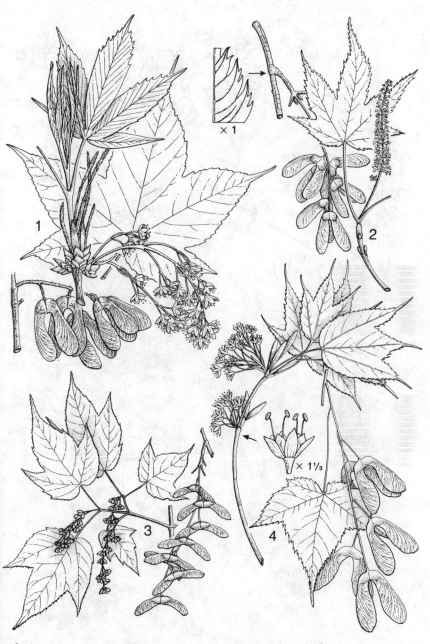

1 *Acer sterculiaceum* 286   3 *Acer pectinatum* 288
2 *Acer caudatum* 289   4 *Acer acuminatum* 287

All × ¹/₃ except where shown

× 3

× ¼

2

× 4

× ²/₃

4

1

3

× ¼

× ²/₃

× ²/₃

5

× 1¹/₃

5

1 *Coriaria napalensis* 297
2 *Cotinus coggygria* 292
3 *Sabia campanulata* 291

4 *Rhus wallichii* 295
5 *Pistacia chinensis* subsp.
  integerrima 296

All × ¹/₃ except where shown

458

× 1

× ²/₃

× 1¹/₃

× 1¹/₃

1

2

3

1 *Lespedeza gerardiana* 320   3 *Flemingia strobilifera* 319
2 *Crotalaria cytisoides* 316

All × ²/₃ except where shown

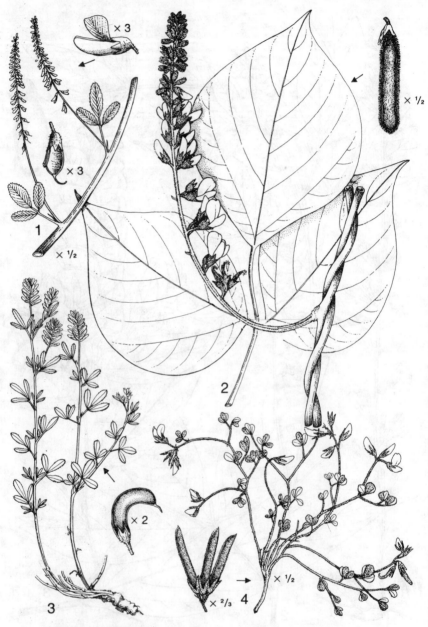

1 *Melilotus officinalis* 327   3 *Medicago falcata* 324
2 *Pueraria tuberosa* 322   4 *Argyrolobium roseum* 323

All × ¹/₃ except where shown

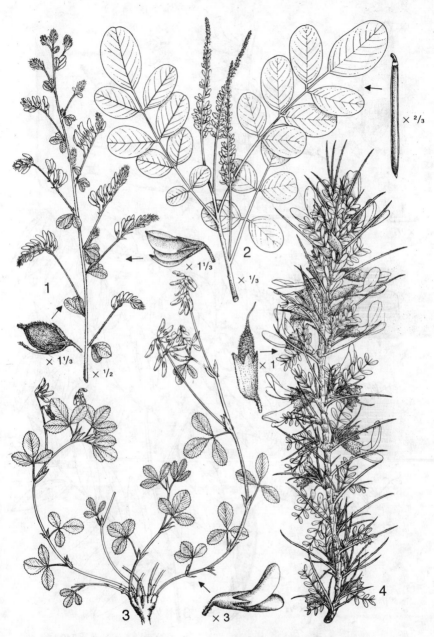

1　*Campylotropis speciosa* 315　　3　*Trigonella emodi* 332
2　*Indigofera atropurpurea* 343　　4　*Caragana gerardiana* 335

All × ²/₃ except where shown

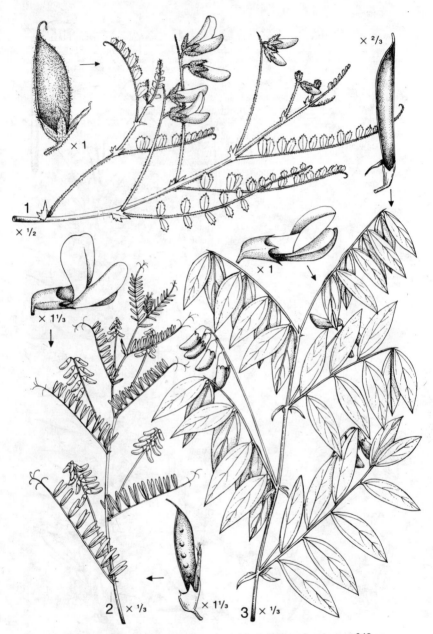

× 2/3

× 1

× 1/2

× 1 1/3

× 1/3

× 1 1/3

2 × 1/3

3 × 1/3

1 *Cicer microphyllum* 348
2 *Vicia bakeri* 355

3 *Lathyrus laevigatus* 349

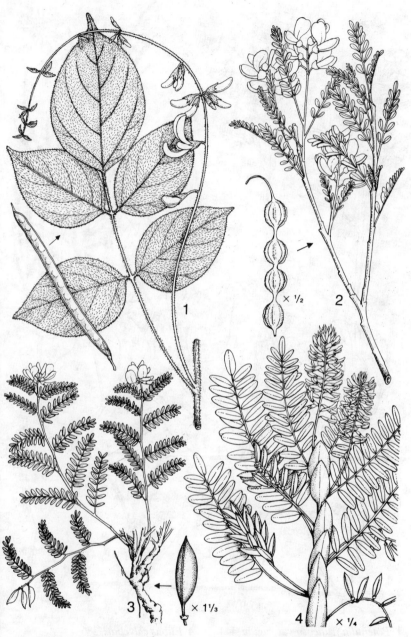

1 *Apios carnea* 357
2 *Sophora mollis* 347
3 *Astragalus himalayanus* 363
4 *Astragalus stipulatus* 371

All × ¹/₃ except where shown

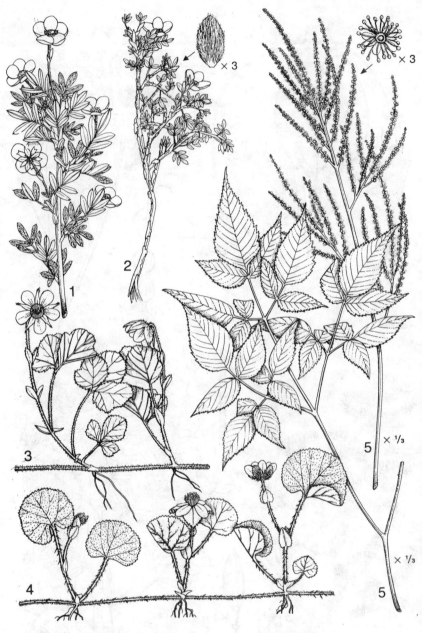

1 *Potentilla fruticosa* var. *rigida* 444   4 *Rubus calycinus* 386
2 *Potentilla fruticosa* var. *pumila* 444   5 *Aruncus dioicus* 461
3 *Rubus napalensis* 387

All × ¹/₂ except where shown

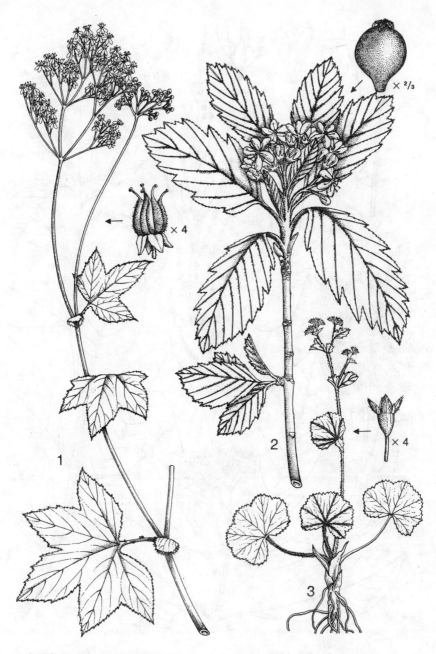

$\times \,^2/_3$

$\times 4$

$\times 4$

1

2

3

1 *Filipendula vestita* 462
2 *Sorbus lanata* 434
3 *Alchemilla trollii* 438

All $\times \,^4/_9$ except where shown

1 *Agrimonia pilosa* 460
2 *Astilbe rivularis* 463
3 *Stranvaesia nussia* 428
4 *Photinia integrifolia* 427

All × ¹/₃ except where shown

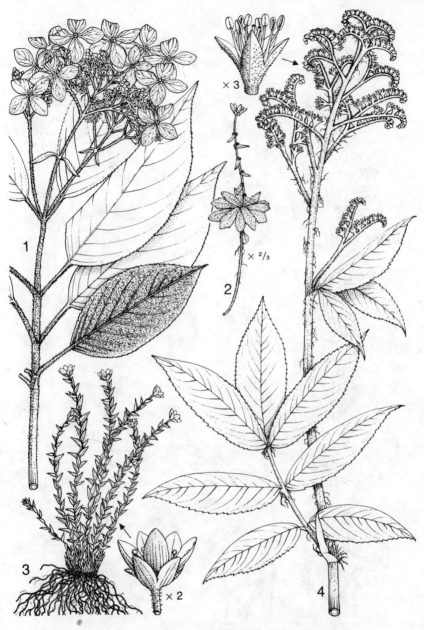

$\times 3$

$\times {}^2/_3$

$\times 2$

1 *Hydrangea heteromalla* 497   3 *Saxifraga brachypoda* 474
2 *Saxifraga strigosa* 487   4 *Rodgersia nepalensis* 464

All × ¹/₃ except where shown

1 *Sedum multicaule* 521   4 *Sedum oreades* 522
2 *Ribes alpestre* 503   5 *Drosera peltata* 523
3 *Rotala rotundifolia* 529

All × ²/₃ except where shown

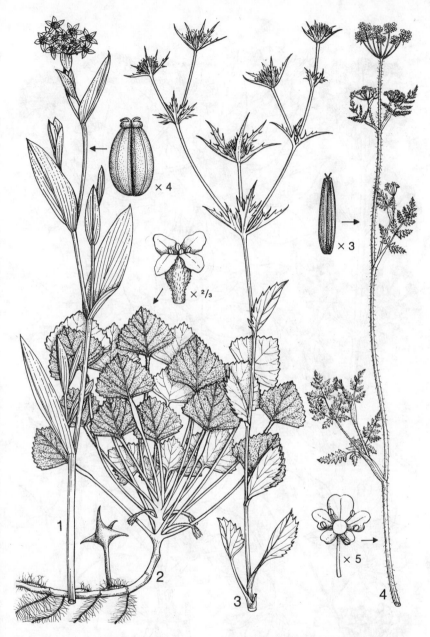

1 *Bupleurum candollii* 549   3 *Eryngium biebersteinianum* 554
2 *Trapa quadrispinosa* 540   4 *Chaerophyllum villosum* 569

All × ⁴/₉ except where shown

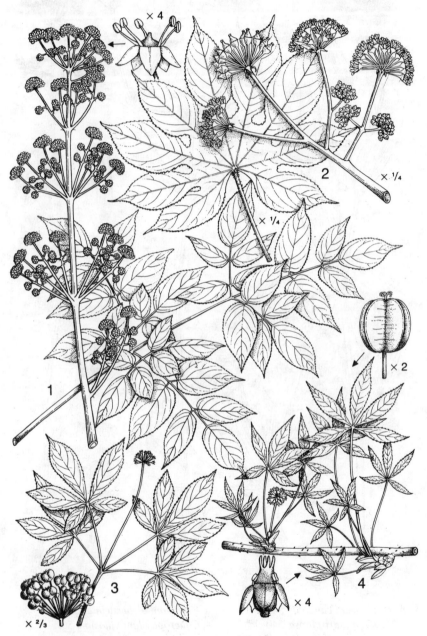

1 *Aralia cachemirica* 573
2 *Trevesia palmata* 576
3 *Panax pseudo-ginseng* 572
4 *Acanthopanax cissifolius* 574

All × ¹/₃ except where shown

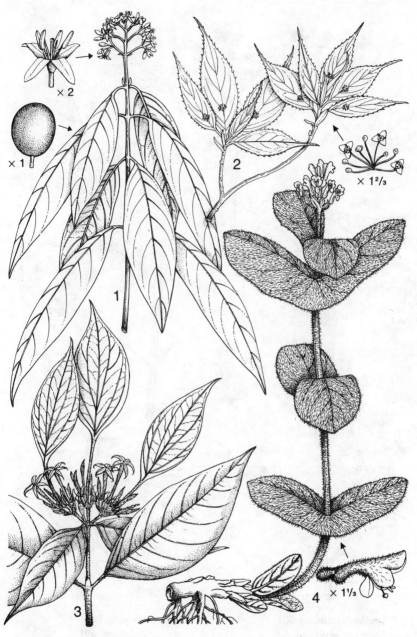

1 *Swida oblonga* 579   3 *Mussaenda treutleri* 615
2 *Helwingia himalaica* 571   4 *Triosteum himalayanum* 581

All × ¹/₃ except where shown

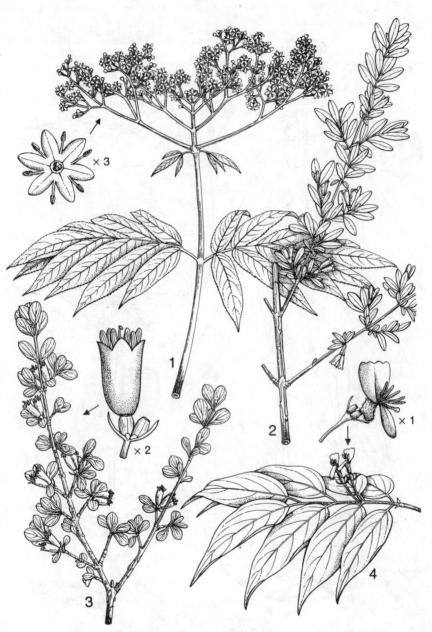

$\times 3$

$\times 2$

$\times 1$

1
2
3
4

1 *Sambucus adnata* 601   3 *Lonicera obovata* 597
2 *Lonicera myrtillus* 526   4 *Lonicera webbiana* 591

All × ¹/₄ except where shown

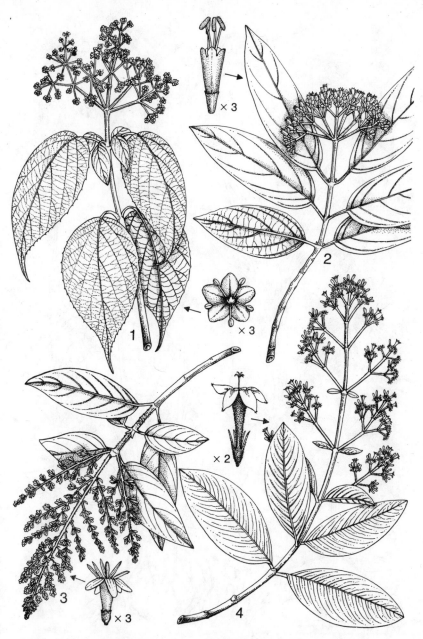

1 *Viburnum mullaha* 608   3 *Wendlandia puberula* 611
2 *Viburnum cylindricum* 603   4 *Spermadictyon suaveolens* 613

All × ⁴/₁₅ except where shown

1 *Nardostachys grandiflora* 620   3 *Valeriana jatamansii* 621
2 *Rubia manjith* 619   4 *Scabiosa speciosa* 629

All × ¹/₃ except where shown

1 *Aster molliusculus* 642
2 *Eupatorium adenophorum* 632
3 *Gnaphalium affine* 647
4 *Tanacetum atkinsonii* 673
5 *Ageratum conyzoides* 631

All × ¹/₂ except where shown

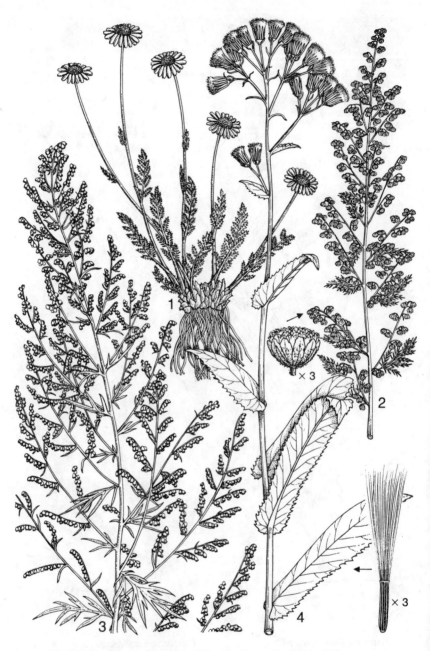

1 *Chrysanthemum pyrethroides* 674 3 *Artemisia dubia* 678

2 *Artemisia gmelinii* 682 4 *Gynura cusimbua* 694

All × ¹/₃ except where shown

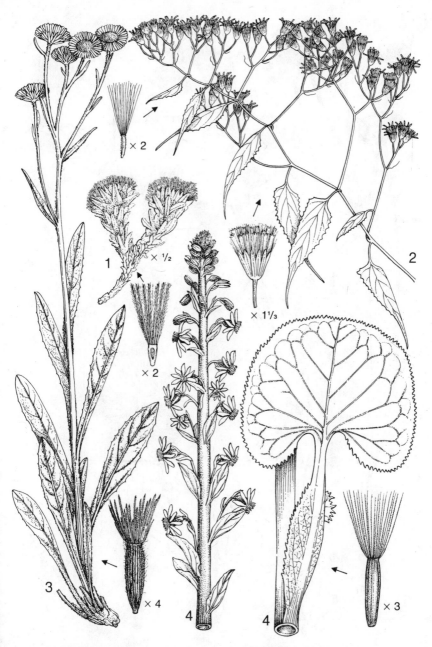

1 *Saussurea gnaphalodes* 714
2 *Senecio scandens* 698
3 *Nannoglottis hookeri* 697
4 *Lingularia fischeri* 707

All × ⁴/₁₅ except where shown

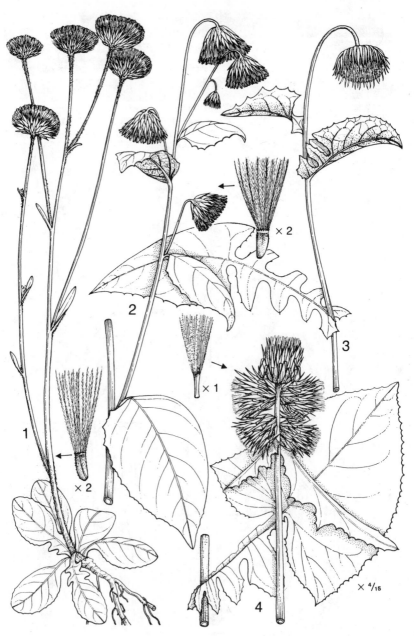

1 *Saussurea heteromalla* 727   3 *Saussurea auriculata* 716

2 *Saussurea deltoidea* 726   4 *Saussurea costus* 725

All × ²/₅ except where shown

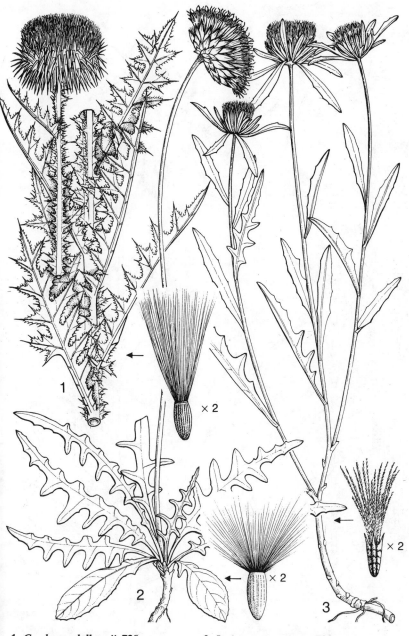

1 *Carduus edelbergii* 730
2 *Serratula pallida* 737
3 *Jurinea ceratocarpa* 729

All × ¹/₂ except where shown

479

1 *Centaurea iberica* 738
2 *Carthamus lanatus* 739
3 *Prenanthes brunoniana* 753

All × ¹/₃ except where shown

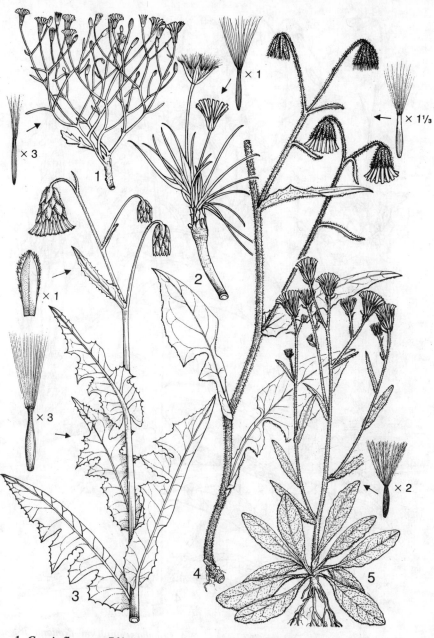

1 *Crepis flexuosa* 761
2 *Tragopogon gracilis* 747
3 *Cicerbita macrantha* 758
4 *Dubyaea hispida* 754
5 *Picris heiracioides* 745

All × ⅓ except where shown

× ²/₃

× 1

× ¼

1 *Campanula modesta* 767
2 *Campanula cashmeriana* 771
3 *Asyneuma thomsonii* 764

4 *Lobelia seguinii* var. *doniana* 785
5 *Campanula sylvatica* 768

All × ¹/₃ except where shown

1 *Rhododendron pendulum* 811
2 *Rhododendron vaccinioides* 808
3 *Lyonia villosa* 818

4 *Rhododendron camelliiflorum* 812
5 *Rhododendron pumilum* 803

All × ¹/₂

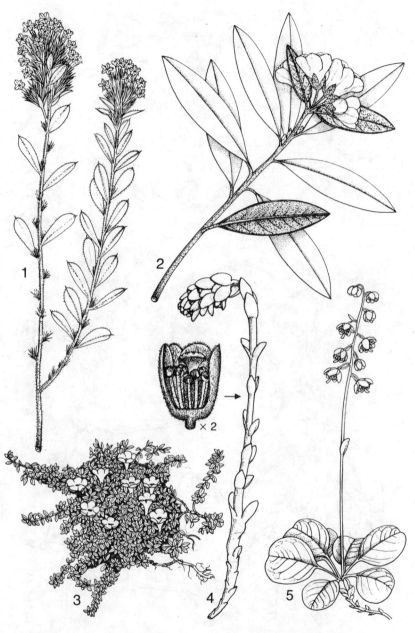

1 *Ceratostigma ulicinum* 835
2 *Rhododendron glaucophyllum* 813
3 *Diapensia himalaica* 833
4 *Monotropa hypopitys* 832
5 *Pyrola karakoramica* 831

All × ⁴/₉ except where shown

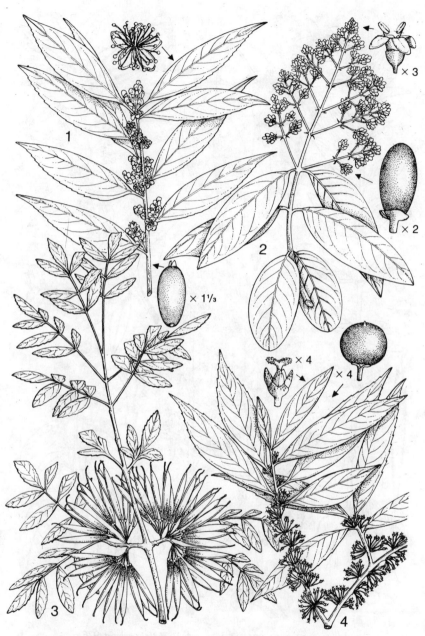

1 *Symplocos theifolia* 893
2 *Ligustrum compactum* 906
3 *Fraxinus xanthoxyloides* 899
4 *Myrsine semiserrata* 892

All × ¹/₃ except where shown

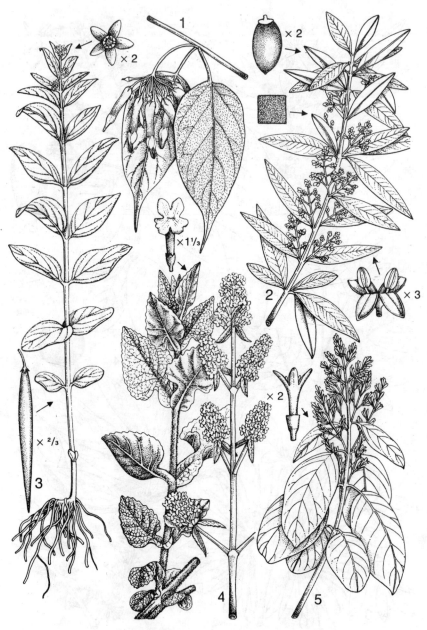

1 *Ceropegia pubescens* 921
2 *Olea ferruginea* 908
3 *Vincetoxicum hirundinaria* 919

4 *Buddleja crispa* 926
5 *Syringa emodi* 905

All × ⁴/₁₅ except where shown

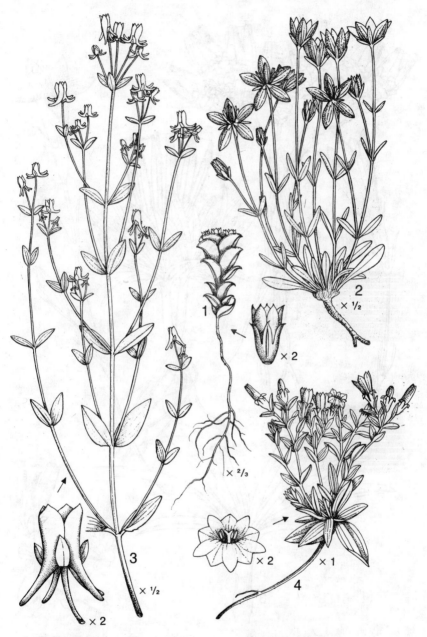

1 *Gentiana capitata* 952   3 *Halenia elliptica* 928
2 *Lomatogonium caeruleum* 930   4 *Gentiana pedicellata* 956

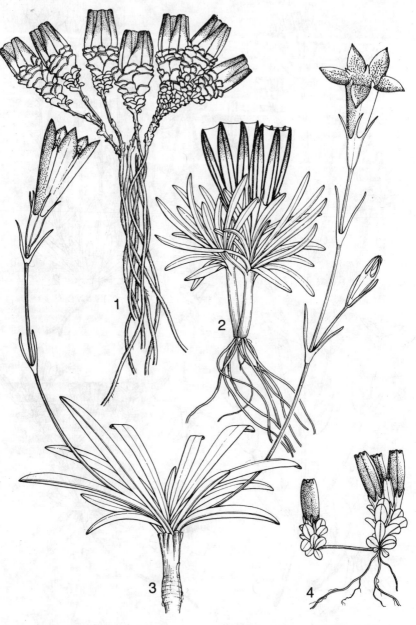

1 *Gentiana urnula* 947
2 *Gentiana algida* 939
3 *Gentiana kurroo* 940
4 *Gentiana venusta* 950

All × ²/₃ except where shown

× 3

2 × ½

× 1⅓

1 *Cynoglossum glochidiatum* 973   3 *Maharanga emodi* 972
2 *Polemonium caeruleum*
  subsp. *himalayanum* 965

All × ⅓ except where shown

1 *Ipomoea quamoclit* 994
2 *Eritrichium nanum* 981
3 *Myosotis alpestris* 988
4 *Chionocharis hookeri* 982
5 *Microula sikkimensis* 987

All × ⁴/₉ except where shown

× ½

× 3

× ⅔

1

3

× ⅓

3

2

3

1 *Scopolia stramonifolia* 1009   3 *Wightia speciosissima* 1015
2 *Scrophularia elatior* 1013

All × ⁴/₁₅ except where shown

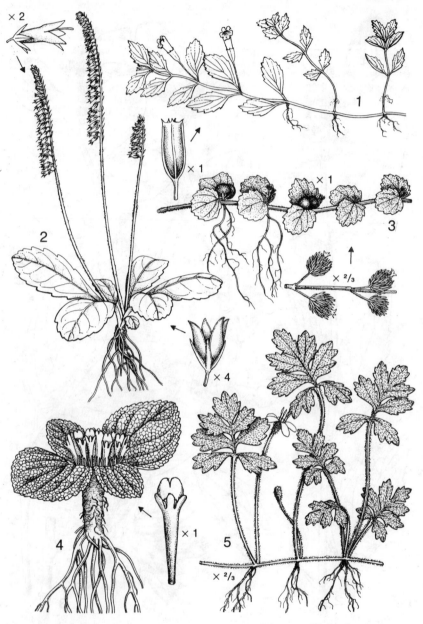

1 *Mimulus nepalensis* 1016
2 *Wulfenia amherstiana* 1027
3 *Hemiphragma heterophyllum* 1022
4 *Oreosolon wattii* 1028
5 *Ellisiophyllum pinnatum* 1021

All × ⁴/₉ except where shown

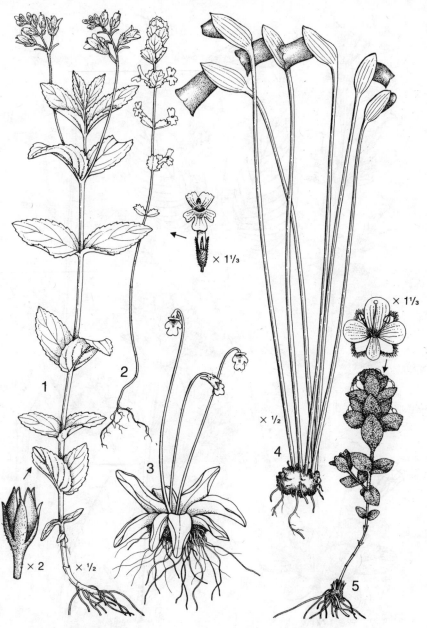

× 1⅓

× ½

× 1⅓

1

2

3

4

5

× 2

× ½

1 *Veronica himalensis* 1033
2 *Euphrasia himalayica* 1037
3 *Pinguicula alpina* 1064

4 *Aeginetia indica* 1059
5 *Veronica lanuginosa* 1034

All × ⅔ except where shown

1 *Aechmanthera gossypina* 1081
2 *Rhynchoglossum obliquum* 1074
3 *Sesamum orientale* 1080
4 *Corallodiscus lanuginosus* 1070
5 *Eranthemum pulchellum* 1089

All × ¹/₃ except where shown

$\times$ ⅓

2

$\times$ ⅔

1

$\times$ 3

3

$\times$ 2

4

5

$\times$ ⅓

1 *Clerodendrum viscosum* 1095
2 *Holmskioldia sanguinea* 1098
3 *Vitex negundo* 1092

4 *Caryopteris odorata* 1099
5 *Goldfussia pentastemonoides* 1083

All $\times$ ⁴/₁₅ except where shown

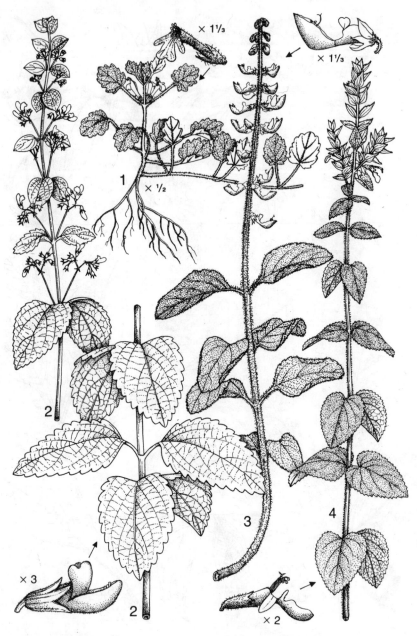

$\times 1\frac{1}{3}$

$\times 1\frac{1}{3}$

$\times \frac{1}{2}$

$\times 3$

$\times 2$

1 *Ajuga lobata* 1103
2 *Rabdosia pharica* 1109
3 *Coleus barbatus* 1107
4 *Teucrium quadrifarium* 1104

All $\times \frac{1}{3}$ except where shown

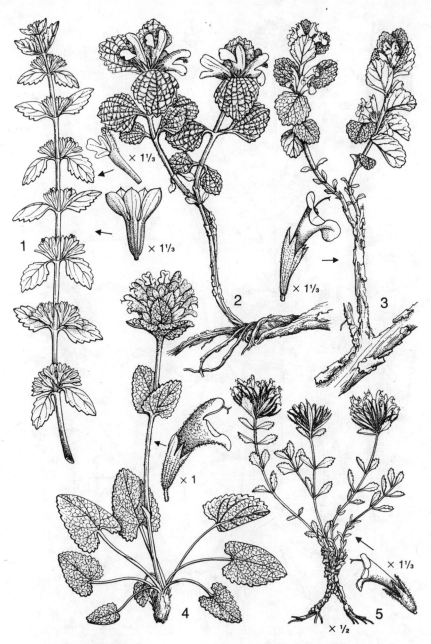

× 1⅓

× 1⅓

1

× 1⅓

2

3

× 1

4

× 1⅓

5

× ½

1 *Roylea cinerea* 1132

2 *Lamium rhomboideum* 1124

3 *Glechoma nivalis* 1150

4 *Dracocephalum wallichii* 1152

5 *Nepeta longibracteata* 1142

All × ⅓ except where shown

497

1 *Origanum vulgare* 1159
2 *Thymus linearis* 1160
3 *Perovskia abrotanoides* 1167
4 *Mentha longifolia* 1161
5 *Micromeria biflora* 1158

All × ¹/₂ except where shown

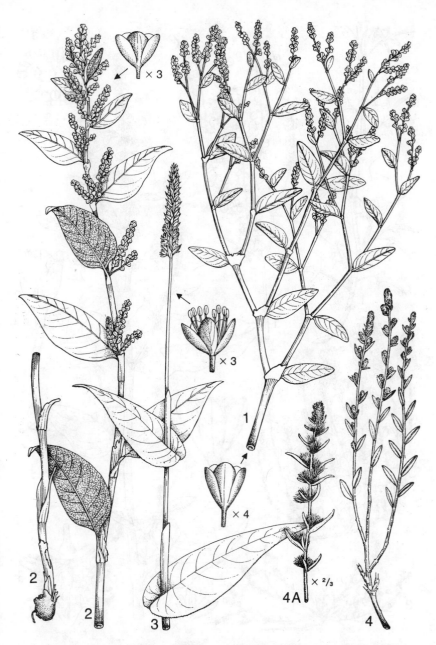

1 *Aconogonum tortuosum* 1179 4 *Krascheninnikovia ceratoides* 1176
2 *Aconogonum rumicifolium* 1182 4A Fruiting stem of 1176
3 *Bistorta amplexicaulis* 1186

All × ¹/₃ except where shown

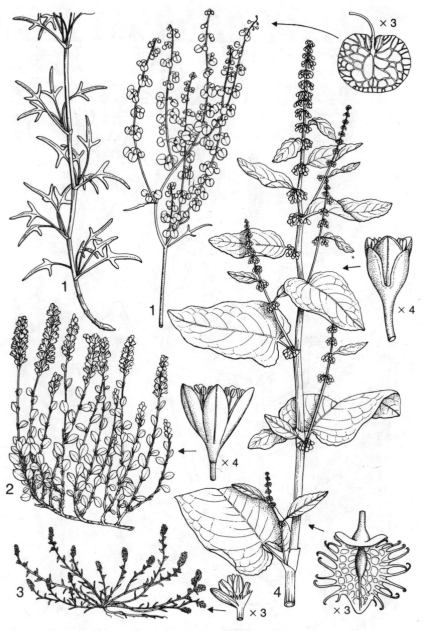

1 *Rumex hastatus* 1204
2 *Bistorta vaccinifolia* 1185
3 *Polygonum plebeium* 1193
4 *Rumex nepalensis* 1205

All × ¹/₃ except where shown

1 *Cinnamomum tamala* 1209 3 *Neolitsea pallens* 1216
2 *Litsea doshia* 1215

All × ⅓ except where shown

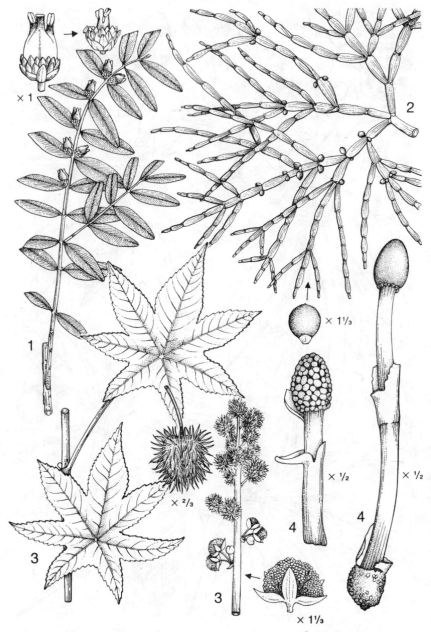

× 1

2

× 1⅓

× ½

× ½

4

4

× ⅔

3

3

× 1⅓

1 *Buxus wallichiana* 1242   3 *Ricinus communis* 1241

2 *Viscum articulatum* 1231   4 *Balanophora involucrata* 1233

All × ⅓ except where shown

$\times 1^2/_3$

$\times 2^2/_3$

1 *Urtica hyperborea* 1252
2 *Lecanthus peduncularis* 1254
3 *Laportea terminalis* 1250
4 *Girardinia diversifolia* 1249
5 *Boehmeria platyphylla* 1246

All × ¹/₃ except where shown

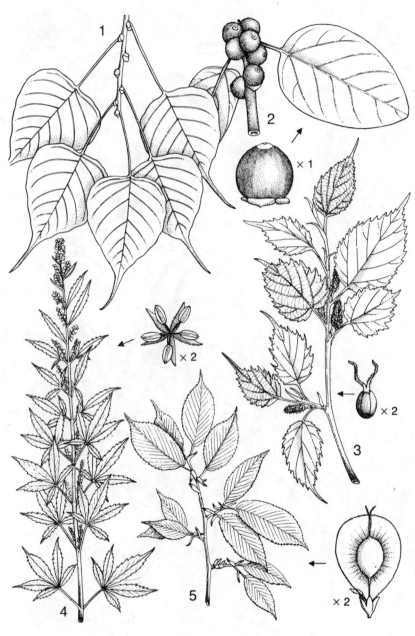

1 *Ficus religiosa* 1259 4 *Cannabis sativa* 1257
2 *Ficus benghalensis* 1258 5 *Ulmus wallichiana* 1256
3 *Morus serrata* 1264

All × ⁴/₁₅ except where shown

1 *Populus ciliata* 1301
2 *Myrica esculenta* 1269
3 *Platanus orientalis* 1266
4 *Engelhardia spicata* 1268
5 *Alnus nepalensis* 1270

All × ²/₉ except where shown

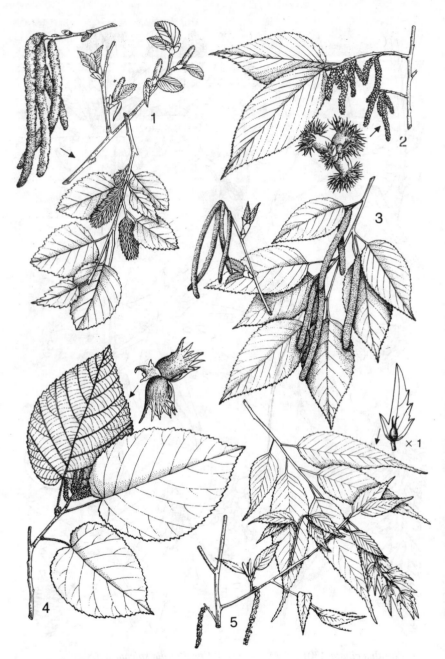

1 *Betula utilis* 1272
2 *Corylus ferox* 1276
3 *Betula alnoides* 1273

4 *Corylus jacquemontii* 1277
5 *Carpinus viminea* 1274

All × ¹/₃ except where shown

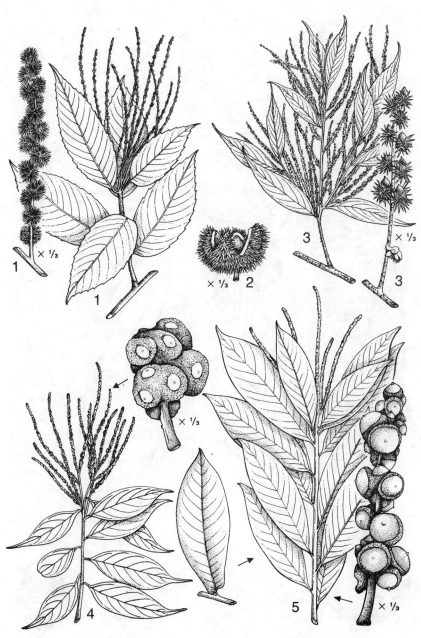

1 *Castanopsis indica* 1278
2 *Castanopsis hystrix* 1279
3 *Castanopsis tribuloides* 1280
4 *Lithocarpus pachyphylla* 1281
5 *Lithocarpus elegans* 1282

All × ²/₉ except where shown

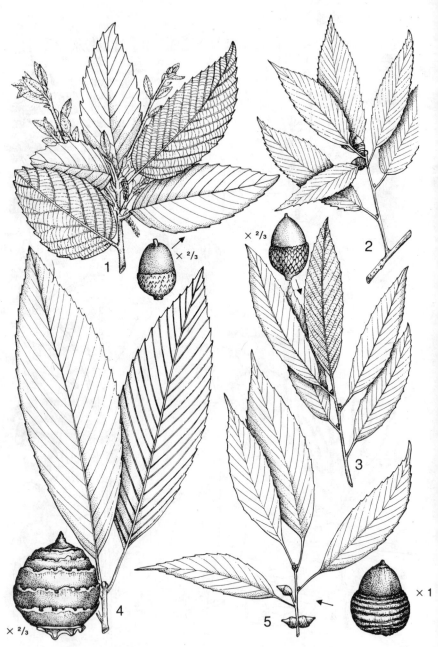

1 *Quercus lanata* 1286
2 *Quercus oxyodon* 1284
3 *Quercus leucotrichophora* 1287
4 *Quercus lamellosa* 1285
5 *Quercus glauca* 1283

All × ¹/₃ except where shown

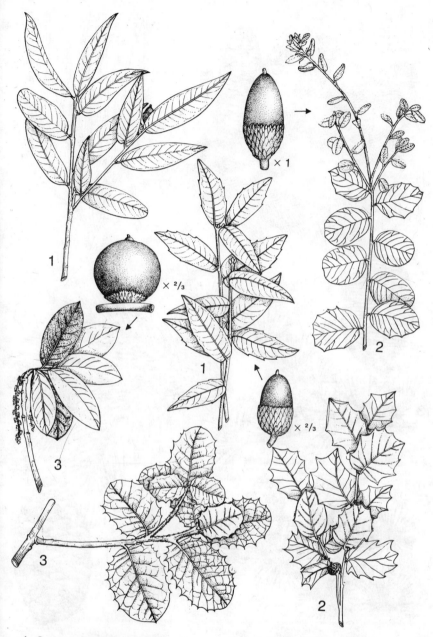

1 *Quercus floribunda* 1290   3 *Quercus semecarpifolia* 1288

2 *Quercus baloot* 1289

All × ¹/₃ except where shown

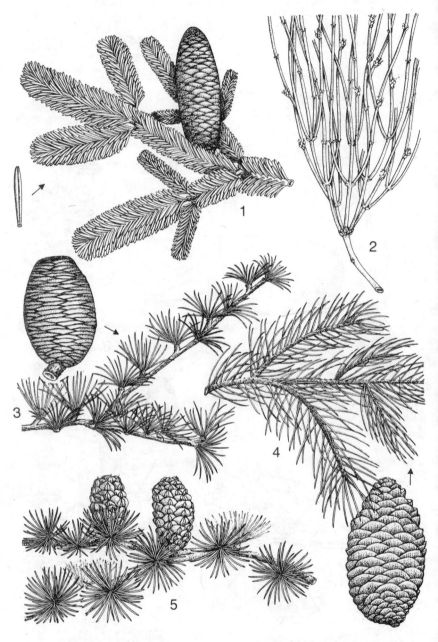

1 *Abies spectabilis* 1308
2 *Ephedra gerardiana* 1305
3 *Cedrus deodara* 1311

4 *Picea smithiana* 1306
5 *Larix himalaica* 1313

All × ¹/₃

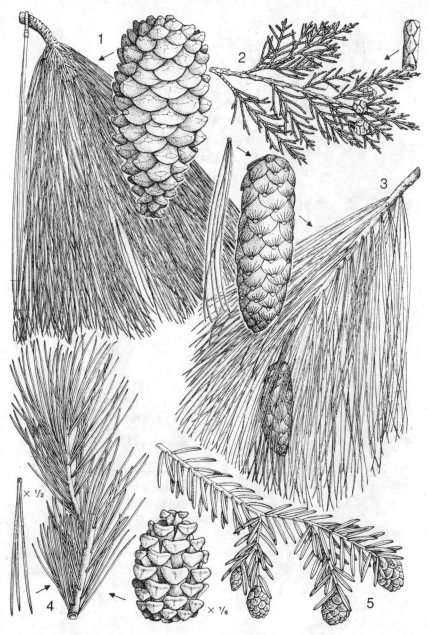

1 *Pinus roxburghii* 1315   4 *Pinus gerardiana* 1316
2 *Cupressus torulosa* 1318   5 *Tsuga dumosa* 1307
3 *Pinus wallichiana* 1314

All × ⅓ except where shown

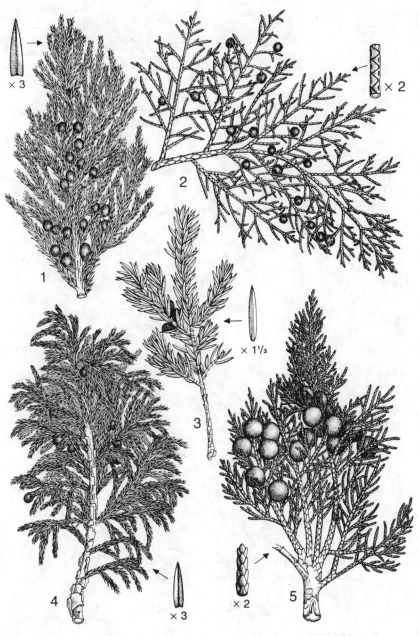

1 *Juniperus squamata* 1321   4 *Juniperus recurva* 1320

2 *Juniperus macropoda* 1323 5 *Juniperus indica* 1322

3 *Juniperus communis* 1319

All × ¹/₂ except where shown

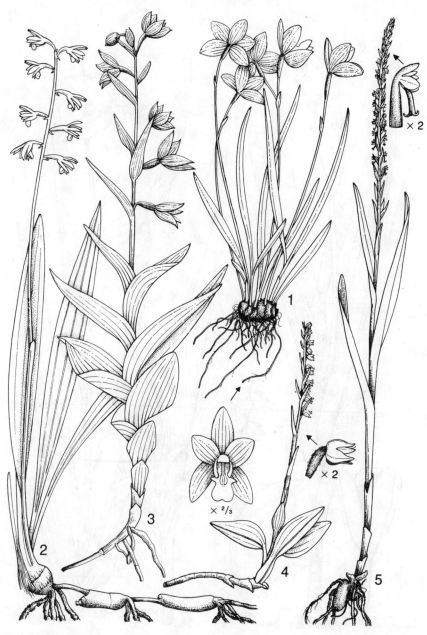

1 *Spathoglottis ixioides* 1355
2 *Oreorchis foliosa* 1349
3 *Epipactis royleana* 1335
4 *Goodyera repens* 1340
5 *Herminium lanceum* 1344

All × ¹/₂ except where shown

1 *Belamcanda chinensis* 1386
2 *Galearis stracheyi* 1338

3 *Cautleya spicata* 1382
4 *Cymbidium hookeranum* 1364

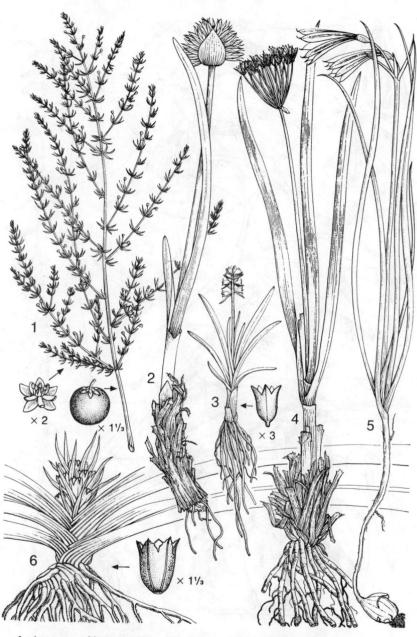

1 *Asparagus filicinus* 1417
2 *Allium semenovii* 1399
3 *Aletris pauciflora* 1415
4 *Allium wallichii* 1405
5 *Ixiolirion karateginum* 1410
6 *Campylandra aurantiaca* 1418

All × ⁴/₉ except where shown

× 1⅓

× 2

1

× 2

2

3

× 2

1 *Disporum cantoniense* 1423   3 *Chlorophytum nepalense* 1420
2 *Smilax ferox* 1450

All × ½ except where shown

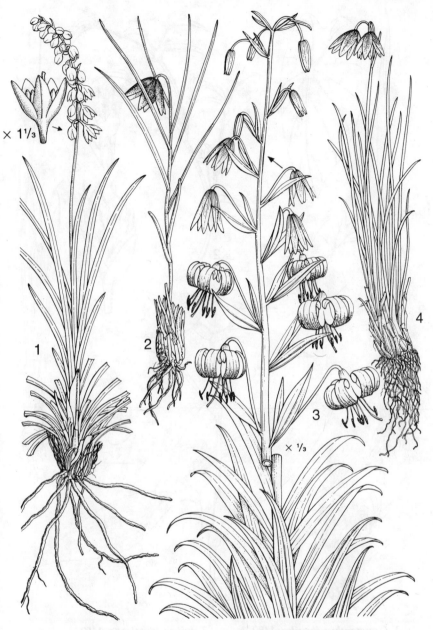

× 1¹⁄₃

× ¹⁄₃

1 *Ophiopogon intermedius* 1441 3 *Lilium polyphyllum* 1433
2 *Lilium nanum* 1430   4 *Lloydia longiscapa* 1437

All × ¹⁄₂ except where shown

1 *Gonatanthus pumilus* 1480
1A Bulbiferous stem of 1480
2 *Cyanotis vaga* 1456

3 *Juncus himalensis* 1459
4 *Commelina paludosa* 1455

× 3

× 3

× 2

× 1⅓

All × ⁴/₉ except where shown

COLOUR PLATES

Aquilegia nivalis Kashmir

8 *Aquilegia pubiflora*
Saraj, Himachal Pradesh

6 *Aquilegia fragrans* × 1½
Lahul, Himachal Pradesh

1

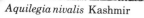

Caltha palustris var. *himalensis*
anesh Himal, C. Nepal

2 *Trollius acaulis* Kashmir

4 *Paraquilegia microphylla* Kashmir

14 *Aconitum hookeri* × 1½ Dolpo, W. Nepal

2

20 *Delphinium cashmerianum*
Bara Lacha, Lahul

21 *Delphinium viscosum* Arun, E. Nepal

17 *Aconitum ferox*
Gatlang, C. Nepal

15 *Aconitum violaceum*
Kashmir

18 *Aconitum spicatum*
Gosainkund, C. Nepal

3

9 *Actaea spicata* var. *acuminata*
Kulu, Himachal Pradesh

12 *Aconitum laeve* Kashmir

26 *Delphinium roylei*
Kashmir

19 *Delphinium brunonianum*
Dolpo, W. Nepal

15 *Aconitum violaceum* var. *robustum* Kashmir

4

23 *Delphinium vestitum*
Ganesh Himal, C. Nepal

10 *Cimicifuga foetida* Kashmir

22 *Delphinium drepanocentrum*
Arun, E. Nepal

30 *Callianthemum pimpinelloides* Gulmarg, Kashmir

38 *Ranunculus brotherusii*
Langtang, C. Nepal

42 *Adonis chrysocyathus* Kashmir

39a *Ranunculus adoxifolius*
Jugal Himal, C. Nepal

39 *Ranunculus hirtellus* Vishensar, Kashmir

27 *Delphinium himalayai*
Kali Gandaki, C. Nepal

31 *Oxygraphis polypetala*
Kashmir

50 *Anemone demissa*
Himal Chuli, C. Nepal

52 *Anemone polyanthes*
Dhaulagiri, C. Nepal

51 *Anemone tetrasepala* Kashmir

6

51 *Anemone tetrasepala* Salnai Sar, Kashmir

48 *Anemone obtusiloba*
Annapurna Himal, C. Nepal

46 *Anemone rupicola*
Lahul, Himachal Pradesh

48 *Anemone obtusiloba* Kashmir

54 *Thalictrum reniforme*
Ganesh Himal, C. Nepal

55 *Thalictrum virgatum* E. Nepal

8

61 *Clematis montana* Dudh Kosi, E. Nepal

58 *Thalictrum cultratum*
Chamba, Himachal Pradesh

68 *Clematis connata*
Trisuli, C. Nepal

66 *Clematis vernayi* Dolpo, W. Nepal

60 *Clematis barbellata*
Silgarhi Dhoti, W. Nepal

70 *Clematis roylei*
Dolpo, W. Nepal

71 *Paeonia emodi* Kashmir

63 *Clematis phlebantha*
Suli Gad, W. Nepal

74 *Michelia doltsopa* Dudh Kosi, E. Nepal

90 *Euryale ferox* Kashmir

10

82 *Berberis aristata* Nawakot, C. Nepal

84 *Berberis wallichiana*
Milke Dara, E. Nepal

86 *Mahonia napaulensis* Darjeeling

89 *Holboellia latifolia*
Dudh Kosi, E. Nepal

Schisandra grandiflora
dh Kosi, E. Nepal

92 *Nymphaea tetragona* Kashmir

11

Berberis angulosa
mjung Himal, C. Nepal

92 *Nymphaea mexicana* hybrid Kashmir

Podophyllum hexandrum
shmir

91 *Nelumbo nucifera* Kashmir

101 *Meconopsis horridula* Jumla, W. Nepal

12

94 *Meconopsis paniculata* Arun, E. Nepal

106 *Meconopsis aculeata*
Umasi La, Zanskar, Ladakh

95 *Meconopsis regia*
Ganesh Himal, C. Nepal

97 *Meconopsis discigera*
Topke Gola, E. Nepal

95a *Meconopsis dhwojii*
Rolwaling, C. Nepal

102 *Meconopsis latifolia*
Trunkal, Kashmir

108 *Papaver macrostomum*
Kashmir

111 *Dicranostigma
lactucoides* Dolpo, W. Nepal

13

103 *Meconopsis grandis* Jumla, W. Nepal

8 *Meconopsis simplicifolia*
Dhaulagiri, C. Nepal

121 *Corydalis juncea*
Langtang, C. Nepal

14

112 *Corydalis cashmeriana*
Kulu, Himachal Pradesh

120 *Corydalis govaniana*
Kulu, Himachal Pradesh

114 *Corydalis rutifolia* Sonamarg, Kashmir   113 *Corydalis crassissima* Kashmir

116 *Corydalis meifolia* Rolwaling, C. Nepal   118 *Corydalis thyrsiflora* Kashmir

122 *Dicentra macrocapnos*
Modi Khola, C. Nepal

132 *Draba setosa*
Zanskar, Ladakh

149 *Erysimum melicentae*
Kashmir

15

153 *Chorispora sabulosa* Kashmir

150 *Cardamine loxostemonoides*
Phoksumdo Tal, W. Nepal

155 *Parrya nudicaulis*
Umasi La, Zanskar, Ladakh

147 *Pycnoplinthopsis bhutanica*
Modi Khola, C. Nepal

138 *Megacarpaea polyandra*
Jumla, W. Nepal

161 *Capparis spinosa*
Indus valley, Ladakh

163 *Viola biflora* Kashmir

164 *Viola wallichiana*
Dudh Kosi, E. Nepal

16

187 *Thylacospermum caespitosum*
Phirtse La, Zanskar, Ladakh

172 *Polygala sibirica*
Langtang, C. Nepal

178 *Arenaria bryophylla*
Dolpo, W. Nepal

180 *Arenaria densissima* Ganesh Himal, C. Nepal

77 *Arenaria griffithii* Dras, Ladakh

186 *Minuartia kashmirica* Kashmir

17

84 *Arenaria glanduligera* Dolpo, W. Nepal

189 *Dianthus anatolicus* Zanskar, Ladakh

88 *Dianthus angulatus* Lahul, Himachal Pradesh

173 *Cerastium cerastioides*
Lahul, Himachal Pradesh

191 *Gypsophila cerastioides* Kashmir

198 *Silene moorcroftiana*
Umasi La, Zanskar, Ladakh

18

175 *Stellaria decumbens* Kashmir

196 *Silene setisperma* Arun, E. Nepal

197 *Silene edgeworthii*
Kashmir

200 *Silene viscosa*
Lahul, Himachal Pradesh

201 *Lychnis coronaria* Kashmir

203 *Myricaria squamosa*
Lahul, Himachal Pradesh

205 *Myricaria rosea*
Naulekh, E. Nepal

218 *Actinidia callosa*
Dudh Kosi, E. Nepal

19

210 *Hypericum oblongifolium*
Saraj, Himachal Pradesh

211 *Hypericum choisianum*
Pindari, Uttar Pradesh

08 *Hypericum elodeoides*
risuli, C. Nepal

219 *Saurauia napaulensis*
E. Nepal

214 *Camellia kissi*
Himal Chuli, C. Nepal

232 *Peganum harmala*
Indus valley, Ladakh

233 *Biebersteinia odora*
Zanskar, Ladakh

237 *Geranium polyanthes*
Ganesh Himal, C. Nepal

20

217 *Schima wallichii* E. Nepal

220 *Stachyurus himalaicus* Dudh Kosi, E. Nepal

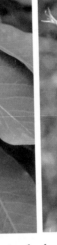

227 *Thespesia lampas*
Malemchi, C. Nepal

224 *Lavatera kashmiriana*
Kashmir

239 *Geranium tuberaria*
Chamba, Himachal Pradesh

234 *Geranium nakaoanum* Ganesh Himal, C. Nepal

235 *Geranium donianum* Langtang, C. Nepal

236 *Geranium wallichianum* C. Nepal

21

230 *Reinwardtia indica* Saraj, Himachal Pradesh

256 *Impatiens stenantha* Darjeeling

254 *Impatiens urticifolia* Talemchi, C. Nepal

250 *Impatiens sulcata* Ganesh Himal, C. Nepal

255 *Impatiens edgeworthii* Kashmir

280 *Aesculus indica*
Garhwal, Uttar Pradesh

22

264 *Zanthoxylum nepalense* Langtang, C. Nepal

259 *Skimmia anquetilia*
Gulmarg, Kashmir

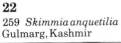

261 *Skimmia laureola* Tonglu, Sikkim

258 *Dictamnus albus*
Sonamarg, Kashmir

249 *Impatiens glandulifera* Sonamarg, Kashmir

269 *Euonymus echinatus*
Mewa Khola, E. Nepal

272 *Euonymus tingens*
Jumla, W. Nepal

290 *Staphylea emodi*
Kishtwar, Kashmir

23

268 *Ilex fragilis* Arun, E. Nepal

267 *Ilex dipyrena* Arun, E. Nepal

94 *Rhus javanica*
Himal Chuli, C. Nepal

275 *Ampelocissus rugosa*
Trisuli, C. Nepal

278 *Vitis jacquemontii*
Tons, Garhwal, Uttar Pradesh

302 *Entada phaseoloides* Tamur, E. Nepal

299 *Albizia chinensis* C. Nepal

24

307 *Cassia fistula* Garhwal, Uttar Pradesh

298 *Albizia julibrissin*
Garhwal, Uttar Pradesh

309 *Cassia tora*
Buri Gandaki, C. Nepal

300 *Mimosa pudica* E. Nepal

306 *Caesalpinia decapetala* Karnali, W. Nepal

25

301 *Mimosa rubicaulis* Trisuli, C. Nepal

311 *Erythrina stricta* Karnali, W. Nepal

310 *Erythrina arborescens* Trisuli, C. Nepal

305 *Bauhinia vahlii*
Narkanda, Himachal Pradesh

26

303 *Bauhinia purpurea* C. Nepal

317 *Desmodium elegans*
Langtang, C. Nepal

329 *Thermopsis barbata* × 1
Garhwal, Uttar Pradesh

341 *Indigofera heterantha*
Manali, Himachal Pradesh

314 *Butea minor* Karnali, W. Nepal

27

321 *Piptanthus nepalensis* E. Nepal

334 *Caragana brevifolia* Zanskar, Ladakh

328 *Parochetus communis* C. Nepal

28

358 *Astragalus candolleanus* Chamba, Himachal Pradesh

369 *Astragalus floridus*
Naulekh, E. Nepal

338 *Caragana brevispina*
Jumla, W. Nepal

366 *Astragalus rhizanthus*
Lahul, Himachal Pradesh

359 *Astragalus grahamianus* Kishtwar, Kashmir

339 *Colutea multiflora* Dudh Kosi, E. Nepal

29

379 *Gueldenstaedtia himalaica* Langtang, C. Nepal

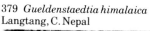

377 *Chesneya cuneata* Dras, Ladakh

378 *Chesneya nubigena* W. Nepal

381 *Hedysarum sikkimense* 373 *Oxytropis williamsii* Jomsom, C. Nepal
Naulekh, E. Nepal

30

383 *Hedysarum campylocarpon*
Langtang, C. Nepal

374 *Oxytropis lapponica* Kashmir

351 *Lathyrus humilis*
Chamba, Himachal Pradesh

382 *Hedysarum cachemirianum* Kashmir

3 *Rubus hoffmeisterianus* Kashmir

391 *Rubus ellipticus* Sikkim

8 *Rubus paniculatus*
ngtang, C. Nepal

409 *Prinsepia utilis* Saraj, Himachal Pradesh

9 *Rubus acuminatus*
rjeeling

395 *Sorbaria tomentosa* Kashmir

398a *Rosa laevigata*
Kathmandu, C. Nepal

32

398 *Rosa sericea* E. Nepal

400a *Rosa foetida*
Lahul, Himachal Pradesh

400 *Rosa webbiana* Lahul, Himachal Pradesh

396 *Rosa brunonii* C. Nepal

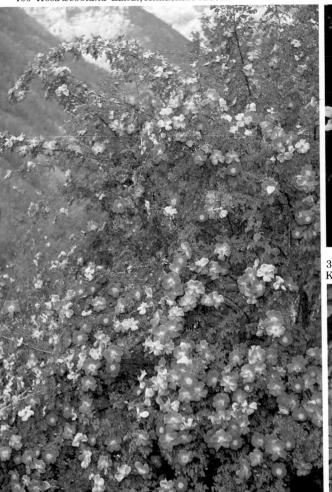

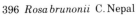

399 *Rosa macrophylla*
Khumbu, E. Nepal

412 *Spiraea arcuata* Khumbu, E. Nepal

404 *Prunus carmesina* Dhailekh, W. Nepal

33

07 *Prunus cornuta*
Kulu, Himachal Pradesh

403 *Prunus cerasoides* Darjeeling

06 *Prunus rufa* W. Nepal

411 *Spiraea canescens* Dudh Kosi, E. Nepal

426 *Pyracantha crenulata* C. Nepal

34

417 *Cotoneaster microphyllus* agg. Garhwal, Uttar Pradesh

420 *Cotoneaster bacillaris* Chamba, Himachal Pradesh

422 *Cotoneaster frigidus* Dudh Kosi, E. Nepal

421 *Cotoneaster affinis* Langtang, C. Nepal

431 *Pyrus pashia* C. Nepal

418 *Cotoneaster nummularia*
Kishtwar, Kashmir

430 *Malus sikkimensis*
Karnali, W. Nepal

410 *Neillia rubiflora* Malemchi, C. Nepal

432 *Sorbus cuspidata* Arun, E. Nepal

439 *Fragaria nubicola* E. Nepal

441 *Geum elatum* Kashmir

36

433 *Sorbus rhamnoides*
Annapurna Himal, C. Nepal

435 *Sorbus microphylla* Khumbu, E. Nepal

52 *Potentilla peduncularis*
anesh Himal, C. Nepal

455 *Potentilla atrosanguinea* Kashmir

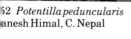

6 *Potentilla curviseta* Kashmir

442 *Geum sikkimense*
Ganesh Himal, C. Nepal

449 *Potentilla anserina* Dolpo, W. Nepal

38

441 *Geum elatum* Jumla, W. Nepal

440 *Fragaria daltoniana* E. Nepal

447 *Potentilla eriocarpa* Arun, E. Nepal

450 *Potentilla coriandrifolia* C. Nepal

451 *Potentilla microphylla* Rolwaling, C. Nepal

456 *Potentilla curviseta* Kashmir

459 *Sibbaldia purpurea* Kulu, Himachal Pradesh

458 *Sibbaldia cuneata* Nichinai, Kashmir

446 *Potentilla cuneata*
Maulekh, E. Nepal

482 *Saxifraga sibirica* Nichinai, Kashmir

465 *Bergenia ciliata*
Saraj, Himachal Pradesh

466 *Bergenia stracheyi*
Kulu, Himachal Pradesh

40

477 *Saxifraga mucronulata*
Sinthan, Kashmir

472 *Saxifraga parnassifolia*
Ganesh Himal, C. Nepal

94 *Parnassia cabulica* Suru, Ladakh

478 *Saxifraga stenophylla*
Bara Lacha, Lahul, Himachal Pradesh

41

75 *Saxifraga lychnitis* Dolpo, W. Nepal

492 *Saxifraga roylei* Ganesh Himal, C. Nepal

480 *Saxifraga saginoides* Langtang, C. Nepal

483 *Saxifraga asarifolia*
Garhwal, Uttar Pradesh

42

468 *Chrysosplenium carnosum*
Naulekh, E. Nepal

479 *Saxifraga jacquemontiana* Nichinai, Kashmir

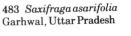

476 *Saxifraga brunonis*
Arun, E. Nepal

489 *Saxifraga pulvinaria* Kanjiroba Himal, W. Nepal

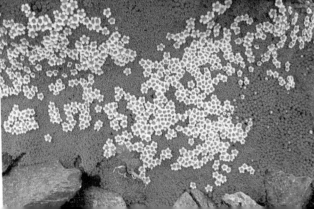

498 *Saxifraga androsacea* Nichinai, Kashmir

501 *Deutzia bhutanensis*
Mewa Khola, E. Nepal

43

500 *Saxifraga andersonii* Kanjiroba Himal, W. Nepal

502 *Dichroa febrifuga*
C. Nepal

481 *Saxifraga engleriana* Naulekh, E. Nepal

485 *Saxifraga pseudo-pallida*
Garhwal, Uttar Pradesh

495 *Hydrangea anomala*
Mewa Khola, E. Nepal

44

498 *Philadelphus tomentosus* Dudh Kosi, E. Nepal

499 *Deutzia staminea*
Trisuli, C. Nepal

500 *Deutzia compacta* Dudh Kosi, E. Nepal

506 *Ribes orientale*
Chamba, Himachal Pradesh

505 *Ribes griffithii*
Arun, E. Nepal

508 *Ribes takare*
Dudh Kosi, E. Nepal

520 *Sedum ewersii*
Zanskar, Ladakh

511 *Rosularia rosulata*
Pindari, Garhwal, Uttar Pradesh

45

509 *Kalanchoe spathulata*
Trisuli, C. Nepal

513 *Rhodiola wallichiana* Kashmir

16 *Rhodiola himalensis*
88 *Cremanthodium decaisnei*

519 *Rhodiola bupleuroides*
Langtang, C. Nepal

517 *Rhodiola heterodonta*
Chamba, Himachal Pradesh

524 *Parrotiopsis jacquemontiana*
Wardwan, Kashmir

525 *Melastoma normale* C. Nepal

526 *Osbeckia stellata* C. Nepal

527 *Osbeckia nepalensis* C. Nepal

531 *Woodfordia fruticosa*
Narkanda, Himachal Pradesh

528 *Oxyspora paniculata* C. Nepal

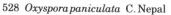

47

535 *Epilobium latifolium* Zanskar, Ladakh

545 *Begonia picta* C. Nepal

539 *Epilobium laxum*
Manali Himachal Pradesh

543 *Thladiantha cordifolia*
Arun, E. Nepal

556 *Pleurospermum govanianum*
Zanskar, Ladakh

48

558 *Pleurospermum candollei* Kashmir

542 *Trichosanthes tricuspidata* C. Nepal

538 *Epilobium wallichianum*
Trisuli, C. Nepal

547 *Begonia rubella*
Godavari, C. Nepal

548 *Opuntia monacantha*
C. Nepal

541 *Herpetospermum pedunculosum* Trisuli, C. Nepal

567 *Cortiella hookeri*
Gosainkund, C. Nepal

49

564 *Heracleum pinnatum* Zanskar, Ladakh

538 *Chaerophyllum reflexum*
Kulu, Himachal Pradesh

566 *Cortia depressa* Nichinai, Kashmir

552 *Ferula jaeschkeana* Dras, Ladakh

555 *Pleurospermum benthamii* Langtang, C. Nepal

50

562a *Heracleum lallii* Jumla, W. Nepal

553 *Prangos pabularia* Dras, Ladakh

577 *Benthamidia capitata* C. Nepal

51

561 *Angelica cyclocarpa* Ganesh Himal, C. Nepal

560 *Selinum tenuifolium*
Ganesh Himal, C. Nepal

609 *Hymenopogon parasiticus* Dudh Kosi, E. Nepal

580 *Toricellia tiliifolia*
W. Nepal

52

590 *Lonicera cyanocarpa* var.
porphyrantha Naulekh, E. Nepal

604 *Viburnum erubescens* Gatlang, C. Nepal

583 *Leycesteria formosa* Bhoté Kosi, Tibet

607 *Viburnum nervosum* Dudh Kosi, E. Nepal

587 *Lonicera rupicola*
Khumbu, E. Nepal

584 *Lonicera glabrata*
Buri Gandaki, C. Nepal

594 *Lonicera hispida*
W. Nepal

599 *Lonicera quinquelocularis*
Lahul, Himachal Pradesh

588 *Lonicera spinosa* Zanskar, Ladakh

606 *Viburnum cotinifolium*
Chamba, Himachal Pradesh

54

582 *Abelia triflora* Suli Gad, W. Nepal

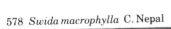

578 *Swida macrophylla* C. Nepal

605 *Viburnum grandiflorum* W. Nepal

614a *Coffea benghalensis*
Sun Kosi, C. Nepal

616a *Xeromphis spinosa* Sun Kosi, C. Nepal

616 *Randia tetrasperma*
Garhwal, Uttar Pradesh

610 *Luculia gratissima* C. Nepal

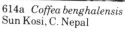

614 *Pavetta indica*
Kamur, E. Nepal

626 *Morina polyphylla* Langtang, C. Nepal

612 *Leptodermis kumaonensis* Dudh Kosi, E. Nepa

56

622 *Valeriana pyrolifolia*
Gangerbal, Kashmir

625 *Morina longifolia* Kulu, Himachal Pradesh

627 *Morina nepalensis* C. Nepal

628 *Pterocephalus hookeri* 1188 *Bistorta macrophylla* Dolpo, W. Nepal

57

630 *Dipsacus inermis* E. Nepal 624 *Morina coulteriana* Sat Sar, Kashmir

634 *Aster albescens* Trisuli C. Nepal

644 *Erigeron multiradiatus* Langtang, C. Nepal

58

638 *Aster falconeri* Kashmir

656 *Inula grandiflora* Kashmir

657 *Inula royleana* Kashmir

59

633 *Solidago virga-aurea* Kashmir

660 *Inula obtusifolia* Zanskar, Ladakh

640 *Aster himalaicus*
Ganesh Himal, C. Nepal

60

639 *Aster flaccidus* Bara Lacha, Lahul, Himachal Pradesh

645 *Erigeron bellidioides* Darjeeling

646 *Psychrogeton andryaloides*
Lahul, Himachal Pradesh

641 *Aster stracheyi* Dolpo, W. Nepal

637 *Aster diplostephioides* Rolwaling, C. Nepal

650 *Leontopodium jacotianum*
Ganesh Himal, C. Nepal

651 *Anaphalis triplinervis* var. *monocephala*
Kashmir

61

759 *Cicerbita macrorhiza*
Ganesh Himal, C. Nepal

649 *Leontopodium himalayanum* Kashmir

648 *Gnaphalium hypoleucum*
Garhwal, Uttar Pradesh

652 *Anaphalis margaritacea*
Malemchi, C. Nepal

651 *Anaphalis triplinervis*
Malemchi, C. Nepal

662 *Guizotia abyssinica* Trisuli, C. Nepal

658 *Inula hookeri* Langtang, C. Nepal

62

714 *Saussurea gnaphalodes*
Sangdah La, C. Nepal

664 *Waldheimia glabra* Kashmir

70a *Tanacetum tibeticum*
hul, Himachal Pradesh

667 *Tanacetum gracile* Ladakh

63

5a *Waldheimia stoliczkai*
nasi La, Zanskar, Ladakh

690 *Cremanthodium ellisii* Zanskar, Ladakh

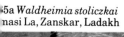

669 *Tanacetum gossypinum* Langtang, C. Nepal

668 *Tanacetum dolichophyllum* Kashmir

64

665 *Waldheimia tomentosa* Umasi La, Zanskar Ladakh

703 *Senecio chrysanthemoides* Kashmir

661 *Inula rhizocephala* var. *rhizocephaloides* Suru, Ladakh

655 *Inula cappa* C. Nepal

2 *Cremanthodium nepalense* Rolwaling, C. Nepal

742 *Gerbera gossypina*
Kulu, Himachal Pradesh

65

1 *Cremanthodium oblongatum*
nesh Himal, C. Nepal

687 *Cremanthodium reniforme* W. Nepal

5 *Senecio graciliflorus* Gosainkund, C. Nepal

696 *Doronicum roylei* Kashmir

686 *Cremanthodium arnicoides*
Dolpo, W. Nepal

66

706 *Ligularia amplexicaulis*
Kulu, Himachal Pradesh

684 *Cremanthodium retusum*
Langtang, C. Nepal

702 *Senecio jacquemontianus* Kashmir

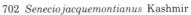

709 *Cousinia thomsonii*
Lahul, Himachal Pradesh

756 *Soroseris hookerana* Dolpo, W. Nepal

728 *Jurinea dolomiaea* Ganesh Himal, C. Nepal

752 *Lactuca lessertiana* Kashmir

712 *Saussurea simpsoniana*
Karakoram, Pakistan

711 *Saussurea gossypiphora* Dolpo, W. Nepal

68

710 *Saussurea graminifolia*
Ganesh Himal, C. Nepal

723 *Saussurea obvallata* Topke Gola, E. Nepal

721 *Saussurea jacea* Suru, Ladakh

719 *Saussurea albescens* Dras, Ladakh

69

722 *Saussurea fastuosa*
Langtang, C. Nepal

717 *Saussurea roylei*
Kulu, Himachal Pradesh

715 *Saussurea nepalensis*
Ganesh Himal, C. Nepal

731 *Cirsium falconeri* Aphorwat, Kashmir   733 *Cirsium verutum* Langtang, C. Nepal

70
740 *Ainsliaea aptera* × ²/₃
Garhwal, Uttar Pradesh   708 *Echinops cornigerus* Kargil, Ladakh

Onopordum acanthium Kashmir

731 *Cirsium falconeri* Langtang, C. Nepal

71

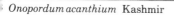

Scorzonera virgata Zanskar, Ladakh

783 *Cyananthus lobatus* Ganesh Himal, C. Nepal

773 *Codonopsis affinis* Arun. E. Nepal

765 *Campanula latifolia* Gulmarg, Kashmir

774 *Codonopsis viridis* C. Nepal

72

772 *Codonopsis convolvulacea* Tukucha, C. Nepal

781 *Cyananthus incanus* Dolpo, W. Nepal

777 *Codonopsis clematidea* Lahul, Himachal Pradesh

778 *Codonopsis ovata* Kashmir

806 *Rhododendron setosum* Naulekh, E. Nepal

787 *Enkianthus deflexus* E. Nepal

73

807 *Rhododendron nivale* Rolwaling, C. Nepal

786 *Cassiope fastigiata*
Chamba, Himachal Pradesh

814 *Rhododendron lowndesii*
Phoksumdo Tal, W. Nepal

805 *Rhododendron lepidotum* Junbesi, E. Nepal

797 *Rhododendron arboreum*
Malemchi, C. Nepal

790 *Rhododendron campanulatum*
Kulu, Himachal Pradesh

74

801 *Rhododendron cinnabarinum*
Dudh Kosi, E. Nepal

796 *Rhododendron barbatum* Malemchi, C. Nepal

9 *Rhododendron dalhousiae* Darjeeling

810 *Rhododendron triflorum* E. Nepal

75

795 *Rhododendron hodgsonii*
Topke Gola, E. Nepal

0 *Rhododendron lindleyi* Hongu Khola, E. Nepal

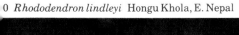

6 *Rhododendron cowanianum*
nesh Himal, C. Nepal

802 *Rhododendron ciliatum*
Dudh Kosi, E. Nepal

792 *Rhododendron wightii* 817 *Lyonia ovalifolia* Annapurna Himal, C. Nepal
Tamur, E. Nepal

76

819 *Pieris formosa* Junbesi, E. Nepal   788 *Rhododendron thomsonii* Dzongri, Sikkim

793 *Rhododendron falconeri*
Tamur, E. Nepal   804 *Rhododendron anthopogon* Rolwaling, C. Nepal

27 *Agapetes incurvata* Arun, E. Nepal

821 *Gaultheria trichophylla* C. Nepal

23 *Gaultheria fragrantissima* Darjeeling

830 *Vaccinium vacciniaceum*
Tamur, E. Nepal

6 *Agapetes serpens* E. Nepal

822 *Gaultheria pyroloides*
Dudh Kosi, E. Nepal

828 *Vaccinium nummularia*
Arun, E. Nepal

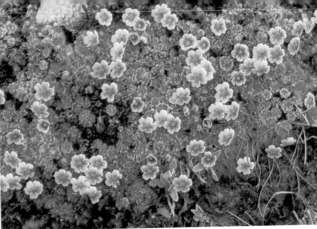

834 *Acantholimon lycopodioides* Ladakh

844 *Androsace globifera* Rolwaling, C. Nepal

840 *Androsace strigillosa* W. Nepal

843 *Androsace delavayi* Dolpo, W. Nepal

836 *Androsace rotundifolia* Kishtwar, Kashmir

842 *Androsace lehmannii* Annapurna Himal, C. Nepal

839 *Androsace primuloides* Kashmir

848 *Androsace mucronifolia* Kashmir

845 *Androsace tapete* 863a *Primula walshii* S. Tibet

838 *Androsace sarmentosa* Garhwal, Uttar Pradesh

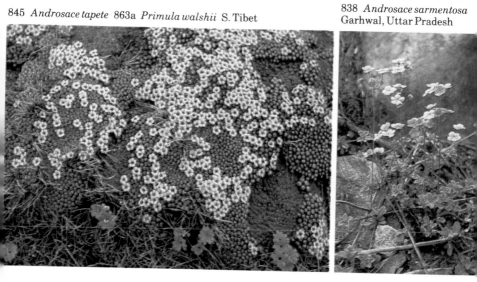

349 *Androsace sempervivoides* Aphorwat, Kashmir

847 *Androsace zambalensis* Dolpo, W. Nepal

870 *Primula calderana* subsp. *strumosa*
Arun, E. Nepal

80

885 *Primula wigramiana* Ganesh Himal, C. Nepal

887 *Cortusa brotheri* Kashmir

851 *Primula geraniifolia* Dudh Kosi, E. Nepal

868 *Primula stuartii* Ganesh Himal, C. Nepal

878 *Primula rotundifolia* Naulekh, E. Nepal

865 *Primula macrophylla*
Kanjiroba Himal, W. Nepal

882 *Primula sikkimensis* Dolpo, W. Nepal

850 *Primula glomerata*
C. Nepal

82

852 *Primula denticulata* Kulu, Himachal Pradesh

870 *Primula calderana* Zongri, Sikkim

867 *Primula obliqua* Barun, E. Nepal

863 *Primula reptans*
Kulu, Himachal Pradesh

864 *Primula primulina* Rolwaling, C. Nepal

Primula involucrata
hmir

860 *Primula rosea* Kashmir

859 *Primula elliptica*
Kashmir

83

Primula irregularis
emchi, C. Nepal

876 *Primula edgeworthii* Kopta, W. Nepal

Primula wollastonii Rolwaling, C. Nepal

853 *Primula atrodentata*
Naulekh, E. Nepal

884 *Primula reidii* W. Nepal

871 *Primula aureata* Gosainkund, C. Nepal

84

882 *Primula sikkimensis* var. *hopeana* E. Nepal 869 *Primula poluninii* Sisne Himal, W. Nepal

880 *Primula caveana* Rolwaling, C. Nepal

888 *Omphalogramma elwesiana* Barun, E. Nepal

894 *Symplocos paniculata* Kashmir

901 *Jasminum humile*
Kishtwar, Kashmir

903 *Jasminum officinale* Jumla, W. Nepal

889 *Maesa chisia*
Godavari, C. Nepal

7 *Marsdenia roylei* C. Nepal

918 *Calotropis gigantea*
C. Nepal

916 *Hoya lanceolata*
Modi Khola, C. Nepal

910 *Beaumontia grandiflora* E. Nepal

912 *Trachelospermum lucidum* C. Nepal

86

911 *Nerium indicum* Garhwal, Utter Pradesh

936 *Swertia alternifolia* Kali Gandaki, C. Nepal

909 *Osmanthus suavis* E. Nepal

941 *Gentiana cachemirica* Kashmir

948 *Gentiana ornata* Khumbu, E. Nepal

929 *Lomatogonium carinthiacum* × 1
Suru, Ladakh

944 *Gentiana phyllocalyx* Arun, E. Nepal

945 *Gentiana tubiflora* Arun, E. Nepal

46 *Gentiana depressa* C. Nepal

938 *Crawfurdia speciosa* Darjeeling

934 *Swertia hookeri*
Hongu, E. Nepal

88

942 *Gentiana tianschanica*
Kashmir

960 *Gentianella paludosa*
Zanskar, Ladakh

963 *Jaeschkea oligosperma*
Kashmir

933 *Swertia petiolata*
Kashmir

927 *Exacum tetragonum*
Ganesh Himal, C. Nepal

954 *Gentiana carinata* Kashmir

961 *Megacodon stylophorus* Khumbu, E. Nepal

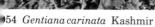

959 *Gentianella moorcroftiana* Zanskar, Ladakh

964 *Nymphoides peltata* Kashmir

966 *Solenanthus circinnatus* 985 *Pseudomertensia moltkioides* Kashmir
Chamba, Himachal Pradesh

90

983 *Pseudomertensia echioides*
Lahul, Himachal Pradesh

984 *Pseudomertensia nemerosa*
Kashmir

971 *Onosma hispidum*
Dras, Ladakh

990 *Trigonotis rotundifolia* E. Nepal

975 *Hackelia uncinata* Kashmir

967 *Arnebia benthamii* Kashmir

91

968 *Arnebia euchroma*
Bara Lacha, Lahul, Himachal Pradesh

978 *Lindelofia stylosa* Zanskar, Ladakh

979 *Lindelofia longiflora*
Kashmir

92

1003 *Cestrum nocturnum*
C. Nepal

996 *Ipomoea purpurea*
C. Nepal

993 *Argyreia hookeri* Bheri, W. Nepal

1007 *Datura stramonium* E. Nepal

1006 *Physochlaina praealta*
Lahul, Himachal Pradesh

997 *Ipomoea carnea* Jammu

992 *Cuscuta europaea* var. *indica*
Kashmir

998 *Porana grandiflora* Trisuli, C. Nepal

1005 *Hyoscyamus niger*
Lahul, Himachal Pradesh

93

1020 *Lindenbergia grandiflora* C. Nepal

1008 *Datura suaveolens* Kathmandu, C. Nepal

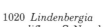

1002 *Mandragora caulescens* Jumla, W. Nepal

1018 *Mazus surculosus* C. Nepal

1019 *Lancea tibetica* Kargil, Ladakh

1049 *Pedicularis pyramidata* Kashmir

1038 *Pedicularis oederi*
Kulu, Himachal Pradesh

1025 *Lagotis kunawurensis* Naulekh, E. Nepal

1040 *Pedicularis scullyana* C. Nepal

1044 *Pedicularis hoffmeisteri* C. Nepal

1039 *Pedicularis bicornuta* Sat Sar, Kashmir

1050 *Pedicularis trichoglossa* Phoksumdo, W. Nepal

1061 *Orobanche cernua* × 1 Lahul, Himachal Pradesh

1043 *Pedicularis longiflora* var. *tubiformis* C. Nepal

96

1048 *Pedicularis pectinata* Kashmir

1060 *Boschniakia himalaica* E. Nepal

1062 *Orobanche alba*
Lahul, Himachal Pradesh

1069 *Chirita urticifolia* Malemchi, C. Nepal

1079 *Martynia annua* C. Nepal

1072 *Didymocarpus primulifolius*
Kali Gandaki, C. Nepal

1077 *Incarvillea mairei*
Phoksumdo Tal, W. Nepal

98

1063 *Lathraea squamaria*
Kulu, Himachal Pradesh

1066 *Aeschynanthus sikkimensis* Tamur, E. Nepal

1024 *Picrorhiza kurrooa*
Kashmir

1067 *Platystemma violoides*
E. Nepal

1090 *Justicia adhatoda*
Kathmandu, C. Nepal

988 *Barleria cristata*
. Nepal

1082 *Asystasia macrocarpa*
Karkani, C. Nepal

1085 *Pteracanthus
urticifolius* Kashmir

99

975 *Incarvillea arguta* Tukucha, C. Nepal

1091 *Phlogacanthus pubinervius* E. Nepal

1097 *Clerodendrum japonicum* Trisuli, C. Nepal

1096 *Clerodendrum indicum*
Kathmandu, C. Nepal

100

1137 *Nepeta podostachys* Lahul, Himachal Pradesh

1126 *Leonurus cardiaca*
Kashmir

1169 *Elsholtzia fruticosa*
Malemchi, C. Nepal

1168 *Elsholzia flava*
Malemchi, C. Nepal

1151 *Dracocephalum nutans*
Kashmir

53 *Salvia moorcroftiana*
shmir

1155 *Clinopodium piperitum*
C. Nepal

1125 *Lamium tuberosum*
Kali Gandaki, C. Nepal

101

55 *Salvia campanulata*
Nepal

1117 *Phlomis rotata* Dolpo, W. Nepal

5 *Nepeta connata* × ½
shmir

1162 *Salvia lanata*
Garhwal, Uttar Pradesh

1113 *Scutellaria prostrata*
× ⅓ Kashmir

1118 *Phlomis cashmeriana* Kashmir   1164 *Salvia hians* Kashmir

102

1120 *Phlomis bracteosa* Sat Sar, Kashmir

1127 *Stachys sericea* Kashmir

1130 *Eriophyton wallichii*
Gosainkund, C. Nepal

103

1133 *Colquhounia coccinea* C. Nepal

1105 *Leucosceptrum canum* Arun, E. Nepal

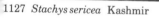

1111 *Scutellaria discolor*
Kathmandu, C. Nepal

104

1119 *Phlomis spectabilis*
Kashmir

1129 *Stachys tibetica*
Dras, Ladakh

1131 *Leucas lanata* C. Nepal

1106 *Orthosiphon incurvus*
Godavari, C. Nepal

1115 *Scutellaria scandens*
Garhwal, Uttar Pradesh

147 *Nepeta erecta* Kashmir

1166 *Salvia nubicola* × ½
Kulu, Himachal Pradesh

1108 *Rabdosia rugosa* C. Nepal

1170 *Elsholtzia eriostachya*
Kashmir

146 *Nepeta clarkei* Kashmir

1141 *Nepeta laevigata* Kashmir

1172 *Amaranthus caudatus*
C. Nepal

1177 *Phytolacca acinosa*
Kulu, Himachal Pradesh

1178 *Aconogonum molle* E. Nepal

106

1183 *Bistorta affinis* Ganesh Himal, C. Nepal

1206 *Oxyria digyna*
Langtang, C. Nepal

1186 *Bistorta amplexicaulis* var
pendula C. Nepal

188 *Bistorta macrophylla* Jumla, W. Nepal

1200 *Rheum australe*
Langtang, C. Nepal

107

190 *Persicaria capitata* C. Nepal 1202 *Rheum nobile* Topke Gola, E. Nepal

198 *Rheum moorcroftianum*
Dolpo, W. Nepal

1219 *Daphne mucronata* Kashmir

1208 *Houttuynia cordata* C. Nepal

108

1217 *Daphne bholua* var. *glacialis* C. Nepal

1222 *Wikstroemia canescens* W. Nepal

1221 *Edgeworthia gardneri*
. Nepal

1214 *Lindera pulcherrima*
E. Nepal

1207 *Aristolochia griffithii*
Dudh Kosi, E. Nepal

109

1218 *Daphne papyracea*
araj, Himachal Pradesh

1213 *Persea duthiei* C. Nepal

1211 *Dodecadenia*
grandiflora C. Nepal

1223 *Stellera chamaejasme*
Kali Gandaki, C. Nepal

1220 *Daphne retusa* Arun, E. Nepal

110

1224 *Elaeagnus parvifolia* C. Nepal

1229 *Scurrula elata* C. Nepal

1228 *Helixanthera parasitica* E. Nepal

1227 *Hippophae tibetana*
Khumbu, E. Nepal

53 *Elatostema platyphyllum*
...un, E. Nepal

1235 *Euphorbia milii*
Pokhara, C. Nepal

1248 *Debregeasia longifolia*
Arun, E. Nepal

111

1243 *Sarcococca saligna*
Saraj, Himachal Pradesh

..8 *Euphorbia cognata* Chamba, Himachal Pradesh

5 *Pilea scripta*
..gtang, C. Nepal

1237 *Euphorbia wallichii* Kashmir

1299 *Salix flabellaris* Kashmir

1234 *Euphorbia royleana* Jammu

112

1298 *Salix calyculata* C. Nepal

1292 *Salix sikkimensis* Topke Gola, E. Nepal

352 *Ponerorchis chusua*
anesh Himal, C. Nepal

1353 *Satyrium nepalense*
C. Nepal

1356 *Spiranthes sinensis*
Pokhara, C. Nepal

113

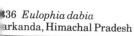

336 *Eulophia dabia*
arkanda, Himachal Pradesh

1329 *Cephalanthera longifolia*
Garhwal, Uttar Pradesh

1325 *Arundina graminifolia*
C. Nepal

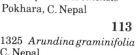

1342 *Habenaria pectinata*
W. Nepal

1333 *Dactylorhiza hatagirea*
Kashmir

1350 *Pecteilis susannae*
Trisuli, C. Nepal

114

1327 *Calanthe brevicornu*
× ⅔ C. Nepal

1328 *Calanthe plantaginea*
Godavari, C. Nepal

1363 *Cryptochilus luteus*
C. Nepal

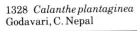

32 *Cypripedium himalaicum* Bheri, W. Nepal

1372 *Pleione praecox* × ⅔ Darjeeling

115

26 *Calanthe tricarinata* Kashmir

1347 *Malaxis muscifera*
Garhwal, Uttar Pradesh

1339 *Galeola lindleyana* Arun, E. Nepal

1358 *Bulbophyllum affine* C. Nepal

116

1330 *Cypripedium cordigerum* W. Nepal

1346 *Herminium monorchis*
Kulu, Himachal Pradesh

1373 *Rhynchostylis retusa*
C. Nepal

1367 *Dendrobium amoenum* × ½ C. Nepal

65 *Cymbidium longifolium*
rjeeling

1374 *Vanda cristata*
Arun, E. Nepal

1357 *Aerides multiflora*
Trisuli, C. Nepal

117

1371 *Pleione humilis*
Darjeeling

66 *Dendrobium densiflorum* E. Nepal

59 *Coelogyne ochracea*
Nepal

1362 *Coelogyne corymbosa*
C. Nepal

1349 *Oreorchis foliosa*
Ganesh Himal, C. Nepal

1376 *Costus speciosus* Pokhara, C. Nepal   1377 *Curcuma aromatica* Arun, E. Nepal

118

1384 *Hedychium ellipticum* C. Nepal   1383 *Hedychium aurantiacum* C. Nepal

85 *Hedychium spicatum* Himal Chuli, C. Nepal   1379 *Roscoea purpurea* Tamur, E. Nepal

80 *Roscoea capitata* Malemchi, C. Nepal   1378 *Roscoea alpina* Khumbu, E. Nepal

1388 *Iris lactea* Kashmir

1392 *Iris milesii*
Saraj, Himachal Pradesh

1393 *Iris goniocarpa*
Langtang, C. Nepal

120

1394 *Iris kemaonensis*
Kulu, Himachal Pradesh

1395 *Iris hookerana* Kashmir

1389 *Iris crocea* Kashmir

Agave sp. Garhwal, Uttar Pradesh

1411 *Zephyranthes carinata* Sikkim

1405 *Allium wallichii* Langtang, C. Nepal

1402 *Allium oreoprasum*
Dolpo, W. Nepal

1404 *Allium carolinianum*
Dolpo, W. Nepal

1406 *Allium caesium*
Lahul, Himachal Pradesh

1401 *Allium humile* Kashmir

1412 *Hypoxis aurea* Arun, E. Nepal

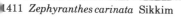

1397 *Crinum amoenum*
Pokhara, C. Nepal

122

1434 *Lilium wallichianum*
Trisuli, C. Nepal

1429 *Hemerocallis fulva*
Manali, Himachal Pradesh

1451 *Streptopus simplex*
Rolwaling, C. Nepal

1422 *Colchicum luteum* Kashmir

1439 *Notholirion thomsonianum*
Kishtwar, Kashmir

1432 *Lilium nepalense*
Kali Gandaki, C. Nepal

1419 *Cardiocrinum giganteum*
W. Nepal

1421 *Clintonia udensis*
C. Nepal

1445 *Polygonatum cirrhifolium*
Lahul, Himachal Pradesh

1452 *Theropogon pallidus*
Ganesh Himal, C. Nepal

1447 *Smilacina purpurea*
Kulu, Himachal Pradesh

1443 *Polygonatum hookeri*
Langtang, C. Nepal

1446 *Polygonatum multiflorum*
Chamba, Himachal Pradesh

1448 *Smilacina oleracea* Sikkim

1440 *Notholirion macrophyllum* W. Nepal

1454 *Tulipa stellata* var. *chrysantha* Kashmir

124

1442 *Paris polyphylla*
Garhwal, Uttar Pradesh

1424 *Eremurus himalaicus*
Lahul, Himachal Pradesh

428 *Gagea elegans* Lahul, Himachal Pradesh

1453 *Trillidium govanianum*
Chamba, Himachal Pradesh

125

1431 *Lilium oxypetalum*
Garhwal, Uttar Pradesh

425 *Fritillaria roylei* Kashmir

1457 *Juncus thomsonii* Langtang, C. Nepal

1462 *Pandanus nepalensis* C. Nepal

126

1481 *Remusatia hookerana*
Sikkim

1467 *Arisaema griffithii* Darjeeling

1481 *Remusatia hookerana*
Sikkim

1477 *Typhonium diversifolium* Jumla, W. Nepal

1474 *Arisaema erubescens* Ganesh Himal, C. Nepal

1464 *Arisaema propinquum* C. Nepal

1463 *Arisaema costatum* C. Nepal

1468 *Arisaema intermedium*
Garhwal, Uttar Pradesh

1471 *Arisaema jacquemontii* Kashmir

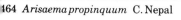

1476 *Sauromatum venosum*
Kulu, Himachal Pradesh

128

1469 *Arisaema flavum*
Kulu, Himachal Pradesh

1478 *Thomsonia napalensis* × ¹/₄
Tamur, E. Nepal

1470 *Arisaema tortuosum* W. Nepal

1472 *Arisaema nepenthoides*
Ganesh Himal, C. Nepal.

Glossary

achene   A one-seeded fruit (a); usually one of many in a fruiting head.   1

acorn   The fruit of *Quercus*. A nut (a) partially surrounded by a cup-like involucre, acorn-cup (b), of bracts or scales.   2

actinomorphic   Flowers regular, radially symmetrical. See 54a, 54b.

adpressed   Pressed flat to a surface as in leaves (a.)   3

alternate   Leaves placed singly, at different heights on a stem.   4

annual   A plant completing its life cycle in one season or one year, cf. biennial.

annulus   A ring of tissue.

anther   The part of the stamen (a) containing the pollen grains.   5

apiculate   With a small broad point at apex (a).   6

appendage   An additional part of a structure or organ (calyx, corolla, or of the spadix, see 124a).

aril   An appendage or outer, often fleshy covering (a), of a seed.   7

auricles   Ear-like extensions (a) at base of leaf-blade.   8

awl-shaped   Broad-based and tapering to a sharp point as in leaves (a).   9

awn   A stiff bristle-like projection (a) borne on an organ.   10

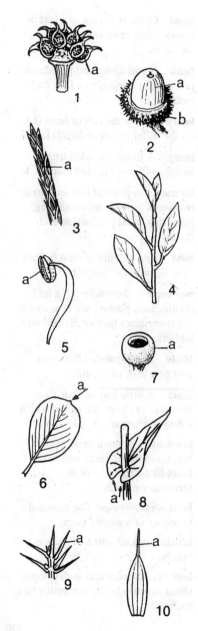

axil, axillary   The angle between the leaf and stem; hence axillary flower, or bud (a).   11

basal   Often referring to a leaf or leaves which arise from the base of the stem.

beak   A stiff elongated projection (a); usually applied to a fruit or ovary.   12

beard   In *Iris* the tuft of hairs (b) on the outer petals or falls (a).   13

berry   A fleshy rounded fruit, usually with hard pips or seeds.   14

biennial   A plant of two seasons or two years growth, from seedling, flowering, seed production and death.

bifid   Divided into two no further than the middle. See 16b

bipinnate = 2-pinnate. Of a leaf divided into pinnae, see 29a, which are themselves further divided into leaflets.

blade   The expanded flattened part of a leaf, petal, etc.

bract   A little leaf or scale-like structure (a) from the axil of which a flower often arises.   15, 127a

bracteole   A small or secondary bract; in Umbelliferae one of the bract-like structures of the secondary umbels (a).   16

branchlets = twigs. The terminal branches of a woody stem.

bristle   A long stiff hair, hence bristly.

bud   A compact and undeveloped shoot surrounded by protective bud scales (a).   17

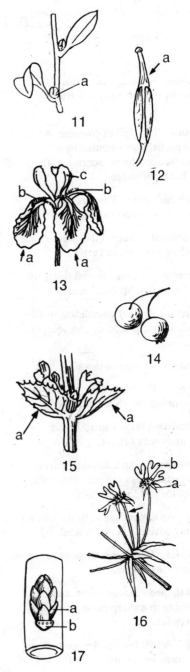

520

bulb Underground storage organ consisting of one or many fleshy scales (a) arising from a basal plate of tissue and encircling a growing point. 18

bulbil A small bulb or swollen bud (a) arising from the stem or inflorescence. 19

bush see shrub.

calyx The sepals collectively; often joined together in a tube, the calyx-tube (a), and with calyx-lobes (b), or calyx-teeth. 20

campanulate Bell-shaped. 21

capitulum = flower-head. A head of small stalkless florets (a) surrounded by an involucre of bracts (b), as in Compositae. 22

capsule A dry fruit formed of two or more fused carpels which splits open when ripe. 23

carpel One of the units (a) of the female part of the flower; they are either separate or fused together into a fruit. 24

cartilaginous Tough and hard: resembling cartilage.

catkin A crowded spike of tiny flowers, usually hanging and tassel-like. 25

ciliate Fringed with hairs along the margin. 26

cladode A green leaf-like lateral shoot (a). 27

claw The lower stalk-like part (b) of some petals (often called a stalk in the descriptions). See 98

cleistogamous Flowers which never open and are self-pollinated.

521

coalesce Growing together to form one whole.

column A structure (a) formed by the fusion of the stamens, styles and stigmas, as in Orchidaceae. 28

compound e.g: compound inflorescence has a branched axis, see 35; a compound leaf has several or many distinct leaflets. 29

cone A distinct rounded or elongated structure often becoming woody and composed of many overlapping scales (a) which bear pollen, or seeds when ripe. 30

connate Organs of the same kind joined together, eg., 2 leaves joined by their bases.

connective The tissue or structure joining the two anthers together. See 50b.

cordate Heart-shaped. 31

corm A swollen underground stem surrounded by a papery covering, the tunic (a), and replaced annually by a new corm. 32

corolla The petals collectively; often joined together into a corolla-tube (a) with corolla-lobes (b). 33

corona = **crown** An additional organ or appendage, arising between the corolla and stamens, or on the corolla (a). 34

crenate = **crenulate** With rounded teeth. See 138.

crest An elevated toothed or irregular ridge of tissue on an organ.

crisped Curled: an extreme form of undulate. See 146

crown = **corona** See 34a.

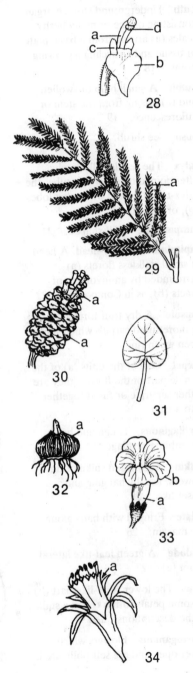

522

cyme A broad more or less flat-topped inflorescence formed of axillary branches terminating in a flower, the central flowers (a) maturing first. 35

deciduous Falling off, as with leaves in the dry season or autumn, and leaving a leaf-scar.

decurrent With the base of the leaf-blade or other organ, continuing as a wing down the stem or stalk (a). 36

decussate Leaves in opposite pairs arranged alternately at right-angles to each other up stem. 37

deflexed = reflexed Bent sharply downwards (a). 38

dehiscent Splitting as in fruits; opening by pores or valves. See 23.

dichotomous Stems branched into 2 equal branches, often equally forked again. 39

digitate Leaves with several finger-like leaflets radiating from a central point and joined only at the base. See 40.

dioecious Having male and female flowers on separate plants.

disk A fleshy, sometimes nectar-secreting, part of the receptacle (a) which surrounds, or surmounts the ovary. 41

disk-florets In Compositae: the central regular 5-lobed florets (a), in contrast to the strap-shaped ray-florets (b). See 52, 107.

dissected Divided into many segments. See 29.

dorsal = back Relating to the back or upper surface of a part or organ, as eg. the upper side of a leaf; the opposite to ventral.

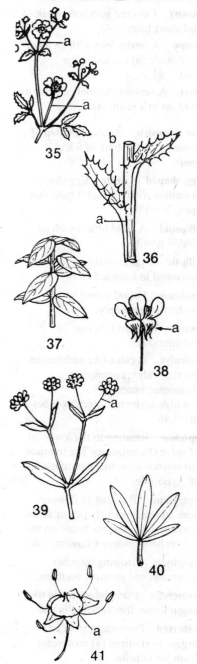

523

downy Covered with soft, weak and short hairs. 42

drupe A fleshy fruit with an inner hard stone (a) enclosing the seed. 43

duct A slender channel carrying a fluid, as in a resin duct.

ear = auricle An ear-like flap of tissue (a) at the base of a leaf blade. 44

egg-shaped = ovate Egg-shaped in outline with a broader base than apex. See 138.

ellipsoid A solid object with an elliptic profile.

elliptic Oval in outline and narrowed to rounded ends. 45

endemic Found native in one country or area only.

entire Without lobes or indentations.

epicalyx A calyx-like addition to the true calyx; common in Rosaceae: hence epicalyx-segments, epicalyx-lobes (a). 46

epichile Relating to the lower lip of some Orchidaceae. The terminal (b) often tongue-like part of the lip; cf. hypochile. 47

epigynous Relating to flowers with an inferior ovary (a), or to the perianth and stamens borne on the ovary: hence inferior flowers. 48

epiphytic Growing on other plants but not parasitic on them.

evergreen Plants which retain their green leaves throughout the year.

exserted Protruding from an organ; as stamens (a) protruding from the corolla. 49

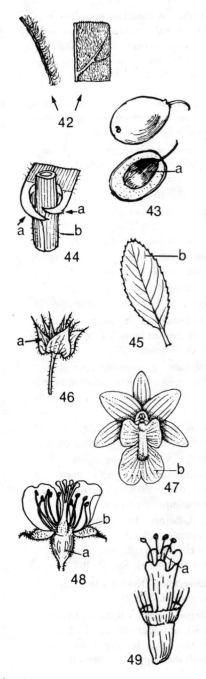

524

eye Relating to the colour of the corolla; the central distinctive and differently coloured part of the flower, as in some *Primula* species.

falls In *Iris*, the outer set of perianth-segments which are down-curved (a), in contrast to the erect standards (c). See 13.

farina A mealy or powdery covering to an organ, as in many *Primula* species: hence farinaceous.

female flowers Flowers with a fertile ovary but without fertile stamens.

ferruginous Rust-coloured.

filament The slender stalk (a) of the stamen, which bears the anthers. 50

flat-topped cluster A branched cluster of flowers, with all flowers terminating at approximately the same level. 51

flexuous Having a wavy or zigzag form: as of some stems.

floret A small flower, usually one of a dense cluster or a head, as in the Compositae; often of two forms: ray-florets (b), see 107, and disk-florets (a). 52

flower-head A densely packed group of flowers or florets. 53, see 22

flowers regular: irregular Regular flowers have equal petals 54a, or a corolla with lobes arranged symmetrically 54b; irregular flowers have petals or a corolla with lobes arranged asymmetrically and divisible in one plane only into 2 equal halves. 54c

foetid Stinking, strong and unpleasant smelling.

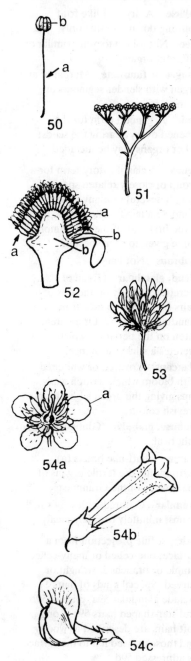

525

follicle   A dry pod-like fruit splitting down one side only.   55

free   Not joined to other similar or different organs.

fringed = fimbriate   Margin of an organ with slender segments or hairs.   56

fruit   A general term for the ripened ovary bearing the seeds; other organs may be included.

genus   A classificatory term for a group of closely related species; usually several or many genera comprise a family. The generic name is the first part of the latin binomal name given to each species.

glabrous   Not hairy.

gland, glandular   Organ of secretion, usually on the tips of hairs (a); hence glandular or glandular-hairy. 57. Or an area, often on the perianth, which secretes a sticky substance.

glaucous   Covered or whitened with bloom which is often waxy, thus giving the organ a bluish or greyish colour.

globose, globular   Globe-shaped, spherical.

glume   Chaff-like bracts (a) surrounding the fertile organs of the flowers of the Gramineae.   58

granular   Covered with very small grains, minutely or finely mealy.

hair   A fine projection from a surface; one-celled or many-celled, simple or branched, straight or curved. Special kinds of hairs include glandular 59a, crisped 59b and star-shaped hairs 59c. Long soft hairs are described as woolly and those pressed flat to the surface as adpressed 59d.

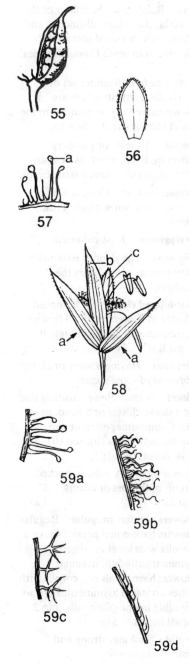

55

56

57

58

59a

59b

59c

59d

half-inferior = perigynous Ovary situated in a hollow with other organs borne on the rim (a). 60

hastate = spear-shaped Shaped like a spear-head with pointed basal lobes projecting outwards at an angle to the blade. 61

head A short dense cluster of flowers, see 53. Also capitulum in Compositae. See 22.

heart-shaped = cordate Ovate in general outline but with 2 rounded basal lobes, particularly referring to the shape of the leaf-blade. See 31.

herbaceous Plants without woody stems, dying down each year or season; also referring to shoots before they become woody.

hermaphrodite With both fertile stamens and ovaries present in the same flower.

hip The fruit of *Rosa*: the receptacle (a) is fleshy and encircles the achenes (b). 62

hispid Coarsely and stiffly hairy.

hoary Covered with very short dense hairs giving a whitish appearance to the organ.

hood The upper petal and/or sepal (a) of certain Orchid and Aconite flowers, hence hooded. 63

hybrid The result of cross-breeding between two different species of plants, the offspring possessing some of the characters of each parent plant.

hypochile The basal often enlarged part of the lip (c) of some Orchidaceae; it is sometimes hinged with the terminal epichile as in *Cephalanthera*. See 28.

60

61

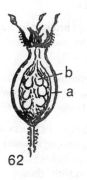

62

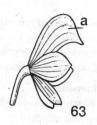

63

hypogynous = superior flower A flower in which the stamens are inserted close to and beside or beneath the base of the ovary, cf. perigynous and epigynous. See 134.

indehiscent Not splitting to release the seeds.

indumentum The hairy coverage of an organ.

inferior ovary = epigynous Ovary situated below the sepals, petals and stamens in the flower. See 48a.

inflorescence Flowering branch, including bracts, flower-stalks and flowers. See 35, 39, 105.

internode The part of the stem (b) between two adjacent nodes (a). 64

interrupted Not continuous, as in an interrupted spike.

introduced A plant brought into a country or region by man, and not native there.

involucel An additional calyx-like structure (b), below the true calyx of each flower, as in Dipsacaceae. 65

involucre A collection of bracts (a) or leaves, surrounding a flower or flower-head, sometimes becoming woody in fruit: thus involucral-bracts. 66

keel A sharp central ridge on an organ resembling the keel of a boat. The lower petal or petals of a flower (b), as in most Leguminosae. 67

kidney-shaped The shape of leaves in which the blade is broader than long and attached at the narrowest part to the leaf-stalk. 68

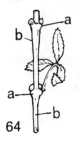

64

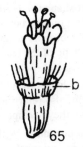

65

66

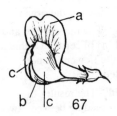

67

68

labellum = lip The lower part of the corolla (a) formed from one or several lobes and differing from the remaining lobes, particularly in Orchidaceae and Labiateae. Both upper (b) and lower lips (a) are frequently present. 69, see 47b

lanceolate Shaped like a lance, with the broadest part nearest the base, regularly narrowed upwards to an acute apex. 70

lateral On or at the side.

latex Milky sap.

lax Loose, not dense, relating in particular to branching.

leaf Usually composed of a leaf-blade or lamina (a), a leaf-stalk or petiole (b), and a leaf-base arising from the node of the stem (c); stipules are sometimes present. See 131a. Leaf-blade may be entire, toothed, lobed or compound. 71

leaflet The individual part (a) of a compound leaf which is usually leaf-like and may possess its own stalk. 72

legume = pod The fruit of the Leguminosae, comprising a single carpel which splits down two sides into two valves. 73

lemma Of grasses: the lower and outer of the two fertile glumes of the floret (b). See 58.

lenticels Corky 'breathing-pores' in bark on twigs, and on some fruit.

liana A woody climber usually associated with tropical forests.

ligule A small flap of tissue (a), or a scale, borne on the surface of a leaf or perianth-segment. 74

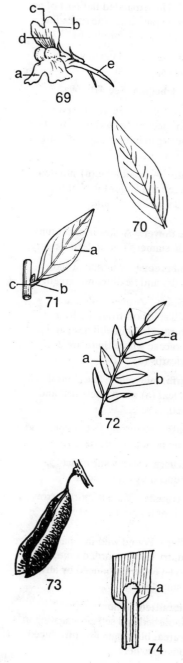

limb The expanded flat part of an organ, in particular the spreading lobes of a fused corolla (b). 75

linear Long and narrow with parallel sides. 76

lip = **labellum**. See 47b, 69a.

lobe A part or segment of an organ, deeply divided from the rest of the organ but not separated from it. 77a

lobule A small lobe (a) situated between the larger lobes of the corolla, as in *Gentiana*. 78

male flower A flower containing fertile stamens but no fertile ovary.

membranous Parchment-like, thin dry and flexible, not green.

mericarp A one-seeded unit which splits off from the fruit at maturity. In Umbelliferae, and Aceraceae where there are 2 mericarps (a). 79

midrib, mid-vein The central vein of a leaf (a) often thickened and raised. 80

mouth Referring to the open end of the perianth. See 69d, 75c.

mucilage Sticky substance produced by some seeds.

mucronate With a short narrow point or **mucro** (a). 81

native Found wild in a particular country since recorded time; not known to be introduced by any human agency.

naturalized Thoroughly established and self-propagating in an area, but originally introduced from elsewhere.

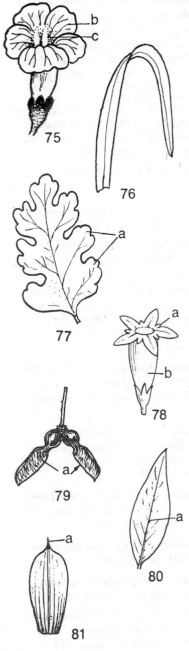

nectary A gland (a) which produces a sugary liquid and serves to attract insects; variously situated on the flower. 82

needle A slender needle-like leaf (a), as found in Pinaceae. 83

netted = reticulate With a net-like arrangement of veins (a). 84

node A point on the stem (a) from which one or more leaves, and or buds arise. 85

notch A v-shaped indentation in an organ (c), as in the apex of a petal or leaf. See 69

nut, nutlet A one-seeded fruit with a hard outer covering 87. See 2a. Nutlets are small nuts, often one of several (a) in a fruit as in Labiateae. 86

ob- Inverted, with the broadest part of the organ near the apex, and not as is more usual near the base. Used as a prefix: thus **oblanceolate, obcordate, obovate, obovoid, obconical**, etc. 88

oblong Longer than broad and with the margins nearly parallel for most of their length. 89

obtuse Blunt or rounded at apex.

ochrea A nodal sheath (a) formed by the fusion of two stipules, as in Polygonaceae. 90

opposite Of two organs; arising at the same level on opposite sides of the stem or structure. 91

orbicular Circular or disk-shaped.

ovary The part of the flower containing the ovules (a) and later the seeds, usually with one or more styles (b) and stigmas (c). 92

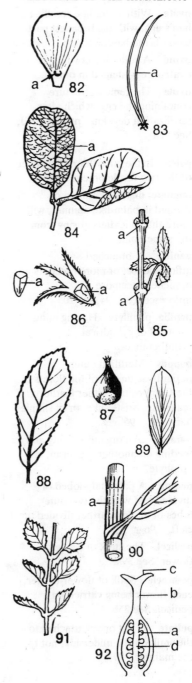

ovate With an outline recalling a hen's egg, with the broadest part towards the base. See 138.

ovoid A solid object which is ovate or egg-shaped in outline.

ovule The female structure containing the egg, which after fertilization develops into the seed. See 92a.

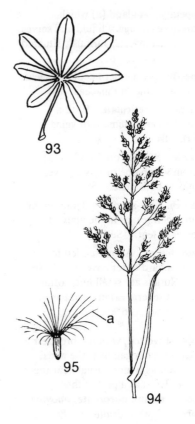

palea In the grass flower the upper of the two fertile bracts. See 58c.

palmate, digitate Lobed or divided in palm-like manner, with more than 3 leaflets arising from the same point. 93

panicle A branched inflorescence, or more precisely a branched racemose inflorescence. 94, see 105

papilla, papillate Having minute rounded or cylindrical protuberances.

pappus Modified outer perianth-segments of the florets of the Compositae, either feathery, scale-like, bristle-like, or of simple hairs (a). 95

parasite An organism living and feeding on another organism, cf. epiphyte.

pedate A palmately-lobed or divided leaf with the two outer lobes or leaflets further divided or cleft. 96

pedicel The stalk of a single flower. See 39a.

peduncle Stalk of flower-cluster, each flower being carried on a pedicel. See 35b.

peltate A flat organ attached to the stalk on the underside, not to the margin. 97

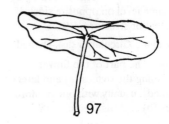

perennial Living for more than 2 years. A herbaceous perennial has non-woody stems which are produced anew every season or year. Many perennials die down above ground in the dry or cold season, and persist below ground until the next growing season.

perianth, perianth-segments The outer non-sexual parts (a) (c) of the flower, usually in two distinct or similar whorls. The segments are the individual organs comprising the two whorls. 98, see 134b.

perianth-tube Tube (b) formed by the fusion of the perianth segments. See 78.

pericarp Wall of the fruit, often with outer part fleshy (a) and inner part stony (b), surrounding the seed (c) or seeds. 99

perigynous Of flowers in which the receptacle forms a ring or cup (a) surrounding the base of the ovary, with the sepals, petals and stamens arising from its rim. The ovary is thus half-inferior. See 60.

petal An individual member (a) of the inner set of sterile organs surrounding the sexual parts of the flower, usually brightly coloured. See 98.

phyllode A green flattened leaf-stalk (a) resembling a leaf. 100

pinnae The primary divisions (a) of a compound leaf bearing leaflets-pinnate- or bearing a second division, with leaflets-bipinnate. 101

pinnate The regular arrangement of leaflets (a) in two rows on either side of the stalk or rachis (b). See 72. Thus pinnately-lobed 102; bi- or 2-pinnate, see 101.

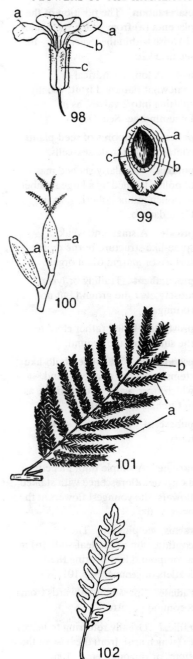

98

99

100

101

102

placentation The position of the placenta (a) from which the ovules (b) arise within the ovary (c). 103, see also 92d.

pod A long cylindrical or somewhat flattened fruit, usually splitting into 2 valves, as in Leguminosae. See 73.

pollen The spores of seed-plants containing the male sex-cells.

pollinia Regularly shaped masses of pollen formed by a large number of pollen grains cohering, as in Orchidaceae.

prickle A small and weak spine-like structure borne on the surface or margin of an organ.

procumbent Trailing or lying loosely over the ground but not rooting.

prostrate Lying rather close to the surface of the ground.

pseudobulb Thickened bulb-like stems (a) borne above ground which store water and nutrient, as in Orchidaceae. 104

pubescent Shortly and softly hairy.

raceme A simple unbranched elongate inflorescence with stalked flowers, the youngest flowers at the apex. 105

rachis, see **pinnate** The continuation of the leaf-stalk (b) in a compound leaf bearing the leaflets or pinnae. See 101, 72b.

radiate Spreading outwards from a common centre.

radical Usually referring to leaves (a) which arise from the base of the stem, or rhizome, etc. 106

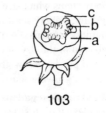

103

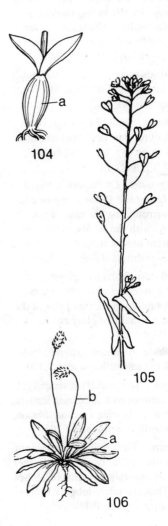

104

105

106

ray The primary (b) and secondary branches (c) of an inflorescence of the Umbelliferae. See 145.

ray-floret One of the strap-shaped florets in the flower-head of many Compositae. 107

receptacle The uppermost, often enlarged part (a) of the flower-stalk from which some or all parts of the flower are borne. 108

recurved Bent or curved downwards or backwards.

reflexed = deflexed Abruptly recurved or bent downwards or backwards. See 38a.

regular flowers See flowers. 54a, 54b

resin A sticky substance often coating bud-scales and which hardens on exposure to the air. Resin is often carried in ducts or canals.

reticulate = netted With a network, usually of veins. See 84a.

rhizome A swollen underground stem (a), with nodes, buds and scale-leaves, and from which the leaves and flowering stems arise each season. 109

rhombic, rhomboid Diamond-shaped, as on playing-cards.

rootstock A general term for an underground stem (a), usually short and erect. 110

rosette An arrangement of leaves radiating from a crown or centre, often spreading over the ground. 111

rostellum A small beak; as in the Orchidaceae. See 28d

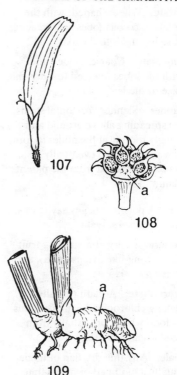

107

108

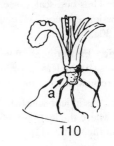

109

110

111

rotate Wheel-shaped; with the petals or corolla-lobes (a) spreading at right-angles to the axis. 112

runcinate Coarsely lobed and with the lobes directed towards the base of the leaf. 113

runner Slender horizontal stems, (a) spreading above ground and usually rooting at the nodes to form new plants which eventually become detached from the parent plant. 114

sac A pouch or baggy cavity to an organ, hence **saccate**.

samara A dry non-splitting fruit with a wing-like extension (a), as in *Fraxinus*. 115

saprophyte A plant (usually lacking chlorophyll) which obtains its food wholly or partially from dead organic matter.

scale Any thin dry flap of tissue; usually a modified or degenetate leaf. Thus bud-scales, see 17a; scale-leaves, see 133a; cone-scales, see 30a.

scape Flowering stem without leaves (b), all leaves arising from the rootstock. See 106.

scar As in leaf-scar (a). The mark left where the leaf-base is detached from the stem. 116

seed Contained within the ovary, produced as a result of fertilization of the ovule: consisting of a seed-coat or testa (a), an embryo (b), and sometimes of food reserves (c). 117

segment A part of an organ; thus leaf-segment (a), perianth-segment, etc. 118

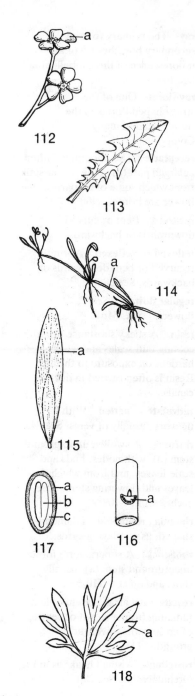

sepal One of the outer set of perianth-segments (c), usually green and protecting in bud, rarely coloured and petal-like. See 98.

septum A partition or cross-wall (a), as in a fruit. 119

serrate Of a margin of a leaf-blade with teeth like a saw. See 45.

sessile Stalkless

sheath A long more or less tubular structure (a) surrounding an organ or part. 120

short-shoot Lateral shoots or branches which bear leaves and have limited growth (a), or no growth. 121

shrub A perennial with woody stems, usually branched, or with several woody branches arising from the base.

shrublet A general term for a small woody-stemmed plant, usually less than about 30 cm tall, often creeping over the ground.

silicula The short fruit of some Cruciferae, less than 3 times as long as broad. 122

siliqua The long fruit of some Cruciferae, more than 3 times as long as broad. 123

silky Having a covering of soft adpressed fine hairs.

simple Of a leaf, not divided into segments; of stems and inflorescences, unbranched.

spadix Fleshy axis bearing clusters of stalkless flowers (b) and often ending in a tail-like or club-shaped appendage (a) as in *Arisaema* 124. Or axis covered to apex with flowers.

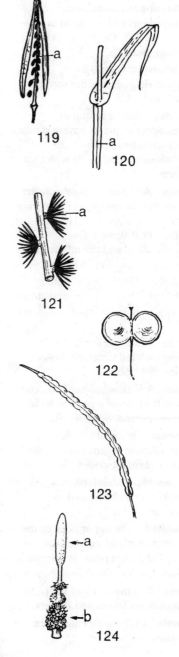

537

spathe   A large bract (a) enclosing a flower-head; sometimes conspicuous and coloured as in Araceae. 125. Or papery as in *Allium*. See 19b.

spathulate   Spoon- or paddle-shaped, broader towards the apex.   126

species   The basic unit of classification; plants having similar characters and in which the individuals breed freely with each other.

spike   A slender elongate cluster or clusters of spikelets of numerous stalkless flowers. 127. Hence spike-like if flowers are short-stalked and the inflorescence is elongated and cylindrical.

spikelet   One or more florets subtended by one or several sterile bracts, as in Gramineae.   128, see 58.

spine   A stiff straight, sharp-pointed structure; hence spiny. See 36b.

spur   A hollow sac-like projection from the base of a petal or sepal; hence spurred. See 69e, 127b.

stamen   One of the male reproductive organs of the flower, which bears the pollen. See 5, 50.

staminode   An infertile stamen (a), often much reduced in structure.   129

standard   The upper petal of the flowers of most Leguminosae. See 67a. The 3 erect petals in *Iris*. See 13c.

stem   The main axis of the plant which bears leaves and flowers.

sterile   Lacking functional sex organs.

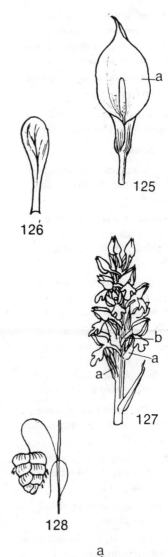

125

126

127

128

129

stigma The part of the female organ (a) which receives the pollen. 130

stipule A scale-like or leaf-like appendage (a) at the base of the leaf-stalk; usually paired; thus stipulate. 131

stolon A creeping stem (a) occurring above or below ground produced by a plant which has either an erect stem or a rosette of leaves. 132, 133

style The more or less elongated part of the female organ (b), which bears the stigmatic surfaces. See 130.

sub-species, or subsp. A group of individuals within a species which have some distinctive characters in common, and often a well-defined geographical range.

subtend To stand below and close to; as when a bract subtends a flower in its axil.

subulate Awl-shaped, narrow, pointed and more or less flattened. See 9a.

succulent Fleshy, juicy, thick.

sucker A shoot originating and spreading below ground, and ultimately appearing above ground, often some distance from the parent plant. 133

superior ovary = hypogynous Ovary (a) situated on the receptacle above the bases of stamens, petals, and sepals. 134

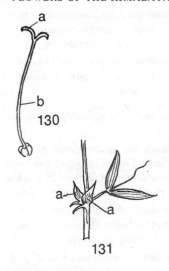

130

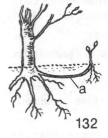

131

132

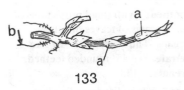

133

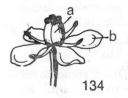

134

symbiotic An association of two dissimilar organisms, usually to their mutual advantage.

synonym A name given to a plant which has been superseded or is no longer used.

taproot The main descending root.

tendril A slender simple or branched organ (a) which can twine round a support, derived from a leaf or part of a leaf, or a stem. 135

terminal At the tip, apical; also an organ borne at the end of a stem and limiting its growth.

ternate Divided into 3 equal parts, these may be further divided into 3 parts, thus 2–ternate, etc. 136

tessellated Chequered.

testa Outer coat of a seed (a). See 117.

thorn A woody sharp-pointed structure formed from a modified branch. 137

throat The opening or orifice of a tubular or funnel-shaped corolla. See 69d, see 75c.

toothed With small projections of the margins of leaves or other organs. Thus triangular saw-like **serrate**. See 45; rounded teethed, **crenate**. 138

tree A long-lived woody plant with a single trunk branching only from above.

trifid Split into three no further than the middle. 139

trifoliate Having 3 leaflets, as in *Trifolium*. 140

tripartite Deeply divided to below the middle into three parts. 141

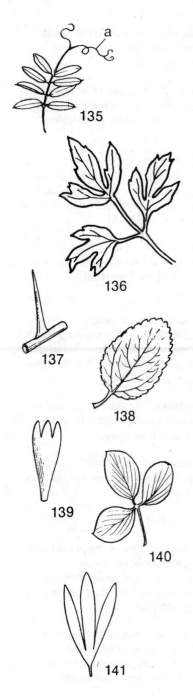

truncate Appearing as if cut-off at the end. 142

tube The fused part of the corolla, see 78b, or calyx (a). 143

tuber A swollen portion of a stem or root (a) arising annually, usually below ground. 144

tunic A dry, usually brown and papery covering round a bulb or corm. See 32a.

twig The youngest woody branches, usually of the present year's growth.

umbel A cluster of flowers whose spreading stalks or rays (b) arise from the apex of the stem, resembling the spokes of an umbrella. Compound umbels have secondary umbels (a) borne at the ends of the rays (b), of the primary umbel. 145

undershrub A general term for a low shrub usually less than 1m tall.

undulate Wavy in a plane at right angles to the surface. 146

unisexual With either male or female fertile sexual organs only.

valve One of the parts (a) into which a capsule splits. 147

variety, or **var.** A subordinate rank to species and subspecies, consisting of individuals differing from the type by one or two characters such as flower colour and hairiness.

vein A strand of strengthening and conducting tissue running through the leaf and other organs; hence mid-vein (a), lateral veins (b). 148

ventral Relating to the front or lower surface of an organ; opposite to the back or dorsal part.

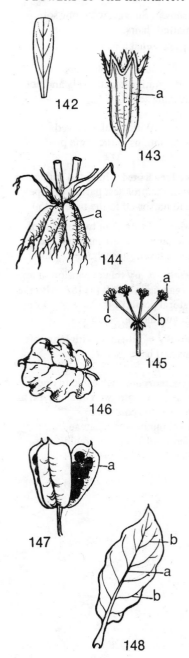

541

villous   Shaggy; with long soft, not matted, hairs.

viscid   Sticky.

viviparous   With flowers proliferating vegetatively and not forming seed.

wavy   With regular curved indentations in the same plane as the surface.   149

wedge-shaped   Inverted triangular with the broadest part at the apex and narrowed to the base.   150

whorl   More than two organs of the same kind arising from the same level; thus whorled.   151

wing   A dry thin expansion of an organ, as in a fruit (a) 152. Also the lateral petals of the flowers of the Leguminosae. See 67c.

woolly = downy   With long soft matted hairs. See 42.

zygomorphic   Divisible in one plane only into two equal halves; as for eg. the corollas of Leguminosae, Labiateae, Scrophulariaceae, Orchidaceae. See 47, 54c, 63, 69, 127.

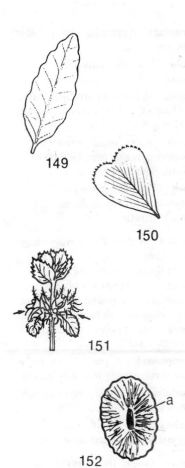

149

150

151

152

Bibliography

The authors have often been asked by people going to the Himalaya what book they should take with them in order to identify plants in the field. The answer is that no such book is available, and it is to remedy the lack that this book has been written.

Much that has been written on the Himalayan flora was published in periodicals, or in books which are now out of print. Those with access to a library may like to refer to some of the works listed below. Those which are still in print have been marked with an asterisk.

Floras

*Hooker, J. D. (1872–97), *The Flora of British India*, 7 vols., L. Reeve, London. Recently reprinted but not revised. It remains the principal Flora of our area, although by now it is very out-of-date. Recent botanical exploration in Nepal has revealed many species which are not included in it. The three Floras next mentioned below are often useful in offering a more contemporary view.

*Nasir, E. & Ali, S. I. (1970–), *Flora of West Pakistan*, Dept. of Botany, University of Karachi.

*Rechinger, K. N. (1963–), *Flora Iranica*, Akademische Druck u. Verlaganstalt, Graz, Austria.

*Komarov, V. L. et al. (1934–64), *Flora of U.S.S.R.*, 31 vols., Botanical Institute of the Academy of Sciences, Leningrad. Translated into English by Israel Program for Scientific Translations (1963–).

*Collett, H. (1902), *Flora Simlensis*, Thacker, Spink, Calcutta and Simla. This is the best of a number of small regional floras. It covers the Simla district.

Other more recent regional floras are as follows:

*R. K. Gupta (1968), *Flora Nainitalensis*, Navayug Traders, New Delhi.

*G. Singh & P. Kachroo (1976), *Forest Flora of Srinagar*, B. Singh & M. P. Singh, Dehra Dun.

*P. Kachroo et al. (1977), *Flora of Ladakh*, ibid.

*M. M. Raizada & H. O. Saxena (1978), *Flora of Mussoorie*, Vol. 1, ibid.

*B. M. Sharma & P. Kachroo (1981), *Flora of Jammu*, ibid.

*Dept. of Medicine Plants (1969), *Flora of Phulchoki & Godawari*, H.M. Govt. of Nepal.

*Ibid. (1976), *Flora of Langtang*, ibid.

Checklists

*Hara, H. et al. (1978–82), *An Enumeration of the Flowering Plants of Nepal*, 3 vols., Trustees of the British Museum (Nat. Hist.), London. This work does much to bring Himalayan nomenclature up-to-date, and is an invaluable guide through the maze of synonyms which surround so many species.

543

FLOWERS OF THE HIMALAYA

*Stewart, R. R. ((1972), An Annotated Catalogue of the Vascular Plants of West Pakistan and Kashmir, Flora of West Pakistan, Dept. of Botany, University of Karachi. This work includes species which occur in Kashmir and Ladakh.
*Department of Medicinal Plants (1976), Catalogue of Nepalese Vascular Plants, H.M. Govt. of Nepal.

Works dealing with trees and shrubs

*Brandis, D. (1906), Indian Trees, Constable, London. Recently reprinted, it remains the standard work on the trees of the whole sub-continent.
*Parker, R. N. (1918), A Forest Flora for the Punjab etc., Lahore. Recently reprinted.
Osmaston, A. E. (1927), A Forest Flora for Kumaon, Allahabad.

Works illustrated by coloured drawings

Royle, J. F. (1833–40), Illustrations of the Botany... of the Himalayan Mountains and of the Flora of Cashemere, 2 vols., W. N. Allen, London.
Hooker, J. D. (1849–51), The Rhododendrons of the Sikkim Himalaya, London.
King, G. & Pantling, R. (1898), Orchids of the Sikkim Himalaya, Ann. Roy. Bot. Gard. Calcutta, 8, p. 1-342.
Duthie, J. F. (1906), Orchids of the N. W. Himalaya, Ann. Roy. Bot. Gard. Calcutta, 9, 2, p. 81-211.
Blatter, E. (1928–9), Beautiful Flowers of Kashmir, 2 vols. John Balc, Danielsson, London.

Works illustrated photographically

Coventry, B. O. (1923–30), Wild Flowers of Kashmir, 3 vols., Raithby, Lawrence, London.
*Hara, H. (1963), Spring Flora of Sikkim Himalaya, Hoikusha, Japan.
*Hara, H. (1968), Photo-Album of Plants of Eastern Himalaya, Inoue Book Co., Tokyo.
*Nakao, S. (1964), Living Himalayan Flowers, Mainichi Newspapers, Tokyo.
Stainton, J. D. A. (1972), Forests of Nepal, John Murray, London.
*Mierow, M. & Shrestha, T. B. (1978), Himalayan Flowers and Trees, Sahayogi Press, Kathmandu.

Ecological descriptions

Many good accounts exist which cover in some depth various limited areas of the Himalaya. They are too numerous to mention individually. To find out whether any particular area has been covered the following works should be consulted.
*Schweinfurth, U. (1957), Die horizontale und vertikale Verbreitung der Vegetation im Himalaya, Ferd. Dummlers Verlag, Bonn.
*Dobremez, J. F. et al. (1972), Bibliographie du Nepal, Vol. 3, Tome 2, Botanique, Centre National de la Recherche Scientifique, Paris.
The work by Schweinfurth contains an ecological map of the whole of the

544

BIBLIOGRAPHY

Himalaya. Dobremez also has published ecological maps covering most of Nepal.

*Mani, M. S. (1978), *Ecology and Phytogeography of High-Altitude Plants of the Northwest Himalaya*, Oxford & IBH Publishing Co., New Delhi. This is the most recent work to appear dealing with alpine ecology in our area.

Index

Accepted names are given in roman type, synonyms in *italic*. Subspecies and varieties are entered under their main species. Numbers given are running numbers, colour plates are indicated by pl., page number of line drawings (pages 445–518), by d. and page numbers by p.

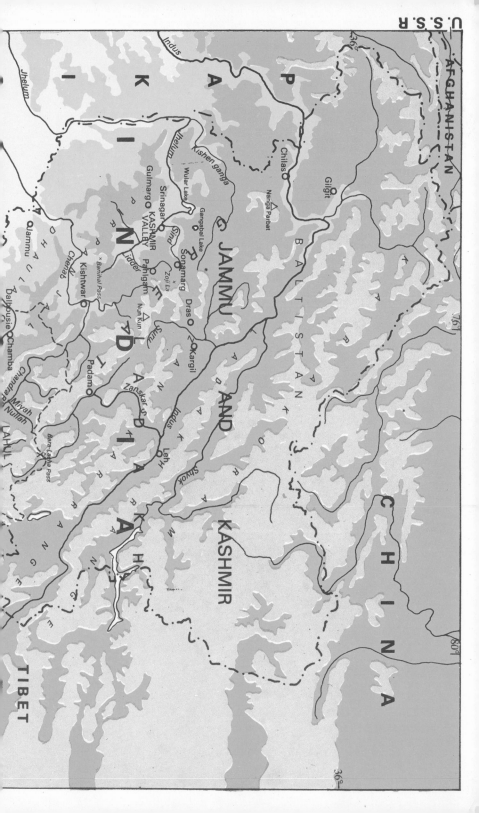

Bistorta amplexicaulis
Parochetus communis
Persicaria capitata

Clover (White/Dutch)
White dead nettle.
Nettles
Docks.
Arum (white fruit)
(Manali valley) 9.9.98